Advances in Synthesis Gas:
Methods, Technologies and Applications

Advances in Synthesis Gas: Methods, Technologies and Applications

Volume 4: Syngas Process Modelling and Apparatus Simulation

Edited by

Mohammad Reza Rahimpour
Department of Chemical Engineering, Shiraz University, Shiraz, Iran

Mohammad Amin Makarem
Methanol Institute, Shiraz University, Shiraz, Iran

Maryam Meshksar
Department of Chemical Engineering, Shiraz University, Shiraz, Iran

ELSEVIER

Elsevier
Radarweg 29, PO Box 211, 1000 AE Amsterdam, Netherlands
The Boulevard, Langford Lane, Kidlington, Oxford OX5 1GB, United Kingdom
50 Hampshire Street, 5th Floor, Cambridge, MA 02139, United States

Notices
Knowledge and best practice in this field are constantly changing. As new research and experience broaden our
understanding, changes in research methods, professional practices, or medical treatment may become
necessary.

Practitioners and researchers must always rely on their own experience and knowledge in evaluating and using
any information, methods, compounds, or experiments described herein. In using such information or methods
they should be mindful of their own safety and the safety of others, including parties for whom they have a
professional responsibility.

To the fullest extent of the law, neither the Publisher nor the authors, contributors, or editors, assume any liability
for any injury and/or damage to persons or property as a matter of products liability, negligence or otherwise, or
from any use or operation of any methods, products, instructions, or ideas contained in the material herein.

ISBN: 978-0-323-91879-4

For information on all Elsevier publications
visit our website at https://www.elsevier.com/books-and-journals

Publisher: Susan Dennis
Acquisitions Editor: Anita Koch
Editorial Project Manager: Zsereena Rose Mampusti
Production Project Manager: Sruthi Satheesh
Cover Designer: Mark Rogers

Typeset by STRAIVE, India
Transferred to Digital Printing 2022

Contents

SECTION 1: Modelling and simulation of syngas production processes

Chapter 1: Thermodynamic and phase equilibrium models of syngas generation through gasification

Soumitra Pati, Dinabandhu Manna, Sudipta De, and Ranjana Chowdhury

Maryam Delshah, Shabnam Yousefi, Mohammad Amin Makarem,
Hamid Reza Rahimpour, and Mohammad Reza Rahimpour

José Antonio Mayoral Chavando, Valter Silva, M. Puig-gamero, João Sousa Cardoso,
Luís A.C. Tarelho, and Daniela Eusébio

SECTION 2: Modelling and simulation of syngas purification processes

Chapter 7: Modeling and simulation of membrane-assisted separation of carbon dioxide and hydrogen from syngas....................................... 199

Nayef Ghasem

Chapter 8: Simulation of cyclone separator for particulate removal from syngas.. 219

Minhaj Uddin Monir, Azrina Abd Aziz, Abu Yousuf, Jafar Hossain, Ahosan Habib, Kuaanan Techato, and Khamphe Phoungthong

Chapter 14: Alcohols synthesis using syngas: Plant design and simulation 401

Mohammad Hasan Khademi, Afshar Alipour-Dehkordi, and Fereshteh Nalchifard

Chapter 15: Modeling, simulation, and optimization of combined heat and power generation from produced syngas 465

Ilenia Rossetti

Contributors

Afshar Alipour-Dehkordi   Department of Chemical Engineering, College of Engineering, University of Isfahan, Isfahan, Iran

Azrina Abd Aziz   Faculty of Civil Engineering Technology, University Malaysia Pahang, Gambang, Malaysia

Ali Behrad Vakylabad   Department of Materials, Institute of Science and High Technology and Environmental Sciences, Graduate University of Advanced Technology, Kerman, Iran

Filippo Bisotti   Dept. CMIC "Giulio Natta", Center for Sustainable Process Engineering Research (SuPER), Politecnico di Milano, Milan, Italy

Enrico Bocci   Marconi University, Rome, Italy

João Sousa Cardoso   Polytechnic Institute of Portalegre, Portalegre; Instituto Superior Técnico, Universidade de Lisboa, Lisboa, Portugal

José Antonio Mayoral Chavando   Polytechnic Institute of Portalegre, Portalegre, Portugal

Ranjana Chowdhury   Chemical Engineering Department, Jadavpur University, Kolkata, West Bengal, India

Sudipta De   Mechanical Engineering Department, Jadavpur University, Kolkata, West Bengal, India

Silvio de Oliveira Junior   Department of Mechanical Engineering, Polytechnic School of University of São Paulo, São Paulo, Brazil

Maryam Delshah   Department of Chemical Engineering, Shiraz University, Shiraz, Iran

Andrea Di Carlo   L'Aquila University, L'Aquila, Italy

Meire Ellen Gorete Ribeiro Domingos   Department of Chemical Engineering, Polytechnic School of University of São Paulo, São Paulo, Brazil

Moisés Teles dos Santos   Department of Chemical Engineering, Polytechnic School of University of São Paulo, São Paulo, Brazil

Daniela Eusébio   Polytechnic Institute of Portalegre, Portalegre, Portugal

Mohammad Farsi   Department of Chemical Engineering, School of Chemical and Petroleum Engineering, Shiraz University, Shiraz, Iran

Matteo Fedeli   Dept. CMIC "Giulio Natta", Center for Sustainable Process Engineering Research (SuPER), Politecnico di Milano, Milan, Italy

Daniel A. Flórez-Orrego   Department of Mechanical Engineering, Polytechnic School of University of São Paulo, São Paulo, Brazil; National University of Colombia, School of Processes and Energy, Medellin, Colombia; École Polytechnic Fédérale de Lausanne, Lausanne, Switzerland

Nayef Ghasem   Department of Chemical and Petroleum Engineering, UAE University, Al-Ain, United Arab Emirates

Ahosan Habib   Geological Survey of Bangladesh, Dhaka, Bangladesh

Jafar Hossain   Department of Petroleum and Mining Engineering, Jashore University of Science and Technology, Jashore, Bangladesh

Mohammad Hasan Khademi   Department of Chemical Engineering, College of Engineering, University of Isfahan, Isfahan, Iran

Hadis Najafi Maharluie   Department of Chemical Engineering, Amirkabir University of Technology, Tehran, Iran

Mohammad Amin Makarem   Methanol Institute, Shiraz University, Shiraz, Iran

Flavio Manenti   Dept. CMIC "Giulio Natta", Center for Sustainable Process Engineering Research (SuPER), Politecnico di Milano, Milan, Italy

Dinabandhu Manna   Chemical Engineering Department, Jadavpur University, Kolkata, West Bengal, India

Vera Marcantonio   Unit of Process Engineering, Department of Engineering, University Campus Bio-Medico di Roma, Rome, Italy

Minhaj Uddin Monir   Department of Petroleum and Mining Engineering, Jashore University of Science and Technology, Jashore, Bangladesh

Fereshteh Nalchifard   Department of Chemical Engineering, College of Engineering, University of Isfahan, Isfahan, Iran

Rafael Nogueira Nakashima   Department of Mechanical Engineering, Polytechnic School of University of São Paulo, São Paulo, Brazil

Soumitra Pati   Mechanical Engineering Department, Jadavpur University, Kolkata, West Bengal, India

Khamphe Phoungthong   Environmental Assessment and Technology for Hazardous Waste Management Research Center, Faculty of Environmental Management, Prince of Songkla University, Songkhla, Thailand

Karen Valverde Pontes   Industrial Engineering Graduate Program (PEI), Federal University of Bahia (UFBA), Street Professor Aristides Novis 02, Bahia, Brazil

M. Puig-gamero   University of Castilla—La Mancha, Ciudad Real, Spain

Poliana P.S. Quirino   Industrial Engineering Graduate Program (PEI), Federal University of Bahia (UFBA), Street Professor Aristides Novis 02, Bahia, Brazil

Hamid Reza Rahimpour   Department of Chemical Engineering, Shiraz University, Shiraz, Iran

Mohammad Reza Rahimpour   Department of Chemical Engineering, Shiraz University, Shiraz, Iran

Mohammad Rahmani   Department of Chemical Engineering, Amirkabir University of Technology, Tehran, Iran

Ilenia Rossetti   Chemical Plants and Industrial Chemistry Group, Dipartimento di Chimica, Università degli Studi di Milano, INSTM Milano Università-Unit, CNR-SCITEC, Milano, Italy

Mohammad Hadi Sedaghat   Department of Mechanical Engineering, Technical and Vocational University (TVU), Tehran, Iran

Sonia Sepahi   Department of Chemical Engineering, Shiraz University, Shiraz, Iran

Valter Silva   Polytechnic Institute of Portalegre, Portalegre; Department of Environment and Planning, Centre for Environmental and Marine Studies (CESAM), University of Aveiro, Aveiro, Portugal

Luís A.C. Tarelho   Department of Environment and Planning, Centre for Environmental and Marine Studies (CESAM), University of Aveiro, Aveiro, Portugal

Kuaanan Techato   Faculty of Environmental Management, Prince of Songkla University, Songkhla, Thailand

Shabnam Yousefi   Department of Chemical Engineering, Shiraz University, Shiraz, Iran

Abu Yousuf   Department of Chemical Engineering and Polymer Science, Shahjalal University of Science and Technology, Sylhet, Bangladesh

About the Editors

Prof. Mohammad Reza Rahimpour is a professor in Chemical Engineering at Shiraz University, Iran. He received his PhD in Chemical Engineering from Shiraz University in cooperation with the University of Sydney, Australia, in 1988. He started his independent career as assistant professor at Shiraz University in September 1998. Prof. Rahimpour was a research associate at the University of California, Davis, from 2012 to 2017. During his stint at the University of California, he developed different reaction networks and catalytic processes such as thermal and plasma reactors for upgrading lignin bio-oil to biofuel with the collaboration of UCDAVIS. He has been a chair of the Department of Chemical Engineering at Shiraz University from 2005 to 2009 and from 2015 to 2020. Prof. Rahimpour leads a research group in fuel processing technology focused on the catalytic conversion of fossil fuels such as natural gas and renewable fuels such as bio-oils derived from lignin to valuable energy sources. He provides young distinguished scholars from developing countries with perfect educational opportunities in both experimental methods and theoretical tools to undertake in-depth research in the various fields of chemical engineering including carbon capture, chemical looping, membrane separation, storage and utilization technologies, novel technologies for natural gas conversion, and improving the energy efficiency in the production and use of natural gas in industries.

Dr. Mohammad Amin Makarem is a research associate at Methanol Institute, Shiraz University. His research interests are focused on gas separation and purification, nanofluids, microfluidics, catalyst synthesis, reactor design, and green energy. In the gas separation field, his focus is on experimental and theoretical investigation and optimization of the pressure swing adsorption process, and in the gas purification field, he is working on novel technologies such as microchannels. Recently, he has investigated methods of synthesizing bio-template nanomaterials and catalysts. He has collaborated in writing and editing various books and book chapters for famous publishers such as Elsevier, Springer, and Wiley, in addition to guest editing journal special issues.

Maryam Meshksar is a research associate at Shiraz University. Her research is focused on gas separation, clean energy, and catalyst synthesis. In the gas separation field, she is working on membrane separation processes, and in the clean energy field, she has worked on different reforming-based processes for syngas production from methane experimentally. She has also synthesized novel catalysts for these processes, which have been tested for the first time. She has reviewed novel technologies such as microchannels for energy production. Recently, she has written various book chapters for famous publishers such as Elsevier, Springer, and Wiley.

Preface

Vol. 4: Syngas process modelling and apparatus simulation

Synthesis gas (syngas) and its products such as hydrogen are indispensable in chemical, oil, and energy industries. They are important building blocks and serve as feedstock for the production of many chemical compounds such as ammonia and methanol. Hydrogen is expected to become a common energy carrier no later than the middle of the 21st century since it offers considerable energy density and releases negligible pollutants. It is also utilized in petroleum refineries for producing clean transportation fuels, and its consumption is expected to increase dramatically in the near future as refineries need to intensify production capacities. Many publications have hitherto focused on syngas production and purification methods, as well as its applications in industrial production units. Despite the fact that extended studies have been undertaken, there is still room for improvement. The four volumes of this book series explain the conventional and state-of-the-art technologies for the production, purification, and conversion of syngas meticulously.

Simulation models are of paramount importance for achieving scenarios closer to the real plant conditions since they provide invaluable information about the operation at a lower cost than empirical studies. Therefore, this book is a promising means to collect any information and investigate the most recent literature on modeling and simulation of syngas production, purification, and conversion processes and units. It will serve as a strong connection between the experts in both academic and industrial sectors aiming to solve operational challenges and instrument designing and will also shed light on the prospects of developing new processes.

In this regard, the three sections of this book are devoted to modeling and simulation of syngas production processes such as gasification and reforming, modeling and simulation of different syngas purification and cleaning processes, and modeling and simulation of processes of syngas transformation to chemicals and/or energy such as alcohols, methane, ammonia, and power generation.

The editors feel obliged to sincerely appreciate the authors of the chapters for their contributions, hard work, and great assistance in this project. Furthermore, the authors, as well as the editors, are grateful to all the Elsevier staff for their invaluable and irreplaceable step-by-step assistance in preparing this book.

Mohammad Reza Rahimpour
Mohammad Amin Makarem
Maryam Meshksar

Reviewer Acknowledgments

The editors feel obliged to appreciate the dedicated reviewers (listed below), who were involved in reviewing and commenting on the submitted chapters and whose cooperation and insightful comments were very helpful in improving the quality of the chapters and books in this series.

Dr. Mohammad Hadi Sedaghat

School of Mechanical Engineering, Shiraz University, Shiraz, Iran

Dr. Ali Bakhtyari

Chemical Engineering Department, Shiraz University, Shiraz, Iran

Dr. Javad Hekayati

Department of Chemical Engineering, Shiraz University, Shiraz, Iran

Ms. Parvin Kiani

Department of Chemical Engineering, Shiraz University, Shiraz, Iran

Ms. Samira Zafarnak

Department of Chemical Engineering, Shiraz University, Shiraz, Iran

Modelling and simulation of syngas production processes

Thermodynamic and phase equilibrium models of syngas generation through gasification

Soumitra Pati[a], Dinabandhu Manna[b], Sudipta De[a], and Ranjana Chowdhury[b]

[a]*Mechanical Engineering Department, Jadavpur University, Kolkata, West Bengal, India* [b]*Chemical Engineering Department, Jadavpur University, Kolkata, West Bengal, India*

1. Introduction

Energy transition based on renewable sources of energy is the key to limiting global warming to 1.5°C by 2050. Substitution of natural gas by renewable gases, such as hydrogen and biogas is expected to play a vital role to ensure the transition toward a zero-emission energy sector [1]. Besides having the potential to be added to gas-grids, syngas also plays a vital role in the generation of liquid fuels for automobile sectors. Ethanol can be produced through direct fermentation of syngas [2,3]. Moreover, higher alcohols such as butanol, hexanol, or heptanol can also be produced through the combination of the syngas fermentation process with a carboxylate platform based on incomplete anerobic digestion of biomass [4]. Only through the incorporation of the gasification step, high lignin lignocellulosic biomass can be utilized for the production of liquid fuels and fuel additives, such as alcohols and other hydrocarbon compounds [5,6]. Any carbonaceous solid can be converted to diesel-grade liquid fuel via the generation of syngas and its conversion through a catalytic chemical process, namely, Fischer-Tropsch synthesis. The processes of liquid fuel generation via this route are known as XTL (feedstock to liquid), where X stands for the particular carbon source of syngas [7]. For example, CTL and BTL signify the conversion of coal and biomass to the liquid fuel, respectively. Syngas is mainly constituted of CO, CO_2, H_2, H_2O, CH_4, and some condensable volatiles (tar), Besides reforming of hydrocarbons, gasification of solids containing C, H, N, O, and S is one of the major routes of production of syngas. The solids are partially oxidized using gasifying agents such as oxygen or air and steam. The solid feedstock, usually used for the production of syngas includes low-grade coal, agricultural biomass, algal biomass, woody biomass, and municipal solid wastes to name a few [8]. A solid stream containing unconverted

Advances in Synthesis Gas: Methods, Technologies and Applications. https://doi.org/10.1016/B978-0-323-91879-4.00007-2

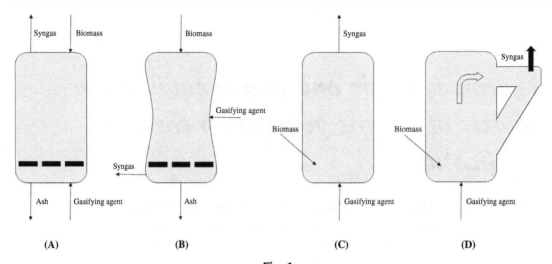

Fig. 1
Different types of gasifiers (A): Updraft; (B) Downdraft; (C) Fluidized bed; (D) Entrained bed.

char and ash is also produced during the process of syngas generation. From the perspective of moving patterns of solid feedstock and the gasifying medium, the gasifiers can be categorized into three types, namely, (a) moving bed, (b) fluidized bed, and (c) entrained bed [9]. Different configurations of syngas reactors are schematically represented in Fig. 1. The moving bed gasifiers can again be classified as updraft and downdraft ones. In both types of moving bed reactors, the solid feedstock is fed from the top of the reactor, and the gasifying medium is fed either from the bottom (updraft) or from the top (downdraft) of the gasifiers [10]. In the updraft and downdraft reactors, the solid-fluid movement is in counter-current and co-current patterns, respectively. The downdraft gasifier is more popularly used because the concentration of condensable volatiles (tar) in the product syngas stream is much less compared to that in the updraft ones [11]. Large-sized particles can be handled in the moving bed gasifiers. In the fluidized bed gasifiers, the superficial velocity of the gasifying medium is sufficient enough to fluidize the solid particles, and hence the gasifier behaves like a continuous stirred tank reactor [12]. The heat and mass transfer between the fluid and solid particles is better in comparison to that of the moving bed versions. The residence time is in the order of minutes to seconds. In the entrained bed gasifiers, micron-sized particles are handled in a co-current pattern with the gasifying medium. Usually, high temperature and very short residence time (in order of seconds) are maintained in the reactor, and the solids are usually in the slagging state [13].

Gasification is a complex process involving physical and chemical changes. Mainly four stages, namely, drying, pyrolysis, combustion, reduction, and gasification are involved in the process. Drying of feedstock occurs at $<150°C$ and causes the removal of moisture [9]. During pyrolysis ($150–700°C$), thermal degradation of feedstock leads to the formation of char and volatiles. While the non-condensable volatile product, that is, the gaseous part is constituted of H_2, CO,

CO_2, CH_4, and some other light hydrocarbon gases, the condensable non-aqueous volatile part mainly comprises high molecular weight hydrocarbons, namely, tar [14]. In the combustion zone (700–1500°C), some of the solid feedstock is combusted to form CO_2 and H_2O. During reduction (800–1100°C), CO_2 reacts with char to form CO (Boudouard reaction). Besides, a further reaction between H_2O with CO (water-shift reaction), CH_4 (methane reforming), and methanation (reaction of char with hydrogen) can also occur [10]. While the combustion reaction is exothermic in nature, reduction and gasification reactions are endothermic. Some tar cracking reactions and the formation of ammonia and H_2S can also occur during gasification. The heat required for endothermic reactions is supplied by the exothermic combustion reaction, and hence the process can be run in a self-sustained way. Different reactions which may occur during gasification are provided in Table 1.

Table 1 Array of reactions in gasifiers.

Reaction	Reaction name	Heat of reaction (ΔH) (kJ/mol)	Reaction number	Ref.
Drying $Biomass_{wet} \rightarrow Biomass_{dry} + H_2O_{(g)}$	–	–	(R1)	[15,16]
Pyrolysis $Biomass_{dry} \rightarrow Gas + Tar + Char$	–	–	(R2)	[15,16]
Heterogeneous reaction $C + O_2 \rightarrow CO_2$	Complete combustion	−394	(R3)	[15,17]
$C + 0.5\,O_2 \rightarrow CO$	Char partial combustion	−111	(R4)	[16–18]
$C + CO_2 \leftrightarrow 2CO$	Boudouard reaction	+172	(R5)	[15–17]
$C + H_2O \leftrightarrow CO + H_2$	Water-gas	+131	(R6)	[15–18]
$C + 2H_2 \leftrightarrow CH_4$	Methane formation	−74.8	(R7)	[15–17]
Homogeneous reactions $CO + 0.5O_2 \rightarrow CO_2$	CO partial combustion	−284	(R8)	[15–17]
$H_2 + 0.5\,O_2 \rightarrow H_2O$	Hydrogen combustion	−242	(R9)	[15–18]
$CO + H_2O \leftrightarrow CO_2 + H_2$	Water-gas shift (WGS)	−41.2	(R10)	[2,5–7]
$CH_4 + H_2O \leftrightarrow CO + 3H_2$	Methane reforming	+206	(R11)	[15–18]
$CH_4 + 1.5\,O_2 \rightarrow CO + 2H_2O$	Methane partial combustion	−520	(R12)	[16]
H_2S and NH_3 formation reactions $H_2 + S \rightarrow H_2S$	H_2S formation	–	(R13)	[17]
$3H_2 + N_2 \rightarrow 2NH_3$	NH_3 formation	–	(R14)	[17]
Tar decomposition $pC_nH_x \rightarrow qC_mH_y + rH_2$	Tar cracking	–	(R15)	[16,17]
$C_nH_x + nH_2O \rightarrow \left(n + \frac{x}{2}\right)H_2 + nCO$	Steam reforming of tar	–	(R16)	[16,17]
$C_nH_x + nCO_2 \rightarrow \left(\frac{x}{2}\right)H_2 + 2nCO$	Dry reforming of tar	–	(R17)	[16,17]
$C_nH_x \rightarrow nC + \left(\frac{x}{2}\right)H_2$	Carbon formation	–	(R18)	[16,17]

The quality and quantity of the syngas are extremely dependent on various operating parameters such as the mass flow rate of feedstock, type of gasifying agents, gasification temperature, pressure inside the gasifier, and equivalence ratio for instance. The thermo-chemical property and elemental composition of the feedstock also affect the production of syngas up to a certain extent [19]. Therefore, it is economically infeasible and sufficiently time-consuming to experimentally determine the optimum condition of the gasification process for any particular feedstock. It is worth mentioning that the operating parameter variation leaves a combined effect on the gasification system. Mathematical modeling of syngas reactors is necessary for (i) a priori prediction of performance with the variation of input parameters such as feedstock properties, temperature, solid-to-gasifying agent ratio, and pressure, for example; (ii) scaling up from laboratory-scale to pilot and industrial scales; (iii) optimization of reactor design and performance [20]. Mathematical modeling has enabled the researchers to virtually model the gasification process and optimize the process parameters, providing a lot of convenience in experimental studies. The main aim of a mathematical model of gasification is to virtually replicate the physical and chemical processes taking place inside a gasifier up to as much extent as possible. As the amount of complicacy of the model is increased, it will provide more realistic results. These complexities are typically based on the chemical reactions of gasification and the dynamic behavior of gas and solid particles inside the reactor [16]. The extent of the implication of such complexity in any mathematical model determines its exactness in turn. However, it is a common practice among researchers to simplify the model up to a certain extent within a certain range of permissible tolerance. These simplifications are based on some assumptions regarding the physical and chemical changes of matter inside the gasifier and regarding the elemental composition of the product [21].

Based on the principles of development, the gasifier models can be broadly categorized as A) kinetic models and B) thermodynamic equilibrium models [19]. However, due to the complexity of gasification, these models are also based on some assumptions. They can predict the overall and spatial response, mostly the composition of product syngas of the gasifier with the variation of input variables both under steady- and unsteady-state conditions. As the name suggests, thermodynamic equilibrium models are entirely based on the idealistic assumption of the attainment of equilibrium and can predict the maximum possible yield of any constituent of product syngas. It is assumed that the reactants and products have stayed inside the reactor for a long enough time to attain chemical equilibrium [22]. Although based on equilibrium assumption, which is very difficult to be attained, particularly at low temperature operation of downdraft gasifier and short residence time observed in fluidized bed gasifiers, this is a very useful tool for the quick assessment of the suitability of different feedstocks at varying operating conditions. Although the thermodynamic equilibrium models have some drawbacks such as overestimating the product composition etc., equilibrium modeling has served as an important tool to estimate and optimize the gasification process for decades [17]. In the present chapter, the principles, applications, limitations, and modifications of thermodynamic models of syngas reactors are discussed in detail.

2. Equilibrium modeling of gasification

As the name suggests, thermochemical equilibrium (TCE) models are based on the principle of attainment of equilibrium. The predictions of the models are independent of the reactor design and fluid dynamics. However, TCE models can predict the maximum possible yield of a gasifier achievable using a particular set of operating parameters. The TCE models can be categorized as (I) **stoichiometric** (based on equilibrium constants of a set of preselected array of reaction) and (II) *non-stoichiometric* ones (based on minimization of Gibbs-free energy). All TCE models are based on the following assumptions [15].

(1)   The gasifier is operated isothermally under atmospheric pressure.
(2)   Residence time of the reactants in the gasifier is infinite and hence steady state is attained.
(3)   The exact mechanism of the pyrolysis reaction and the accounting of intermediates are not considered.
(4)   The reactor is well-mixed and transient behavior and spatial gradient of temperature and pressure are negligible.
(5)   No unreacted oxygen escapes the gasifier.
(6)   Nitrogen behaves almost as an inert gas during the gasification.
(7)   There is no change in kinetic and potential energy within the reactor.
(8)   The oxidizing agent is sufficient to convert all carbon entering through biomass or coal fed to the gasifier.
(9)   Gas phase behaves ideally.
(10)   Energy content of ash, that is, the mixture of unconverted carbon and mineral matter escaping through the effluent solid stream is negligible.
(11)   The condensable volatiles, that is, tar formed during pyrolysis is in vapor state and behave as an ideal gas as the temperature inside the gasifier is very high.
(12)   Heat loss to the environment is negligible.

In the case of the non-stoichiometric equilibrium model, the volatile pyro-products are only to be specified according to the quantity of different elements present. Under stoichiometric equilibrium, model distribution of moles of different components is necessary. Typically, the components such as CO, CO_2 CH_4, H_2, H_2O, and N_2 are considered. The equivalent quantities of C, H_2, N_2, O_2, Cl_2, and S fed to the gasifier appear in the product syngas. In both stoichiometric and non-stoichiometric equilibrium models, char is considered to be inert carbon during pyrolysis. Although the approaches are different, it has already been established that both types of equilibrium modeling yield the same results [11]. The principles of the two types of modeling are described in the following sections.

Table 2 Standard properties of gases in the ideal-gas state.

Chemical species	Standard enthalpy of formation	Standard Gibbs-free energy of formation	Standard absolute entropy	Constant coefficients of molar heat capacity (C_p) (kJ/mol/K)			
	Reference state $T = 298.15\,K$, $P = 0.1\,MPa$ [23]			$C_p = a + bT + cT^2 + dT^3$ (T in °C) [24]			
	(kJ/mol)	(kJ/mol)	(J/kmol/K)	$a \times 10^3$	$b \times 10^5$	$c \times 10^8$	$d \times 10^{12}$
CO	−110.257	−137.163	197.653	28.95	0.411	0.3548	−2.22
H_2	0	0	130.680	28.84	0.00765	0.3288	−0.8698
CO_2	−393.522	−394.389	213.795	36.11	4.233	−2.887	7.464
CH_4	−74.873	−50.768	186.251	34.31	5.469	0.3661	−11.00
$H_2O_{(g)}$	−241.826	−228.582	188.834	33.46	0.6880	0.7604	−3.593
N_2	0	0	191.609	29.00	0.2199	0.5723	−2.871

2.1 Stoichiometric equilibrium model

The product syngas stream comprises CO, H_2, CO_2, CH_4, H_2O, and N_2. This method of modeling is based on the equilibrium constants of a selected array of reactions. Table 2 provides the standard properties of all the gaseous components participating in the gasification process [23,24].

Chemical equilibrium in a single-phase system is considered. One of the main criteria of chemical equilibrium is that the entropy, S and Gibbs energy, G are maximum and minimum. Respectively [25]. This leads to

$$\sum_{i=1}^{c} \vartheta_i \overline{G}_i = 0 \tag{1}$$

where, $\overline{G}_i$, ϑ_i, and c are partial molar Gibbs energy and stoichiometric coefficient of ith component and total number of components, respectively.

The above-mentioned equation along with the atom balance equations and the constraint equations of state variables can be used to identify the equilibrium state. The total Gibbs energy for a gaseous system can be represented as follows:

$$G = \sum_{i=1}^{c} N_i \overline{G}_i(T,P,y) = \sum_{i=1}^{c} N_i \overline{\overline{G}}_i(T,P) + \sum_{i=1}^{c} N_i \left(\overline{G}_i(T,P,y) - \overline{\overline{G}}_i(T,P) \right) \tag{2}$$

$\overline{\overline{G}}_i(T,P)$ and $\overline{G}_i(T,P,y)$ are the pure and partial molar Gibbs energy of the ith component at the condition of reaction. Eq. (2) can also be written as

$$G = \sum_{i=1}^{c} N_i \overline{\overline{G}}_i(T,P) + \sum_{i=1}^{c} N_i RT \ln \frac{\overline{f}_i(T,P,y)}{\overline{\overline{f}}(T,P)} \tag{3}$$

$$= \sum_{i=1}^{c} N_i \bar{\bar{G}}_i(T,P) + \sum_{i=1}^{c} N_i RT ln \frac{(y_i P)^{\frac{\bar{f}_i(T,P,y)}{(y_i P)}}}{(P)^{\frac{\bar{f}_i(T,P)}{(P)}}} \tag{4}$$

$\frac{\bar{f}_i(T,P)}{(P)}$ and $\frac{\bar{f}_i(T,P,y)}{(y_i P)}$ are the fugacity coefficients of pure component i and when in the reaction mixture, respectively. Considering the ideal gas behavior, all fugacity coefficients can be set at unity. Therefore,

$$G = \sum_{i=1}^{c} N_i \bar{\bar{G}}_i(T,P) + RT \sum_{i=1}^{c} N_i \; ln \, (y_i) \tag{5}$$

The first term on the RHS represents the summation of Gibbs energy of all pure components and the second term represents the Gibbs-free energy change. As the reactions proceed, the number of moles and hence the mole fraction of any component change. They are guided by the extent of reactions, ξ and the stoichiometric coefficients, ϑ of the components in different reactions.

For any component, i, participating in a single reaction,

$$N_i = N_{i0} + \vartheta_i \xi$$

or

$$\xi = \frac{N_i - N_{i0}}{\vartheta_i} \tag{6}$$

For multiple reactions, for the ith component participating in several reactions,

$$N_i = N_{i0} + \sum_j \vartheta_{ij} \xi_j \tag{7}$$

where j represents any reaction.

The value of ϑ_i is positive for reaction products and negative for reactants. Mathematically, the equilibrium state can be identified by the criterion that G will be minimum, that is,

$$G = minimum \; or, \; dG = 0 \tag{8}$$

or,

$$\left(\frac{\partial G}{\partial \xi} \right)_{T,P} = 0 \tag{9}$$

This equation is solved to determine the value of ξ at equilibrium, that is, ξ^*. From that the equilibrium mole fraction of each species is determined.

In general, the following equation is valid under equilibrium:

$$-\frac{\sum_{i=1}^{c} \vartheta_i \bar{\bar{G}}_i}{RT} = \sum_{i=1}^{c} ln \, y_i^{\vartheta_i} = ln \prod_{i=1}^{c} y_i^{\vartheta_i} \tag{10}$$

Again, under equilibrium,

$$\sum_{i=1}^{c} \overline{G}_i(T, P, \vartheta_i) = 0 \tag{11}$$

$$G = \sum_{i=1}^{c} N_i \overline{\overline{G}}_i(T, P) + \sum_{i=1}^{c} N_i RT ln \frac{\overline{f}_i(T, P, y)}{\overline{\overline{f}}(T, P)} \tag{12}$$

$$= \sum_{i=1}^{c} N_i \overline{\overline{G}}_i(T, P) + \sum_{i=1}^{c} N_i RT ln \frac{(y_i P)^{\frac{\overline{f}_i(T,P,y)}{(y_i P)}}}{(P)^{\frac{\overline{f}(T,P)}{(P)}}} \tag{13}$$

For a general single-phase reaction,

$$\overline{G}_i(T, P, \xi_i) = \overline{G}_i^{\,o} + \left[\overline{G}_i(T, P, \xi_i) - \overline{G}_i^{\,o} \right] \tag{14}$$

$\overline{G}_i^{\,o}$ is the standard Gibbs energy of species, i. ($T = 298$ K; $P = 1$ bar and initial ξ_i, $\xi_i^{\,o}$).

Again,

$$\overline{G}_i(T, P, \xi_i) = \overline{G}_i^{\,o}(T = 298K, P = 1bar, \xi_i^{\,o}) + RT ln \frac{\overline{f}_i(T, P, \xi_i)}{\overline{f}_i^{\,o}(T, P, \xi_i^{\,o})} \tag{15}$$

$$\frac{\overline{f}_i(T, P, \xi_i)}{\overline{f}_i^{\,o}(T, P, \xi_i^{\,o})} = a_i = \text{activity of } i^{\text{th}} \text{ species} \tag{16}$$

Therefore,

$$a_i = \exp \left(\frac{\overline{G}_i(T, P, \xi_i) - \overline{G}_i^{\,o}(T = 298K, P = 1bar, \xi_i^{\,o})}{RT} \right) \tag{17}$$

Under equilibrium,

$$-\frac{\Delta_{rxn}G^o}{RT} = ln \prod_{i=1}^{c} a_i^{\vartheta_i} \tag{18}$$

The equilibrium constant is defined as follows:

$$K_a(T) = \prod_{i=1}^{c} a_i^{\vartheta_i} = \exp \left(-\frac{\Delta_{rxn}G^o}{RT} \right) \tag{19}$$

This equation can be used to predict the values of variables under the equilibrium state.

$$\Delta_{rxn}G^o = \sum_{i=1}^{c} \vartheta_i \Delta G_{fi}^{\,o} \tag{20}$$

$\Delta G_{fi}^{\,o} = \text{standard heat of formation of } i^{\text{th}} \text{ component}$

$$\frac{\partial}{\partial T}\left(\frac{\overline{G}_i}{T}\right)_P = \frac{1}{T}\left(\frac{\partial \overline{G}_i}{\partial T}\right)_P - \frac{\overline{G}_i}{T^2} = -\frac{\overline{S}_i}{T} - \frac{\overline{H}_i}{T^2} + \frac{\overline{S}_i}{T} = -\frac{\overline{H}_i}{T^2} \tag{21}$$

$$lnK_a = -\frac{\sum_{i=1}^{c}\vartheta_i\Delta_f G_i{}^o}{RT} \tag{22}$$

Therefore,

$$\left(\frac{dlnK_a}{dT}\right)_P = -\frac{1}{R}\frac{d}{dT}\left(\frac{\sum_{i=1}^{c}\vartheta_i\Delta G_{fi}{}^o}{T}\right) = \frac{1}{RT^2}\sum_{i=1}^{c}\vartheta_i\Delta_f H_i{}^o = \frac{\Delta_{rxn}H^o}{RT^2} \text{ (van't Hoff equation)}$$

$$\tag{23}$$

$$\Delta_{rxn}H^o = \sum_{i=1}^{c}\vartheta_i\Delta_f H_i{}^o \tag{24}$$

Functionality of constant pressure heat capacity on temperature is as follows:

$$c_{pi} = a_i + b_i T + c_i T^2 + d_i T^3 \tag{25}$$

The values of the constants are listed in Table 2.

Now, the relationship between the heat of reaction at any temperature T and that at the reference temperature T_R is as follows:

$$\Delta_{rxn}H^o(T) = \Delta_{rxn}H^o(T_R) + \Delta a(T - T_R) + +\frac{\Delta b}{2}\left(T^2 - T_R{}^2\right) + \frac{\Delta c}{2}\left(T^3 - T_R{}^3\right)$$
$$+ \frac{\Delta d}{2}\left(T^4 - T_R{}^4\right) - \Delta e\left(\frac{1}{T} - \frac{1}{T_R}\right) \tag{26}$$

$$\ln\frac{K_a(T)}{K_a(T_R)} = -\frac{\Delta_{rxn}H^o(T_R)}{R}\left(\frac{1}{T} - \frac{1}{T_R}\right) + \frac{\Delta_{rxn}c_p}{R}\ln\frac{T}{T_R} + \frac{\Delta_{rxn}c_p}{R}\left(\frac{T_R}{T} - 1\right) \tag{27}$$

$$K_a(T) = \prod_{i=1}^{c}a_i{}^{\vartheta_i} \tag{28}$$

$$a_i = \frac{y_i P}{1\ bar} \tag{29}$$

Therefore,

$$K_a(T) = \prod_{i=1}^{c}\left(\frac{y_i P_{rxn}}{1\ bar}\right)^{\vartheta_i} \tag{30}$$

For any reaction, $\alpha A + \beta B \rightarrow \gamma C + \delta D$

$$K_a(T) = \frac{y_C^\gamma y_D^\delta}{y_A^\alpha y_B^\beta} \left(P^{\gamma + \delta - \alpha - \beta} \right) \tag{31}$$

2.1.1 Combined phase and chemical equilibrium

The total number of moles of species I in all P phases is given by,

$$\sum_{k=1}^{P} N_i^k = N_{i0} + \sum_{j=1}^{M} \vartheta_{ij} \xi_j \tag{32}$$

Total Gibbs-free energy is given by,

$$G = \sum_{k=1}^{P} \sum_{i=1}^{C} N_i^k \overline{G}_i^k \tag{33}$$

The condition of phase equilibrium should be satisfied for all species in each phase, and chemical reactions are in equilibrium. Therefore, the following equation is valid:

$$\frac{\partial G}{\partial \xi_j} = 0 = \sum_{k=1}^{P} \sum_{i=1}^{C} N_i^k \left(\frac{\partial \overline{G}_i^k}{\partial \xi_j} \right) - \sum_{i=1}^{C} a_i \vartheta_{ij} = 0 \tag{34}$$

Again,

$$\sum_{i=1}^{C} \vartheta_{ij} \overline{G}_i^k = 0 \tag{35}$$

This signifies that all reactions are at equilibrium in each phase.

Phase rule is applied in the PVT system to obtain the number of independent variables to be specified, that is, the number of degrees of freedom, F of a system under equilibrium [23]. The phase rule is stated as follows:

$$F = 2 - \pi + N - r \tag{36}$$

where F=degrees of freedom; π=number of phases; r=number of independent reactionsThis rule is applied to determine the unknown composition.

For the application of the abovementioned principles in case of synthesis gas production, the following steps are followed:

2.1.2 Step-I

The generalized stoichiometric equation representing the syngas production is usually written as the first step of the stoichiometric modeling:

$$CH_aO_bN_cS_d + \vartheta_wH_2O + \vartheta_o(O_2 + \varepsilon N_2) \rightarrow \vartheta_{H_2}H_2 + \vartheta_{co}CO + \vartheta_{CO_2}CO_2 + \vartheta_{CH_4}CH_4$$
$$+ \vartheta_{H_2O}H_2O + \vartheta_{N_2}N_2 + \vartheta_cC + \vartheta_tTar \tag{37}$$

where the molecular formula of the biomass is based on C -mole and hence

$$\alpha = \frac{w_H(MW)_C}{w_C(MW)_H}; b = \frac{w_O(MW)_C}{w_C(MW)_O}; c = \frac{w_N(MW)_C}{w_C(MW)_N}; d = \frac{w_S(MW)_C}{w_C(MW)_S} \tag{38}$$

w_C, w_H, w_O, w_N. and w_S are the weight fraction of C, H, O, N, and S, respectively, determined through the elemental analysis of the biomass.

Similarly, $(MW)_C$, $(MW)_H$, $(MW)_O$, $(MW)_N$, *and* $(MW)_S$ are the molar mass of C, H, O, N, and S. The molar ratio of nitrogen to oxygen in air is $\varepsilon\left(= \frac{y_{N_2}}{y_{O_2}} = \frac{0.79}{0.21} = 3.76\right)$.

The phase rule (Eq. 36) is applied to determine the unknown composition. Under this situation,

$$F = 2 - \pi + N - r = 2 - 3 + 11 - 1 = 9 \tag{39}$$

Therefore, specification of temperature and pressure in three atom balance equations [Step II], three equilibrium constants [Step III], and one energy balance equation [Step IV]leads to the evaluation of equilibrium composition of syngas.

2.1.3 Step II

Based on the stoichiometric equation, I, the atom balance equations are written to determine the number of moles of product components, that is, H_2, CO, CO_2, CH_4, H_2O, N_2, C, and Tar. Tar is a complex combination of different hydrocarbons and is generally represented by $C_xH_yO_z$. For simplification, in some literature, tar is considered to be C_6H_6 [26] and thus, $x=6$, $y=6$ and $z=0$. In another reported article, $CH_{1.003}O_{0.33}$ has been used to represent tar (i.e., $x=1$, $y=1.003$, $z=0.33$) [27]. Tar formation is a function of temperature and other input parameters.

In terms of the generalized representation of chemical formula, $C_xH_yO_z$ of tar Eq. (37) can be represented as follows:

$$CH_aO_bN_cS_d + \vartheta_wH_2O + \vartheta_o(O_2 + \varepsilon N_2) \rightarrow \vartheta_{H_2}H_2 + \vartheta_{co}CO + \vartheta_{CO_2}CO_2$$
$$+ \vartheta_{CH_4}CH_4 + \vartheta_{H_2O}H_2O + \vartheta_{N_2}N_2 + \vartheta_cC + \vartheta_tC_xH_yO_z \tag{40}$$

Atom balance equations for C, H, and O are as follows:

$$C : 1 = \vartheta_{co} + \vartheta_{CO_2} + \vartheta_{CH_4} + \vartheta_c + x\vartheta_t$$

Or,

$$\vartheta_{co} + \vartheta_{CO_2} + \vartheta_{CH_4} + \vartheta_c + x\vartheta_t - 1 = 0 \tag{41}$$

$$H : a + 2\vartheta_w = 2\vartheta_{H_2} + 4\vartheta_{CH_4} + 2\vartheta_{H_2O} + y\vartheta_t$$

Or,

$$2\vartheta_{H_2} + 4\vartheta_{CH_4} + 2\vartheta_{H_2O} + y\vartheta_t - a - 2\vartheta_w = 0 \tag{42}$$

$$O : b + \vartheta_w + 2\vartheta_o = \vartheta_{co} + 2\vartheta_{CO_2} + \vartheta_{H_2O} + z\vartheta_t$$

$$\vartheta_{co} + 2\vartheta_{CO_2} + \vartheta_{H_2O} + z\vartheta_t - b - \vartheta_w - 2\vartheta_o = 0 \tag{43}$$

2.1.4 Step III

Since the reactions involving oxygen are expected to be completed under gasification condition, most of the reported articles consider that the attainment of equilibrium of two or three of the following reactions (Table 3) influences the exit product composition [16]:

According to Eq. (19), $K(T) = \prod\limits_{i=1}^{c} a_i^{\vartheta_i} = \exp\left(-\frac{\Delta_{rxn}G^o}{RT}\right)$

Since according to Eq. (29), $a_i = \frac{y_i P}{1\,bar}$,

$$K_a(T) = \prod_{i=1}^{c}\left(\frac{y_i P_{rxn}}{1\,bar}\right)^{\vartheta_i} = \exp\left(-\frac{\Delta_{rxn}G^o}{RT}\right) \tag{44}$$

Applying Eq. (36),

$$K_{al}(T) = \frac{(y_{CO})^2}{y_{CO_2}^{1}}\left(\frac{P_{rxn}}{P = 1bar}\right)^{2-1} = \frac{(y_{CO})^2}{y_{CO_2}}\frac{P_{rxn}}{P = 1bar} \tag{45}$$

Therefore,

$$\frac{(y_{CO})^2}{y_{CO_2}}\frac{P_{rxn}}{P = 1bar} = \exp\left(-\frac{\Delta_{rxn,I}G^o}{RT}\right) \tag{46}$$

$$K_{all}(T) = \frac{y_{CO}^{1} y_{H_2}^{1}}{y_{H_2O}^{1}}\left(\frac{P_{rxn}}{P = 1bar}\right)^{1+1-1} = \frac{y_{CO} y_{H_2}}{y_{H_2O}}\left(\frac{P_{rxn}}{P = 1bar}\right) \tag{47}$$

Therefore,

$$\frac{y_{CO} y_{H_2}}{y_{H_2O}}\left(\frac{P_{rxn}}{P = 1bar}\right) = \exp\left(-\frac{\Delta_{rxn,II}G^o}{RT}\right) \tag{48}$$

Table 3 Reactions considered for attainment of equilibrium.

Boudouard reaction:	$C + CO_2 \leftrightarrow 2CO$ (R5)
Char reforming:	$C + H_2O \leftrightarrow CO + H_2$ (R6)
Methane forming:	$C + 2H_2 \leftrightarrow CH_4$ (R7)
Water gas shift reaction:	$CO + H_2O \leftrightarrow CO_2 + H_2$ (R10)
Methane reforming:	$CH_4 + H_2O \leftrightarrow CO_2 + 3H_2$ (R11)

$$K_{aIII}(T) = \frac{y_{CH_4}{}^1}{y_{H_2}{}^2} \left(\frac{P_{rxn}}{P = 1bar} \right)^{1-2} = \frac{y_{CH_4}{}^1}{y_{H_2}{}^2} \left(\frac{P_{rxn}}{P = 1bar} \right)^{-1} \tag{49}$$

Therefore,

$$\frac{y_{CH_4}}{y_{H_2}{}^2} \left(\frac{P_{rxn}}{P = 1bar} \right)^{-1} = \exp\left(-\frac{\Delta_{rxn,III}G^o}{RT} \right) \tag{50}$$

$$K_{aIV}(T) = \frac{y_{CO_2}{}^1 y_{H_2}{}^1}{y_{CO}{}^1 y_{H_2O}{}^1} \left(\frac{P_{rxn}}{P = 1bar} \right)^{1+1-1-1} = \frac{y_{CO_2} y_{H_2}}{y_{CO} y_{H_2O}} \tag{51}$$

Therefore,

$$\frac{y_{CO_2} y_{H_2}}{y_{CO} y_{H_2O}} = \exp\left(-\frac{\Delta_{rxn,IV}G^o}{RT} \right) \tag{52}$$

$$K_{aV}(T) = \frac{y_{CO_2}{}^1 y_{H_2}{}^3}{y_{CH_4}{}^1 y_{H_2O}{}^1} \left(\frac{P_{rxn}}{P = 1bar} \right)^{1+3-1-1} = \frac{y_{CO_2} y_{H_2}}{y_{CH_4} y_{H_2O}} \left(\frac{P_{rxn}}{P = 1bar} \right)^2 \tag{53}$$

$$\frac{y_{CO_2} y_{H_2}}{y_{CH_4} y_{H_2O}} \left(\frac{P_{rxn}}{P = 1bar} \right)^2 = \exp\left(-\frac{\Delta_{rxn,V}G^o}{RT} \right) \tag{54}$$

According to Eq. (20), $\Delta_{rxn}G^o = \sum_{i=1}^{c} \vartheta_i \Delta G_{fi}^o$

Where, ΔG_{fi}^o = standard heat of formation of ith component.

Chemical equilibria of three reactions, namely, methane formation, water gas shift reaction, and char reforming reaction are the most common ones which are considered by the researchers in the stoichiometric equilibrium modeling of syngas production [28,29]. Some of the researchers have also considered water gas shift reaction and methane reforming [30]. Methane formation and methane reforming have also been considered by some other researchers [31].

2.1.5 Step IV-energy balance

The generalized energy balance equation for a flow system is as follows:

$$\Delta \dot{H} + \Delta \dot{E}_k + \Delta \dot{E}_p = \dot{Q} - \dot{W}_s \tag{55}$$

where,

$$\Delta \dot{H} = \text{rate of change of enthalpy;}$$

$$\Delta \dot{E}_k = \text{rate of change of kinetic energy;}$$

$$\Delta \dot{E}_p = \text{rate of change of potential energy;}$$

$$\dot{Q} = \text{rate of heat transfer between the system and the surroundings;}$$

$$\dot{W}_s = \text{shaft work}$$

For the reactor,

$$\Delta \dot{E}_k = \Delta \dot{E}_p = \dot{W}_s = 0 \tag{56}$$

Therefore,

$$\Delta \dot{H} = \dot{Q} \tag{57}$$

When non-stoichiometric amounts of reactants and products enter and leave respectively at different temperatures, the following steps have to be followed:

- Declare a reference state (usually 25°C and 1 atm)
- Calculate enthalpies of each stream. For each stream, three components of enthalpy may be involved:
- (a) heat of formation; (b) sensible heat relative to the reference state; (c) latent heat if phase change is involved

Therefore, if no phase change is considered,

$$\sum_{i=reactant} \dot{n}_i \left(\hat{h}^o_{f,i} + \Delta \hat{h}_i(T_{i,in}) \right) + \dot{Q}_{in} = \sum_{p=product} \dot{n}_p \left(\hat{h}^o_{f,p} + \Delta \hat{h}_p(T_{j,rxn}) \right) + \dot{Q}_{out} \tag{58}$$

$\hat{h}^o_{f,i}$ and $\hat{h}^o_{f,p}$ are the heats of formation per mole of reactant, i and product, p, respectively. $\Delta \hat{h}_i$ and $\Delta \hat{h}_p$ are the specific enthalpy changes associated with sensible heat relative to the reference state for reactant, i and product, p.

Under adiabatic condition, there is no heat exchange with surroundings and hence,

$$\dot{Q}_{in} = \dot{Q}_{out} = 0$$

Therefore,

$$\sum_{i=reactant} \dot{n}_i \left(\hat{h}^o_{f,i} + \Delta \hat{h}_i(T_{i,in}) \right) = \sum_{p=product} \dot{n}_p \left(\hat{h}^o_{f,p} + \Delta \hat{h}_p(T_{j,rxn}) \right) \tag{59}$$

If inert components are also fed to the reactor involved,

$$\sum_{in} \dot{n}_{in} \left(\hat{h}^o_{f,in} + \Delta \hat{h}_{in}(T_{i,in}) \right) = \sum_{out} \dot{n}_{out} \left(\hat{h}^o_{f,out} + \Delta \hat{h}_{out}(T_{j,rxn}) \right) \tag{60}$$

$\hat{h}^o_{f,in}$ and $\hat{h}^o_{f,out}$ are the specific heat of formation of inlet and outlet components, respectively. $\Delta \hat{h}_{in}$ and $\Delta \hat{h}_{out}$ are the specific enthalpy changes associated with sensible heat relative to reference temperature, T_R for inlet and outlet components, respectively. Standard heats of formation of most of the conventional compounds are available in the literature [23].

For multiple reactions, the energy balance equation for an adiabatic reactor reduces to

$$\sum_{reactions} \sum \dot{\xi}_j \Delta H^o_{rxn,j} + \sum_{out} \dot{n}_{out} \left(\Delta \hat{h}_{out}(T_{j,rxn}) \right) - \sum_{in} \dot{n}_{in} \left(\Delta \hat{h}_{in}(T_{i,in}) \right) = 0 \tag{61}$$

Sensible enthalpy $\left(\sum\limits_{out} \dot{n}_{out}\left(\Delta \hat{h}_{out}(T_{rxn}) \right) and \sum\limits_{in} \dot{n}_{in}\left(\Delta \hat{h}_{in}(T_{in}) \right) \right)$ relative to reference temperature T_R are as follows:

$$\sum_{out} \dot{n}_{out}\left(\Delta \hat{h}_{out}\left(T_{j,rxn} \right) \right) = \sum_{out} \dot{n}_{out} c_{p,out} \int_{T_R}^{T_{rxn}} c_{p,out} dT \tag{62}$$

$$\sum_{in} \dot{n}_{in}\left(\Delta \hat{h}_{in}\left(T_{i,in} \right) \right) = \sum_{out} \dot{n}_{in} c_{p,in} \int_{T_R}^{T_{in}} c_{p,in} dT \tag{63}$$

The correlation of molar-specific heat capacity at constant pressure with temperature (in K) is as follows:

$$c_{pi} = a_i + b_i T + c_i T^2 + d_i T^3 \tag{64}$$

Standard heats of formation for most of the conventional compounds are available in the literature [23].

Following *Step I*, the generalized stoichiometric equation of the overall gasification reaction is represented by Eq. (40) as follows:

$$CH_a O_b N_c S_d + \vartheta_w H_2 O + \vartheta_o (O_2 + \varepsilon N_2) \rightarrow \vartheta_{H_2} H_2 + \vartheta_{co} CO + \vartheta_{CO_2} CO_2 +$$
$$\vartheta_{CH_4} CH_4 + \vartheta_{H_2O} H_2 O + \vartheta_{N_2} N_2 + \vartheta_c C + \vartheta_t Tar$$

Therefore, according to Eq. (54), the energy balance equation reduces to the following one:

$$\hat{h}^o_{f,Biomass} + \Delta \hat{h}_{Biomass}(T_{i,in}) + \vartheta_w \left(\hat{h}^o_{f,w} + \Delta \hat{h}_w(T_{i,in}) \right) + \vartheta_o \left(\hat{h}^o_{f,O_2} + \Delta \hat{h}_{O_2}(T_{i,in}) \right)$$
$$+ \vartheta_o \varepsilon \left(\hat{h}^o_{f,N_2} + \Delta \hat{h}_{N_2}(T_{i,in}) \right)$$
$$= \vartheta_{H_2}\left(\hat{h}^o_{f,H_2} + \Delta \hat{h}_{H_2}(T_{i,in}) \right) + \vartheta_{co}\left(\hat{h}^o_{f,co} + \Delta \hat{h}_{co}(T_{i,in}) \right) + \vartheta_{CO_2}\left(\hat{h}^o_{f,CO_2} + \Delta \hat{h}_{CO_2}(T_{i,in}) \right)$$
$$+ \vartheta_{CH_4}\left(\hat{h}^o_{f,CH_4} + \Delta \hat{h}_{CH_4}(T_{i,in}) \right) + \vartheta_{H_2O}\left(\hat{h}^o_{fH_2O} + \Delta \hat{h}_{H_2O}(T_{i,in}) \right) + \vartheta_{N_2}\left(\hat{h}^o_{f,N_2} + \Delta \hat{h}_{N_2}(T_{i,in}) \right)$$
$$+ \vartheta_c\left(\hat{h}^o_{f,c} + \Delta \hat{h}_c(T_{i,in}) \right) + \vartheta_t\left(\hat{h}^o_{f,t} + \Delta \hat{h}_t(T_{i,in}) \right)$$

$$\tag{65}$$

Since the standard heat of formation of O_2, N_2, and H_2 are zero.

$$\hat{h}^o_{f,Biomass} + \Delta \hat{h}_{Biomass}(T_{i,in}) + \vartheta_w \left(\hat{h}^o_{f,w} + \Delta \hat{h}_w(T_{i,in}) \right) +$$
$$= +\vartheta_{co}\left(\hat{h}^o_{f,co} + \Delta \hat{h}_{co}(T_{i,in}) \right) + \vartheta_{CO_2}\left(\hat{h}^o_{f,CO_2} + \Delta \hat{h}_{CO_2}(T_{i,in}) \right)$$
$$+ \vartheta_{CH_4}\left(\hat{h}^o_{f,CH_4} + \Delta \hat{h}_{CH_4}(T_{i,in}) \right) + \vartheta_{H_2O}\left(\hat{h}^o_{fH_2O} + \Delta \hat{h}_{H_2O}(T_{i,in}) \right) + \vartheta_c\left(\hat{h}^o_{f,c} + \Delta \hat{h}_c(T_{i,in}) \right)$$
$$+ \vartheta_t\left(\hat{h}^o_{f,t} + \Delta \hat{h}_t(T_{i,in}) \right)$$

$$\tag{66}$$

Heat of formation of biomass is calculated from the lower heating value (LHV) of biomass.

The combustion reaction of the biomass can be represented as follows:

$$CH_aO_bN_cS_d + \left(1 + \frac{a}{4} + c + d - \frac{b}{2}\right)O_2 \rightarrow CO_2 + \frac{a}{2}H_2O(v) + cNO_2 + dSO_2 \tag{67}$$

$$LHV = -\Delta H^o_{c,Biomass}\widehat{h}^o_{f,Biomass} - \widehat{h}^o_{f,CO_2} - \frac{a}{2}\widehat{h}^o_{fH_2O}(v) - c\widehat{h}^o_{f,NO_2} - d\widehat{h}^o_{f,SO_2} \tag{68}$$

Or,

$$\widehat{h}^o_{f,Biomass} = LHV + \widehat{h}^o_{f,CO_2} + \frac{a}{2}\widehat{h}^o_{fH_2O}(v) + c\widehat{h}^o_{f,NO_2} + d\widehat{h}^o_{f,SO_2} \tag{69}$$

The value of $\widehat{h}^o_{f,Biomass}$ obtained using Eq. (63) is substituted in Eq. (56).

2.1.6 Solution of stoichiometric equilibrium model

In the simplest models without considering the yield of char and tar, three mass balance equations (Eqs. 33–35), two equilibrium reaction equations selected from Eqs. (38), (42), (44), (46), and (48), and one energy balance equation (Eq. 60) are solved. When tar and char yields are considered, more equilibrium reactions are chosen from reactions III to VII. All equations are solved using the Newton-Raphson method. Two strategies are followed. In Strategy-I, the equivalence ratio (ER) is the input. All equations are solved simultaneously using the Newton-Raphson method, and the temperature is determined. In Strategy-II, temperature is the input, and all the mass balance equations and the equilibrium equations are solved. In the next step, the energy balance equation is solved to determine the temperature. The input value of temperature and the determined value are compared until their convergence.

2.2 Non-stoichiometric equilibrium model

The Gibbs-free energy minimization approach is a suitable option for modeling when the product of gasification is known, but their chemical reactions are unknown.

For the closed system, the energy balance equation is as follows [25]:

$$\frac{dU}{dt} = \dot{Q} + \dot{W}_s - P\frac{dV}{dt} \tag{70}$$

and

$$\frac{dS}{dt} = \frac{\dot{Q}}{T} + \dot{S}_{gen}$$

For the multicomponent non-reactive system,

$$U = \sum_{i=1}^{c} N_i \overline{U}_i(T, P, y) \tag{71}$$

$$S = \sum_{i=1}^{c} N_i \bar{s}_i(T,P,y) \tag{72}$$

Equilibrium criteria are again the maximum and minimum values of entropy and Gibbs energy, respectively.

Gibbs energy for the reactive system can be written as follows:

$$G = \sum_{i=1}^{c} N_i \overline{G}_i = \sum_{i=1}^{c} (N_{i0} + \vartheta_i \xi) \overline{G}_i \tag{73}$$

At constant temperature and pressure, the extent of reaction, ξ is the only variable.

Hence, the equilibrium criterion is

$$\left(\frac{\partial G}{\partial \xi}\right)_{T,P} = 0 \tag{74}$$

This leads to

$$\sum_{i=1}^{c} \vartheta_i \overline{G}_i = 0 \tag{75}$$

This correlation is also valid when the temperature and pressure are not constant.

For a system undergoing M number of multiple reactions, number of moles of the *i*th component can be written as

$$N_i = \sum_{j=1}^{M} \left(N_{i0} + \vartheta_{ij} \xi_j\right) \tag{76}$$

Total Gibbs energy can be written as follows:

$$G = \sum_{i=1}^{c} N_i \overline{G}_i = \sum_{i=1}^{c} \left(N_{i0} + \sum_{i=1}^{M} \vartheta_{ij}\xi_j\right) \overline{G}_i = \sum_{i=1}^{c} N_{i0}\overline{G}_i + \sum_{i=1}^{c}\sum_{j=1}^{M} \left(\vartheta_{ij}\xi_j\overline{G}_i\right) \tag{77}$$

If chemical equilibrium is attained,

$G = $ minimumor, $dG = 0$ or, $\left(\frac{\partial G}{\partial \xi_j}\right)_{T,P,y_{i \neq j}} = 0, j = 1, 2, \ldots .M$
Or,

$$\sum_{i=1}^{c} \vartheta_{ij}\overline{G}_i + \sum_{i=1}^{c} N_i \left(\frac{\partial G}{\partial \xi_j}\right)_{T,P,y_{i \neq j}} = 0 \tag{78}$$

According to Gibbs-Duhem equation,

$$\sum_{i=1}^{c} N_i \left(\frac{\partial G}{\partial \xi_j}\right)_{T,P,y_{i \neq j}} = 0 \tag{79}$$

Therefore,

$$\sum_{i=1}^{c} \vartheta_{ij}\overline{G}_i = 0, j = 1, 2....M \tag{80}$$

Eq. (80) implies that if chemical equilibrium is achieved, all simultaneous reactions are in equilibrium.

According to the Gibbs-Duhem equation,

$$\sum_{i=1}^{c} N_i \left(\frac{\partial G}{\partial \xi_j}\right)_{T,P,y_{i\neq j}} = 0 \tag{81}$$

Therefore,

$$\sum_{i=1}^{c} \vartheta_{ij}\overline{G}_i = 0, \ j = 1, 2....M \tag{82}$$

$$G = H - TS \tag{83}$$

$$d(nG) = -nSdT + nVdP + \sum_i \mu_i dn_i \tag{84}$$

The quantity μ_i is called chemical potential of species I, and it plays an important role in phase and chemical equilibria. This is actually partial molar Gibbs energy. Therefore,

$$\mu_i = \left(\frac{\partial nG}{\partial n_i}\right)_{T,P,n_j} \tag{85}$$

$$\left(\frac{\partial \mu_i}{\partial P}\right)_{T,n} = \overline{v}_i; \tag{86}$$

$$\left(\frac{\partial \mu_i}{\partial T}\right)_{P,n} = -\overline{S}_i \tag{87}$$

where μ_i is the chemical potential of ith species.

$$\mu = f(T, P) \tag{88}$$

$$d\mu_i = \overline{v}_i dP - \overline{S}_i dT = d\overline{G}_i \tag{89}$$

$$G = \sum_i \mu_i n_i \tag{90}$$

n_i is number of moles.

$$\sum_i n_i a_{ik} = A_k \ (k = 1, 2, ...W); \tag{91}$$

a_{ik} = number of atoms of the kth element in each molecule of i

A_k is the total number of atomic masses of the kth element

W = No.of atoms present in the system

Chemical potential is calculated using the following equation

$$\mu_i = RT \left[ln \left(\frac{f_i P}{P_0} \right) + ln\, y_i + G_i^o(T, P_0) \right] \tag{92}$$

The values of standard free energy of formation are utilized for the prediction of chemical change through minimization of Gibbs-free energy.

$$\Delta G^o = \Delta H^o - T \Delta S^o \tag{93}$$

2.2.1 Solution of the non-stoichiometric model

The solution of this model involves minimization of the Gibbs-free energy function. The minimization is carried out by various optimization methods such as Lagrange Multipliers method, Morley method, Gordon and McBride method, and RAND method to name a few [32].

2.2.2 Lagrange multipliers method

Most of the non-stoichiometric equilibrium model equations are solved using the Lagrange Multipliers method. The equation obtained by applying Lagrangean multipliers method and using Eqs. (81)–(84) is as follows:

$$\sum_k \lambda_k \left(\sum_k n_i a_{ik} - A_k \right)_i = 0 \tag{94}$$

$$\mu_i + \sum_k n_i a_{ik} = 0 \tag{95}$$

The minimization technique involves the minimization of the function G with respect to n_i. For this purpose, the Lagrangian function L is formed with the help of mass balance constraints as [25].

$$L = G - \sum_k \lambda_k \left(\sum_k n_i a_{ik} - A_k \right)_i \tag{96}$$

The minimum point of this function is calculated by setting the partial derivative of L with respect to n_i equal to zero.

$$\frac{\delta L}{\delta n_i} = 0 \tag{97}$$

This Eq. (14) creates a nonlinear equation for each species in the reactor. For multiple species, a set of equations is created and solved by an iterative technique using the Newton-Raphson method.

Morley method

This optimization technique is basically a modification of the Newton-Raphson method proposed by Morley, 2005 [33]. In this technique, the next value of the function $\frac{\delta L}{\delta n_i}$ is approximated by a first-order Taylor development of $\frac{\delta L}{\delta n_i}$ function. This method is limited to gas species only, and solid materials cannot be handled by this method.

Gordon and McBride method Gordon and McBride [34] proposed this method which is much more improved than the Morley method. In this method, they used a first-order Taylor expansion of $C_i = \ln(n_i)$ and $D = \ln \left(\sum_i^{NG} n_i \right)$ to fit the data much better with the evolution of G.

RAND algorithm **The** RAND algorithm which has been widely used in various non-stoichiometric equilibrium modeling has been proposed by Gautam et al. [35]. According to this algorithm, a quadratic Taylor development is approximated to use it as a vector of mole numbers for next iterations. This technique is also used in Aspen plus software for calculating Gibbs-free energy minimization at RGibbs block [26,36].

Along with these algorithms, linear programming, genetic algorithm, and Monte Carlo method can also be utilized for the minimization of Gibbs-free energy [32].

2.3 Aspen plus simulation of equilibrium models

Equilibrium models of gasification have been developed using Aspen Plus for a long time now. Aspen plus is a software that simulates various chemical processes with inbuilt unit operation blocks or user defined blocks with the help of its large quantity of physical and chemical data library of various components. Gasification is not present in Aspen plus as a single-unit operation block. Therefore, it is necessary to connect various unit operations and transfer and manipulator blocks to form a consolidated simulation. The whole process is completed in a number of steps. The flow diagram is provided in Fig. 2.

2.3.1 Declaration of components and property methods

First, it is necessary in Aspen Plus to declare the components participating in the gasification process. It is known that the biomass/coal/municipal wastes are not defined by a standard molecular formula. Therefore, the gasification feedstock is declared as a non-conventional component in Aspen Plus. As the component is non-conventional, it is necessary to provide its

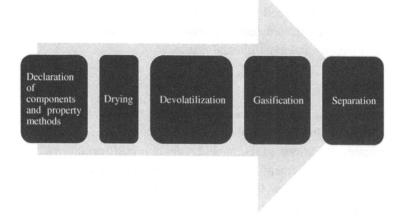

Fig. 2
Methodical representation of Aspen plus equilibrium modeling.

ultimate and proximate analysis along with an enthalpy and density model. Enthalpy and density models chosen in most of the literatures are HCOALGEN and DCOALIGT [37]. Other conventional components are also specified from the Aspen plus database. A suitable property method stating the equation of state for the gaseous and liquid elements is then selected to calculate the thermodynamic parameters of various elements. In gasification modeling IDEAL method [38], Peng-Robinson cubic equation of state (PENG-ROB) [26,39,40], Peng-Robinson cubic equation of state with Boston-Mathias alpha function (PR-BM) [41–44], Redlich-Kwong cubic equation of state, Redlich-Kwong cubic equation of state with Boston-Mathias alpha function (RK-BM), Redlich-Kwong-Soave cubic equation of state, and Redlich-Kwong-Soave cubic equation of state with Boston-Mathias alpha function (RKS-BM) [45,46] are the most utilized property methods. Using these methods, the properties of gaseous elements are needed to be determined by running the property analysis.

2.3.2 Drying

The simulation is conducted first by passing the biomass through a stoichiometric block (RStoic) for drying. Here, with the help of dry air, a pseudo-stoichiometric reaction is assumed and moisture is released from the feedstock at a temperature range of 100–150°C. The equation for drying is as follows.

$$\text{Biomass} \rightarrow \text{dry biomass} + 0.0555 H_2O \qquad (98)$$

2.3.3 Devolatilisation

The dry biomass is then transferred to a decomposer block (RYield) to convert the non-conventional biomass into conventional elements according to its ultimate analysis at a temperature range of 350–500°C. A calculator block is used to specify the yields of the products. This process is basically a simulative assumption of the pyrolysis process where most of the researchers have approached it with breaking the biomass according to its ultimate analysis [47–49]. However, in some cases, researchers have also tried to specify the yields of pyrolysis products from actual experimental results.

2.3.4 Gasification

Furthermore, RStoic or Rgibbs blocks are used to carry out the oxidation and reduction equations of gasification. In the Rstoic block, reaction equilibrium is performed with the help of equilibrium constants of various oxidation and reduction gasification reactions, whereas the RGibbs block performs the elemental equilibrium by minimizing the Gibbs-free energy of the reactant and product mixture. Now, the simulation of gasification reactions can be performed in two methods. The first one is to separate the solid char from the pyrolysis mixture and performing the heterogeneous reactions separately. This process can be represented as the Eq-sep procedure. The second one is to consider all the products at the same time and then send them to an RGibbs block for simultaneous reaction simulation. This process can be represented as the Eq-sing procedure.

2.3.5 Separation

After the gasification is complete, the gas-solid mixture is sent to a separator block for the removal of ash and any other solid impurities present after gasification.

Fig. 3 represents a typical process flow diagram of Aspen plus modeling.

2.4 Literatures in Aspen plus equilibrium modeling

Many researchers have developed equilibrium gasification models with the help of Aspen plus which are majorly focused on the Gibbs-free energy minimization technique [50]. Table 4 shows usage of different unit operation blocks by various authors along with their selected property methods and modeling procedure. It is quite self-explanatory from the table that in most *Aspen plus equilibrium modeling*, the non-stoichiometric method is followed using the RYield and RGibbs blocks.

3. Application of equilibrium models

Equilibrium models, both stoichiometric and non-stoichiometric ones, are used to analyze the sensitivity of the syngas composition toward the gasifying agent (oxygen/air and steam), reaction temperature, C:H ratio in the feedstock, steam to solid feedstock ratio, and equivalence ratio (ER). Equivalence ratio, ER, is defined as the ratio of the actual number of moles of oxygen to the number of moles of oxygen required for complete combustion of the biomass or

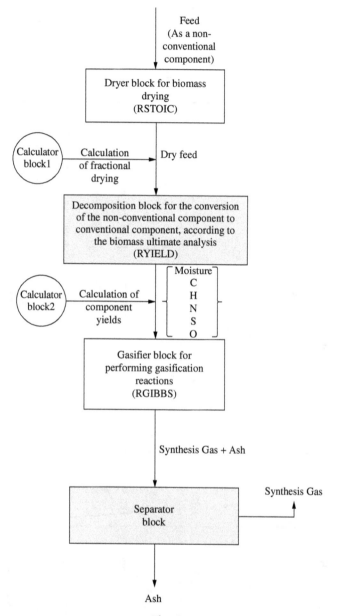

Fig. 3
Process flow diagram of Aspen plus modeling.

coal. If n_{O_2} be the number of moles of oxygen fed per unit mole of feedstock ($CH_aO_bN_cS_d$) for gasification, the ER can be represented by the following equation [16]

$$ER = \frac{n_{O_2}}{1 + \frac{a}{4} + c + d - \frac{b}{2}} \tag{99}$$

Table 4 Modeling procedure of Aspen plus equilibrium models.

Authors	Aspen Plus blocks used	Property method	Tar formation	Modeling procedure	Validation
Sharmina Begum et al. [45]	RYield, RGibbs	RKS-BM	–	Gibbs-free energy minimization with two-stage gasification (separate modeling of oxidation and reduction reactions)	Literature
Tungalag et al. [38]	RYield, RGibbs	IDEAL	C_6H_6O, $C_{16}H_{10}$	Gibbs-free energy minimization with two-stage gasification (separate modeling of oxidation and reduction reactions)	Self-experiment
Gagliano et al. [26]	RYield, RGibbs	PENG-ROB	–	Gibbs-free energy minimization	Self-experiment
Zhai et al. [41]	RYield, RGibbs	PR-BM	C_6H_6	Staging pyrolysis with Gibbs-free energy minimization	Self-experiment
Doherty et al. [47]	RYield, RGibbs	PR-BM	–	Gibbs-free energy minimization with approach	Literature
Adnan et al. [39]	RYield, RGibbs, REquil	PENG-ROB	C_4H_4O	Gibbs-free energy minimization along with tar reformer	Literature
Wan [42]	RYield, RGibbs	PR-BM	C_6H_6, C_7H_8, $C_{10}H_8$	Gibbs-free energy minimization	Literature
Safarian et al. [44]	RYield, RGibbs	PR-BM	–	Gibbs-free energy minimization	Literature
Tavares et al. [46]	RYield, RGibbs	RKS-BM	–	Gibbs-free energy minimization	Literature
Okolie et al. [40]	RYield, RGibbs	PENG-ROB	–	Hydrothermal gasification with Gibbs-free energy minimization	Self-experiment
Ramzan et al. [36]	RYield, RGibbs	PR-BM	–	Gibbs-free energy minimization Assuming complete chemical equilibrium	Self-Experiment
Rupesh et al. [51]	RYield, RGibbs	–	–	Gibbs-free energy minimization with devolatilization modeling using empirical relations.	Literature
Han et al. [49]	RYiled, RGibbs	IDEAL	–	Gibbs-free energy minimization with restricted chemical equilibrium	Literature
Indrawan et al. [52]	RYiled, RGibbs	–	–	Gibbs-free energy minimization with two-stage gasification (separate modeling of oxidation and reduction reactions)	Literature

As ER affects the extent of combustion of the feedstock and hence the equilibrium composition, and it is used as one of the input parameters in most of the articles reported on modeling of syngas generation.

The following research articles have been reported in the literature on the application of stoichiometric and non-stoichiometric models.

3.1 Application of stoichiometric equilibrium modeling

This section deals with the most fundamental models of biomass gasification developed by various researchers for syngas production considering thermodynamic and chemical equilibrium inside the reactor (Table 5).

3.2 Application of non-stoichiometric equilibrium modeling

This section includes the non-stoichiometric equilibrium models developed following the Gibbs-free energy minimization approach (Table 6).

It is observed that non-stoichiometric equilibrium modeling has been mostly explored using Aspen plus as simulating software. The modeling procedure has been discussed in the next part.

3.3 Limitations of equilibrium models

The equilibrium models are able to predict the hypothetically possible yields of all components in product syngas. As one can expect, there always lies the deviation between model prediction and experimental results. The syngas yields predicted by the equilibrium models often differ from the experimental values in the real gasification process. This occurs majorly because the residence time of the gaseous species inside the gasifier is not as long as it is assumed, and char conversion is limited and various other products such as tars, hydrocarbons, and unconverted carbon also come out of the gasifier which are generally neglected by most of the researchers. The notable deviations are observed in (i) methane concentration (under-prediction) and (ii) char obtained (under-prediction). It has been observed by Loha et al. [57] that in a steam fed gasifier, the real reactor shows 4.2% methane in the product syngas, compared with the model prediction of 2%. Accurate predictions are possible through the introduction of empirical correlations.

4. Modification of equilibrium models to address deviations

To rectify the predictions of the models, different modifications are introduced in the equilibrium model by many researchers. Some of them are discussed below:

Table 5 Modeling procedures of various stoichiometric equilibrium models.

Authors	Modeling procedure	Reactions considered	validation
Zainal et al. [28]	Combined mass and energy balance along with stoichiometric equilibrium. All equations solved simultaneously using the Newton-Raphson algorithm	R5, R6, R7, R10	Literature
Jayah et al. [53]	Lumped zone modeling with special emphasis on the flaming pyrolysis zone	–	Self-experiment
Jarungthammachote and Dutta [29]	Combined mass and energy balance along with stoichiometric equilibrium. Assuming an initial temperature to calculate equilibrium constants and subsequently using the Newton-Raphson method to calculate new value of temperature by solving mass and energy balance equations, finally repeating the procedure to converge the temperatures	R5, R6, R7, R10	Literature
Sharma [54]	Stoichiometric equilibrium modeling of reduction reactions	R5, R6, R7, R10, R11	Literature
Ramanan et al. [55]	Mass and energy balance of components along with stoichiometric and non-stoichiometric equilibrium considerations	R3, R5, R6, R7, R10,	Self-experiment
Huang and Ramaswamy [56]	Stoichiometric equilibrium modeling of the reduction zone in the gasifier	R6, R10, R11	Literature
Loha et al. [57]	Mass and energy balance of components along with stoichiometric equilibrium of heterogeneous reactions.	R5, R6, R7	Literature
Barman et al. [30]	Stoichiometric equilibrium modeling with tar consideration and temperature-dependent empirical equations of equilibrium constants	R7, R10, R11	Literature
Mendiburu et al. [31]	Stoichiometric equilibrium modeling by introducing equilibrium constants as a function of Gibbs-free energy	R5, R6, R7, R10, R11	Literature
Sharma and Seth [58]	Stoichiometric equilibrium model considering char formation	R7, R10, R11	Literature and self-experiment literature
Aydin et al. [59]	Stoichiometric equilibrium model based on empirical correlations of equilibrium constants including tar formation	R5, R6, R7, R10	
Pradhan et al. [60]	Semi-empirical approach using equilibrium constants derived from five other equilibrium models. Both Newton-Raphson and Gauss-Newton algorithms are used	R5, R6, R7, R10, R11	Literature

Table 6 Modeling procedure of various non-stoichiometric equilibrium models.

Authors	Modeling procedure	validation
Altafani et al. [61]	Gibbs-free energy minimization of each component at the output. Model equations solved with the help of the Newton-Raphson algorithm	Literature
Li et al. [62]	Gibbs-free energy minimization using the RAND algorithm	Self-experiment
Jarungthammachote and Dutta [63]	Gibbs-free energy minimization of each component at the output. Model equations solved with the help of the Newton-Raphson algorithm	Literature
Yan et al. [64]	Non-stoichiometric equilibrium based on Gibbs-free energy minimization using the RAND algorithm [65]	Literature
Buragohain et al. [66]	Non-stoichiometric equilibrium based on Gibbs-free energy minimization using the SOLGASMIX algorithm introduced by Eriksson [67]	Literature
Mendiburu et al. [68]	Non-stoichiometric equilibrium based on Gibbs-free energy minimization	literature
Gambrotta et al. [69]	Non-stoichiometric equilibrium based on Gibbs-free energy minimization using the RAND algorithm	Literature
Kashyap et al. [70]	Simple non-stoichiometric equilibrium model based on Gibbs-free energy minimization	Literature and self-experiment

4.1 Tar and char inclusion

For the ease of modeling, most of the researchers seem to omit the formation of tar and unreacted char while equilibrium modeling of syngas production. This results in over prediction of the other gasification products. To avoid that problem, various researchers have taken different ways to include tar or unconverted char or both in the equilibrium model. In the works of Yamazaki et al. [71] Barman et al. [30], and Gagliano et al. [72], they considered fixed tar content of 4.5% of dry biomass for their models. Tar yield as a function of temperature has also been considered in various literatures as a modeling input. This correlation was presented by Sadaka et al. [73].

$$Y_{tar} = 35.98e^{-0.0029(T-273.15)} \tag{100}$$

Phase rule (Eq. 36) is applied in the PVT system to obtain the number of independent variables to be specified, that is, the number of degrees of freedom, F of a system under equilibrium. In the case of gasification, described by stoichiometric reaction I, the number of phases is three (solid, liquid, and gas), number of unknown coefficients is 11, and number of overall reactions is one. Thus, the phase rule reduces to the following:

$$F = 2 - \pi + N - r = 2 - 3 + 11 - 1 = 9 \tag{101}$$

This rule is applied to determine the unknown composition. Three atom balance equations, three equilibrium constants, specification of temperature and pressure, and one energy balance equation leads to the evaluation of equilibrium composition of syngas. When the molecular formula of tar is unknown, additional equilibrium constant, for example, that of methane reforming reaction and the empirical correlation between the syngas to hydrocarbon ratio (SHR) and temperature and equivalence ratio are used. Some typical correlation, reported in the literature, are as follows:

$$SHR = 8*10^{-8}T^2 - 2*10^{-4}T + 0.3172 \tag{102}$$

$$SHR = -0.894(ER)^2 + 0.4835ER + 0.1296 \tag{103}$$

In Aspen plus equilibrium modeling, tar has been considered an inert compound in most of the literatures. Compounds such as benzene (C_6H_6), phenol, etc. are considered tar compounds in these modeling [74]. Similarly constant char yield has also been proposed by many researchers. Kirsanovs et al. presented an empirical equation for predicting the char yield as a function of temperature, equivalence ratio, moisture content, and ash [75].

4.2 Adjustments on equilibrium constants

The accuracy of equilibrium model predictions can also be increased if multiplication factors are applied to the equilibrium constants of different gasification reactions. The process is to dislocate the equilibrium position of a particular reaction to predict accurate model yields. These models are thus called quasi-equilibrium models. Most of the researchers have made adjustments to the methane formation and water-gas shift reaction [16]. The multiplication correlation is set according to the experimental data obtained. Some literatures suggest considering a fixed multiplication factor whereas some suggest correlations as a function of temperature, equivalence ratio, etc. The literatures suggest that model predictions have been considerably improved with the inclusion of these adjustments.

4.3 Mole fraction of methane as input

Some researchers have also taken the liberty to compensate the under prediction of methane in equilibrium models by using the mole fraction of methane derived from various experiments as input. Jayah et al. [53] adjusted the predicted amount of methane in their model to be equal to the experimentally obtained values. Anukam et al. [76] assumed a fixed methane concentration of 2% of the total syngas as an input to their model. Wu et al. [77] used an empirical correlation for the prediction of methane formation in their air-steam gasification modeling. The empirical equation is given in Eq. (104).

$$\alpha_{CH_4} = 0.0052e^{\frac{3960}{T}}P^{0.149} \tag{104}$$

4.4 Molar distribution of products

Researchers have also tried to modify the equilibrium model by applying empirical correlation of CO/CO_2 or CO/H_2 ratio. Mendiburu et al. [31] have applied empirical correlations for these ratios, whereas Ratnadhariya and Channiwala [78] experimentally determined these ratios and used them in their equilibrium models.

4.5 Heat loss as input

Generally, the heat loss from the gasifier is neglected and adiabatic operation is assumed. However, for a more realistic representation of the system, the effects of heat loss to the ambient can be incorporated. Anukam et al. [76] considered 12.8% of heat loss to the ambient. Srinivas et al. [79] and Shayan et al. [80] assumed 5% of the total heat input to be lost to the ambient.

Table 7 provides some of the modifications in equilibrium models, as reported in the literature.

Table 7 Modifications in equilibrium models.

Authors	Types of modification made	Improvement achieved
Jarungthammachote and Dutta [29]	Equilibrium constant of methanation reaction multiplied by a factor of 11.28 and that of the water-gas shift reaction multiplied by a factor of 0.91	Better agreement with experimental results in terms of CO_2 and CH_4 production and improved HHV of syngas
Huang and Ramaswamy [56]	Equilibrium constants of methane reforming reaction and water-gas shift reaction multiplied by constant factors. Char included in one of the models	Improvement in matching CH_4 and CO production with experimental data while considering char
Barman et al. [30]	Consideration of tar as $CH_{1.003}O_{0.33}$ [27] and modification of methanation reaction according to Jarungthammachote and Dutta [29]	Improvement in CH_4 prediction and prediction of other gasification products
Mendiburu et al. [31]	Three modified models made First, modification of methanation and water-gas shift reaction with constant factors Second, implication of the modified CO/CO_2 ratio Third, implication of empirical equations of equilibrium constants	Improvement of prediction in case of considering constant factors for methanation and water-gas shift reactions
Li et al. [62]	Experimental data on carbon conversion and methane yield implemented	Improved prediction of syngas
Costa et al. [81]	Multi-objective optimization based on model modification factors used by various researchers to minimize the difference between predicted and experimental yield values.	Set of correction factors for different biomass achieved.
Gagliano et al. [72]	Correction coefficient for methanation and water-gas shift reaction used and a new correlation between biomass moisture content and equivalence ratio proposed	CH_4 prediction in calibrated model is more accurate
Shayan et al. [80]	NO_x and SO_x formation included in the model	Problem of over prediction of CO and H_2 in the model solved

4.6 Bi-equilibrium models

Usually, in both stoichiometric and non-stoichiometric equilibrium models, it is considered that all types of pyro-products, that is, char and volatiles enter the gasifier together with oxygen and steam. These are categorized as Equi-Sing models and the scheme is described in Fig. 4. As the concentration of methane and CO_2 are under-predicted by the Equi-Sing models, bi-equilibrium models, namely, Equi-Sep models considering partitioning of either the char stream or that of volatiles are developed [17]. The Equi-Sep models considering partitioning of char can be called Equi-Sep-Ch and those considering the partitioning of volatiles matter can be called Equi-Sep-VM.

4.6.1 Equi-Sep-Ch

In the Equi-Sep-Ch models, represented in Fig. 5, pyro-char is bifurcated into two streams. One stream enters an equilibrium reactor for combustion along with oxygen. The other char stream enters the equilibrium-gasifier along with condensable and non-condensable volatile pyro-products, that is, tar and gas, respectively, oxygen and steam.

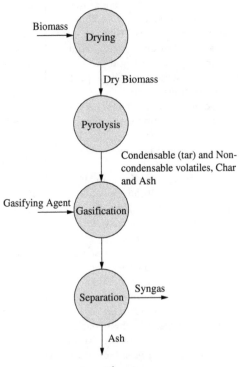

Fig. 4
Schematic of Equi-Sing model.

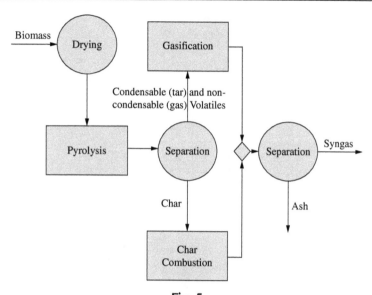

Fig. 5
Schematic of Equi-Sep-Ch models.

Based on the participation of volatiles in the gasification step, two options are considered in the bi-equilibrium Equi-Sep-VM models. These are as follows:

4.6.2 Equi-Sep-VM1

In Equi-Sep-VM1, a repartition factor λ_1 is introduced as a parameter. Volatile product stream is bifurcated into two fractions, γ_1 and $(1-\gamma_1)$. According to this model, γ_1 fractionenters the gasification zone at the peak temperature along with moisture, converted char, and air. On the other hand, the residual part, that is, $1-\gamma_1$ fraction directly joins the exit stream of the reactor (Fig. 6).

4.6.3 Equi-Sep-VM2

In Equi-Sep-VM2, a repartition factor γ_2 is introduced as a parameter. The combined stream of overall pyrolysis products (char+volatiles) and moisture is split into (γ_2) and $(1-\gamma_2)$ fractions. The fraction γ_2 enters the equilibrium reactor at T_{onset} (low temperature). The residual part, that is, $1-\gamma_2$ fraction contacts gasifying air and enters the equilibrium reactor at T_{Peak}. The combined stream of two equilibrium reactors represents the product syngas. Onset temperature is nothing but the pyrolysis temperature and has been declared as 473 K in the literature [82]. The peak temperature is the temperature of the gasifier under equilibrium.

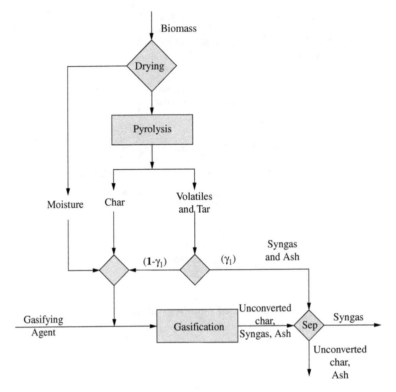

Fig. 6
Schematic diagram of Equi-Sep-VM1 models.

For Equi-Sep-VM1 and Equi-Sep-VM2, iterative solutions of model equations are determined up to the convergence level of 1 K using the brute force method varying the temperature in the range of 1000–2500 K at fixed values of γ (Fig. 7).

Optimization with respect to repartition coefficient, γ.

The optimum value of γ, that is, γ^* minimizing the sum of fractional deviations of model prediction of mole fraction of constituent compounds of product syngas, volumetric flow rate, and heating value of syngas

$$Sum = \sum_i \left[\frac{y_i^{exp} - y_i^p}{y_i^{exp}} \right] + \sum_i \left[\frac{q_i^{exp} - q_i^p}{q_i^{exp}} \right] + \sum_i \left[\frac{LHV_i^{exp} - LHV_i^p}{LHV_i^{exp}} \right] \qquad (105)$$

The only hydrocarbon appearing in the product syngas is assumed to be methane. The concentration of methane is determined considering the equivalence of C_2H_4, C_2H_6, C_2H_2 using the following correlation:

$$y_{CH_4,equivalent} = y_{CH_4} + y_{C_2H_4} * \frac{LHV_{C_2H_4}}{LHV_{CH_4}} + y_{C_2H_6} * \frac{LHV_{C_2H_6}}{LHV_{CH_4}} + y_{C_2H_2} * \frac{LHV_{C_2H_2}}{LHV_{CH_4}} \qquad (106)$$

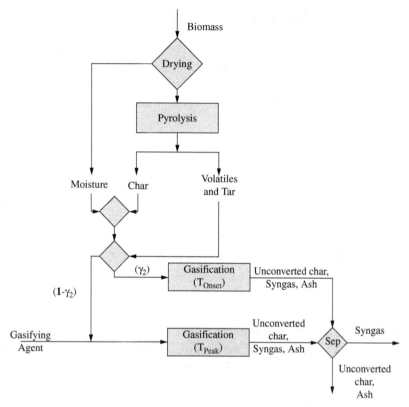

Fig. 7
Schematic diagram of Equi-Sep-VM2 models.

The parameter, γ_1^* is inversely proportional to ER. T_{peak} is proportional to ER. Heating time is directly proportional with the material parameter

$$mp = \frac{d_{p,eq}^2 \rho}{V_b} = \frac{\sum_j d_{p,j}^2 y_j \rho^2}{M_b} \qquad (107)$$

For different feedstock, linear correlations between γ and mp has been reported [82].

The linear correlation between the following parameter, namely, equivalence ratio parameter, ERP, and ER has also been reported:

$$ERP = V_b \sqrt{\frac{\rho}{d_{p,eq}}} \qquad (108)$$

V_b, ρ and $d_{p,eq}$ are the bulk volume, density, and equivalent diameter of solid feedstock.

4.7 Comparison between Equi-Sing and Equi-Sep models

Equi-Sing models are a perfect choice when the aim is just to know the thermodynamic limit of the yields of different components under equilibrium—a hypothetical condition. To reduce the deviation between the experimental yields of different components, particularly the outlet concentrations of methane and CO_2 which are underpredicted by Equi-Sing models, the Equi-Sep models can be used.

5. Strategy for choice of models

From the overall discussion presented in the chapter, the strategy on model choice can be developed using the following flow diagram of decision-making (Fig. 8). However, one should be cautious of the fact although the Equi-Sing model is based on the thermodynamic equilibrium consideration whereas Equi-Sep models are empirical in nature [11].

6. Conclusion

The chapter provides multiple modeling techniques for the generation of syngas based on the basic criteria of thermodynamic and phase equilibria. With known gasification reactions and equilibrium constant stoichiometric equilibrium model can be used, whereas the Gibbs-free

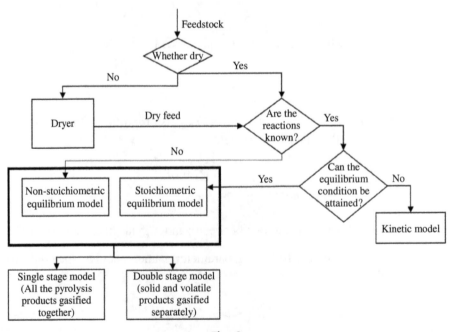

Fig. 8
Decision diagram of syngas production modeling.

energy minimization technique is used when gasification reactions are unknown. It is discussed in this chapter that the equilibrium models can serve as useful tools for the overall prediction of composition of syngas generated through gasification of carbonaceous feedstock. Thus, the models can be utilized for the comparison of different gasifiers without performing the actual experiments. This, in turn, will help in drawing the strategic decision in setting up of gasification plants using new non-conventional feedstock such as waste biomass. However, due to the simplistic assumptions used in equilibrium models, they are associated with various limitations such as underprediction of CH_4, tar, and char compounds. Some modification methods involving inclusion of char and tar compounds as products, adjustment of equilibrium constants, insertion of methane formation, and inclusion of empirical correlation of CO/CO_2 ratio have also been discussed in the chapter.

Abbreviations and symbols

$=$	quantity at pure state
$-$	quantity per unit mole
$\cdot$	rate of change with respect to time
A_k	total number of atomic masses of the kth element
a_i	activity of ith component
a_{ik}	number of atoms of the kth element in each molecule of i
c	total number of components
c_{pi}	constant pressure heat capacity of ith component
$\hat{}$	specific value
$\Delta \dot{E}_k$	rate of change of kinetic energy
$\Delta \dot{E}_p$	rate of change of potential energy
$\Delta G_{fi}{}^o$	standard heat of formation ith component
$\Delta \dot{H}$	rate of change of enthalpy
$\Delta \hat{h}$	specific enthalpy change associated with sensible heat
$\Delta_{rxn}G^o$	total Gibbs energy of reaction
$\Delta_{rxn}H^o$	total heat of reaction
ER	equivalence ratio
F	degrees of freedom
f	formation
f_i	fugacity of ith component
$\bar{f}_i$	fugacity coefficients of pure component i in reaction mixture
$\bar{\bar{f}}$	fugacity coefficients of pure component
G_i	Gibbs-free energy of ith component
$\bar{G}_i$	partial molar Gibbs energy of ith component
$\bar{\bar{G}}_i{}^o$	Gibbs energy of species at standard temperature and pressure
G_i	pure molar Gibbs energy of ith component
$\bar{G}_i{}^k$	partial molar Gibbs energy of ith component in kth phase
H_i	enthalpy of ith component
$\hat{h}_f{}^o$	molar heat of formation
i	ith gas species
j	jth chemical element
K_a	equilibrium constant
L	Lagrangian function

LHV	lower heating value
λ	Lagrangian multiplier
M	mass
MW	molecular weight
μ_i	chemical potential of ith species
N_i	total number of moles in ith component
N_i^k	total number of moles of ith component in kth phase
o	standard reference state
P	total pressure
π	number of phases
$\dot{Q}$	rate of heat transfer between the system and the surroundings
R	gas constant
R	reference state
r	number of independent reactions
rxn	reaction
$\dot{S}_{gen}$	specific entropy of generation
S_i	entropy of ith component
$\overline{S}_i$	partial molar entropy of ith component
SHR	syngas to hydrocarbon ratio
T	temperature
T_R	reference temperature
U	internal energy
V	volume
ϑ_i	stoichiometric coefficient of ith component
W	number of atoms present in the system
$\dot{W}_s$	shaft work
w	weight fraction
ξ	extent of reaction
y_i	mole fraction of ith component

References

[1] I.G. Jensen, F. Wiese, R. Bramstoft, M. Münster, Potential role of renewable gas in the transition of electricity and district heating systems, Energ. Strat. Rev. 27 (2020), 100446.

[2] C.M. Drapcho, N.P. Nhuan, T.H. Walker, Biofuels Engineering Process Technology, first ed., Focus on Catalysts, vol. 2008, McGraw-Hill, New York, 2008.

[3] S. Michailos, D. Parker, C. Webb, Design, sustainability analysis and multiobjective optimisation of ethanol production via syngas fermentation, Waste Biomass Valoriz. 10 (4) (2019) 865–876.

[4] R. Chowdhury, S. Ghosh, D. Manna, S. Das, S. Dutta, S. Kleinsteuber, et al., Hybridization of sugar-carboxylate-syngas platforms for the production of bio-alcohols from lignocellulosic biomass (LCB)—a state-of-the-art review and recommendations, Energ. Conver. Manage. 200 (June) (2019), 112111.

[5] H. Richter, S.E. Loftus, L.T. Angenent, Integrating syngas fermentation with the carboxylate platform and yeast fermentation to reduce medium cost and improve biofuel productivity, Environ. Technol. 34 (13–14) (2013) 1983–1994.

[6] F.C.F. Baleeiro, S. Kleinsteuber, A. Neumann, H. Sträuber, Syngas-aided anaerobic fermentation for medium-chain carboxylate and alcohol production: the case for microbial communities, Appl. Microbiol. Biotechnol. 103 (21 – 22) (2019) 8689–8709.

[7] D. Selvatico, A. Lanzini, M. Santarelli, Low temperature Fischer-Tropsch fuels from syngas: kinetic modeling and process simulation of different plant configurations, Fuel 186 (2016) 544–560.

[8] R. Chowdhury, S. Ghosh, B. Debnath, D. Manna, Indian agro-wastes for 2G biorefineries: strategic decision on conversion processes, Green Energy Technol. (2018) 353–373. 978c9811071874.

[9] M. Puig-Arnavat, J.C. Bruno, A. Coronas, Review and analysis of biomass gasification models, Renew. Sustain. Energy Rev. 14 (9) (2010) 2841–2851.

[10] T.K. Patra, P.N. Sheth, Biomass gasification models for downdraft gasifier: a state-of-the-art review, Renew. Sustain. Energy Rev. 50 (2015) 583–593.

[11] P. Sharma, B. Gupta, M. Pandey, K. Singh Bisen, P. Baredar, Downdraft biomass gasification: a review on concepts, designs analysis, modelling and recent advances, Mater. Today: Proc. 46 (2020) 5333–5341.

[12] M.B. Nikoo, N. Mahinpey, Simulation of biomass gasification in fluidized bed reactor using ASPEN PLUS, Biomass Bioenergy 32 (12) (2008) 1245–1254.

[13] X. Ku, J. Wang, H. Jin, J. Lin, Effects of operating conditions and reactor structure on biomass entrained-flow gasification, Renew. Energy 139 (2019) 781–795.

[14] R. Chowdhury, A. Sarkar, Reaction kinetics and product distribution of slow pyrolysis of Indian textile wastes, Int. J. Chem. React. Eng. 10 (2012).

[15] M. La Villetta, M. Costa, N. Massarotti, Modelling approaches to biomass gasification: a review with emphasis on the stoichiometric method, Renew. Sustain. Energy Rev. 74 (February) (2017) 71–88.

[16] I.P. Silva, R.M.A. Lima, G.F. Silva, D.S. Ruzene, D.P. Silva, Thermodynamic equilibrium model based on stoichiometric method for biomass gasification: a review of model modifications, Renew. Sustain. Energy Rev. 114 (July) (2019), 109305.

[17] S. Safarian, R. Unnþórsson, C. Richter, A review of biomass gasification modelling, Renew. Sustain. Energy Rev. 110 (November 2018) (2019) 378–391.

[18] V. Marcantonio, A.M. Ferrario, C.A. Di, Z.L. Del, D. Monarca, E. Bocci, Biomass steam gasification: a comparison of syngas composition between a 1-d matlab kinetic model and a 0-d aspen plus quasi-equilibrium model, Comput. Des. 8 (4) (2020) 1–15.

[19] D. Baruah, D.C. Baruah, Modeling of biomass gasification: a review, Renew. Sustain. Energy Rev. 39 (2014) 806–815.

[20] M.L. de Souza-Santos, Solid Fuels Combustion and Gasification: Modeling, Simulation, and Equipment Operations, second ed., CRC Press, 2010, pp. 1–487.

[21] P. Basu, Biomass Gasification and Pyrolysis: Practical Design and Theory, Academic Press, 2010.

[22] K. Jana, S. De, Biomass integrated gasification combined cogeneration with or without CO2 capture—a comparative thermodynamic study, Renew. Energy 72 (2014) 243–252.

[23] R. Perry, D. Green, Perry's Chemical Engineers' Handbook, eight ed., McGraw-Hill, New York, 2007.

[24] R.M. Felder, R.W. Rousseau, Elementary Principles of Chemical Processes, third ed., John Wiley and Sons, 2005.

[25] S. Sandler, Chemical Biochemical and Engineering Thermodynamics, fourth ed., Jhon Wiley and Sons, 2006.

[26] A. Gagliano, F. Nocera, M. Bruno, G. Cardillo, Development of an equilibrium-based model of gasification of biomass by Aspen Plus, Energy Procedia 111 (2017) 1010–1019.

[27] F.V. Tinaut, A. Melgar, J.F. Pérez, A. Horrillo, Effect of biomass particle size and air superficial velocity on the gasification process in a downdraft fixed bed gasifier. An experimental and modelling study, Fuel Process. Technol. 89 (11) (2008) 1076–1089.

[28] Z.A. Zainal, R. Ali, C.H. Lean, K.N. Seetharamu, Prediction of performance of a downdraft gasifier using equilibrium modeling for different biomass materials, Energ. Conver. Manage. 42 (12) (2001) 1499–1515.

[29] S. Jarungthammachote, A. Dutta, Thermodynamic equilibrium model and second law analysis of a downdraft waste gasifier, Energy 32 (9) (2007) 1660–1669.

[30] N.S. Barman, S. Ghosh, S. De, Gasification of biomass in a fixed bed downdraft gasifier—a realistic model including tar, Bioresour. Technol. 107 (2012) 505–511.

[31] A.Z. Mendiburu, J.A. Carvalho, C.J.R. Coronado, Thermochemical equilibrium modeling of biomass downdraft gasifier: stoichiometric models, Energy 66 (2014) 189–201.

[32] Q. Liu, C. Proust, F. Gomez, D. Luart, C. Len, The prediction multi-phase, multi reactant equilibria by minimizing the Gibbs energy of the system: review of available techniques and proposal of a new method based on a Monte Carlo technique, Chem. Eng. Sci. 216 (2020).

[33] C. Morley, gaseq.co.uk, 2005.

[34] S. Gordon, B.J. McBride, Computer Program for Calculation of Complex Chemical Equilibrium Compositions Rocket Performance Incident and Reflected Shocks, and Chapman-Jouguet Detonations, 250, Scientific and Technical Information Office, National Aeronautics and Space Administration, 1971.

[35] R. Gautam, W.D. Seider, Computation of phase and chemical equilibrium: part I. local and constrained minima in Gibbs free energy, AICHE J. 25 (6) (1979) 991–999.

[36] N. Ramzan, A. Ashraf, S. Naveed, A. Malik, Simulation of hybrid biomass gasification using Aspen Plus: a comparative performance analysis for food, municipal solid and poultry waste, Biomass Bioenergy 35 (9) (2011) 3962–3969.

[37] S. Pati, S. De, R. Chowdhury, Process modelling and thermodynamic performance optimization of mixed Indian lignocellulosic waste co-gasification, Int. J. Energy Res. 45 (12) (2021) 17175–17188.

[38] A. Tungalag, B.J. Lee, M. Yadav, O. Akande, Yield prediction of MSW gasification including minor species through ASPEN Plus simulation, Energy 198 (2020), 117296.

[39] M.A. Adnan, H. Susanto, H. Binous, O. Muraza, M.M. Hossain, Feed compositions and gasification potential of several biomasses including a microalgae: a thermodynamic modeling approach, Int. J. Hydrogen Energy 42 (27) (2017) 17009–17019.

[40] J.A. Okolie, S. Nanda, A.K. Dalai, J.A. Kozinski, Hydrothermal gasification of soybean straw and flax straw for hydrogen-rich syngas production: experimental and thermodynamic modeling, Energ. Conver. Manage. 208 (February) (2020), 112545.

[41] M. Zhai, L. Guo, Y. Wang, Y. Zhang, P. Dong, H. Jin, Process simulation of staging pyrolysis and steam gasification for pine sawdust, Int. J. Hydrogen Energy 41 (47) (2016) 21926–21935.

[42] W. Wan, An innovative system by integrating the gasification unit with the supercritical water unit to produce clean syngas: effects of operating parameters, Int. J. Hydrogen Energy 41 (33) (2016) 14573–14582.

[43] W. Doherty, A. Reynolds, D. Kennedy, Aspen Plus simulation of biomass gasification in a steam blown dual fluidised bed, Mater. Process Energy (2013) 212–220.

[44] S. Safarian, C. Richter, R. Unnthorsson, Waste biomass gasification simulation using Aspen Plus: performance evaluation of wood chips, sawdust and mixed paper wastes, J. Power Energy Eng. 07 (06) (2019) 12–30.

[45] S. Begum, M.G. Rasul, D. Akbar, A numerical investigation of municipal solid waste gasification using Aspen Plus, Procedia Eng. 90 (2014) 710–717.

[46] R. Tavares, E. Monteiro, F. Tabet, A. Rouboa, Numerical investigation of optimum operating conditions for syngas and hydrogen production from biomass gasification using Aspen Plus, Renew. Energy 146 (2020) 1309–1314.

[47] W. Doherty, A. Reynolds, D. Kennedy, Simulation of a circulating fluidised bed biomass gasifier using ASPEN Plus—A performance analysis, in: ECOS 2008—Proceedings of the 21st International Conference on Efficiency, Cost, Optimization, Simulation and Environmental Impact of Energy Systems, 2008, pp. 1241–1248.

[48] R. Barrera, C. Salazar, J.F. Pérez, Thermochemical equilibrium model of synthetic natural gas production from coal gasification using Aspen Plus, Int. J. Chem. Eng. 2014 (2014).

[49] J. Han, Y. Liang, J. Hu, L. Qin, J. Street, Y. Lu, et al., Modeling downdraft biomass gasification process by restricting chemical reaction equilibrium with Aspen Plus, Energ. Conver. Manage. 153 (2017) 641–648.

[50] S. Pati, S. De, Model development and thermodynamic analysis of biomass co-gasification using Aspen Plus®, Indian Chem. Eng. 63 (2) (2021) 172–183.

[51] S. Rupesh, C. Muraleedharan, P. Arun, ASPEN plus modelling of air-steam gasification of biomass with sorbent enabled CO2 capture, Resour. Eff. Technol. 2 (2) (2016) 94–103.

[52] N. Indrawan, S. Mohammad, A. Kumar, R.L. Huhnke, Modeling low temperature plasma gasification of municipal solid waste, Environ. Technol. Innov. 15 (2019), 100412.

[53] T.H. Jayah, L. Aye, R.J. Fuller, D.F. Stewart, Computer simulation of a downdraft wood gasifier for tea drying, Biomass Bioenergy 25 (4) (2003) 459–469.

[54] A.K. Sharma, Equilibrium and kinetic modeling of char reduction reactions in a downdraft biomass gasifier: a comparison, Sol. Energy 82 (10) (2008) 918–928.

[55] M. Venkata Ramanan, E. Lakshmanan, R. Sethumadhavan, S. Renganarayanan, Performance prediction and validation of equilibrium modeling for gasification of cashew nut shell char, Braz. J. Chem. Eng. 25 (3) (2008) 585–601.

[56] H.J. Huang, S. Ramaswamy, Modeling biomass gasification using thermodynamic equilibrium approach, Appl. Biochem. Biotechnol. 154 (1–3) (2009) 193–204.

[57] C. Loha, H. Chattopadhyay, P.K. Chatterjee, Thermodynamic analysis of hydrogen rich synthetic gas generation from fluidized bed gasification of rice husk, Energy 36 (7) (2011) 4063–4071.

[58] S. Sharma, P.N. Sheth, Air-steam biomass gasification: experiments, modeling and simulation, Energ. Conver. Manage. 110 (2016) 307–318.

[59] E.S. Aydin, O. Yucel, H. Sadikoglu, Development of a semi-empirical equilibrium model for downdraft gasification systems, Energy 130 (2017) 86–98.

[60] P. Pradhan, A. Arora, S.M. Mahajani, A semi-empirical approach towards predicting producer gas composition in biomass gasification, Bioresour. Technol. 272 (October 2018) (2019) 535–544.

[61] C.R. Altafini, P.R. Wander, R.M. Barreto, Prediction of the working parameters of a wood waste gasifier through an equilibrium model, Energ. Conver. Manage. 44 (17) (2003) 2763–2777.

[62] X.T. Li, J.R. Grace, C.J. Lim, A.P. Watkinson, H.P. Chen, J.R. Kim, Biomass gasification in a circulating fluidized bed, Biomass Bioenergy 26 (2) (2004) 171–193.

[63] S. Jarungthammachote, A. Dutta, Equilibrium modeling of gasification: Gibbs free energy minimization approach and its application to spouted bed and spout-fluid bed gasifiers, Energ. Conver. Manage. 49 (6) (2008) 1345–1356.

[64] Q. Yan, L. Guo, Y. Lu, Thermodynamic analysis of hydrogen production from biomass gasification in supercritical water, Energ. Conver. Manage. 47 (11 – 12) (2006) 1515–1528.

[65] W. Smith, M.R. Smith WR, R.W. Missen, Chemical Reactions and Equilibrium Analysis, second ed., Wiley, New York, 1982.

[66] B. Buragohain, P. Mahanta, V.S. Moholkar, Performance correlations for biomass gasifiers using semi-equilibrium non-stoichiometric thermodynamic models, Int. J. Energy Res. 36 (4) (2012) 590–618.

[67] G. Eriksson, Thermodynamic studies of high temperature equilibria—XII: SOLGAMIX, a computer program for calculation of equilibrium composition in multiphase systems, Chem. Scr. 8 (1975) 100–103.

[68] A.Z. Mendiburu, J.A. Carvalho, R. Zanzi, C.R. Coronado, J.L. Silveira, Thermochemical equilibrium modeling of a biomass downdraft gasifier: constrained and unconstrained non-stoichiometric models, Energy 71 (2014) 624–637.

[69] A. Gambarotta, M. Morini, A. Zubani, A non-stoichiometric equilibrium model for the simulation of the biomass gasification process, Appl. Energy 227 (July 2017) (2018) 119–127.

[70] P.V. Kashyap, R.H. Pulla, A.K. Sharma, P.K. Sharma, Development of a non-stoichiometric equilibrium model of downdraft gasifier, Energy Sources, Part A 00 (00) (2019) 1–19.

[71] T. Yamazaki, H. Kozu, S. Yamagata, N. Murao, S. Ohta, S. Shiya, et al., Effect of superficial velocity on tar from downdraft gasification of biomass, Energy Fuel 19 (3) (2005) 1186–1191.

[72] A. Gagliano, F. Nocera, F. Patania, M. Bruno, D.G. Castaldo, A robust numerical model for characterizing the syngas composition in a downdraft gasification process, C. R. Chim. 19 (4) (2016) 441–449.

[73] S.S. Sadaka, A.E. Ghaly, M.A. Sabbah, Two phase biomass air-steam gasification model for fluidized bed reactors: part I—model development, Biomass Bioenergy 22 (6) (2002) 439–462.

[74] P.C. Font, Modelling of tar formation and evolution for biomass gasification: a review, Appl. Energy 111 (2013) 129–141.

[75] V. Kirsanovs, A. Žandeckis, C. Rochas, Biomass gasification thermodynamic model including tar and char, Agron. Res. 14 (4) (2016) 1321–1331.

[76] A. Anukam, S. Mamphweli, E. Meyer, O. Okoh, Computer simulation of the mass and energy balance during gasification of sugarcane bagasse, J. Energy 2014 (2014) 1–9.

[77] Y. Wu, W. Yang, W. Blasiak, Energy and exergy analysis of high temperature agent gasification of biomass, Energies 7 (4) (2014) 2107–2122.

[78] J.K. Ratnadhariya, S.A. Channiwala, Three zone equilibrium and kinetic free modeling of biomass gasifier—a novel approach, Renew. Energy 34 (4) (2009) 1050–1058.

[79] T. Srinivas, A.V.S.S.K.S. Gupta, B.V. Reddy, Thermodynamic equilibrium model and exergy analysis of a biomass gasifier, J. Energy Resour. Technol. 131 (September) (2009) 1–7.

[80] E. Shayan, V. Zare, I. Mirzaee, Hydrogen production from biomass gasification; a theoretical comparison of using different gasification agents, Energ. Conver. Manage. 159 (January) (2018) 30–41.

[81] M. Costa, M. La Villetta, N. Massarotti, Optimal tuning of a thermo-chemical equilibrium model for downdraft biomass gasifiers, Chem. Eng. Trans. 43 (November 2017) (2015) 439–444.

[82] E. Biagini, F. Barontini, L. Tognotti, Development of a bi-equilibrium model for biomass gasification in a downdraft bed reactor, Bioresour. Technol. 201 (2016) 156–165.

CHAPTER 2

Process modeling and apparatus simulation for syngas production

Filippo Bisotti[a],*, Matteo Fedeli[a], Poliana P.S. Quirino[b], Karen Valverde Pontes[b], and Flavio Manenti[a]

[a]Dept. CMIC "Giulio Natta", Center for Sustainable Process Engineering Research (SuPER), Politecnico di Milano, Milan, Italy [b]Industrial Engineering Graduate Program (PEI), Federal University of Bahia (UFBA), Street Professor Aristides Novis 02, Bahia, Brazil

1. Introduction

Syngas (mixture of H_2, CO, and CO_2) is one of the most important chemical products in the industrial sector, due to its wide use as a raw material for energy generation and synthesis of many products. Depending on its composition, the synthesis gas has different applications such as the manufacture of ammonia, hydrogen, methanol, and synthetic fuels by the Fischer-Tropsch (FT) process to quote a few [1,2]. Based on feedstock, the global syngas market is segmented into coal, natural gas, biomass/waste, and others. Among these segments, coal and natural gas hold an important share in the global synthesis gas market [3]. Depending on the technology used in the production of synthesis gas, this market can be classified into two main groups: gasification and reforming. Gasification combines partial oxidation of solid fossil fuels, such as coal, biomass, waste with high carbon content, such as coke, and urban waste with the steam treatment of coal, as described by the (R1) and (R2) [4]:

$$C + 1/2O_2 \rightarrow CO \quad \Delta H^{298 \text{ K}} = -123 \frac{kJ}{mol} \tag{R1}$$

$$C + H_2O \leftrightarrows CO + H_2 \quad \Delta H^{298 \text{ K}} = 131 \frac{kJ}{mol} \tag{R2}$$

The use of these solid fuels is advantageous, as it reduces dependence on oil, and some of them are sustainable and renewable, such as biomass. Research in this area has been developed to reduce the cost of collection, transportation, and pretreatment, to make the gasification of these

---

* Current address: Department of Process Technology, SINTEF Industry, Torgarden, Trondheim, Norway

Advances in Synthesis Gas: Methods, Technologies and Applications. https://doi.org/10.1016/B978-0-323-91879-4.00004-7

feedstock feasible [5]. These processes generally result in an inadequate synthesis gas composition for direct use in processes downstream. For satisfactory methanol production, for example, the H_2/CO and H_2/CO_2 ratios in the synthesis gas must be around 2 and 3, respectively. Therefore, some additional treatments to adjust this composition and achieve this goal are needed, which means adding cost and complexity to the overall process. This type of coal-based technology is commonly used in countries such as China and India, due to the large availability of coal reserves in the Asia-Pacific region. Most companies established in such countries have invested great efforts in an attempt to improve the gasification process and its technologies, making them more attractive and environmentally sustainable.

The reforming uses natural gas as the main source of raw material, which is commonly represented by methane, its main component. The preference for natural gas over coal in these units is mainly due to economic and environmental reasons. In general, the capital cost of a coal-based plant is higher than that of a methane-based plant. Furthermore, when reforming methane, the proportion of carbon converted to CO_2 is much lower than when coal is gasified [6]. Natural gas is converted to synthesis gas by different routes, such as steam reforming, partial oxidation reforming; autothermal reforming, and dry reforming. Among these, steam reforming is undoubtedly the most widely used and studied catalytic route for the synthesis of gas production [5,7–10]. The steam reforming segment is expected to have the fastest growth and reach up to US $24,708.1 million by 2027, with an increase from USD 16,435.0 million in 2019 [11]. Its popularity is due in part to its greater efficiency compared with other competing processes.

The steam methane reforming (SMR) is usually described by two reactions: steam reforming of methane (R3) and the water-gas shift reaction (R4), which interconverts CO and CO_2 [8,12–16]. Other authors further consider the reverse methanation reaction (R5), which is the sum of (R3) and (R4) [17–26]. Carbon formation is a parallel and undesirable reaction, which occurs less frequently and therefore is often neglected [17,21,27–29].

$$CH_4 + H_2O \rightleftarrows CO + 3H_2 \quad \Delta H^{298\ K} = 206\frac{kJ}{mol} \tag{R3}$$

$$CO + H_2O \leftrightarrows CO_2 + H_2 \quad \Delta H^{298\ K} = -41\frac{kJ}{mol} \tag{R4}$$

$$CH_4 + 2H_2O \rightleftarrows CO_2 + 4H_2 \quad \Delta H^{298\ K} = 165\frac{kJ}{mol} \tag{R5}$$

Such steam reforming reactions have a strongly endothermic character and require excess steam to avoid the catalyst deactivation by carbon deposition. Typically, the industrial steam methane reforming units operate at pressures around 15–35 bar [30–33], with feed gas temperature ranging from 723 to 923 K. The products leave at a temperature of 1073–1223 K, depending on the application of the synthesis gas, for example, the methanol and ammonia production [34–36]. The H_2/CO ratio produced by steam reforming methane is equal to 3, so it is the most suitable route for the production of hydrogen.

In the partial oxidation reforming, natural gas reacts with a limited amount of oxygen, producing H_2 and CO, through the (R6) reaction, which is moderately exothermic. In addition to these two components, the synthesis gas also contains smaller amounts of CO_2 and H_2O due to side reactions such as the oxidation reactions of H_2 and CO (R7 and R8) and the complete methane oxidation reaction (R9), which is highly exothermic [4]:

$$CH_4 + 1/2O_2 \rightarrow CO + 2H_2 \quad \Delta H^{298\,K} = -36\frac{kJ}{mol} \tag{R6}$$

$$H_2 + 1/2O_2 \rightarrow H_2O \quad \Delta H^{298\,K} = -242\frac{kJ}{mol} \tag{R7}$$

$$CO + 1/2O_2 \rightarrow CO_2 \quad \Delta H^{298\,K} = -283\frac{kJ}{mol} \tag{R8}$$

$$CH_4 + 2O_2 \rightarrow CO_2 + 2H_2O \quad \Delta H^{298\,K} = -880\frac{kJ}{mol} \tag{R9}$$

Partial oxidation of methane (POM) produces a synthesis gas with an H_2/CO ratio of around 2, less than the steam reforming, but still ideal for the methanol or Fischer-Tropsch synthesis [6,37]. However, the oxidation of H_2 and CO (R7 and R8) can also result in H_2/CO ratios less than 2 in the synthesis gas [38]. POM, however, is considered advantageous for small-scale operations, where efficiency is not such an issue [39]. Therefore, the main motivation for using this technology is its low overall cost, more compact reactor, rapid start-up, and responsiveness to load change [40–43]. In general, this process operates at a temperature of 1473–1773 K [44], which results in large temperature gradients, leading to the formation of hotspots on the catalyst surface if the O_2 is not evenly distributed along the length of the reactor [45].

Endothermic reforming reactions can be combined with exothermic (partial) oxidation to simplify heat management in the synthesis gas production process and thus achieve a thermodynamically neutral system [4,38]. The autothermal reforming (ATR) results from the combination of conventional steam methane reforming with the partial oxidation of methane, then the main reactions involved are (R3), (R4), and (R6) [46]. Natural gas reacts simultaneously with steam and oxygen within a single reactor, reducing the cost of the process. This reactor generally includes a burning zone, a combustion chamber, and a catalytic zone [45]. The combustion chamber usually operates at high temperatures, around 2200 K, while the temperature of the catalytic zone operates around 1200–1400 K [37]. As the endothermic (R3 and R4) and exothermic (R6) reactions should ideally be conducted at different temperatures and pressures, a two-reactor arrangement can also be used, in which the effluent from steam reforming feeds a partial oxidation reactor. This configuration results in a smaller amount of residual methane in the synthesis gas since unreacted methane from the steam reformer can be oxidized in the second reactor [4,38].

ATR is mainly applied to small-scale hydrogen production, as this technology is characterized by compact reactors with internal energy generation [47–49]. Some works investigate the ATR for the production of hydrogen, using as raw material higher-molecular-weight hydrocarbons, instead of natural gas. Rice and Mann [50], for example, utilize naphtha and pyrolyzed biomass in a laboratory-scale ATR test unit, using a combination of cracking, reforming, and water-gas shift chemistry in a compact, integrated system. Other examples of ATR application involving heavy hydrocarbons as raw material are reported in Dreyer et al. [51], Williams and Schmidt [52], Gould et al. [53], Dorazio and Castaldi [54], Shi et al. [55], and Faheem et al. [56].

Dry methane reforming (DMR) consists of reacting methane with carbon dioxide, according to the reaction (R10), to obtain synthesis gas with a low H_2/CO ratio (≤ 1) [57]. In this case, CO_2 is used to replace steam, and this reaction is normally conducted at temperatures around 1073–1273K [4]. The water-gas shift reaction (R4) in the reverse direction, as well as the CO_2 methanation reaction (R5), can also occur simultaneously with the (R10) reaction.

$$CH_4 + CO_2 \rightleftarrows 2CO + 2H_2 \quad \Delta H^{298 \text{ K}} = 247 \, \frac{kJ}{mol} \tag{R10}$$

One of the main disadvantages of dry reforming is that it tends to form and deposit coke on the catalyst surface, which drastically reduces its useful life. This unwanted formation of carbon may be due to the thermal decomposition reaction of methane (R11), considered the simplest form of carbon formation; the Boudouard reaction (R12), in which carbon monoxide is disproportionate to carbon dioxide; the carbon monoxide reduction reaction (R13); among others [10].

$$CH_4 \rightleftarrows C_{(s)} + 2H_2 \quad \Delta H^{298 \text{ K}} = 75 \text{ kJ/mol} \tag{R11}$$

$$2CO \rightleftharpoons C_{(s)} + CO_2 \quad \Delta H^{298 \text{ K}} = -172 \text{ kJ/mol} \tag{R12}$$

$$CO + H_2 \rightleftharpoons C_{(s)} + H_2O \quad \Delta H^{298 \text{ K}} = -131 \text{ kJ/mol} \tag{R13}$$

Although the (R10) is more endothermic than the steam reforming of methane (R3), the industrial application of this technology could contribute to reducing the concentration of CO_2 in the atmosphere, mitigating the environmental impacts resulting from global warming; therefore, the study and application of this technology are being encouraged. The benefit from the CO_2 consumption and the cost of capture and conversion, though, must be weighed. Biogas, derived from biomass, is a clean and renewable energy source and has been highlighted as an alternative raw material to replace natural gas in the synthesis gas production process. Its use as fuel results in a carbon-neutral process, as the carbon dioxide produced from this process can be captured via photosynthesis during biomass growth [58]. Therefore, biomass emerges as an alternative both to reduce the emission of greenhouse gases and to gradually replace fossil fuels. One of the main problems of the biogas dry reforming is the coke formation through the methane decomposition (R11) and/or Boudouard reaction (R12) [59,60]. The steam reforming reaction (without the CO_2 separation step) consumes less energy compared with the dry biogas reforming. The addition of steam to the dry biogas reforming process may result in a significant reduction in coke deposition on the catalyst [61]. These authors performed an extensive review of the latest reforming technologies that produce synthesis gas from biogas. Other studies specifically focused on dry reforming are reported in Corthals et al. [62], Fan et al. [63], Muraza and Galadima [64], Medeiros et al. [65], Rezaei et al. [66], Aouad et al. [67], and Wittich et al. [68].

New pathways have been studied to improve the synthesis gas production process. Most of the technologies presented above, especially dry reforming, often result in a synthesis gas composition unsuitable for direct use in downstream processes. Additional treatment to tailor this composition for a specific application, then, is necessary, but it means adding cost and complexity to the overall process [64,69]. To solve this inconvenience, Olah et al. [70] combined steam reforming and dry reforming methane into a single step called bireforming as per the reaction (R14) [69]. These authors aimed to produce synthesis gas with an H_2/CO ratio of exactly 2:1, which was called "metgas" to differentiate it from the synthesis gas obtained by traditional technologies, with varying proportions of H_2/CO [70].

$$3CH_4 + 2H_2O + CO_2 \rightleftarrows 4CO + 8H_2 \quad \Delta H^{298 \text{ K}} = 660.3 \frac{kJ}{mol} \tag{R14}$$

The bireforming has the advantage of suppressing coke formation, due to the presence of steam in the process, and is preferably conducted at temperatures from about 1073 to 1223 K and pressures ranging from 5 to 40atm, with a feed composition of CH_4, steam, and CO_2 in a ratio of 3/2/1 [4,69,70]. Furthermore, the bireform allows higher aliphatic hydrocarbons to be used as raw material for the production of synthesis gas, according to the (R15) reaction. Thus, shale gas can be used without the need to separate the higher alkanes [4]. Some applications of bireforming for the production of synthesis gas can also be seen in Kumar et al. [71], Singh et al. [72], Cunha et al. [73], Mallikarjun et al. [74], Santos et al. [75], and Mohanty et al. [76].

$$3C_nH_{2n+2} + (3n - 1)H_2O + CO_2 \rightleftarrows (3n + 1)CO + (6n + 2)H_2 \tag{R15}$$

The trireforming combines three reforming processes commonly used in synthesis gas production: (i) steam methane reforming (R3), partial oxidation of methane (R6), and DMR (R10). In addition to these reactions, the complete methane oxidation reaction can also occur (R9). Therefore, similarly to ART, there is a combination of endothermic reactions (R3 and R10) and exothermic reactions (R2 and R4) to obtain a thermoneutral process [77]. The trireform is proposed to mitigate CO_2 emissions since it can be consumed in the (R10) reaction [78]. Thus, the combination of reactions (R3), (R4), (R5), and (R10) has advantages over simple reforming reactions, such as lower energy consumption compared with the reforming reaction, due to the presence of exothermic reactions; greater catalyst stability, as O_2 and H_2O are strong oxidizers and help to minimize carbon formation and deposit on the catalyst surface [79,80].

Studies conducted by Nguyen and Zondervan [81] demonstrated that, for methanol production, the methane trireforming technology results in lower total annual costs and CO_2 emissions, when compared with the bireforming. Therefore, a good option for methanol synthesis from CO_2 and CH_4 is the trireforming process [82]. The trireforming approach has been the subject of some lab-scale studies [83,84]. This technology was applied in a pilot plant and demonstration unit in Korea, using natural gas as a raw material to obtain a synthesis gas with an H_2/CO ratio around 1.2 to 1.5, which is suitable for DME synthesis [85].

2. *Prereforming for the syngas production*

Steam reforming processes commonly use a prereformer (adiabatic catalytic reactor), positioned upstream of the main steam reformer, to completely convert heavier hydrocarbons contained in the raw material into a mixture of carbon oxides, hydrogen, and methane [86]. The main reactions involved in the prereformer are (R3), (R4), (R5). The reforming of higher-molecular-weight hydrocarbons is described according to the (R16) reaction [87]:

$$C_nH_m + nH_2O \rightleftharpoons nCO + \frac{(n + m)}{2}H_2 \qquad (R16)$$

Depending on the raw material used, the temperature profile can be endothermic or exothermic. For a typical natural gas feed, the overall reforming process is endothermic, resulting in a temperature drop. On the other hand, when heavier raw materials such as naphtha are used, the overall process is exothermic or thermoneutral [87,88]. The inclusion of a prereforming unit provides some advantages, namely [4,50,81,82,85,87,88]:

- Reduction of the reformer's capital cost
- Fuel savings over standalone primary reformer
- Greater raw material flexibility, which can range from natural gas to naphtha
- Possibility of operating with very low steam-to-carbon ratio, which is interesting in units where the main objective is to have more CO, as well as a specific ratio of H_2/CO in the syngas
- Increased hydrogen production capacity (about 10%–15%) without additional energy costs
- Prevents coke formation, protecting the downstream reformer catalyst in case of any poison slip across the purification section

The inclusion of the prereformer in the steam reforming units, on the downside, results in an increase in pressure drop. In addition, the catalyst useful life in the prereformer is much shorter compared with the main reformer's catalyst and therefore needs more frequent replacements [89].

3. Thermodynamic equilibrium

Thermodynamic is a key aspect in syngas production modeling since it enables estimation of the final mixture composition or, at least, predicts the maximum allowable content in H_2. Very recently, many researchers focus attention and both theoretical and experimental studies on the equilibria composition in the syngas production. As previously remarked, this renewed interest lies in the use of new potential feedstock to be exploited to produce syngas and the several feasible technologies and chemical treatments available to convert them. For instance, plastic wastes and less recently biomass gasification have acquired a growing interest. Finally, dry reforming (DR) is gaining more and more attention since it enables the consumption of greenhouse gases (methane and CO_2) converting them into a syngas mixture suitable for several fuel and chemicals production. Table 1 lists some of the most recent publications in the field grouped according to the analyzed reforming technology. Further details and comments are reported in the note column.

Kakoee and Gharehghani [90] proposed a complete model including SMR, DR, RWGS, Boudouard reaction, and soot formation. They evaluated the equilibrium constants as a function of the temperature. Accordingly, almost all reactions show a sign inversion at around 900 K (630°C). Reforming processes are favored at higher temperatures ($T > 650$°C), in parallel also methane decomposition occurs; however, steam and CO_2 rapidly removed consume the soot. Finally, CO_2 activation is a heavily endothermic process due to CO_2 stability. Hence, reactions such as RWGS and DR are shifted toward syngas production for temperatures higher than 1200 K (930°C) as proven in Zhou et al. [92]. This means that both reforming reactions are energetically starved as shown in Carapellucci and Giordano [45].

Temperature impacts the extent of the reaction and the reactant conversion. As depicted in this recent publication [91], reforming reactions are ineffective below 550 K, while SMR process is active at around 550–1100 K due to its lower endothermicity with regard to other methane reforming reactions. For temperatures higher than 1100 K also DR and RWGS reactions prevail consuming CO_2. Similar results are reported in Zhou et al. [92]. This different reaction activation is compliant with the CO_2 conversion as a function of temperature: the minimum is due to WGS reaction (exothermic), which consumes CO and steam producing CO_2 and H_2. Hence, a net negative CO_2 conversion is registered as proven in other equilibrium analyses [104]. In the light of this consideration, considering the typical industrial reformer temperature range of 700–950°C [109–112], trends in H_2/CO ratio in the final syngas mixture are straightforward:

- In bireforming (BR) process, the SMR guarantees H_2/CO>3; however, the activation of the RWGS, which consumes both CO_2 and the produced hydrogen, decreases such ratio, and independently from the pressure (for $T > 850$°C), H_2/CO is slightly larger than 2 [104]. Similar considerations are present in Zhou et al. [92]. The steam amount in the feed stream impacts the final syngas quality. The rule of thumb is that the higher the steam content, the

Table 1 Overview of recent publication on syngas production thermodynamic assessment.

Syngas production technology	Reference	Notes and comments
General overview	[90]	This work offers a complete overview of the equilibrium constants for a large reactions scheme including soot formation and methanation
	[91]	Both kinetic and thermodynamic studies for several technologies: DR, SMR, POX, ATR, BR and TR (including the combination of these). The final remarkable result is the surrogate model able to predict syngas composition at different operating conditions. Authors also compare their results with thermodynamic predictions showing that thermodynamic affects the product mixture compositions
	[45]	Complete thermodynamic assessment for SMR, DR and ATR. Authors compare in silico studies with experimental data available in the literature
	[92]	Thermodynamic assessments for DR, BR, TR and ATR
	[93]	Complete work on the SMR and DR (including BR). Sensitivity analysis on the final mixture composition according to the operating conditions and the feed composition
	[94]	The paper mainly deals with the DR thermodynamic assessment in presence of coke formation. However, this work accounts for SMR and POX
Dry reforming (DR)	[95]	Evaluation (including experimental tests) of CO_2/CH_4 effect on the syngas composition and H_2 or CO co-feeding impact
	[96]	Analysis of DR and the effect of O_2 as co-fed specie
	[97]	The work is mainly devoted to DR as a potential reforming process for emissions reduction. It presents a complete analysis of DR and CO_2 conversion. The authors includes also the SMR, POX and ATR in their analysis to be combined with DR to optimize the process yield
Reverse water gas shift (RWGS)	[98]	Effect of H_2/CO_2 on the syngas composition and CO_2 conversion. The work also deals with the performances of several commercial and laboratory catalyst
	[99]	Assessment for RWGS equilibrium. The article deals with reactor modeling in presence of transport limitations
Partial oxidation (POX)	[100]	Thermodynamic study of biogas partial oxidation with particular emphasis on soot formation and modeling using experimental observations
	[101]	Thermodynamic analysis on POX in O_2 and air. The paper deals with the effect of the feed composition and comburent on the products mixture including the soot formation
Steam methane reforming (SMR) and bireforming (BR)	[102]	Investigation of bi-reforming reactions on Ni-based catalyst. In this work, the thermodynamic assessment on CH_4 and CO_2 assessment

Table 1 Overview of recent publication on syngas production thermodynamic assessment—cont'd

Syngas production technology	Reference	Notes and comments
Trireforming (TR)	[103]	An exhaustive assessment of equilibrium constants for bireformer including soot formation and removal
	[104]	Starting from previous experimental studies, the paper reports a complete thermodynamic assessment of the bi-reformer system for syngas-to-methanol
	[105]	Complete work on thermodynamic equilibria involved in methane reforming in presence of CO_2, steam and O_2. The work deals with the bi-reforming and oxy-bi-reforming
	[106]	The paper proofs the overlapping of thermodynamic and kinetic in bi-reformer systems at high temperature ($T > 650°C$)
	[92]	In this work, a section is dedicated to thermodynamic analysis of tri-reforming reactions and the comparison with other syngas production technologies such as POX, BR and ATR
	[107]	The work deals with methane and coal gasification using flue gas. A complete thermodynamic study of the thermodynamic involved in TR is provided
Biomass gasification	[108]	Thermodynamic assessment of biomass gasification in presence of steam. The work shows the impact of steam/biomass ratio on the H_2 content in the produced syngas. Experimental results corroborate the in-silico study

higher is the H_2 amount in the syngas mixture, though, the asymptotic trend pushes the system toward $H_2/CO = 2$ independently from the initial steam/methane ratio [104,113,114].

- The DR is a very promising greenhouse gases emissions cutting technology; in any case the final expected $H_2/CO < 1.5$ (likely H_2/CO close to 1) due to parasite reactions such as soot formation and RWGS. However, it is possible to reach a higher H_2/CO ratio by properly tuning the inlet temperature, residence time, and CH_4/CO_2 ratio. For instance, oxidant deficiency ($CH_4/CO_2 < 1$) enables to increase the H_2 final content [95], but the final hydrogen content does satisfy the minimum requirement for many chemicals syntheses. Hence, generally, DR is coupled with SMR to increase the hydrogen content [97].
- In autothermal processes, also including trireforming (TR), POX occurs in the presence of BR. Mild POX increases the content of oxidized compounds and partially mitigates the energy requirement due to endothermic reactions (RWGS, SMR, and DR); hence, the final syngas content is $1 < H_2/CO < 2$ [107]. As a matter of fact, the O_2/CH_4 ratio manages the final syngas composition [105]. For $O_2/CH_4 > 0.5$, the syngas quality drops. Hydrogen content is progressively reduced since CO_2 and steam are the most stable compounds, and they are favored when combustion occurs.

As the temperature, the pressure influences the syngas quality. More in general, reforming processes cause an increment of moles in the system. For this reason, pressure reduction would favor syngas production. However, this is not possible due to several reasons: (1) volume required to reform the methane or the feedstock; (2) final pressure would not be suitable for chemical productions, for instance, both methanol and derivatives and Fischer-Tropsch synthesis require pressurized reactors; (3) syngas composition refinement is operated in pressurized absorber or adsorption unit [38]. More in general, the hydrogen yield benefits on a pressure reduction; however, this effect is negligible over 1400°C [113,114]. The pressure reduction mitigates the soot formation. As discussed in the next bullet, steam and/or oxygen injection prevents the system from coke depositing [101].

Soot formation is industrially unavoidable as emphasized in Haldor-Topsøe published research [112], and coking occurs in the hotspot of the reformer since it is mainly generated during methane or more in general feedstock thermal decomposition. The coke formation is also confirmed in recent publications [90,96,100,107]. The same work highlights that steam and oxygen (in the case of TR) mitigate the soot formation, and at $T > 700°C$ coking is suppressed.

Several researchers proposed kinetic model for SMR, DR, and RWGS [18,24,115–118]. Zeppieri et al. [118] demonstrated that the syngas composition is far from the thermodynamic predictions [118]. Conversely, several more recent publications in the field of catalysis demonstrated that at high temperature $(T > 700°C)$, thermodynamic equilibrium affects the syngas mixture composition [99,102,106,119] meaning that the kinetic rapidly drives the system toward equilibrium conditions. Chein and collaborators further verified this conclusion in two different works [94,120] by comparing kinetic-based reactor results and thermodynamic predictions as also remarked in ref. [104]. Other researchers clearly demonstrated that thermodynamic equilibrium mirrors the experimental observation on the produced syngas mixtures justifying the use of equilibria reactor in simulation software as well [120–123]. Hence, at the typical industrial reformer operating temperature range $(700 < T < 950°C)$, the syngas final composition is likely in the equilibrium conditions.

4. Kinetic models

The quality of the syngas strongly depends on the catalytic reactions taking place inside the tubular reactors. In the past, the investigations on reforming reactions were mainly focused on catalyst preparation and process analysis, neglecting kinetics and reaction mechanisms [124]. So, due to the scarce information on the subject, in recent decades, many efforts have been expended to propose reaction mechanisms and carbon formation, as well as to determine the kinetic parameters and estimate more precisely the chemical reaction rates [18–20,24,25,125–135]. Despite the interest on the DMR [128,134,135], the autothermal methane reforming [132], the methane trireforming [127], and the biogas dry reforming [130], the kinetics of the steam methane reforming is the most frequently studied in the literature and several kinetic models have been proposed.

The kinetic model from Xu and Froment [24,25] is the most widely employed. It considers three reactions (R3), (R4), and (R5), whose reaction rates are given by:

$$r_3 = \frac{\frac{k_1}{P_{H_2}^{2.5}} \left[P_{CH_4} P_{H_2O} - \frac{P_{CO} P_{H_2}^3}{K_1} \right]}{\left(1 + K_{CH_4} P_{CH_4} + K_{CO} P_{CO} + K_{H_2} P_{H_2} + K_{H_2O} \frac{P_{H_2O}}{P_{H_2}} \right)^2} \tag{1}$$

$$r_4 = \frac{\frac{k_2}{P_{H_2}} \left[P_{CO} P_{H_2O} - \frac{P_{CO_2} P_{H_2}}{K_2} \right]}{\left(1 + K_{CH_4} P_{CH_4} + K_{CO} P_{CO} + K_{H_2} P_{H_2} + K_{H_2O} \frac{P_{H_2O}}{P_{H_2}} \right)^2} \tag{2}$$

$$r_5 = \frac{\frac{k_3}{P_{H_2}^{3.5}} \left[P_{CH_4} P_{H_2O}^2 - \frac{P_{CO_2} P_{H_2}^4}{K_3} \right]}{\left(1 + K_{CH_4} P_{CH_4} + K_{CO} P_{CO} + K_{H_2} P_{H_2} + K_{H_2O} \frac{P_{H_2O}}{P_{H_2}} \right)^2} \tag{3}$$

where r_3, r_4, and r_5 are given in $kmol \cdot kg_{cat}^{-1} \cdot h^{-1}$; P is the partial pressure of the component (bar); k_1 $(kmol \cdot bar^{0.5} \cdot kg_{cat}^{-1} \cdot h^{-1})$, k_2 $(kmol \cdot bar^{-1} \cdot kg_{cat}^{-1} \cdot s^{-1})$, and k_3 $(kmol \cdot bar^{0.5} \cdot kg_{cat}^{-1} \cdot h^{-1})$ are the kinetic constants; K_{CH_4}, K_{CO}, $K_{H_2}(bar^{-1})$, and K_{H_2O} (−) are the adsorption constants; the chemical equilibrium constants are given by K_1, K_3 (bar^{-2}), and K_2 (−). The adsorption on the catalytic sites follows the Langmuir-Hinshelwood (LH) hypothesis and neglects CO_2 adsorption.

Hou and Hughes [18–20] studied the kinetics of steam methane reforming in an integral reactor, considering six possible reaction mechanisms. These mechanisms were investigated in detail and the intrinsic rate equations were suggested and tested under different conditions of temperature, pressure, and steam-to-methane ratio. As Xu and Froment [24,25], Hou and Hughes [18–20] consider the three reactions (R3, R4, and R5), which are modeled as:

$$r_3 = \frac{k_4 \left[\frac{P_{CH_4} P_{H_2O}^{0.5}}{P_{H_2}^{1.25}} \right] \left[P_{CH_4} P_{H_2O} - \frac{P_{CO} P_{H_2}^3}{K_4} \right]}{\left(1 + K_{CO} P_{CO} + K_{H_2} P_{H_2}^{0.5} + K_{H_2O} \frac{P_{H_2O}}{P_{H_2}} \right)^2} \tag{4}$$

$$r_4 = \frac{k_5 \left[\frac{P_{CO} P_{H_2O}^{0.5}}{P_{H_2}^{0.5}} \right] \left[P_{CO} P_{H_2O} - \frac{P_{CO_2} P_{H_2}}{K_5} \right]}{\left(1 + K_{CO} P_{CO} + K_{H_2} P_{H_2}^{0.5} + K_{H_2O} \frac{P_{H_2O}}{P_{H_2}} \right)^2} \tag{5}$$

$$r_5 = \frac{k_6 \left[\frac{P_{CH_4}P_{H_2O}}{P_{H_2}^{1.75}}\right]\left[P_{CH_4}P_{H_2O}^2 - \frac{P_{CO_2}P_{H_2}^4}{K_6}\right]}{\left(1 + K_{CO}P_{CO} + K_{H_2}P_{H_2}^{0.5} + K_{H_2O}\frac{P_{H_2O}}{P_{H_2}}\right)^2} \tag{6}$$

where r_3, r_4, and r_5 are given in $kmol \cdot kg_{cat}^{-1} \cdot s^{-1}$; k_4 $(kmol \cdot kPa^{0.25} \cdot kg_{cat}^{-1} \cdot s^{-1})$, k_5 $(kmol \cdot kPa^{-1} \cdot kg_{cat}^{-1} \cdot s^{-1})$, and k_6 $(kmol \cdot kPa^{0.25} \cdot kg_{cat}^{-1} \cdot h^{-1})$ are the kinetic constants; K_{CH_4}, K_{CO}, $K_{H_2}(kPa^{-1})$, and K_{H_2O} (−) are the adsorption constants; the chemical equilibrium constants are given by K_4, K_6 (kPa^{-2}), and K_5 (−); P is the partial pressure of the component (kPa). The adsorption on the catalytic sites combines the LH theory with the Freundlich's concept of nonideal adsorption and neglects not only the adsorption of CO_2, but also methane, which is hindered due to the high steam concentration, especially at high temperatures.

Singh and Saraf [131] consider that (R5) is quickly completed at the reactor inlet; therefore, their kinetic model comprehends only (R3) and (R4) and follows a power law [136,137], according to:

$$r_3 = \frac{k_7}{P^{0.5}P_{H_2O}}\left[P_{CH_4}P_{H_2O} - \frac{P_{CO}P_{H_2}^3}{K_7}\right] \tag{7}$$

$$r_4 = \frac{k_8}{P^{0.5}P_{H_2O}}\left[P_{CO}P_{H_2O} - \frac{P_{CO_2}P_{H_2}}{K_8}\right] \tag{8}$$

where r_3 and r_4 are given in $lbmol \cdot lb_{cat}^{-1} \cdot h^{-1}$, as functions of total system pressure and partial pressure of component, which are represented by P (atm); k_7 and k_8 $(lbmol \cdot atm^{-0.5} \cdot lb_{cat}^{-1} \cdot h^{-1})$ are the kinetic constants; K_7 (atm^{-2}) and K_8 (−) are the chemical equilibrium constants. The authors assume the hypothesis that all higher-molecular-weight hydrocarbons are promptly hydrocracked to methane [28,136]. The influence of the total pressure on the apparent reaction rate is incorporated to account with the high pore diffusion resistance.

The kinetic model of Numaguchi and Kikuchi [129] was originally developed as a general model for reforming processes using only five experiments. As Singh and Saraf [131], it only considers reactions (R3) and (R4) whose kinetic expressions follow a power law:

$$r_3 = \frac{k_9}{P_{H_2O}^{1.596}}\left[P_{CH_4}P_{H_2O} - \frac{P_{CO}P_{H_2}^3}{K_9}\right] \tag{9}$$

$$r_4 = \frac{k_{10}}{P_{H_2O}}\left[P_{CO}P_{H_2O} - \frac{P_{CO_2}P_{H_2}}{K_{10}}\right] \tag{10}$$

where r_3 and r_4 are given in $kmol \cdot m_{cat}^{-3} \cdot h^{-1}$; k_9 and $k_{10}(kmol \cdot bar^{0.404} \cdot m_{cat}^{-3} \cdot h^{-1})$ are the kinetic constants; the chemical equilibrium constants are given by K_9, K_{10} (bar^{-2}), and K_5 (−); P is the partial pressure of the component (bar).

Numaguchi and Kikuchi [129], as Hou and Hughes [18–20], combine the LH theory with the Freundlich's concept of nonideal adsorption, but further simplify the model after considering that the steam adsorption equilibrium constant is much larger than the other ones. The authors proposed and compared the intrinsic and apparent kinetic rates, which neglects interphase and intraphase diffusions. After validating the models using five experimental runs, they observed that the diffusion on the catalyst surface significantly influences the axial profiles and, therefore, might not be neglected. This model has been more frequently applied for ATR reactors [40,55,138,139], which operate at higher pressures and temperatures due to the exothermic combustion reactions, which occur simultaneously with the reforming reactions.

The kinetic rates presented so far differ regarding the partial pressure of steam and hydrogen. The Numaguchi and Kikuchi [129] model exhibits an inverse dependence on the steam partial pressure, probably due to the competitive adsorption of methane and steam on the catalytic active sites. Singh and Saraf [131] consider that their kinetic model is valid for any operating condition, as well as for the case of significant pore diffusion resistance. The Xu and Froment [24,25] and Hou and Hughes [18–20] models tend to present numerical instability problems because of the negative order of the H_2 partial pressure. Xu and Froment [24,25] advocate, though, that their model is suitable for industrial reformers because the reactor is usually fed with hydrogen or higher hydrocarbons, which are rapidly converted to CO and H_2.

Table 2 presents the operating conditions used to develop the kinetic models discussed so far, as well as the most important assumptions made when formulating the kinetic rates. Xu and Froment [24,25] conducted their experiments at temperatures much lower than those usually applied industrially [173] to avoid reaching the kinetic equilibrium. Despite that, most authors who investigate reforming processes, especially industrial reformers, use the Xu and Froment [24,25] model without any previous investigation. Singh and Saraf [131] and Numaguchi and Kikuchi [129], on the other hand, apply operating conditions more similar to the industrial unit.

Table 2 further summarizes the studies that employ these kinetic models when approaching reforming reactors. Some studies attempt to compare the kinetic models, but they are mainly concerned with noncommercial reactors. Vaccaro and Malangone [155] compared the Xu and Froment [24,25] model with the Hou and Hughes [18–20] model when studying a micro-scale reactor. They observed that the latter outperformed, probably because of the presence of hydrogen in the feed stream. Leonzio [146] compared the Xu and Froment [24,25] model with the Numaguchi and Kikuchi [129] model, when modeling a membrane reactor. The author concluded that the Numaguchi and Kikuchi [129] model was more consistent with the data from literature. More recently, Quirino et al. [173] made a more comprehensive study, comparing the Xu and Froment [24,25] model with the models from Numaguchi and Kikuchi [129], Hou and Hughes [18–20], and Singh and Saraf [131], when modeling an industrial top-fired methane reformer. They observed that the kinetic models presented very similar temperature and composition profiles, except at the very first section of the reactor, as Fig. 1 illustrates.

Table 2 Kinetic model proposed in the literature for modeling the steam reforming process (author's table, 2021) [20,21,24,27,32,129,131,140,141].

Reference	Operating conditions[a]	Works that consider this kinetic model
Singh and Saraf [131]	Modeling approach: PL Temperature (K): 644–1317 Pressure (bar): 13.60–34.8 $[H_2O/CH_4]_{in}$: 3.84–6.73 $[H_2/CO_2]_{in}$: 0 Catalyst/support: Ni/Al_2O_4 Ni wt (%): NA MSA (m^2/g): NA CSS: raschig rings (mm); $16 \times 16 \times 6$; pellets (in) − $1/4 \times 1/4$ BDC (kg/m^3): 1202; 1225; 1384 Diffusion limitations: Yes	Singh and Saraf [142] Assaf et al. [143] Ravi et al. [144] Acuña et al. [145] Rydén and Lyngfelt [14] Vakhshouri and Hashemi [16] Shayegan et al. [15]
Numaguchi and Kikuchi [129]	Modeling approach: LH/F Temperature (K): 674–1160 Pressure (bar): 1.2–25.5 $[H_2O/CH_4]_{in}$: 1.44–4.50 $[H_2/CO_2]_{in}$: NA Catalyst/support: Ni/Al_2O_3 Ni wt (%): 8.7 MSA (m^2/g): 3.6 CSS: cylindrical (in) − $(5/8 \times 1/4)$ and $(5/8 \times 5/8)$ BDC (kg/m^3): 1970 Diffusion limitations: Yes	De Smet et al. [40] Shi et al. [55] Azarhoosh et al. [138] Leonzio [13] Leonzio [146] Aloisi et al. [147] Aloisi et al. [12] Spallina et al. [148] Leonzio [8] Cherif and Nebbali [139]
Xu and Froment [24,25]	Modeling approach: LH Temperature (K): 573–848 Pressure (bar): 3–15 $[H_2O/CH_4]_{in}$: 3–5 $[H_2/CO_2]_{in}$: 0.5–1.0 Catalyst/support: $Ni/MgAl_2O_4$ Ni wt (%): 15.2 MSA (m^2/g): 9.3 CSS (mm): raschig rings; 0.18–0.25 BDC (kg/m^3): 1870 Diffusion limitations: No	De Smet et al. [40] Xiu et al. [149] Gallucci et al. [150] Hoang et al. [151] Yu et al. [152] Zamaniyan et al. [26] Oliveira et al. [153] Silva et al. [154] Pantoleontos et al. [22] Vaccaro and Malangone [155] Sadooghi and Rauch [156] Lee et al. [157] Farniaei et al. [158] Qi et al. [159] Abid and Jassim [160] Pret et al. [23] Lao et al. [21] Vidal Vázquez et al. [161] Kumar et al. [162]

Table 2 Kinetic model proposed in the literature for modeling the steam reforming process (author's table, 2021) [20,21,24,27,32,129,131,140,141]—cont'd

Reference	Operating conditions	Works that consider this kinetic model
Hou and Hughes [18–20]	Modeling approach: LH/F Temperature (K): 598–823 Pressure (bar): 1.2–6 $[H_2O/CH_4]_{in}$: 4–7 $[H_2/CO_2]_{in}$: 0.5; 0.75 Catalyst/Support: Ni/α-Al$_2$O$_3$ Ni wt (%): 11.8–13.4 MSA (m^2/g): 14.3 CSS (mm): cylindrical; 0.105–0.18 BDC (kg/m^3): 2355 Diffusion limitations: No	Darvishi and Zareie-Kordshouli [17] Abbas et al. [163] Wu et al. [164] Dixon et al. [165] Nijemeisland et al. [166] Karimi-Golpayegani et al. [167] Dixon et al. [168] Klein et al. [169] Silva et al. [154] Vaccaro et al. [170] Vaccaro and Malangone [155] Kuznetsov and Kozlov [171] Vidal Vázquez et al. [161] Lu et al. [172]

[a]*PL*, power law; *LH*, Langmuir-Hinshelwood model; *F*, Freundlich model; *[x/y]in*, component molar ratio in feed stream; *CSS*, catalyst shape and size; *BDC*, bulk density of catalysts; *MSA*, metal surface area; *NA*, not available.

The absence of hydrogen in the reforming gas inlet artificially created disturbances on the Xu and Froment [24,25] and Hou and Hughes [18–20] models, unlike the simpler Singh and Saraf [131] and Numaguchi and Kikuchi [129] models. The prediction of the Hou and Hughes [18–20] model overlaps the prediction of the Xu and Froment [24,25] model because of the high similarity of both kinetic rates. The authors further concluded that the reforming reactor operates close to equilibrium, as evidenced in Fig. 2; therefore, simpler kinetic model as Singh and Saraf [131] and Numaguchi and Kikuchi [129] might be considered without loss of accuracy, to represent the SMR unit and further present the advantage of better numerical stability.

5. Reactors modeling

Steam reforming processes involve complex transport phenomena such as molecular diffusion, solid and/or fluid phase heat transfer, in addition to the complex flow profile developed within the catalytic bed [175]. The heat released by the burners in the furnace is transferred to the interior of the reforming tubes and refractory surface by radiation, convection, and conduction mechanisms. The chemical reactions create radial and axial temperature and concentration gradients in the tube and its surroundings, as well as inside the catalyst porous particles. Thus,

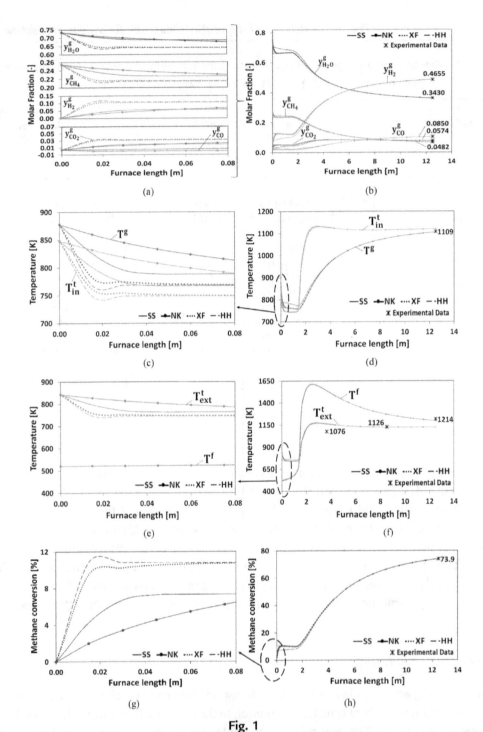

Fig. 1
Influence of the kinetic models on molar composition of components (A and B); process gas (T^g) and inner tube wall (T^t_{in}) temperatures (C and D); furnace (T^f);and outer tube wall (T^t_{ext}) temperatures (E and F); and methane conversion (G and H) along the length of the reformer. Experimental data reported by Latham [32] at the output of the tubular reactor and at two intermediate points on the outer tube wall. Focus is given at the initial section of reformer, where the kinetic models show the greatest discrepancies [173].

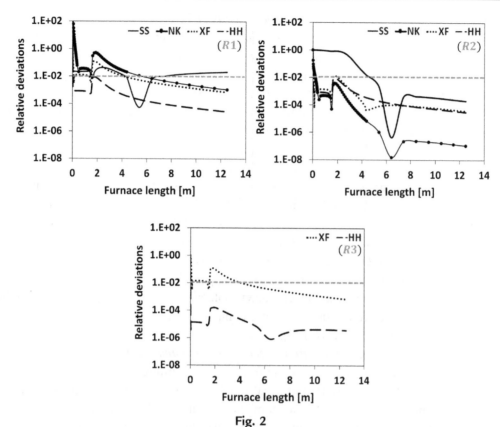

Fig. 2

Deviations of chemical equilibrium along the reformer length for different kinetic models. The Hou and Hughes [20], Xu and Froment [24], Singh and Saraf [131] deviations are calculated by comparing the equilibrium constants, computed at the reaction temperature, with the reaction quotients, i.e., the ratios between the relative amounts of products and reagents. If the equilibrium constant is equal to the reaction quotient, the chemical equilibrium is achieved. The *red (dark gray in the printed version) dotted line* refers to a 1% deviation from equilibrium [13,20,24,129,131,146,174]. *Adapted from Quirino, P.P.S., Amaral, A., Pontes, K.V., Rossi, F., Manenti, F. (2021). Impact of kinetic models in the prediction accuracy of an industrial steam methane reforming unit. Comput. Chem. Eng. 152, 107379. https://doi.org/10.1016/j.compchemeng.2021.107379.*

the precise analytical determination of all phenomena that occur in a reformer through highly complex mathematical models is a challenging task. For practical purposes, some simplifying hypotheses are established; however, excessive simplification may limit the model's adherence to the real system.

Depending on the application, several approaches for modeling and simulation of SMR processes are reported in the literature regarding:

- *Steady state or dynamic models.* Most authors model the reform process in steady state ([15,17,23,26,30,32,157,162,176]; to cite just a few), and some describe the reforming process in transient conditions [21,22,27,140,141,163,177].

- *Description of the catalytic bed by homogeneous or heterogeneous approach.* The main difference between the pseudo-homogeneous and heterogeneous models is the evaluation of the temperature and concentration profiles inside the catalyst particles, besides in the gas phase. Heterogeneous models make explicit reference to the catalyst, using effectivity factors multiplied by kinetic reaction rates. These effectivity factors vary along the length of the reactor and are calculated through the Thiele module [15,17,26,30]. On the other hand, in the pseudo-homogeneous models, the effectiveness factors are constant parameters (≤ 1), as reviewed by Pantoleontos et al. [22]. This approach does not explicitly consider diffusion limitations or is related to the assumption of "pseudo homogeneity" due to insignificant mass transport resistances because of the small catalyst particles used [18–20,28,149,162,171,176]. Quirino et al. [173], when comparing four kinetic models of reforming reactions [18–20,24,25,129,131], concluded that diffusion effects are negligible because the system quickly reaches the equilibrium. Both time-dependent models are commonly formulated using a pseudo-homogeneous approach due to its greater simplicity and versatility for different reforming process applications. The heterogeneous model, on the other hand, might present convergence problems due to the abrupt gradients in the catalyst particle, therefore requires greater computational effort [18–20,28,136,149,162,171,176,178].
- *Description of the furnace (radiation mechanisms, combustion models, and heat flow patterns).* Some authors model only the furnace, disregarding the heat and mass transport inside the tubes [177,179–184], and usually only the radiation mechanism is modeled, neglecting the other heat transfer phenomena. The precise description of the flaming radiation or hot gases is not straightforward because the gaseous medium participates in the emission and absorption of radiation. Lobo and Evans [185] pioneered in the first model for calculating the radiant heat exchange in furnaces, being referenced for several studies, as Roesler [182] and Hottel et al. [180]. The approach of Hottel et al. [180] is the most frequently used because it offers lower complexity to estimate the heat exchange and the outlet temperature of the flue gases [23,157,177]. Their approach subdivides the furnace volume and surface into a finite number of isothermal zones, as carried out in computational fluid dynamics (CFD) [23,157,177]. Quirino et al. [2], on the other hand, believe that the furnace discretization only in the axial direction is sufficient, as it avoids the complex determination of the radiation heat exchange areas between the radiating zones in the furnace. This hypothesis provides a trade-off between model complexity, computational cost, and accuracy.
- *Number of control volumes representing the system.* The models reported in the literature usually consider three control volumes to represent the industrial reformer, even the most detailed approaches. They consider the tubular reactor, the tube wall, and the furnace, but neglect the refractory wall. Quirino et al. [2], on the other hand, model all the four control volumes, including the refractory wall temperature. Predicting the large temperature gradients inside the furnace, though, would allow to take corrective actions avoiding

damages to the refractory material and reduce energy losses to the environment. Furthermore, the knowledge of the wall temperature could be used to estimate thermomechanical properties of the refractory material as mechanical strength, conductivity, and porosity, useful for maintenance purposes.

- *Determination of the furnace gas emissivity* [2]. The gray gas hypothesis assumes that the absorption coefficient solely depends on temperature and gas composition. Thanks to its simplicity, several authors assume the gray gas hypothesis, as Xue et al. [186], Al-Omari [187], Deshmukh et al. [188], and Narayanan and Trouvé [189]. More complex models have also been reported, further taking into account the furnace geometry and pressure effects on the gas emissivity [190].

- *Heat release profile in the furnace.* The combustible gas is assumed to be completely and isothermally burnt at the furnace inlet. To avoid including the distributed energy balance for the tube wall, some authors simplify the reform model considering empirical profiles for the heat release along the reformer length [26,28,162], or for the temperature of the tube outer wall [22,191], or even for the furnace temperature [17]. Experimental data of flame length and total heat of combustion are strictly required to identify the correlations of heat profile release. Roesler [182] estimated a parabolic profile to represent the heat release along the furnace, while Hyde et al. [192] considered an exponential function. The required experimental information, though, is usually scarce [17,26,193]. Considering the kinetics of the combustion reactions in the furnace would provide more flexibility to the model. Therefore, Lee et al. [157], Pret et al. [23], and Tran et al. [177] describe the rate of the combustion reactions, which are limited by the mixing effect among fuel, oxygen, and products, so that slower temperature gradients are predicted within a Computational Fluid Dynamics (CFD) framework. Quirino et al. [2] also modeled the combustion reactions, but without considering the mixing effects, in a more straightforward approach than the CFD, yet reasonably precise when compared with industrial data [2,177].

- *Prediction of gaseous mixture properties.* The temperature gradient across the reformer reaches up to 1000 K; therefore, the physical properties of the gaseous mixture might significantly change. Some authors assume ideal mixture, while other authors state that the physical properties vary with temperature and pressure, without showing the correlations used [18–20,30,145,153,159].

- *Consideration of different heat transfer effects.* Some simplifying hypotheses are frequently adopted in the modeling of the steam methane reformer. Some heat transfer mechanisms are neglected, such as the convection at the furnace, the reflection of the radiated heat by the reforming tubes, the role of the gas in the absorption of the radiation, and the shading effects caused by the other tubes [22,30,131,141,152,162,176]. The degree of sophistication of the reformer model primarily depends on the purpose for which this model is developed. If the model is intended to model-based control, for example, the usual CFD approach is impractical because its computational cost is incompatible with the real-time computation of control actions. This is confirmed by several authors who proposed optimal control applications in reforming units [194,195]. Quirino et al. [2], then, formulated a first-principles stationary model for practical purposes, i.e., to help operators in monitoring the most important output variables of industrial reformers.

6. Process description and mathematical model

Fig. 3 shows a typical industrial top-fired SMR unit, which comprehends a furnace equipped with one to several rows of tubes. The process gas flows down through tubes filled with catalyst. The burners are located among the tube rows at the top of the furnace. The tubes are heated by the radiation of the flames, combustion gases, and some convective effects. The flows of process gas and flue gas are cocurrents and come out at the bottom of the furnace. The feeding of the tubular reactors consists of a mixture composed of natural gas, steam, and CO_2, and at the end of these reactors is obtained the syngas (reformed gas). The energy for the reforming reactions is supplied by combustion of the fuel gas [2].

Fig. 4 shows the schematic flowchart of the control volumes and heat transfer mechanisms in an industrial reformer, as presented by Quirino et al. [2] for the purpose of developing a stationary mathematical model. The reformer is subdivided into four control volumes: (i) the tubular reactor, (ii) the tube wall, (iii) the furnace, and (iv) the refractory. For a more detailed description of the heat transfer mechanisms considered, as well as the main variables used in the model, the reader is addressed to Quirino et al. [2].

The mixture fed to the reforming tubes (natural gas + steam + CO_2) contains different components, and methane is the main hydrocarbon. Higher-molecular-weight hydrocarbons,

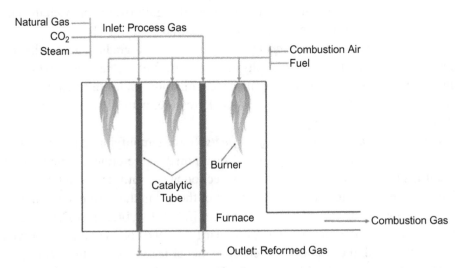

Fig. 3

Schematic flow diagram of the industrial SMRU (Steam Methane Reforming Unit) [2].

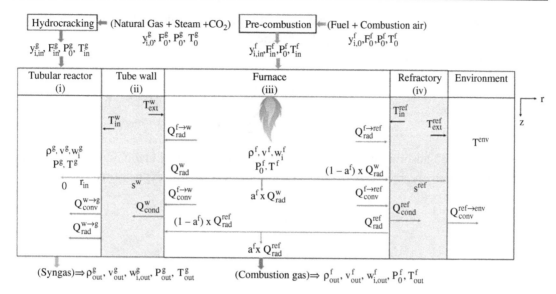

Fig. 4

Schematic flowchart of the control volumes, the heat transfer mechanisms, and main variables of the industrial reformer model [2].

such as ethane and propane, are present in lower composition. For this reason, to limit the number of material balances and simplify the reaction kinetics model, all the higher alkanes are considered to be irreversibly hydrocracked at the entrance of the tubular reactor. Hyman [136] admits that these superior hydrocarbons are completely converted to methane, according to the generic reaction:

$$C_nH_{2n+2} + (n-1)H_2 \rightarrow n\,CH_4 \; per \; n = 1, 2, 3, 4, \dots. \tag{R17}$$

However, when considering this reaction, if the hydrogen fed to the tubular reactors is not sufficient to completely break up the upper alkanes, they will remain unreacted in the fed stream. In this situation, most of the kinetic models that describe the reforming reactions cannot be used, since such models refer only to hydrocarbon methane. Moreover, accounting for unbroken upper alkanes will require many additional species and new equations should be added to the model to adequately describe the reforming system. To avoid these undesirable effects, Quirino et al. [2] considered that the hydrocracking of the higher alkanes occurs according to the reaction (R18) [32]:

$$C_nH_{2n+2} + \frac{n-1}{3}H_2O \rightarrow \frac{n-1}{3}CO + \frac{(2n+1)}{3}CH_4 \; per \; n > 1 \tag{R18}$$

At the furnace, all higher alkanes are considered to be completely burned, according to the general reaction:

$$C_nH_{2n+2} + \frac{3n+1}{2}O_2 \rightarrow n\,CO_2 + (n+1)H_2O \; if \; n = 2, 3, 4, \dots \tag{R19}$$

The new equivalent streams after the hydrocranking at the tubes inlet and the precombustion at the burners inlet are respectively computed by the following mass and energy balances, according to Quirino et al. [173]:

$$F_{i,eq} = F_{i,0} + \sum v_{i,n} r_n \tag{11}$$

$$\left(\sum F_i H_i\right)_{eq} = \left(\sum F_i H_i\right)_0 \tag{12}$$

The kinetic (k_i) and adsorption (K_i) constants used to compute the kinetic rates (r_4, r_5, r_6) are calculated, respectively, by the Arrhenius law and the Van't Hoff expression:

$$k_i = A_i \exp\left(\frac{-E_i}{RT}\right) \quad i = 1, 2, 3 \tag{13}$$

$$K_i = A_i \exp\left(\frac{-\Delta H_i^0}{RT}\right) \quad i = CH_4, CO, H_2O, H_2 \tag{14}$$

The kinetic parameters applied in Eqs. (13) and (14) are reported by the corresponding authors of the kinetic reaction rates [18–20,24,25,129,131]. The equilibrium constants (K_i) are calculated according to Twigg [33]:

$$K_1 = c \cdot \exp \exp\left(\sum d^i \alpha^i\right) \tag{15}$$

$$K_2 = \exp \exp\left(\sum b^i \alpha^i\right) \tag{16}$$

$$K_3 = K_1 K_2 \tag{17}$$

$$\alpha = \frac{1000}{T^g} - 1 \tag{18}$$

where the parameter c is a conversion from atm to Pa, T is temperature (K), and the parameters a^i and b^i are summarized in Table 3.

Table 4 presents the material, momentum, and energy balances [2]. The global mass balances for the tubular reactor and the furnace are expressed through Eqs. (19) and (24), applying the ideal gas equation (Eq. 33) and the mass flow concept (Eq. 34) to define the boundary conditions. Differential (Eqs. 20 and 25) and algebraic (Eqs. 21 and 26) equations compute the component mass fractions in the tube and in the furnace. The model assumes the hypothesis of

Table 3 Parameters used by Twigg [33] for the calculation of equilibrium constants [173].

i	0	1	2	3	4
d^i	−3.2770	27.1337	−0.58101	−0.3665	0.2513
b^i	0.31688	4.1778	0.63508	−0.29353	–

Tubular reactor		(19)

$$v^g \frac{d\rho^g}{dz} + \rho^g \frac{dv^g}{dz} = 0$$

$$\rho^g(0) = \rho^g_{in}; \quad v^g(0) = v^g_{in} \tag{20}$$

$$\frac{\partial w^g_i}{\partial z} = \frac{\rho^c \cdot (1-\phi) \cdot A^t}{\dot{m}^g} M_i \sum_k \eta_k v^g_{i,k} \cdot R^g_k \quad i = 1, \ldots, Ni - 1$$

$$w^g_i(0) = w^g_{i,in} \tag{21}$$

$$\sum w^g_i = 1 \quad i = 1, \ldots, Ni \tag{22}$$

$$\frac{dT^g}{dz} = -\frac{\rho^c \cdot (1-\phi) \cdot A^t}{\dot{m}^g \cdot \hat{c}^g_p} \sum_k \eta_k R^g_k \Delta H_{kt} + \frac{Q^{w \to g}_{conv}}{\dot{m}^g \cdot \hat{c}^g_p \cdot \Delta z} + \frac{Q^{w \to g}_{rad}}{\dot{m}^g \cdot \hat{c}^g_p \cdot \Delta z}$$

$$T^g(0) = T^g_{in} \tag{23}$$

$$\frac{dP^g}{dz} = -150 \frac{(1-\phi)^2}{\phi^3} \frac{\mu^g \cdot v^g}{(D^c)^2} - 1.75 \frac{(1-\phi)}{\phi^3} \frac{\dot{m}^g \cdot v^g}{D^c \cdot A^t}$$

$$P^g(0) = P^g_{in} \tag{24}$$

Furnace		

$$v^f \frac{dP^f}{dz} + \rho^f \frac{dv^f}{dz} = 0$$

$$\rho^f(0,z) = \rho^f_{in}; \quad v^f(0) = v^f_{in} \tag{25}$$

$$\frac{dw^f_i}{dz} = \frac{A^f}{\dot{m}^f} M^f_i \sum_k \eta_k v^f_{i,k} \cdot R^f_k \quad i = 1, \ldots, Ni - 1$$

$$w^f_i(0) = w^f_{i,\,in}$$

$$\sum w^f_i = 1 \quad i = 1, \ldots, Ni \tag{26}$$

$$\frac{dP^f}{dz} = 0 \quad P^f(0) = P^f_{in} \tag{27}$$

$$\frac{dT^f}{dz} = -\frac{A^f}{\dot{m}^f \cdot \hat{c}^f_p} \sum_k R^f_k \Delta H_k + \frac{1}{\dot{m}^f \cdot \hat{c}^f_p \cdot \Delta z} \cdot \left(-Q^{f \to w}_{conv} - Q^{f \to ref}_{conv} - Q^{f \to w}_{rad} - Q^{f \to ref}_{rad} + d^f \cdot Q^w_{rad} + d^f \cdot Q^{ref}_{rad} \right)$$

$$T^f(0) = T^f_{in} \tag{28}$$

Tube wall		

$$\frac{d}{dr} \cdot \left(r \cdot \frac{\partial T^w}{\partial r} \right) = 0$$

$$Q^w_{cond}(r_{in}) = -k^w \cdot A^t_{L,in} \cdot \frac{dT^w}{dr}\bigg|_{r=r_{in}} = Q^{w \to g}_{conv} + Q^{w \to g}_{rad}$$

$$Q^w_{cond}(r_{in} + s^w) = -k^w \cdot A^t_{L,ext} \cdot \frac{dT^w}{dr}\bigg|_{r=r_{in}+s^w} = Q^{f \to w}_{conv} + Q^{f \to w}_{rad} + (1-d^f) \cdot Q^{ref}_{rad} - Q^w_{rad} \tag{29}$$

Refractory		

$$\frac{d^2 T^{ref}}{dz^2} = 0$$

$$Q^{ref}_{cond}\left(s^{ref}_0\right) = -k^{ref} \cdot A^{ref}_r \cdot \frac{dT^{ref}}{dr}\bigg|_{r=s^{ref}_0} = Q^{f \to ref}_{conv} + Q^{f \to ref}_{rad} + (1-d^f) \cdot Q^w_{rad} - Q^{ref}_{rad}$$

$$Q^{ref}_{cond}\left(s^{ref}\right) = -k^{ref} \cdot A^{ref}_r \cdot \frac{dT^{ref}}{dr}\bigg|_{r=s^{ref}} = Q^{ref \to env}_{conv} \tag{30}$$

insignificant mass transport resistances; hence, the effectiveness factors (η_k) of the kinetic rates, Eqs. (20) and (22), are equal to 1 [18–20,149,150,154,171].

The component mass balances are formulated in mass terms, unlike usually reported in the literature, which is on a molar basis. Most authors do not provide any explanation for the use of this approach or admit that the variation in the total number of moles is negligible along the reactor. This variation is commonly compensated by the variation in the number of moles of the inert components, in order to close the molar balances [196]. The use of molar balances, however, does not confer greater benefits to the simulation compared with mass balances, as it does not simplify the reactor modeling procedure nor the numerical integration [197]. On the other hand, the formulation of balances in mass terms allows us to account for the variation in the total number of moles, which is important when aiming to assess more accurately the composition of the components along the axial coordinate of the reactor and not just at its output. Consequently, substitution methods, which can cause numerical instabilities, do not need to be used to continuously update the molar fractions along the axial coordinate of the reactor, in order to compensate for the total molar variation. Therefore, the balance in mass terms is more advantageous, as they result in systems with high numerical stability [194,196,197].

The reforming gas pressure is modeled according to Ergun equation (Eq. 23), and no pressure drop is considered for the flue gas (Eq. 27). The energy balances (Eqs. 22, 28, 29, and 30) provide the temperature profiles of the tubular reactor, the tube wall, the furnace, and the refractory wall, respectively. The boundary conditions for these differential equations are determined from the molar fractions of the individual components in the equivalent streams fed to the tubes and the furnace (Eq. 31) and are summarized in Table 5. This table also presents other constitutive relations applied to compute the parameters used in the model.

Table 6 presents the equations used to calculate the heat transfer rates by convection and radiation in the energy balances in Table 4. The convective coefficients on the tube and furnace sides, h^t [27,141,145,198–200] and h^f are, respectively, the Beek's correlation (1962), and the Dittus and Boelter's equation (1930) [17,28,152], while the convective coefficient to the external environment is a constant. The physical properties of the process gases in the reactor and in the furnace are computed using correlations from the literature, assuming ideal mixing

Table 5 Boundary conditions and other correlations of the steady-state model of the reformer (author's table, 2021).

$y_i = M\frac{w_i}{M_i}$	(31)	$M = \frac{1}{\sum \frac{w_i}{M_i}}$	(32)	$\rho = \frac{PM}{RT}$	(33)
$G = \rho v = \frac{\dot{m}}{A}$	(34)	$A^t = \pi r_{in}^2$	(35)	$A^t_{L,in} = 2\pi r_{in}\Delta z$	(36)
$A^t_{L,ext} = 2\pi . (r_{in} + s^w). \Delta z$	(37)	$A^f = \frac{LW}{n^t} - \pi r_{ext}^2$	(38)	$A^{ref}_r = \frac{2.\Delta z.(L+W)}{n^t}$	(39)

Table 6 Equations for heat transfer rates by convection and radiation (author's table, 2021).

Convection		Radiation	
$Q_{conv}^{w\to g}=h^t.\ 2\pi r_{in}.\ \Delta z.\ (T_{in}^w-T^g)$	(40)	$Q_{rad}^{f\to w}=\sigma.\ 2\pi(r_{in}+s^w).\ \Delta z.\ \varepsilon^f.\ (T^f)^4$	(41)
$Q_{conv}^{f\to w}=h^f.\ 2\pi(r_{in}+s^w).\ \Delta z.\ (T^f-T_{ext}^w)$	(42)	$Q_{rad}^{f\to ref}=\sigma.\ A_L^f.\ \varepsilon^f.\ (T^f)^4$	(43)
$Q_{conv}^{f\to ref}=h^f.\ A_L^f.\ (T^f-T_{in}^{ref})$	(44)	$Q_{rad}^{w\to g}=\sigma.\ 2\pi r_{in}.\ \Delta z.\ \varepsilon^w.\ (T_{in}^w)^4$	(45)
$Q_{conv}^{ref\to env}=h^{env}.\ A_L^f.\ (T_{ext}^{ref}-T^{env})$	(46)	$Q_{rad}^{ref}=\beta_1.\ \sigma.\ A_L^f.\ \varepsilon^{ref}.\ (T_{in}^{ref})^4$	(47)
		$Q_{rad}^{w}=\beta_2.\ \sigma.\ 2\pi(r_{in}+s^w).\ \Delta z.\ \varepsilon^w.\ (T_{ext}^w)^4$	(48)

rule: thermal conductivity and viscosity, according to Yaws [201] and Poling et al. [202] specific heat of the pure components, from NASA correlations [203].

The mathematical model presented in Table 4 results in a stiff differential algebraic equation (DAE) system, which comprehends equations with different dynamics: while the reforming reactions occur more slowly, the combustion reactions are pretty fast. The reformer model proposed by Quirino et al. [2] was validated against industrial and literature data [32] with reasonable accuracy [2,32]. The temperature and composition profiles predicted by Quirino et al. [2] are compared with the simulated data from Latham [32] in Fig. 5. Remarkable deviations are noticed only at the first section of the reactor, probably because Quirino et al. [2] modeled the combustion reactions in the furnace, unlike Latham [32], who assumed a heat release profile [2,32]. There is a lack in the literature on experimental data for the reforming reactor, especially along the tube length, but both models present very similar deviations from the experimental data provided by Latham [32]. Any experimental or simulated data for the temperature at the inner and outer surfaces of the refractory were available to validate the model, but the profiles presented by Quirino et al. [2] were consistent. H_2 is steadily produced as steam and methane are consumed. The CO concentration is higher than the CO_2 concentration at the reactor outlet, indicating that the reverse water-gas shift reaction (R2) takes place at high temperatures due to its exothermic nature [2].

Quirino et al. [2] concluded that the radiation inside the tubes, which is usually disregarded, changes the model predictions, while heat transfer by convection inside the furnace might be neglected, as indicated in Fig. 6, probably because the radiation dominates the other heat transfer mechanisms. Cremer and Olver [190] demonstrated that the refractory emissivity, when varied from 0.4 to 0.9 using the gray gas model, had no influence on the process variables. Then using a spectral gas line model (spectral-line-weighted, SLW), though, the efficiency of the radiant furnace increased by 2% and the arc temperature decreased by 30°C [2,177,179–181,183,184].

To identify which of the variables are most relevant to the SMR unit, a screening factorial design 2_{IV}^{8-4} is performed using data obtained by simulating the model in ref. [2]. The inlet variables considered are: feed gas temperature (T_{in}^g); (I_2) inlet pressure (P_{in}^g); (I_3)

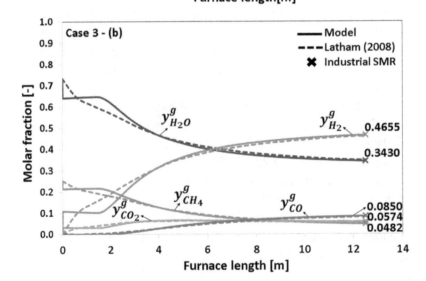

Fig. 5

Temperature profiles of process gas (T^g), inner tube wall (T^t_{in}), outer tube wall (T^t_{ext}), and furnace (T^f) (case 3-a); and composition profiles (case 3-b) predicted by the models from Quirino et al. [2] and Latham [32]. Experimental data from industrial SMR from Latham [32] at the output of the tubular reactor and at two intermediate points on the outer tube wall. *Adapted from Quirino, P.P.S., Amaral, A., Pontes, K.V., Rossi, F., Manenti, F. (2020). Modeling and simulation of an industrial top-fired methane steam reforming unit. Ind. Eng. Chem. Res. 59 (24), 11250–11264. https://doi.org/10.1021/acs.iecr.0c00456.*

steam-to-carbon ratio (SCR); (I_4) total feed flow to the tube (Fw^g_{in}); (I_5) combustion air temperature (T^f_{air}); (I_6) fuel gas temperature (T^f_{fuel}); (I_7) air/fuel ratio (AFR); (I_8) molar composition of methane in fuel gas ($y^f_{CH_4}$). The effects of these variables are investigated in four responses (output variables): tubular reactor outlet temperature (T^g_{out}), tube outer wall temperature at one-third of the furnace length $\left(T^w_{\frac{1}{3}}\right)$, methane conversion ($X_{CH_4}$), and H$_2$/CO

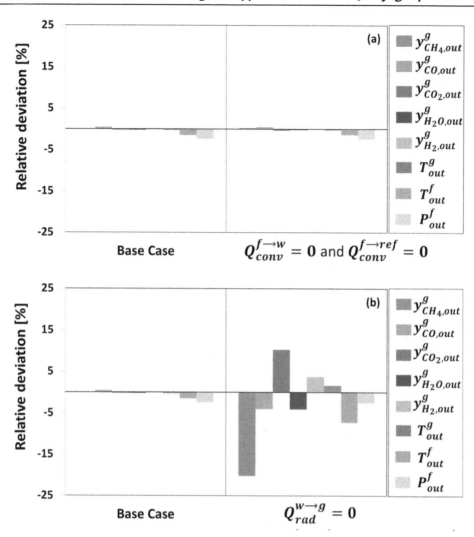

Fig. 6

Effect of heat transfer by convection from the furnace to the tube wall ($Q_{conv}^{f \to w}$) and to the refractory ($Q_{conv}^{f \to ref}$) (A) and by radiation in the tube side ($Q_{rad}^{w \to g}$) (B) on the model accuracy. *Adapted from Quirino, P.P.S., Amaral, A., Pontes, K.V., Rossi, F., Manenti, F. (2020). Modeling and simulation of an industrial top-fired methane steam reforming unit. Ind. Eng. Chem. Res. 59 (24), 11250–11264. https://doi.org/10. 1021/acs.iecr.0c00456.*

ratio in the reformed gas. The selection of inlet variables and process responses is based on works available in the literature on methane steam reforming [152,163,204–213]. Fig. 7 illustrates the main effects, through the Pareto diagram at a 90% confidence level. The results obtained from the 2_{IV}^{8-4} Fractional Factorial Design show that six ($SCR, Fw_{in}^{g}, T_{in}^{g}, T_{air}^{f}, AFR, y_{CH_4}^{f}$) of the eight factors significantly affect at least one of the monitored responses. The first three

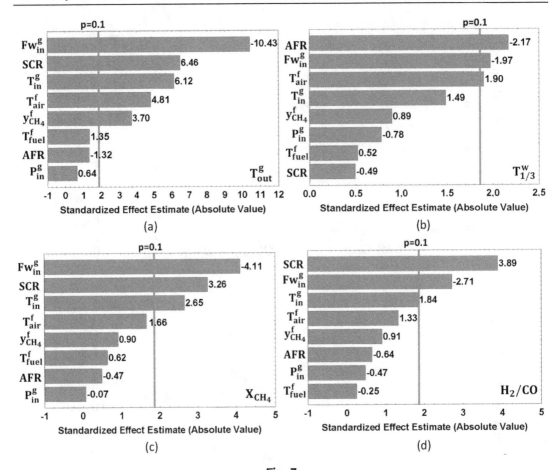

Fig. 7
Pareto charts of absolute standardized effects at a 90% confidence level ($p = 0.1$) for the responses (A) T^g_{out}, (B) $T^w_{\frac{1}{3}}$, (C) X_{CH_4}, and (D) H_2/CO for the 2^{8-4}_{IV} fractional factorial design (author's figure, 2021).

factors stand out among the others, playing an important role in the performance of the steam reforming process.

7. Solid wastes gasification and pyrolysis modeling—GasDS package

The modeling of solid fuels such as tars, coal, and biomasses [214–221] and/or wastes as in case of urban wastes and more recently plastics and bio-chars [222–225] pyrolysis/ gasification-to-syngas require specific numerical tools able to catch both syngas formation and residual char quantity [226,227]. The interest in using solids wastes as alternative feedstock for syngas production is continuously growing in the light of conventional source shortage and the possibility to consume wastes that otherwise would be discharged into landfills or burnt into thermal generation plants [38,214,225]. Nevertheless, the comprehensive description of solids particles combustion is a challenging task due to a multilayer complexity: (1) multicomponent problem related also to natural and seasonal oscillations in the feedstock composition, (2) multiphase problem since combustion involves different phases; (3) multiscale approach since intra- and interphase transport phenomena may occur, and these are not negligible; finally, (4) multidimensional since different order of magnitude may be considered: the reactor scale, the solid particle surface, and the solid porous as well. Moreover, also the time (i.e., residence time is a variable to be accounted). This complexity is further enhanced considering the request for transport phenomena and kinetic modeling coupling to acquire a discrete accuracy in modeling gasification and pyrolysis.

GasDS (developed at Politecnico di Milano) allows the model of both gasification and pyrolysis processes for solids fuels [226]. The characterization of the solid fuel composition is a key step to model the kinetic scheme involved in the solid degradation and the consequent syngas quality. The multilayer structure is briefly described here.

7.1 Biomass characterization and multistep kinetic model

The van Krevelen diagram is able to determine the elemental relative abundance of the relative quantitative ratio for each element as depicted in Fig. 8. If only the elemental analysis in terms of C, H, and O content is available, then a suitable combination of the reference species is simply derived from the three atoms balance. In the absence of biochemical analysis, bare estimates can be done. Hence, three mixtures of the reference component (cellulose, hemicellulose, and lignin) are proposed, and biomasses are characterized as a linear combination of these reference mixtures. The concept is naturally extended to plastic wastes, for instance, polyethylene (PE) is constituted by H and C; hence, oxygen is not present [226,227]. Finally, an additional point to consider is that the heating value is a function of the ratio of the relative elements: increasing the H/C ratio and decreasing the oxygen content is reflected in a larger heating value.

Taking advantage of several published works (refer to Ranzi and coworkers cited works for a full list and detailed description of each work), different kinetic mechanisms, detailed scheme, and the possibility to adopt the lumped approach, pyrolysis, and gasification have been modeled. At this step, it is necessary to account for (1) the heterogeneous reactions of char gasification and combustion, and (2) the secondary reactions occurring in the gas phase. The first considers that in thermally thick particles, the heat is transferred from the external surface

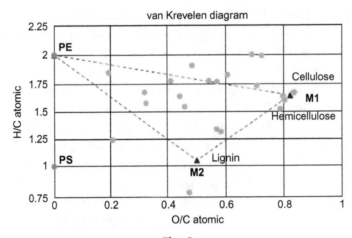

Fig. 8

Van Krevelen diagram [226].

toward the inner part of the particle. Char heterogeneous reactions are initially inhibited by the diffusion of volatile pyrolysis products. Similarly, char gasification and/or combustion reactions take place once biomass pyrolysis is accomplished. The secondary gas-phase reactions involving the released volatile species (tar and gas) are described using a general and detailed kinetic scheme of pyrolysis and combustion of hydrocarbons and oxygenated fuels. The number of species included in the kinetic scheme is always a good compromise between the accuracy and computational efforts.

7.2 Multiscale modeling

Two different scales are involved in the pyrolysis processes: the reactor scale and solid particle scale. According to prior works, a convenient way to present mass and energy balances is to distinguish these different scales [227]. The particle scale should be able to predict temperature profile and product distribution as a function of time. This model requires not only reaction kinetics but also reliable rules for estimating the transport properties such as thermal conductivity, material diffusion, and so forth to account for the morphological changes during the pyrolysis process. The intraparticle mass and heat transfer resistances are described assuming the solid particle as an isotropic sphere. The sphere is discretized into a series of concentric shells to dynamically characterize the temperature profile, exposed surface (biomass particles shrink as much as 50% during the process), and concentration profile under pyrolysis, gasification, and combustion regime. The gradients of temperature and volatile species inside the particles are evaluated employing the energy and continuity equations, respectively.
N sectors are assumed to discretize the particles. The implemented equations and more details

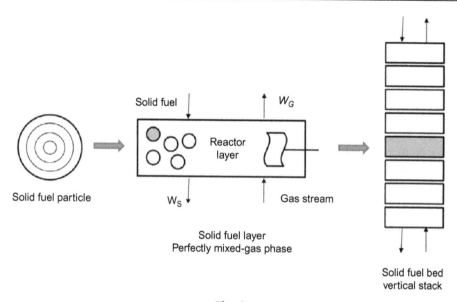

Fig. 9
Particle, layer, and equipment (gasifier, pyrolysis chamber) scales implemented in GasDS [227].

are reported in Ranzi and coworkers article [227]. The mathematical model of the fluidized bed or entrained bed reactor takes can directly refer to the particle model: the modeling of the fixed-bed reactor takes advantage from the definition of an elemental reactor layer describing the gas-solid interactions. The solid bed is then simulated as a series of NR interconnected layers as shown in Fig. 9, which highlights the interactions between adjacent layers. Boundary conditions such as feed temperature, composition, and so on are provided by the user. The height of each layer is of the same order of magnitude as the size of the biomass particle, accounting for vertical dispersion phenomena. The complete mixing inside the layer both for solid and gas phases is assumed. Balance equations are reported in the already cited work.

The GasDS has been successfully applied for different applications both on pilot scale and industrial scale showing a good agreement between model solution and experimental observations. Additional details on the experiments setup, model validation, and model assumptions are reported in the literature [227–231].

8. Reformer CFD modeling

As far the chapter dealt with the presentation of conventional modeling for the simulation and design of reformer for syngas production starting from natural gas, naphtha, and solid fuels such as coal or nonconventional fuel (i.e., municipal wastes, plastics, and so forth). In the field of reformer modeling and simulation, the Computational Fluid Dynamics (CFD) deserves a dedicated paragraph. The CFD has been receiving an increasing interest in the last decades to the its potential for detailed process analysis, process optimization, and transport phenomena (mass and heat transfer, diffusion limitations, microkinetics) investigation [232–234]. From a technical and engineering perspective, the coupling between process simulation and CFD detailed analysis is an interesting combination showing promising results and future outlooks [235]. Such kinds of analysis are continuously growing thanks to the increasing computational power, which lead to less costly and time-consuming CFD coding [232,233]. The CFD simulation requires the solution of Partial Differential Equations (PDEs) and specific numeral algorithms are devoted to solve such a class of numerical problem and partially mitigate the numerical solution stiffness. Considering (i) the potential high dimensions (as in industrial applications), (ii) the necessity to optimize the mesh grid for 2D and 3D problem, and (iii) the mathematical tools already available in the literature, it is also common practice to adopt commercial software suit to perform CFD simulations [234]. It is important to highlight that CFD simulations are time costly and CPU demanding; hence, a good trade-off is requested between several parameters such as numerical accuracy, mesh grid size, and computational time [236].

Despite the reformer dimensions and the complexity of huge reactors such as the reformer unit, the CFD simulation is progressively covering a key role also in the reforming simulation. In the last 20 years, many specific publications appeared. For instance, some researchers exploit ANSYS Fluent software to investigate the SMR [177,237], others analyzed the impact of the CFD to model the fluidized reactor for Coal Gasification (CG) [238,239], and more recently German researchers worked on DR. They coupled the CFD with microkinetics to evaluate their impact on the industrial data analysis and predictions just working at the microscale [240]. As a future outlook, the CFD simulation demonstrates that microchannel reactors appear as promising alternatives where it is possible to better manage the temperature profile, hence, the reaction rates and final syngas quality [241]. This brief overview remarks that the CFD applications are broadened, and it is a versatile approach enabling to achieve different targets according to the final task. Recent publications show that the CFD in the reforming reactions modeling is mainly devoted to (i) determine the specie distribution, (ii) temperature profile and gradient within the tubes and around the burners, (iii) optimize the catalyst shape to avoid hotspots and large pressure drops, (iv) optimize the overall process performance in terms of yield and also process controllability, and (v) better understand the transfer phenomena involved in the process. As mentioned above, hence, the CFD is a powerful approach to find applications for optimization and deep process analysis purposes.

Several works proved that the CFD simulations implemented in commercial software suits provide reliable mirrors of the process performance and final syngas quality regardless the kind of reforming (i.e., SMR, DR, CG, and so forth). These also compared experimental observations and industrial data showing a good agreement [177,237–240,242]. The CFD simulation allows one to better understand the flame temperature close to the burner, gas velocity, product distribution, and the impact of the flue/air ratio on such variables [243]. These results are useful to accurately predict radiative chamber temperature and the thermal exchange between the external wall tube and the surrounding environment. Other works focused within the tube bundle, and they demonstrated the potential of the CFD simulation to verify the optimal operating conditions. For instance, the CFD coding enables it to correctly predict the heat flux close to the inner tube surface, the inner temperature gradient, and the temperature hotspots within the catalyst particles. In this case, the results of the CFD simulation are meant to optimize the control system to reach the syngas specifications without wasting time [21]. Finally, the optimization of the catalyst particles is a great enhancement reached thanks to the CFD simulation. Dixon's research group is really active in this branch. They studied the species, temperature, and gas velocity distributions around industrial catalyst particles [168]. Their works proved that the CFD simulation allows to visualize the flow and gas velocity around the catalyst, the channeling effect close to the tube wall and to estimate the real the solids heat transfer coefficient [166]. These pieces of information are really useful to avoid any temperature hotspot, mass transfer limitations, and damage to the catalysts. These preliminary studies enable us to define the optimal catalyst configuration, increasing the catalyst performance in terms of productivity, life, heat transfer coefficients, and reducing the expected pressure drops [244]. Hence, it is possible to state that the interest in the CFD simulation is justified looking at the improvements that it brought to the process modeling.

9. Simulation methods for syngas production

Syngas is one of the leading mixture compounds in the chemical processes and in its chemicals production. Syngas differs chemically from gasses normally generated by gasification processes at low temperatures, including fluidized-bed reactors [245]. The product gas is defined as a fuel gas with H_2 and CO as well as with substantial amounts of hydrocarbons, such as methane. Product gas contains CO_2 and H_2O, and often N_2, also inevitably. Throughout the chemical industry, syngas is a substantial intermediate product. Each year, around 6 EJ of syngas is manufactured globally, which is almost 2% of the world's current primary energy consumption [222]. Once produced, the main utilization of syngas involved the production of ammonia, methanol, dimethyl-ether, and synthetic fuels with the Fischer-Tropsch process. The current global market distribution of syngas utilization is reported in Fig. 10.

Only ammonia and methanol capacity production cover 500 million metric tons/year worldwide as highlighted in Sheldon review [247]. Being the reforming operations fundamental in the chemical process industry is important for the branch of process simulation modeling, studying, and analyzing the behavior of this unit in the process. This section treats two examples of process simulation based on bireforming unit, in the first case a sensitivity analysis is conducted to study the perturbations output depending on the main variables such as temperature, pressure of the reactor. The second example deals with the study of the dynamic

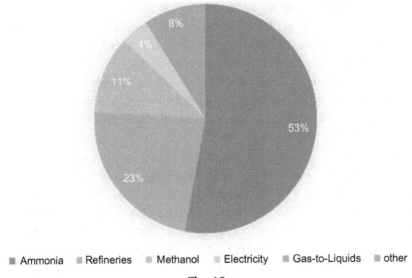

Ammonia   **Refineries**   **Methanol**   **Electricity**   **Gas-to-Liquids**   **other**

Fig. 10
Actual global syngas market [246].

simulation of a bireformer focusing on the control loop and the regulation of the control. Both examples are focused with modeling simulation of bio-refinery: biogas feedstock is used to produce bio-syngas. This definition is in accordance with the SYNBIOS-conference definition. Biogas production has an important potential of continuous growth because it allows conferring an economic value to agricultural by-products, which otherwise are considered wastes. Biogas is a mixture of methane (CH_4) and carbon dioxide (CO_2) produced by anaerobic digestion of organic matter. Ammonia, water, hydrogen sulfide are the main impurities of biogas, and their concentration is usually lower than 1%–2%, their exact value depends on the biomass material, the condition of anaerobic digestion plant, and the different feedstock [248]. Despite the main exploitation of biogas in the cogeneration or biomethane upgrading, to challenge the end of life of incentivization politics and high investment cost is necessary to introduce a third way for biogas valorization: the production of chemicals, as methanol and acetic acid, through the transformation in bio-syngas.

Bireforming unit is implemented in each process simulation, the reaction active is described in Eqs. (R20)–(R23).

$$\text{Dry reforming} \quad CH_4 + CO_2 \rightarrow 2CO + 2H_2 \tag{R20}$$

$$\text{Steam reforming} \quad CH_4 + H_2O \rightarrow 2CO + 3H_2 \tag{R21}$$

$$\text{Water gas shift} \quad CO + H_2O \rightarrow CO_2 + H_2 \tag{R22}$$

$$\text{Methanation} \quad CO_2 + 4H_2 \rightarrow CH_4 + 2H_2O \tag{R23}$$

9.1 Bireforming case study

The first design represents the process simulation to produce bio-syngas starting from biogas feedstock. The model is steady-state and is useful to understand some critical parameters of the bireforming reactor. Fig. 11 reports the flowsheet implemented in Aspen Hysis V11.

Biogas, upgraded from traces of H_2S and NH_3 is sent to a compressor before entering in the bireformer reactor R-100. The compressed feedstock is mixed with a stream of water, reactant for the steam reforming reaction. The mixture passes through an economizer with a preheating function, the enthalpy required is provided by the hot gas products. R-100 is modeled in Aspen as a plug flow reactor, the reactants source is heterogeneous catalytic-type, and the kinetic model is taken from the work of Balasubramanian et al. [91]. Reactor data are reported in Table 7, size and quantity of catalyst are strictly related to the kinetics and performance of the unit. The hot products exiting from the reactor, stream HOT_SYNGAS, passes in a process-process heat exchanger where the reactants are heated up. After this process, cold syngas enters in a separator to remove the humidity, the accumulation of water is recycled back as reagent for steam reforming. To keep the ratio steam/carbon constant is necessary a stream of make-up water. In Table 7, are reported the properties of the main process streams.

Useful parameters and results of a bireformer reactor are the carbon conversion, the quantity of CO_2 and CH_4 consumed in the reactions, and H_2/CO ratio, which identify the quality of syngas.

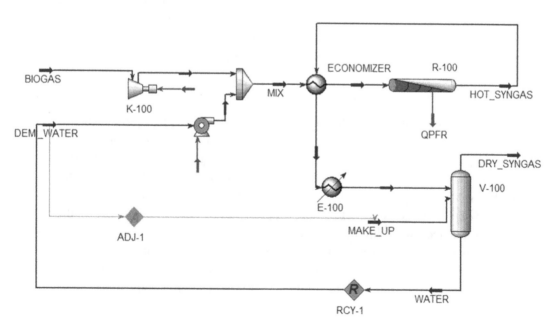

Fig. 11
Aspen Hysis steady-state process simulation.

Table 7 Molar flow rates of the main process streams.

Stream	H_2	CO	CO_2	CH_4	H_2O	UOM
Biogas	0	0	55.77	167.30	0	kmol/h
Mix	0	0	55.77	167.30	111.27	kmol/h
Hot-syngas	368.90	184.00	9.98	29.11	18.87	kmol/h
Dry syingas	368.90	184.00	9.98	29.11	4.51	kmol/h

Typical value of these parameters is fixed in a range between 1 and 2 depending on the leading reaction in the bireformer. Once the process simulation is assessed, a global sensitivity analysis is made perturbing the pressure, the temperature of the reactor R-100 and the CO_2/Steam ratio reactants inlet. In Fig. 12 is depicted the sensitivity analysis based on main parameters perturbations.

In Fig. 12A, the chosen parameters are the pressure of the reformer unit, the output results show that the conversions of CH_4 and CO_2 are very similar and are higher with low pressure. This is quite reasonable because steam reforming and dry reforming reactions proceed with the increase in the number of moles. Despite this optimal performance at atmospheric pressure, the industrial operations of reforming are operated at higher pressure [109] to reduce the dimension of the unit. The H_2/CO ratio seems to oscillate between 1.997 and 2.006, pressure doesn't have an impact for this value. Bireformer reactor is designed as an isothermal reactor; the temperature of the catalytic tubes is perturbed from 700°C to 1000°C, in Fig. 12B. At 700°C, the conversion of CO_2 is negative because the direct water gas shift is active, this is remarkable also for the higher presence of H_2 than CO. Increasing the temperature, the conversions rise up, converging at 85% at 1000°C.

The trend of H_2/CO ratio (Fig. 12C) is representative of the dualism between steam and dry reforming: the first produces, stoichiometrically, a syngas with a ratio of 3 while the carbon dioxide reforming sets the ratio to 1. In this perturbation analysis, it is possible to conclude that at lower temperatures steam reforming is more active, CO_2 is produced through WGS and not active in the reforming, while the dry reforming reactions are more active at elevated temperatures. Industrially, heating a catalytic system until 1000°C is quite complex, this is the main reason for seldom application for the dry reformer unit. CO_2/Steam ratio is another critical parameter in the analysis of bireforming reactions. It's clearly statable that analysis is made varying the steam flow rate, because CO_2 entering in the reactor is intrinsic characteristics of the feedstock. When the quantity of CO_2 is higher than steam flow rate, dry reforming reaction is dominating the kinetics; this is depictable from the H_2/CO ratio, which increase when the ratio is lower than 1 [73].

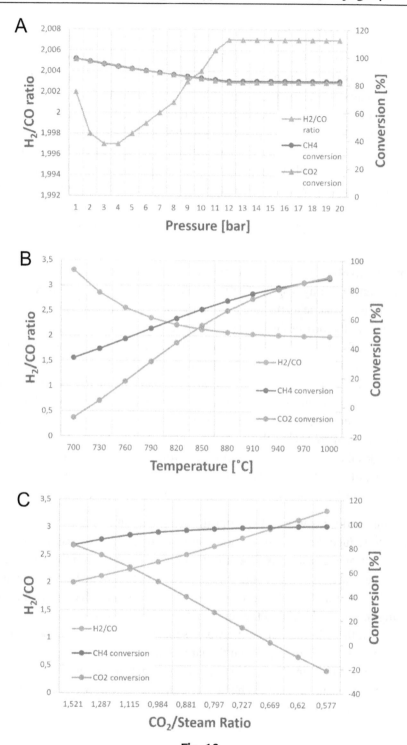

Fig. 12

Carbon conversion and H_2/CO trends with different perturbations: (A) reformer temperature, (B) reformer pressure, (C) CO_2/steam ratio at the inlet condition.

9.2 Steam-methane reforming digital twin

The second proposed study deals with the dynamic simulation and plantwide control of biogas to liquid process; this process is patented as BigSquid™. Mainly focused on biogas to syngas production, for syngas with properties suitable for the production of methanol, on a further stage of the process. The process has been studied and developed by the SUPER (Sustainable Process Engineering Research) group of the Department of Chemistry, Materials and Chemical Engineering "Giulio Natta" of Politecnico di Milano. The scope of this work is developing its dynamic simulation, plantwide control, and the start-up procedure for the plant. The control scheme was proposed, the controllers were tuned and the sequences of operation for the start-up were shown to be effective and stable in its transition state, reaching a steady state without difficulties. In Fig. 13, is depicted the Dynsim simulation flowsheet of the syngas production from biogas.

The components involved in the simulation are water, methane, carbon dioxide, carbon monoxide, methanol, and hydrogen. The thermodynamic method used is SRK for the production section (the same as the one used in the steady-state simulation in Pro/II). The feed stream that enters to the process is biogas without impurities, with a composition of 67% CH_4 and 33% CO_2 at 35°C, and superheated steam (SRC1) at 207°C. Both streams are previously mixed, preheated (in the E201), and fed to the reformer (furnace) at 17 barg (approx.) and around 500°C. In the furnace, the main reactions take place, methane CO_2 reforming, Methane steam reforming and water gas shift. It is important to notice that not all the biogas entering the system is fed to the furnace, before mixing the biogas with the superheated steam, part it is split and burnt in the combustor where the combustion reaction occurs and the hot stream, through radiation, provides the necessary energy for the reformer to work so as to achieve the energy self-sustainability of the BIGSQUID process. The reformer works at a pressure of 16.6 barg and a temperature of 735°C, Table 8 shows the main physical parameters relating to the unit.

These operating conditions were taken after the optimization of the conceptual design, which were oriented to reduce the coke formation and to increase the process yield, taking into account the operational constraints. The hot stream coming out from the reformer, containing traces of methane, water, hydrogen, and carbon oxides, is fed to the heat exchanger mentioned before, which has the function of heat recovery by preheating the feed stream to the reformer. The preheating of the feed reduces the temperature of the product to 450°C, but there is still need to cool down the mixture before the water separates. Therefore, another heat exchanger (E202) has the function of cooling down the stream to the specified temperature that allows further separation of the water from the stream in the next unit operation, a flash (V201) that operates at 16.5 barg with no heat exchange. The syngas is produced on the top of the flash while water is produced from the bottom.

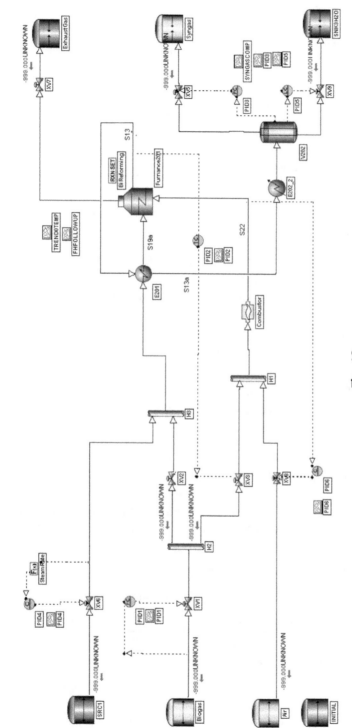

Fig. 13

Dynsim process flowsheet simulation.

Table 8 Geometrical parameter of the reformer unit.

Parameter	Value	UOM
Number of tubes	5500	–
Particle diameter	5.47	mm
Heat capacity	5	$kJ\ Kg^{-1}\ K^{-1}$
Reactor length	7.022	m
Bed void fraction	0.39	–
Catalyst bed density	1140	$kg\,m^{-3}$
Tube ID	38	mm
Tube OD	42	mm

The syngas plant with a fixed molar flow rate of natural gas has 6 degrees of freedom, Table 9 shows the control strategies.

The DYNSIM tool used for developing and analyzing the start-up procedure was "Scenarios" tool, which allows to program a sequence of actions such as setting parameters (controllers' set-points), open/close a valves, start/stop a motors, set timers, program malfunctions, change the feed conditions, and so on. This simulation assumes as initial conditions for the start-up that all proper nitrogen flow has been established as in many cases in the industry in which, for fluids with narrow flammability limits, an inert compound is contained in the equipment before the start-up, once the sequence has started, the inert compound will be gradually purged from the system. With reference to Fig. 13, each step for the start-up procedure developed and implemented in this work by the corresponding DYNSIM scenario command is described as follows:

(a) Loading initial conditions.
(b) Starting fans: for the proper required furnace flow. During start-up, the PID6 that controls the oxygen concentration is operated manually and with a valve opening of 79% (in this case, the valve represents the fans that feed air into the combustion chamber).
(c) Burner ignition: Slowly introduce fuel into the combustion chamber. Two valves need to be opened: XV1 and XV3 controlled by PID1 and PID2 respectively. Both PID are set to manual and both valves are set on 10%.

Table 9 Control strategies of dynamic simulation.

Manipulated variables	UOM	Control
Biogas inlet flow rate	kmol/h	PID1
Outlet temperature of fired heater	°C	PID2
Pressure at syngas section	bar	PID3
Steam flow rate	kmol/h	PID4
Liquid level of the flash separator	%	PID5
Air flow rate	kmol/h	PID6

(d)   Heating the reformer with steam.

Before biogas is fed to the reformer, two (main) conditions must be met: the replacement of the nitrogen by steam and the achievement of the minimum temperature required for the reforming reactions to occur without forming considerably high amounts of coke. To speed up this process, the steam inlet valve is opened widely by setting the PID4 manually output on 70%, this enables the furnace tubes to be filled by steam and at the same time speed up the heating process.

(e)   Keeping the pressure constant during the start-up is important on the flow-driven simulators, to minimize the flow variations of the feed, the pressure control on the flash is set manually and opened at a previously known value.

(f) Run.

(g)   Temperature to start feed introduction: Hydrocarbon feed should be introduced when the process temperature downstream the reformer is in the range of 538–650°C. If introduced below the lower limit, there is a risk of carbon formation on the reformer catalyst [46]. The actual feed temperature will depend on the feed composition and shall be confirmed with the plant design.

(h)   Minimum steam flow: It is recommended that the steam-to-carbon ratio is greater than 5 during feed introduction.

(i)   Start feed introduction: Once the minimum temperature is reached and there is steam in the furnace, the feed is increased manually with a ramp on the PID1 that enables the valve XV1 to open from 10% until 40%. After this moment, the PID1 and PID6 work automatically.

(j)   Temperature changes resulting from feed introduction: Once the reactions start, the temperature of the furnace will decrease due to the endothermic reforming reaction. As a result, additional firing is required, and the PID2 will oversee controlling the temperature automatically (no need for code, PID 2 is already in automatic).

(k)   Firing control: Once the temperature is reached, and the PID2 is working automatically, the fuel flow rate needed to keep the temperature will be significant than the one used to reach the reaction temperature, for which, the oxygen concentration on the combustion chamber would change if no proper control is implemented, for which the valve is manually closed through a ramp from 80% to 30% and kept it for some minutes before turning in automatic, while the oxygen concentration stabilizes.

(l)   Pressure control: Since the inlet flow in the dynamic simulation is very sensitive to pressure, the pressure control is kept open until the inlet flow of biogas stabilizes.

(m)   Steam-methane ratio (PID4) and oxygen concentration control (PID6): these two controllers can be set to automatic since all the start-up sequence has been done.

(n)   Level control: The level is set automatically during all the start-up processes.

To simulate the procedure described above, a scenario with the steps (automatic) has been written down to run and analyze the results. Before the whole sequence described above, a huge work has been performed checking every equipment, its dynamic behavior and its interaction with other process units. There is need to highlight that the vent gas from the furnace, during the start-up, since contains a considerable amount of hydrocarbon species besides nitrogen, must be abated before being vented in the atmosphere to avoid discharge of explosive mixtures. In the same way, it is evident that the syngas produced during the start-up is out of specs; therefore, they cannot be sent to products holding, thus they should be withdrawn through a blowdown line or better they could be recycled to the feed storage tank to avoid products waste.

Fig. 14A shows the PID1 for the simulated inlet flow rate of the biogas. The valve is opened at minimum value until second 200 because the biogas entering is being used for heating up the furnace and not for feeding the tubes for the reactions. In parallel, since only steam is being fed to the furnace the values for the steam methane ratio in the Fig. 14D are not representative, because it would be 100%/0% until biogas starts being fed. As a consequence of the combustion and the steam passing through the furnace tubes in the Fig. 14B, the temperature behavior of the furnace is expected to be increasing gradually; however, it seems that before 100s, it does not start increasing, this can be explained together with the Fig. 14F where the oxygen concentration or air quantity in the combustion chamber is shown to be too high, high enough to not allow the spark to occur and therefore the combustion to start. After this point, the oxygen reaches the optimum concentration for the combustion to be carried out, and the fuel is efficiently used to heat up the furnace, reflected on an increase in its temperature and an optimal concentration of O_2. In Fig. 14B, it can be observed that around the second 200, the minimum temperature required to feed the fuel to the reformer is reached, and as observed in Fig. 15 the steam has almost completely cleaned the N_2 in the system. Since the minimum temperature is reached and almost at the same time the inert gas has been displaced from the system, the conditions are appropriate to start feeding biogas to the furnace to start the syngas production. Consequently, in Fig. 14A it can be observed at around 200s, the ramp increase of the feed introduction. This change has several effects on the rest of the controlled variables, for example, the most notable is the oxygen concentration, Fig. 14F reflects a huge change in the oxygen concentration immediately after the feed flow rate starts to increase. This effect is due to the opening of the valve that feeds biogas to the furnace, because since it is flow-driven and there is more pressure gradient in the line of the tubes of the furnace, the flow initially is directed to the furnace, decreasing the fuel directed to the combustor and suddenly causing the increment on oxygen concentration, however, in matter of seconds, the PID2 for furnace temperature and PID6 for oxygen concentration manage to stabilize both, temperature and oxygen concentration, that is why just after this high perturbation moment, the automatic control of this parameters is turned on. On the other hand, the Fig. 14D shows how the steam methane ratio was higher than 5 (as recommended) before feed introduction and reaches a stable value after it is set in automatic in the second 300. This procedure is considered successful since, among

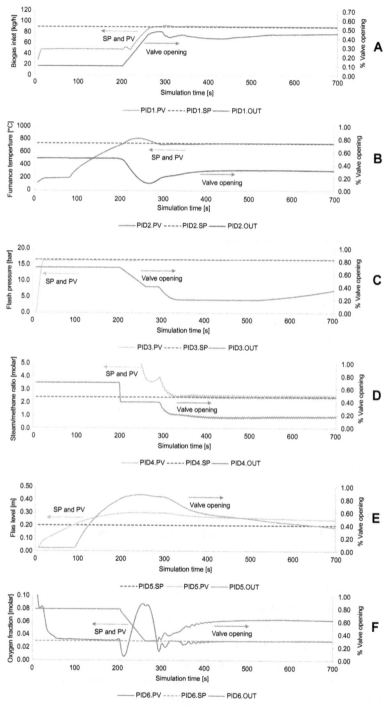

Fig. 14
Manipulated variables start-up.

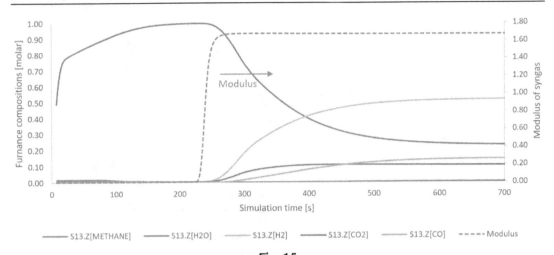

Fig. 15
Composition and modulus of the produced syngas at the outlet conditions.

reaching the products specifications, a maximum of 80°C of temperature over the setpoint during the whole startup process was reached. After the sequence has been completed, the system is left to operate under the regulation of the control loops described in Table 9. Finally, in Fig. 15 it is observable that once the biogas is fed to the reactor, the modulus (dry base) starts to increase and in about 50 s, it reaches a constant value. However, it is remarkable that the modulus dry base in the outlet of the furnace, from which between the second 600 and 700 the water produced is constant with the assumption of water removal from the syngas and the actual production starts to produce methanol on a further stage.

10. Conclusion

The present chapter showed the application of the process modeling for industrial syngas production. We especially focused our efforts on the syngas production starting from natural gas even though alternative feedstock has been reported and the corresponding model to simulate their gasification/pyrolysis. After a brief introduction where key concepts concerning feedstock, kinetics, and thermodynamic equilibria, it has been proven that the feedstock nature influences the tools to simulate syngas production. For instance, any process simulator suitcase or any standard programming language may be adopted to reproduce the reforming unit. Conversely, the presence of nonconventional feedstock such as biomasses, plastic, solid wastes, and similar requires different numerical tools. In the present work, we presented GasDS as an effective tool for the process simulation and its validation has been proven in different works. GasDS enables the description of solid particle gasification using a complex kinetic scheme and considering also potential diffusional intraparticle limitations. Finally, the second part demonstrated the impact of the process simulation on the syngas composition:

- First principle models based on energy and material balances integration are powerful tools to simulate methane reforming. However, they may require a huge numerical effort. The sensitivity analysis on the kinetic model highlighted that (1) the kinetic model does not impact the output results accuracy; however, internal temperature profile may be different; hence, (2) it is possible to complete that you outlet syngas composition is almost close to the equilibrium condition. These results are in line with studies published in the literature.
- Reformer unit may be easily modeled as an equilibrium reactor as highlighted in the previous point. This leads to a more simple process simulation flowsheet since all the equilibria calculations are in charge of the process simulation, which generally adopted robust Gibbs free-energy minimization methods. This is a simplified scheme but enables to roughly estimate the syngas composition and composition. However, in the present work, it has been proposed that the kinetic-based reactor. The reformer unit is designed as a plug-flow reactor where the fuel combustion sustains the endothermic reactions.
- Finally, we proposed the digital twin of the steam methane reforming for biogas. The digital twin (equipment and corresponding control system) has been implemented in the Dynsim. The digital twin is a very useful tool since it allows (1) to verify the series of operations to operate reforming, and (2) in the meanwhile, it allows the dynamic response of the system both in start-up and in case of disturbances. The proposed case studies show that a proper control loop implementation and start-up procedure make it possible to reach nominal operating conditions in a relatively short time, moreover, even in presence of process variables fluctuations, controllers can manage these disturbances driving the system toward a new stable steady-state condition.

Abbreviations and symbols

0	input condition before hydrocracking and precombustion or y coordinate origin
A	preexponential factor or area (m^2)
a	absorptivity factor
α	parameter for the calculation of the equilibrium constant
β	corrective radiation factor
c	as superscript: catalyst particle
c	conversion factor for the pressure
cond	conduction
conv	convection
cp	specific heat $(J. kg^{-1}. K^{-1})$
D	diameter (m)
ΔH	standard chemical adsorption energy $(J. kmol^{-1})$ or reaction enthalpy $(J. kmol^{-1})$
Δz	length of each control volume element (m)
E	reaction activation energy $(J. kmol^{-1})$
env	external environment

ε	emissivity
η	effectiveness factor
F	molar flow (kmol. s^{-1})
f	furnace
g	process gas inside the tubular reactor
γ	Beek correlation adjustment parameter
H	SMRU height (m) or molar enthalpy (J. kmol^{-1})
h	convective coefficient (W. m^{-2}. K^{-1})
i	chemical species
in	input condition after hydrocracking and precombustion or inner tube wall
J	fuel gas or combustion air stream
K_{sub}	adsorption constant or thermal conductivity (W. m^{-1}. K^{-1}); sub: substance
K_n	equilibrium constant; n: numerical value
k_n	kinetic constant; n: numerical value
k	as subscript: reforming or combustion reactions number
L	as subscript: side furnace
L	SMRU length (m)
M	molecular weight (kg. kmol^{-1})
$\dot{m}$	total mass flow (kg. s^{-1})
μ	dynamic viscosity (Pa. s^{-1})
nt	tubes number
out	output condition or outer wall of the tube
P	pressure (Pa)
ϕ	porosity of the catalytic bed
Q	heat transfer rate (J. s^{-1})
R	chemical reaction rate of reaction or ideal gas constant (m^3. Pa. kmol^{-1}. K^{-1})
r	radius (m) or radial coordinate (m)
rad	radiation
ref	refractory
ρ	density (kg. m^{-3})
S	thickness (m)
σ	Stefan-Boltzmann constant (s^{-1}. m^{-2}. K^{-4})
T	temperature (K)
υ	stoichiometric coefficient
v	flow velocity (m. s^{-1})
W	SMRU width (m)
w	mass fraction
w	as superscript: tube wall
y	molar fraction or coordinate of the flat wall (m)
z	axial coordinate (m)

References

[1] F. Bisotti, M. Fedeli, K. Prifti, A. Galeazzi, A.D. Angelo, M. Barbieri, C. Pirola, G. Bozzano, F. Manenti, Century of technology trends in methanol synthesis: any need for kinetics refitting? Ind. Eng. Chem. Res. (2021), https://doi.org/10.1021/acs.iecr.1c02877.

[2] P.P.S. Quirino, A. Amaral, K.V. Pontes, F. Rossi, F. Manenti, Modeling and simulation of an industrial top-fired methane steam reforming unit, Ind. Eng. Chem. Res. 59 (24) (2020) 11250–11264, https://doi.org/10.1021/acs.iecr.0c00456.

[3] Research Dive, Syngas Market Report, 2021. RA08407.

[4] A. Goeppert, M. Czaun, J.-P. Jones, G.K.S. Prakash, G.A. Olah, Recycling of carbon dioxide to methanol and derived products - closing the loop, Chem. Soc. Rev. 43 (23) (2014) 7957–8194, https://doi.org/10.1039/C4CS00122B.

[5] J.H. Ghouse, D. Seepersad, T.A. Adams, Modelling, simulation and design of an integrated radiant syngas cooler and steam methane reformer for use with coal gasification, Fuel Process. Technol. 138 (2015) 378–389, https://doi.org/10.1016/j.fuproc.2015.05.035.

[6] M.E. Dry, The fischer—tropsch process : 1950–2000, Catal. Today 71 (2002) 227–241, https://doi.org/10.1016/s0920-5861(01)00453-9.

[7] R. Balzarotti, A. Beretta, G. Groppi, E. Tronconi, A comparison between washcoated and packed copper foams for the intensification of methane steam reforming, React. Chem. Eng. 4 (8) (2019) 1387–1392, https://doi.org/10.1039/c9re00125e.

[8] G. Leonzio, ANOVA analysis of an integrated membrane reactor for hydrogen production by methane steam reforming, Int. J. Hydrog. Energy 44 (23) (2019) 11535–11545, https://doi.org/10.1016/j.ijhydene.2019.03.077.

[9] M. Mbodji, J.M. Commenge, L. Falk, D. Di Marco, F. Rossignol, L. Prost, S. Valentin, R. Joly, P. Del-Gallo, Steam methane reforming reaction process intensification by using a millistructured reactor: experimental setup and model validation for global kinetic reaction rate estimation, Chem. Eng. J. 207–208 (2012) 871–884, https://doi.org/10.1016/j.cej.2012.07.117.

[10] J.R. Rostrup-Nielsen, J. Sehested, J.K. Nørskov, Hydrogen and synthesis gas by steam- and C02 reforming, Adv. Catal. 47 (2002) 65–139, https://doi.org/10.1016/s0360-0564(02)47006-x.

[11] Research Dive, Syngas Market Report, 2021 (June 2021).

[12] I. Aloisi, A. Di Giuliano, A. Di Carlo, P.U. Foscolo, C. Courson, K. Gallucci, Sorption enhanced catalytic steam methane reforming: experimental data and simulations describing the behaviour of bi-functional particles, Chem. Eng. J. 314 (2017) 570–582, https://doi.org/10.1016/j.cej.2016.12.014.

[13] G. Leonzio, Mathematical modeling of an integrated membrane reactor for methane steam reforming, Int. J. Res. Eng. Technol. 05 (10) (2016) 262–274, https://doi.org/10.15623/ijret.2016.0510043.

[14] M. Rydén, A. Lyngfelt, Using steam reforming to produce hydrogen with carbon dioxide capture by chemical-looping combustion, Int. J. Hydrog. Energy 31 (10) (2006) 1271–1283, https://doi.org/10.1016/j.ijhydene.2005.12.003.

[15] J. Shayegan, M.M.Y. Motamed Hashemi, K. Vakhshouri, Operation of an industrial steam reformer under severe condition: a simulation study, Can. J. Chem. Eng. 86 (4) (2008) 747–755, https://doi.org/10.1002/cjce.20082.

[16] K. Vakhshouri, M.M.Y.M. Hashemi, Simulation study of radial heat and mass transfer inside a fixed bed catalytic reactor, Int. J. Chem. Biol. Eng. 1 (10) (2008) 1–8.

[17] P. Darvishi, F. Zareie-Kordshouli, A rigorous mathematical model for online prediction of tube skin temperature in an industrial top-fired steam methane reformer, Chem. Eng. Res. Des. 126 (2017) 32–44, https://doi.org/10.1016/j.cherd.2017.08.005.

[18] K. Hou, R. Hughes, Intrinsic kinetics of methane steam reforming over a Ni/α-Al2O3 catalyst, Chem. Eng. J. 82 (2001) 311–328.

[19] K. Hou, R. Hughes, The kinetics of methane steam reforming over a Ni/a-Al2O catalyst, Chem. Eng. J. 82 (1385–8947) (2001) 311–328.

[20] K. Hou, R. Hughes, The kinetics of methane steam reforming over a Ni/α-Al2O catalyst, Chem. Eng. J. 82 (1–3) (2001) 311–328, https://doi.org/10.1016/S1385-8947(00)00367-3.

[21] L. Lao, A. Aguirre, A. Tran, Z. Wu, H. Durand, P.D. Christofides, CFD modeling and control of a steam methane reforming reactor, Chem. Eng. Sci. 148 (2016) 78–92, https://doi.org/10.1016/j.ces.2016.03.038.

[22] G. Pantoleontos, E.S. Kikkinides, M.C. Georgiadis, A heterogeneous dynamic model for the simulation and optimisation of the steam methane reforming reactor, Int. J. Hydrog. Energy 37 (21) (2012) 16346–16358, https://doi.org/10.1016/j.ijhydene.2012.02.125.

[23] M.G. Pret, D. Ferrero, A. Lanzini, M. Santarelli, Thermal design, modeling and validation of a steam-reforming reactor for fuel cell applications, Chem. Eng. Res. Des. 104 (2015) 503–512, https://doi.org/10.1016/j.cherd.2015.09.016.

[24] J. Xu, G.F. Froment, Methane steam reforming, methanation and water-gas shift: I. Intrinsic kinetics, AICHE J. 35 (1) (1989) 88–96, https://doi.org/10.1002/aic.690350109.

[25] J. Xu, G.F. Froment, Methane steam reforming: intrisic kinetics, AICHE J. 35 (1) (1989) 88–96.

[26] A. Zamaniyan, H. Ebrahimi, J.S.S. Mohammadzadeh, A unified model for top fired methane steam reformers using three-dimensional zonal analysis, Chem. Eng. Process. Process Intensif. 47 (5) (2008) 946–956, https://doi.org/10.1016/j.cep.2007.03.005.

[27] J.H. Ghouse, T.A. Adams, A multi-scale dynamic two-dimensional heterogeneous model for catalytic steam methane reforming reactors, Int. J. Hydrog. Energy 38 (24) (2013) 9984–9999, https://doi.org/10.1016/j.ijhydene.2013.05.170.

[28] D.A. Latham, K.B. McAuley, B.A. Peppley, T.M. Raybold, Mathematical modeling of an industrial steam-methane reformer for on-line deployment, Fuel Process. Technol. 92 (8) (2011) 1574–1586, https://doi.org/10.1016/j.fuproc.2011.04.001.

[29] D. Wang, X. Feng, Simulation and multi-objective optimization of an integrated process for hydrogen production from refinery off-gas, Int. J. Hydrog. Energy 38 (29) (2013) 12968–12976, https://doi.org/10.1016/j.ijhydene.2013.04.077.

[30] R.A. Abid, A.A. Jassem, Thermodynamic equilibrium analysis of natural gas steam reforming in Basra fertilizer plant, J. Environ. Sci. Comput. Sci. Eng. Technol. 3 (4) (2014) 2208–2218 (4).

[31] Copenor, Personal Communication, 2014. Contact: Mr. John Kennedy Fernandes and Mr. Alan Costa, Brasil.

[32] D. Latham, Mathematical Modelling of an Industrial Steam Methane Reformer, MS Thesis, Queen's University Kingston, 2008, pp. 1–279. December.

[33] M.V. Twigg, Catalyst Handbook, second ed., Taylor & Francis, 1989.

[34] L. Basini, Issues in H2 and synthesis gas technologies for refinery, GTL and small and distributed industrial needs, Catal. Today 106 (1–4) (2005) 34–40, https://doi.org/10.1016/j.cattod.2005.07.179.

[35] P. Ferreira-Aparicio, M.J. Benito, J.L. Sanz, New trends in reforming technologies: from hydrogen industrial plants to multifuel microreformers, Catal. Rev. Sci. Eng. 47 (4) (2005) 491–588, https://doi.org/10.1080/01614940500364958.

[36] N.D. Vasconcelos, Reforma a vapor do metano em catalisadores à base de níquel promovidos com nióbia, Dissertação de Mestrado, UFF, 2006, p. 94.

[37] M.A. Peña, J.P. Gómez, J.L.G. Fierro, New catalytic routes for syngas and hydrogen production, Appl. Catal. A Gen. 144 (1–2) (1996) 7–57, https://doi.org/10.1016/0926-860X(96)00108-1.

[38] G. Bozzano, F. Manenti, Efficient methanol synthesis: perspectives, technologies and optimization strategies, Prog. Energy Combust. Sci. 56 (2016) 71–105, https://doi.org/10.1016/j.pecs.2016.06.001.

[39] Y.M.A. Welaya, M.M. El Gohary, N.R. Ammar, Steam and partial oxidation reforming options for hydrogen production from fossil fuels for PEM fuel cells, Alex. Eng. J. 51 (2) (2012) 69–75, https://doi.org/10.1016/j.aej.2012.03.001.

[40] C.R.H. De Smet, M.H.J.M. De Croon, R.J. Berger, G.B. Marin, J.C. Schouten, Design of adiabatic fixed-bed reactors for the partial oxidation of methane to synthesis gas. Application to production of methanol and hydrogen-for-fuel-cells, Chem. Eng. Sci. 56 (16) (2001) 4849–4861, https://doi.org/10.1016/S0009-2509(01)00130-0.

[41] K.H. Hofstad, J.H.B.J. Hoebink, A. Holmen, G.B. Marin, Partial oxidation of methane to synthesis gas over rhodium catalysts, Catal. Today 40 (1998) 157–170.

[42] T.D. Kusworo, A.R. Songip, N.A.S. Amin, Optimization of partial oxidation of methane for hydrogen production on NiO-CoO/MgO catalyst using design of experiment, Int. J. Eng. Technol. 10 (1) (2010) 1–8.

[43] S.Z. Ozdogan, A. Ersoz, H. Olgun, Simulation study of a PEM fuel cell system fed by hydrogen produced by partial oxidation, J. Power Sources 155 (2006).

[44] A. Pavone, Mega Methanol Plants, Report 43D, Process Economics Program, SRI Consulting, Menlo Park, 2003.

[45] R. Carapellucci, L. Giordano, Steam, dry and autothermal methane reforming for hydrogen production: a thermodynamic equilibrium analysis, J. Power Sources 469 (June) (2020), https://doi.org/10.1016/j.jpowsour.2020.228391, 228391.

[46] K. Aasberg-Petersen, T.S. Christensen, C.S. Nielsen, I. Dybkjær, Recent developments in autothermal reforming and pre-reforming for synthesis gas production in GTL applications, Fuel Process. Technol. 83 (1-3 SPEC) (2003) 253–261, https://doi.org/10.1016/S0378-3820(03)00073-0.

[47] O.L. Ding, S.H. Chan, Autothermal reforming of methane gas-modelling and experimental validation, Int. J. Hydrog. Energy 33 (2) (2008) 633–643, https://doi.org/10.1016/j.ijhydene.2007.10.037.

[48] S.H.D. Lee, D.V. Applegate, S. Ahmed, S.G. Calderone, T.L. Harvey, Hydrogen from natural gas: part I—autothermal reforming in an integrated fuel processor, Int. J. Hydrog. Energy 30 (8) (2005) 829–842, https://doi.org/10.1016/j.ijhydene.2004.09.010.

[49] M. Simeone, L. Salemme, C. Allouis, Reactor temperature profile during autothermal methane reforming on Rh/Al2O3 catalyst by IR imaging, Int. J. Hydrog. Energy 33 (18) (2008) 4798–4808, https://doi.org/10.1016/j.ijhydene.2008.05.089.

[50] S.F. Rice, D.P. Mann, Autothermal reforming of natural gas to synthesis gas, in: Sandia report, vol. 125, 2007. Issue April). https://doi.org/DOE/AL/85000-2007-2331.

[51] B.J. Dreyer, I.C. Lee, J.J. Krummenacher, L.D. Schmidt, Autothermal steam reforming of higher hydrocarbons : n -Decane, n -hexadecane, and JP-8, Appl. Catal. A Gen. 307 (2006) 184–194, https://doi.org/10.1016/j.apcata.2006.03.039.

[52] K.A. Williams, L.D. Schmidt, Catalytic autoignition of higher alkane partial oxidation on Rh-coated foams, Appl. Catal. A Gen. 299 (1–2) (2006) 30–45, https://doi.org/10.1016/j.apcata.2005.09.039.

[53] B.D. Gould, X. Chen, J.W. Schwank, Dodecane reforming over nickel-based monolith catalysts, J. Catal. 250 (2) (2007) 209–221, https://doi.org/10.1016/j.jcat.2007.06.020.

[54] L. Dorazio, M.J. Castaldi, Autothermal reforming of tetradecane (C14H30): a mechanistic approach, Catal. Today 136 (3–4) (2008) 273–280, https://doi.org/10.1016/j.cattod.2008.01.017.

[55] L. Shi, D.J. Bayless, M. Prudich, A model of steam reforming of iso-octane: the effect of thermal boundary conditions on hydrogen production and reactor temperature, Int. J. Hydrog. Energy 33 (17) (2008) 4577–4585, https://doi.org/10.1016/j.ijhydene.2008.06.017.

[56] H.H. Faheem, H.U. Tanveer, S.Z. Abbas, F. Maqbool, Comparative study of conventional steam-methane-reforming (SMR) and auto-thermal-reforming (ATR) with their hybrid sorption enhanced (SE-SMR & SE-ATR) and environmentally benign process models for the hydrogen production, Fuel 297 (February) (2021) 1–16, https://doi.org/10.1016/j.fuel.2021.120769.

[57] M.C.J. Bradford, M.A. Vannice, Catalytic reforming of methane with carbon dioxide over nickel catalysts II. Reaction kinetics, Appl. Catal. A Gen. 142 (1) (1996) 97–122, https://doi.org/10.1016/0926-860X(96)00066-X.

[58] G. Guan, M. Kaewpanha, X. Hao, A. Abudula, Catalytic steam reforming of biomass tar: prospects and challenges, Renew. Sust. Energ. Rev. 58 (2016) 450–461, https://doi.org/10.1016/j.rser.2015.12.316.

[59] N.D. Charisiou, S.L. Douvartzides, G.I. Siakavelas, L. Tzounis, K. Polychronopoulou, M.A. Goula, The relationship between reaction temperature and carbon deposition on nickel catalysts based on Al2O3, ZrO2 or SiO2 supports during the biogas dry reforming reaction, Catalysts 9 (8) (2019) 676.

[60] K. Jabbour, A. Saad, L. Inaty, A. Davidson, P. Massiani, N. El Hassan, Ordered mesoporous Fe-Al2O3 based-catalysts synthesized via a direct "one-pot" method for the dry reforming of a model biogas mixture, Int. J. Hydrog. Energy 44 (29) (2019) 14889–14907, https://doi.org/10.1016/j.ijhydene.2019.04.110.

[61] X. Zhao, B. Joseph, J. Kuhn, S. Ozcan, Biogas reforming to syngas: a review, IScience 23 (5) (2020), https://doi.org/10.1016/j.isci.2020.101082, 101082.

[62] S. Corthals, T. Witvrouwen, P. Jacobs, B. Sels, Development of dry reforming catalysts at elevated pressure: D-optimal vs. full factorial design, Catal. Today 159 (1) (2011) 12–24, https://doi.org/10.1016/j.cattod.2010.06.021.

[63] M.S. Fan, A.Z. Abdullah, S. Bhatia, Hydrogen production from carbon dioxide reforming of methane over Ni-Co/MgO-ZrO2 catalyst: process optimization, Int. J. Hydrog. Energy 36 (8) (2011) 4875–4886, https://doi.org/10.1016/j.ijhydene.2011.01.064.

[64] O. Muraza, A. Galadima, A review on coke management during dry reforming of methane, Int. J. Energy Res. 39 (9) (2015) 1196–1216, https://doi.org/10.1002/er.

[65] R.L.B.A. Medeiros, H.P. Macedo, V.R.M. Melo, Â.A.S. Oliveira, J.M.F. Barros, M.A.F. Melo, D.M.A. Melo, Ni supported on Fe-doped MgAl2O4 for dry reforming of methane: use of factorial design to optimize H2 yield, Int. J. Hydrog. Energy 41 (32) (2016) 14047–14057, https://doi.org/10.1016/j.ijhydene.2016.06.246.

[66] R. Rezaei, G. Moradi, S. Sharifnia, Dry reforming of methane over Ni-Cu/Al2O3 catalyst coatings in a microchannel reactor: modeling and optimization using design of experiments (Research-article), Energy Fuel 33 (7) (2019) 6689–6706, https://doi.org/10.1021/acs.energyfuels.9b00692.

[67] S. Aouad, M. Labaki, S. Ojala, P. Seelam, E. Turpeinen, C. Gennequin, J. Estephane, E.A. Aad, A Review on the Dry Reforming Processes for Hydrogen Production: Catalytic Materials and Technologies (Issue December), 2018, https://doi.org/10.2174/9781681087580118020007.

[68] K. Wittich, M. Krämer, N. Bottke, S.A. Schunk, Catalytic dry reforming of methane: insights from model systems, ChemCatChem 12 (8) (2020) 2130–2147, https://doi.org/10.1002/cctc.201902142.

[69] W.C. Liu, J. Baek, G.A. Somorjai, The methanol economy: methane and carbon dioxide conversion, Top. Catal. 61 (7–8) (2018) 530–541, https://doi.org/10.1007/s11244-018-0907-4.

[70] G.A. Olah, A. Goeppert, M. Czaun, G.K.S. Prakash, Bi-reforming of methane from any source with steam and carbon dioxide exclusively to metgas (CO-2H2) for methanol and hydrocarbon synthesis, J. Am. Chem. Soc. 135 (2) (2013) 648–650, https://doi.org/10.1021/ja311796n.

[71] N. Kumar, M. Shojaee, J.J. Spivey, Catalytic bi-reforming of methane: from greenhouse gases to syngas, Curr. Opin. Chem. Eng. 9 (2015) 8–15, https://doi.org/10.1016/j.coche.2015.07.003.

[72] S. Singh, M.B. Bahari, B. Abdullah, P.T.T. Phuong, Q.D. Truong, D.V.N. Vo, A.A. Adesina, Bi-reforming of methane on Ni/SBA-15 catalyst for syngas production: influence of feed composition, Int. J. Hydrog. Energy 43 (36) (2018) 17230–17243, https://doi.org/10.1016/j.ijhydene.2018.07.136.

[73] A.F. Cunha, T.M. Mata, N.S. Caetano, A.A. Martins, J.M. Loureiro, Catalytic bi-reforming of methane for carbon dioxide ennoblement, Energy Rep. 6 (2020) 74–79, https://doi.org/10.1016/j.egyr.2019.08.022. ScienceDirect.

[74] G. Mallikarjun, T.V. Sagar, S. Swapna, N. Raju, P. Chandrashekar, N. Lingaiah, Hydrogen rich syngas production by bi-reforming of methane with CO 2 over Ni supported on CeO 2 -SrO mixed oxide catalysts, Catal. Today 356 (December 2019) (2020) 1–7, https://doi.org/10.1016/j.cattod.2020.01.005.

[75] D.B.L. Santos, F.B. Noranha, C.E. Hori, Bi-reforming of methane for hydrogen production using LaNiO 3 CexZr1-x O2 as precursor material, Int. J. Hydrog. Energy 45 (2020) 13947–13959, https://doi.org/10.1016/j.ijhydene.2020.03.096.

[76] U.S. Mohanty, M. Ali, M.R. Azhar, A. Al-Yaseri, A. Keshavarz, S. Iglauer, Current advances in syngas (CO + H2) production through bi-reforming of methane using various catalysts: a review, Int. J. Hydrog. Energy 46 (65) (2021) 32809–32845, https://doi.org/10.1016/j.ijhydene.2021.07.097.

[77] Y. Zhang, S. Zhang, J.L. Gossage, H.H. Lou, T.J. Benson, Thermodynamic analyses of tri-reforming reactions to produce syngas, Energy Fuel 28 (4) (2014) 2717–2726, https://doi.org/10.1021/ef500084m.

[78] A.J. Majewski, J. Wood, Tri-reforming of methane over Ni@SiO2 catalyst, Int. J. Hydrog. Energy 39 (24) (2014) 12578–12585, https://doi.org/10.1016/j.ijhydene.2014.06.071.

[79] A.M. Borreguero, F. Dorado, M. Capuchino-Biezma, L. Sánchez-Silva, J.M. García-Vargas, Process simulation and economic feasibility assessment of the methanol production via tri-reforming using experimental kinetic equations, Int. J. Hydrog. Energy 45 (51) (2020) 26623–26636, https://doi.org/10.1016/j.ijhydene.2020.07.013.

[80] M. Fekri Lari, M. Farsi, M.R. Rahimpour, Modification of a tri-reforming reactor based on the feeding policy to couple with methanol and GTL units, Chem. Eng. Res. Des. 144 (2014) (2019) 107–114, https://doi.org/10.1016/j.cherd.2019.01.029.

[81] T.B.H. Nguyen, E. Zondervan, Methanol production from captured CO2 using hydrogenation and reforming technologies- environmental and economic evaluation, J. CO$_2$ Util. 34 (May) (2019) 1–11, https://doi.org/10.1016/j.jcou.2019.05.033.

[82] E. Rezaei, L.J.J. Catalan, Evaluation of CO 2 utilization for methanol production via tri-reforming of methane, J. CO$_2$ Util. 42 (June) (2020), https://doi.org/10.1016/j.jcou.2020.101272, 101272.

[83] J. Hong-tao, L.I. Hui-quan, Z. Yi, Tri-reforming of methane to syngas over Ni/Al 2 O 3—thermal distribution in the catalyst bed, J. Fuel Chem. Technol. 35 (1) (2007) 72–78, https://doi.org/10.1016/s1872-5813(07)60012-7.

[84] S. Lee, W. Cho, W. Ju, B. Cho, Y. Lee, Y. Baek, Tri-reforming of CH 4 using CO 2 for production of synthesis gas to dimethyl ether, Catal. Today 87 (2003) 133–137, https://doi.org/10.1016/j.cattod.2003.10.005.

[85] W. Cho, T. Song, A. Mitsos, J.T. McKinnon, G.H. Ko, J.E. Tolsma, D. Denholm, T. Park, Optimal design and operation of a natural gas tri-reforming reactor for DME synthesis, Catal. Today 139 (4) (2009) 261–267, https://doi.org/10.1016/j.cattod.2008.04.051.

[86] K. Ahmed, K. Föger, Fuel processing for high-temperature high-efficiency fuel cells, Ind. Eng. Chem. Res. 49 (16) (2010) 7239–7256, https://doi.org/10.1021/ie100778g.

[87] K.R. Ramakumar, S. Orman, I.B. Kara, Pre-reformer catalyst in a hydrogen plant, in: Digital Refining Processing, Operations and Maintenance, 2020.

[88] Linde Engineering, Hydrogen, 2021.

[89] IETD, Using an Adiabatic Pre-Reformer, Industrial Efficiency Technology Database, 2021.

[90] A. Kakoee, A. Gharehghani, Carbon oxides methanation in equilibrium; a thermodynamic approach, Int. J. Hydrog. Energy 45 (55) (2020) 29993–30008, https://doi.org/10.1016/j.ijhydene.2020.08.073.

[91] P. Balasubramanian, I. Bajaj, M.M.F. Hasan, Simulation and optimization of reforming reactors for carbon dioxide utilization using both rigorous and reduced models, J. CO$_2$ Util. 23 (July 2017) (2018) 80–104, https://doi.org/10.1016/j.jcou.2017.10.014.

[92] C. Zhou, L. Zhang, A. Swiderski, W. Yang, W. Blasiak, Study and development of a high temperature process of multi-reformation of CH4 with CO2 for remediation of greenhouse gas, Energy 36 (9) (2011) 5450–5459, https://doi.org/10.1016/j.energy.2011.07.045.

[93] K. Jabbour, Tuning combined steam and dry reforming of methane for "metgas" production: a thermodynamic approach and state-of-the-art catalysts, J. Energy Chem. 48 (2020) 54–91, https://doi.org/10.1016/j.jechem.2019.12.017.

[94] R.Y. Chein, Y.C. Chen, C.T. Yu, J.N. Chung, Thermodynamic analysis of dry reforming of CH4 with CO2 at high pressures, J. Nat. Gas Sci. Eng. 26 (2015) 617–629, https://doi.org/10.1016/j.jngse.2015.07.001.

[95] R.Y. Chein, W.H. Hsu, C.T. Yu, Parametric study of catalytic dry reforming of methane for syngas production at elevated pressures, Int. J. Hydrog. Energy 42 (21) (2017) 14485–14500, https://doi.org/10.1016/j.ijhydene.2017.04.110.

[96] C. Jensen, M.S. Duyar, Thermodynamic analysis of dry reforming of methane for valorization of landfill gas and natural gas, Energ. Technol. 9 (7) (2021) 1–12, https://doi.org/10.1002/ente.202100106.

[97] M.M.B. Noureldin, N.O. Elbashir, K.J. Gabriel, M.M. El-Halwagi, A process integration approach to the assessment of CO2 fixation through dry reforming, ACS Sustain. Chem. Eng. 3 (4) (2015) 625–636, https://doi.org/10.1021/sc5007736.

[98] M. González-Castaño, B. Dorneanu, H. Arellano-García, The reverse water gas shift reaction: a process systems engineering perspective, React. Chem. Eng. 6 (6) (2021) 954–976, https://doi.org/10.1039/d0re00478b.

[99] A. Wolf, A. Jess, C. Kern, Syngas production via reverse water-gas shift reaction over a Ni-Al2O3 catalyst: catalyst stability, reaction kinetics, and modeling, Chem. Eng. Technol. 39 (6) (2016) 1040–1048, https://doi.org/10.1002/ceat.201500548.

[100] H. Nourbakhsh, J. Rahbar Shahrouzi, A. Zamaniyan, H. Ebrahimi, M.R. Jafari Nasr, A thermodynamic analysis of biogas partial oxidation to synthesis gas with emphasis on soot formation, Int. J. Hydrog. Energy 43 (33) (2018) 15703–15719, https://doi.org/10.1016/j.ijhydene.2018.06.134.

[101] A.C.D. Freitas, R. Guirardello, Oxidative reforming of methane for hydrogen and synthesis gas production: thermodynamic equilibrium analysis, J. Nat. Gas Chem. 21 (5) (2012) 571–580, https://doi.org/10.1016/S1003-9953(11)60406-4.

[102] N. Kumar, A. Roy, Z. Wang, E.M. L'Abbate, D. Haynes, D. Shekhawat, J.J. Spivey, Bi-reforming of methane on Ni-based pyrochlore catalyst, Appl. Catal. A Gen. 517 (2016) 211–216, https://doi.org/10.1016/j.apcata.2016.03.016.

[103] B.A.V. Santos, J.M. Loureiro, A.M. Ribeiro, A.E. Rodrigues, A.F. Cunha, Methanol production by bi-reforming, Can. J. Chem. Eng. 93 (3) (2015) 510–526, https://doi.org/10.1002/cjce.22068.

[104] N. Entesari, A. Goeppert, G.K.S. Prakash, Renewable methanol synthesis through single step bi-reforming of biogas, Ind. Eng. Chem. Res. 59 (22) (2020) 10542–10551, https://doi.org/10.1021/acs.iecr.0c00755.

[105] A.C.D. Freitas, R. Guirardello, Thermodynamic analysis of methane reforming with CO2, CO 2 + H2O, CO2 + O2 and CO2 + air for hydrogen and synthesis gas production, J. CO$_2$ Util. 7 (2014) 30–38, https://doi.org/10.1016/j.jcou.2014.06.004.

[106] W.J. Lee, C. Li, J. Patel, Upgrading of bio-syngas via steam-CO$_2$ reforming using Rh/alumina monolith catalysts, Catalysts 11 (2021) 180, https://doi.org/10.3390/catal11020180.

[107] R.Y. Chein, W.H. Hsu, Thermodynamic analysis of syngas production via tri-reforming of methane and carbon gasification using flue gas from coal-fired power plants, J. Clean. Prod. 200 (2018) 242–258, https://doi.org/10.1016/j.jclepro.2018.07.228.

[108] E.S. Aydin, O. Yucel, H. Sadikoglu, Numerical and experimental investigation of hydrogen-rich syngas production via biomass gasification, Int. J. Hydrog. Energy 43 (2) (2018) 1105–1115, https://doi.org/10.1016/j.ijhydene.2017.11.013.

[109] P.M. Mortensen, I. Dybkjær, Industrial scale experience on steam reforming of CO2-rich gas, Appl. Catal. A Gen. 495 (2015) 141–151, https://doi.org/10.1016/j.apcata.2015.02.022.

[110] J.H. Oakley, A.F.A. Hoadley, Industrial scale steam reforming of bioethanol: a conceptual study, Int. J. Hydrog. Energy 35 (16) (2010) 8472–8485, https://doi.org/10.1016/j.ijhydene.2010.05.003.

[111] A.S.K. Raju, C.S. Park, J.M. Norbeck, Synthesis gas production using steam hydrogasification and steam reforming, Fuel Process. Technol. 90 (2) (2009) 330–336, https://doi.org/10.1016/j.fuproc.2008.09.011.

[112] J.R. Rostrup-Nielsen, Activity of nickel catalysts for steam reforming of hydrocarbons, J. Catal. 31 (2) (1973) 173–199, https://doi.org/10.1016/0021-9517(73)90326-6.

[113] Ş. Özkara-Aydnolu, Thermodynamic equilibrium analysis of combined carbon dioxide reforming with steam reforming of methane to synthesis gas, Int. J. Hydrog. Energy 35 (23) (2010) 12821–12828, https://doi.org/10.1016/j.ijhydene.2010.08.134.

[114] J. Zhu, D. Zhang, K.D. King, Reforming of CH4 by partial oxidation: thermodynamic and kinetic analyses, Fuel 80 (7) (2001) 899–905, https://doi.org/10.1016/S0016-2361(00)00165-4.

[115] D. Chen, R. Lødeng, H. Svendsen, A. Holmen, Hierarchical multiscale modeling of methane steam reforming reactions, Ind. Eng. Chem. Res. 50 (5) (2011) 2600–2612, https://doi.org/10.1021/ie1006504.

[116] J.T. Richardson, S.A. Paripatyadar, Carbon dioxide reforming of methane with supported rhodium, Appl. Catal. 61 (1) (1990) 293–309, https://doi.org/10.1016/S0166-9834(00)82152-1.

[117] X.E. Verykios, Catalytic dry reforming of natural gas for the production of chemicals and hydrogen, Chem. Ind. Chem. Eng. Q. 8 (2) (2002) 238–255, https://doi.org/10.2298/hemind0206238v.

[118] M. Zeppieri, P.L. Villa, N. Verdone, M. Scarsella, P. De Filippis, Kinetic of methane steam reforming reaction over nickel- and rhodium-based catalysts, Appl. Catal. A Gen. 387 (1–2) (2010) 147–154, https://doi.org/10.1016/j.apcata.2010.08.017.

[119] C. Wang, C. Liu, W. Fu, Z. Bao, J. Zhang, W. Ding, K. Chou, Q. Li, The water-gas shift reaction for hydrogen production from coke oven gas over Cu/ZnO/Al2O3 catalyst, Catal. Today 263 (2016) 46–51, https://doi.org/10.1016/j.cattod.2015.09.044.

[120] R.Y. Chein, W.H. Hsu, Analysis of syngas production from biogas via the tri-reforming process, Energies 11 (5) (2018), https://doi.org/10.3390/en11051075.

[121] A. Sunny, P.A. Solomon, K. Aparna, Syngas production from regasified liquefied natural gas and its simulation using Aspen HYSYS, J. Nat. Gas Sci. Eng. 30 (2016) 176–181, https://doi.org/10.1016/j.jngse.2016.02.013.

[122] E. Terrell, C.S. Theegala, Thermodynamic simulation of syngas production through combined biomass gasification and methane reformation, Sustain. Energy Fuels 3 (6) (2019) 1562–1572, https://doi.org/10.1039/c8se00638e.

[123] M. Zahedi nezhad, S. Rowshanzamir, M.H. Eikani, Autothermal reforming of methane to synthesis gas: modeling and simulation, Int. J. Hydrog. Energy 34 (3) (2009) 1292–1300, https://doi.org/10.1016/j.ijhydene.2008.11.091.

[124] A. Iulianelli, S. Liguori, J. Wilcox, A. Basile, Advances on methane steam reforming to produce hydrogen through membrane reactors technology : a review, Catal. Rev. 58 (1) (2016) 1–35, https://doi.org/10.1080/01614940.2015.1099882.

[125] N.M. Bodrov, L.O. Apelbaum, M.I. Temkin, Kinetics of the reaction of methane with water vapour, catalyzed nickel on a porous carrier, Kinet. Catal. 8 (1967) 821–828.

[126] N.M. Bodrov, L.O. Apelbaum, M.I. Temkin, Kinetics of the reactions of methane with steam on the surface of nickel at 400-600°C, Kinet. Catal. 9 (1968) 1065–1071.

[127] J.M. García-Vargas, J.L. Valverde, J. Díez, F. Dorado, S. Paula, Catalytic and kinetic analysis of the methane tri-reforming over a Ni e Mg/b -SiC catalyst, Int. J. Hydrog. Energy 40 (2015) 0–10, https://doi.org/10.1016/j.ijhydene.2015.05.032.

[128] M. Maestri, D.G. Vlachos, A. Beretta, G. Groppi, E. Tronconi, Steam and dry reforming of methane on Rh : microkinetic analysis and hierarchy of kinetic models, J. Catal. 259 (2) (2008) 211–222, https://doi.org/10.1016/j.jcat.2008.08.008.

[129] T. Numaguchi, K. Kikuchi, Intrinsic kinetics and design simulation in a complex reaction network; steam-methane reforming, Chem. Eng. Sci. 43 (8) (1988) 2295–2301, https://doi.org/10.1016/0009-2509(88)87118-5.

[130] P.V. Ponugoti, V.M. Janardhanan, Mechanistic kinetic model for biogas dry reforming, Ind. Eng. Chem. Res. (2020), https://doi.org/10.1021/acs.iecr.0c02433.

[131] C.P.P. Singh, D.N. Saraf, Simulation of side fired steam-hydrocarbon reformers, Ind. Eng. Chem. Process. Des. Dev. 18 (1) (1979) 1–7, https://doi.org/10.1021/i260069a001.

[132] A.E.A.M. Souza, L.J.L. Maciel, V.O. Cavalcanti-Filho, N.M.L. Filho, C.A.M. Abreu, Kinetic-operational mechanism to autothermal reforming of methane, Ind. Eng. Chem. Res. 50 (5) (2011) 2585–2599, https://doi.org/10.1021/ie100637b.

[133] C. Sprung, B. Arstad, U. Olsbye, Methane steam reforming over a Ni/NiAl 2 O 4 model catalyst—kinetics, ChemCatChem 6 (2014) 1–15, https://doi.org/10.1002/cctc.201402017.

[134] J. Wei, E. Iglesia, Isotopic and kinetic assessment of the mechanism of reactions of CH 4 with CO 2 or H 2 O to form synthesis gas and carbon on nickel catalysts, J. Catal. 224 (2004) 370–383, https://doi.org/10.1016/j.jcat.2004.02.032.

[135] D. Zambrano, J. Soler, J. Herguido, M. Menéndez, Kinetic study of dry reforming of methane over Ni—Ce/Al 2 O 3 catalyst with deactivation, Top. Catal. 62 (2019) 456–466, https://doi.org/10.1007/s11244-019-01157-2.

[136] M.H. Hyman, Simulate methane reformer reactions, Hydrocarb. Process. 47 (7) (1968) 131–137.

[137] Haldor Topsoe, Aspects of catalytic cracking. The institution of gas engineers, in: 31st Autumn Res. Meeting, London, UK, 1965.

[138] M.J. Azarhoosh, H. Ale Ebrahim, S.H. Pourtarah, Simulating and optimizing hydrogen production by low-pressure autothermal reforming of natural gas using non-dominated sorting genetic algorithm-II, Chem. Biochem. Eng. Q. 29 (4) (2015) 519–531, https://doi.org/10.15255/CABEQ.2014.2158.

[139] A. Cherif, R. Nebbali, Numerical analysis on autothermal steam methane reforming: effects of catalysts arrangement and metal foam insertion, Int. J. Hydrog. Energy 44 (39) (2019) 22455–22466, https://doi.org/10.1016/j.ijhydene.2018.12.203.

[140] A. Aguirre, Computational Fluid Dynamics Modeling and Simulation of Steam Methane Reforming Reactors and Furnaces, University of California, 2017.

[141] A.D. Nandasana, A.K. Ray, S.K. Gupta, Dynamic model of an industrial steam reformer and its use for multiobjective optimization, Ind. Eng. Chem. Res. 42 (17) (2003) 4028–4042, https://doi.org/10.1021/ie0209576.

[142] C.P.P. Singh, D.N. Saraf, Process simulation of Ammonia plant, Ind. Eng. Chem. Process. Des. Dev. 20 (3) (1981) 425–433, https://doi.org/10.1021/i200014a003.

[143] E.M. Assaf, C.D.F. Jesus, J.M. Assaf, Mathematical modelling of methane steam reforming in a membrane reactor: an isothermic model, Braz. J. Chem. Eng. 15 (2) (1998) 160–166, https://doi.org/10.1590/S0104-66321998000200010.

[144] K. Ravi, S.C. Dhingra, B.K. Guha, Y.K. Joshi, Simulation of primary and secondary reformers for improved energy performance of an ammonia plant, Chem. Eng. Technol. 12 (1) (1989) 358–364, https://doi.org/10.1002/ceat.270120151.

[145] A. Acuña, C. Fuentes, C.A. Smith, Dynamic simulation of a furnace of steam reforming of natural gas, C. T. F. Cien. Tecnol. Futuro 1 (5) (1999) 35–44.

[146] G. Leonzio, The process simulation of natural gas steam reforming in an integrated membrane reactor, Int. J. Res. Eng. Technol. 10 (2016) 314–323.

[147] I. Aloisi, N. Jand, S. Stendardo, P.U. Foscolo, Hydrogen by sorption enhanced methane reforming: a grain model to study the behavior of bi-functional sorbent-catalyst particles, Chem. Eng. Sci. 149 (2016) 22–34, https://doi.org/10.1016/j.ces.2016.03.042.

[148] V. Spallina, B. Marinello, F. Gallucci, M.C. Romano, M. Van Sint Annaland, Chemical looping reforming in packed-bed reactors: modelling, experimental validation and large-scale reactor design, Fuel Process. Technol. 156 (2017) 156–170, https://doi.org/10.1016/j.fuproc.2016.10.014.

[149] G. Xiu, J.L. Soares, P. Li, A.E. Rodrigues, Simulation of five-step one-bed sorption-enhanced reaction process.pdf, AICHE J. 48 (12) (2002) 2817–2832, https://doi.org/10.1002/aic.690481210.

[150] F. Gallucci, L. Paturzo, A. Basile, A simulation study of the steam reforming of methane in a dense tubular membrane reactor, Int. J. Hydrog. Energy 29 (6) (2004) 611–617, https://doi.org/10.1016/j.ijhydene.2003.08.003.

[151] D.L. Hoang, S.H. Chan, O.L. Ding, Kinetic and modelling study of methane steam reforming over sulfide nickel catalyst on a gamma alumina support, Chem. Eng. J. 112 (1–3) (2005) 1–11, https://doi.org/10.1016/j.cej.2005.06.004.

[152] Z. Yu, E. Cao, Y. Wang, Z. Zhou, Z. Dai, Simulation of natural gas steam reforming furnace, Fuel Process. Technol. 87 (8) (2006) 695–704, https://doi.org/10.1016/j.fuproc.2005.11.008.

[153] E.L.G. Oliveira, C.A. Grande, A.E. Rodrigues, Steam methane reforming in a Ni/Al2O3 catalyst: kinetics and diffusional limitations in extrudates, Can. J. Chem. Eng. 87 (6) (2009) 945–956, https://doi.org/10.1002/cjce.20223.

[154] L.C. Silva, V.V. Murata, C.E. Hori, A.J. Assis, Hydrogen production from methane steam reforming: parametric and gradient based optimization of a Pd-based membrane reactor, Optim. Eng. 11 (3) (2010) 441–458, https://doi.org/10.1007/s11081-010-9106-2.

[155] S. Vaccaro, L. Malangone, Influence of the kinetic sub-model on the performance of a comprehensive model of a micro-scale catalytic reactor, Int. J. Chem. React. Eng. 10 (1) (2012), https://doi.org/10.1515/1542-6580.2949.

[156] P. Sadooghi, R. Rauch, Pseudo heterogeneous modeling of catalytic methane steam reforming process in a fixed bed reactor, J. Nat. Gas Sci. Eng. 11 (2013) 46–51, https://doi.org/10.1016/j.jngse.2012.12.002.

[157] J.S. Lee, J. Seo, H.Y. Kim, J.T. Chung, S.S. Yoon, Effects of combustion parameters on reforming performance of a steam-methane reformer, Fuel 111 (2013) 461–471, https://doi.org/10.1016/j.fuel.2013.04.078.

[158] M. Farniaei, M. Abbasi, H. Rahnama, M.R. Rahimpour, A. Shariati, Syngas production in a novel methane dry reformer by utilizing of tri-reforming process for energy supplying: modeling and simulation, J. Nat. Gas Sci. Eng. 20 (2014) 132–146, https://doi.org/10.1016/j.jngse.2014.06.010.

[159] Y. Qi, Z. Cheng, Z. Zhou, Steam reforming of methane over Ni catalysts prepared from hydrotalcite-type precursors: catalytic activity and reaction kinetics, Chin. J. Chem. Eng. 23 (1) (2015) 76–85, https://doi.org/10.1016/j.cjche.2013.11.002.

[160] R.A. Abid, A.A. Jassim, Effect of operating conditions on the performance of primary steam reformer in Basra fertilizer plant, J. Chem. Pharm. Res. 7 (3) (2015) 2333–2346.

[161] F. Vidal Vázquez, P. Pfeifer, J. Lehtonen, P. Piermartini, P. Simell, V. Alopaeus, Catalyst screening and kinetic modeling for CO production by high pressure and temperature reverse water gas shift for Fischer-Tropsch applications, Ind. Eng. Chem. Res. 56 (45) (2017) 13262–13272, https://doi.org/10.1021/acs.iecr.7b01606.

[162] A. Kumar, M. Baldea, T.F. Edgar, A physics-based model for industrial steam-methane reformer optimization with non-uniform temperature field, Comput. Chem. Eng. 105 (2017) 224–236, https://doi.org/10.1016/j.compchemeng.2017.01.002.

[163] S.Z. Abbas, V. Dupont, T. Mahmud, Kinetics study and modelling of steam methane reforming process over a NiO/Al2O3 catalyst in an adiabatic packed bed reactor, Int. J. Hydrog. Energy 42 (5) (2017) 2889–2903, https://doi.org/10.1016/j.ijhydene.2016.11.093.

[164] Z. Wu, A. Aguirre, A. Tran, H. Durand, D. Ni, P.D. Christofides, Model predictive control of a steam methane reforming reactor described by a computational fluid dynamics model, Ind. Eng. Chem. Res. 56 (20) (2017) 6002–6011, https://doi.org/10.1021/acs.iecr.7b00390.

[165] A.G. Dixon, M. Nijemeisland, H. Stitt, CFD simulation of reaction and heat transfer near the wall of a fixed bed, Int. J. Chem. React. Eng. 1 (A22) (2003), https://doi.org/10.2202/1542-6580.1069.

[166] M. Nijemeisland, A.G. Dixon, E.H. Stitt, Catalyst design by CFD for heat transfer and reaction in steam reforming, Chem. Eng. Sci. 59 (22−23) (2004) 5185–5191, https://doi.org/10.1016/j.ces.2004.07.088.

[167] M. Karimi-Golpayegani, A. Akhavan-Abdolahian, N. Mostoufi, Simulation of a Fluidized-Bed Steam Reformer, Industrial Fluidization South Africa, 2005, pp. 323–329. November.

[168] A.G. Dixon, M. Ertan Taskin, E. Hugh Stitt, M. Nijemeisland, 3D CFD simulations of steam reforming with resolved intraparticle reaction and gradients, Chem. Eng. Sci. 62 (18–20) (2007) 4963–4966, https://doi.org/10.1016/j.ces.2006.11.052.

[169] J.M. Klein, Y. Bultel, S. Georges, M. Pons, Modeling of a SOFC fuelled by methane: from direct internal reforming to gradual internal reforming, Chem. Eng. Sci. 62 (6) (2007) 1636–1649, https://doi.org/10.1016/j.ces.2006.11.034.

[170] S. Vaccaro, L. Malangone, P. Ciambelli, Micro-scale catalytic reactor for syngas production, Ind. Eng. Chem. Res. 49 (21) (2010) 10924–10933, https://doi.org/10.1021/ie100464b.

[171] V.V. Kuznetsov, S.P. Kozlov, Modeling of methane steam reforming in a microchannel subject to multicomponent diffusion, J. Eng. Thermophys. 20 (3) (2011) 229–239, https://doi.org/10.1134/S1810232811030015.

[172] N. Lu, F. Gallucci, T. Melchiori, D. Xie, M. Van Sint Annaland, Modeling of autothermal reforming of methane in a fluidized bed reactor with perovskite membranes, Chem. Eng. Process. Process Intensif. 124 (2018) 308–318, https://doi.org/10.1016/j.cep.2017.07.010.

[173] P.P.S. Quirino, A. Amaral, K.V. Pontes, F. Rossi, F. Manenti, Impact of kinetic models in the prediction accuracy of an industrial steam methane reforming unit, Comput. Chem. Eng. 152 (2021), https://doi.org/10.1016/j.compchemeng.2021.107379, 107379.

[174] I. Severinsen, A. Herritsch, M. Watson, Modeling kinetic, thermodynamic, and operational effects in a steam methane reformer. Part A: reformer output, Ind. Eng. Chem. Res. 60 (5) (2021) 2041–2049, https://doi.org/10.1021/acs.iecr.0c04909.

[175] J.R. Rostrup-Nielsen, Production of synthesis gas, Catal. Today 18 (4) (1993) 305–324, https://doi.org/10.1016/0920-5861(93)80059-A.

[176] A. Olivieri, F. Vegliò, Process simulation of natural gas steam reforming: fuel distribution optimisation in the furnace, Fuel Process. Technol. 89 (6) (2008) 622–632, https://doi.org/10.1016/j.fuproc.2007.12.001.

[177] A. Tran, A. Aguirre, H. Durand, M. Crose, P.D. Christofides, CFD modeling of a industrial-scale steam methane reforming furnace, Chem. Eng. Sci. 171 (2017) 576–598, https://doi.org/10.1016/j.ces.2017.06.001.

[178] M.H. Wesenberg, H.F. Svendsen, Mass and heat transfer limitations in a heterogeneous model of a gas-heated steam reformer, Ind. Eng. Chem. Res. 46 (3) (2007) 667–676, https://doi.org/10.1021/ie060324h.

[179] H. Ebrahimi, A. Zamaniyan, J.S. Soltan Mohammadzadeh, A.A. Khalili, Zonal modeling of radiative heat transfer in industrial furnaces using simplified model for exchange area calculation, Appl. Math. Model. 37 (16–17) (2013) 8004–8015, https://doi.org/10.1016/j.apm.2013.02.053.

[180] H.C. Hottel, A.F. Sarofim, E.J. Fahimian, The role of scatter in determining the radiative properties of surfaces, Sol. Energy 11 (1) (1967) 2–13, https://doi.org/10.1016/0038-092X(67)90012-6.

[181] E.P. Keramida, H.H. Liakos, M.A. Founti, A.G. Boudouvis, N.C. Markatos, Radiative heat transfer in natural gas-fired furnaces, Int. J. Heat Mass Transf. 43 (10) (2000) 1801–1809, https://doi.org/10.1016/S0017-9310(99)00244-6.

[182] F.C. Roesler, Theory of radiative heat transfer in co-current tube furnaces, Chem. Eng. Sci. 22 (10) (1967) 1325–1336, https://doi.org/10.1016/0009-2509(67)80023-X.

[183] N. Selcuk, R.G. Siddall, J.M. Beer, Prediction of the effect of flame length on temperature and radiative heat flux distribution in a process heater, J. Inst. Fuel 48 (1975) 89–96.

[184] N. Selcuk, R.G. Siddall, J.M. Beér, A comparison of mathematical models of the radiative behaviour of a large-scale experimental furnace, Symp. Combust. 16 (1) (1977) 53–62, https://doi.org/10.1016/S0082-0784(77)80313-5.

[185] W.E. Lobo, J.E. Evans, Heat transfer in radiant section of petroleum heaters, Trans. Am. Inst. Chem. Eng. 35 (1939) 748–778.

[186] H. Xue, J.C. Ho, Y.M. Cheng, Comparison of different combustion models in enclosure fire simulation, Fire Saf. J. 36 (1) (2001) 37–54, https://doi.org/10.1016/S0379-7112(00)00043-6.

[187] S.A.B. Al-Omari, On the sensitivity of soot and thermal radiation simulation results to the adopted PDF for temperature underhighly sooting combustion conditions, Int. Commun. Heat Mass Transfer 33 (10) (2006) 1273–1280, https://doi.org/10.1016/j.icheatmasstransfer.2006.08.009.

[188] K.V. Deshmukh, D.C. Haworth, M.F. Modest, Direct numerical simulation of turbulence-radiation interactions in homogeneous nonpremixed combustion systems, Proc. Combust. Inst. 31 (1) (2007) 1641–1648, https://doi.org/10.1016/j.proci.2006.07.139.

[189] P. Narayanan, A. Trouvé, Radiation-driven flame weakening effects in sooting turbulent diffusion flames, Proc. Combust. Inst. 32 (1) (2009) 1481–1489, https://doi.org/10.1016/j.proci.2008.06.056.

[190] M. Cremer, J. Olver, Impact of high-emissivity coatings on process furnace heat transfer, in: AIChE Spring Meeting and Global Congress on Process Safety, 2015.

[191] M.N. Pedernera, J. Pina, D.O. Borio, V. Bucalá, Use of a heterogeneous two-dimensional model to improve the primary steam reformer performance, Chem. Eng. J. 94 (1) (2003) 29–40, https://doi.org/10.1016/S1385-8947(03)00004-4.

[192] D. Hyde, J.S. Truelve, J. Sykes, 1984. Zone 3 User's Manual, AERE-R 11652.

[193] B.F. Oechsler, J.C.S. Dutra, R.C.P. Bittencourt, J.C. Pinto, Simulation and control of steam reforming of natural gas—reactor temperature control using residual gas, Ind. Eng. Chem. Res. 56 (10) (2017) 2690–2710, https://doi.org/10.1021/acs.iecr.6b03665.

[194] F. Manenti, G. Buzzi-Ferraris, S. Pierucci, M. Rovaglio, H. Gulati, Process dynamic optimization using ROMeo, in: Computer Aided Chemical Engineering, vol. 29, Elsevier B.V., 2011, https://doi.org/10.1016/B978-0-444-53711-9.50091-2.

[195] F. Rossi, M. Rovaglio, F. Manenti, Model predictive control and dynamic real-time optimization of steam cracking units, in: Computer Aided Chemical Engineering, first ed., vol. 45, Elsevier B.V, 2019, https://doi.org/10.1016/B978-0-444-64087-1.00018-8.

[196] F. Manenti, S. Cieri, M. Restelli, Considerations on the steady-state modeling of methanol synthesis fixed-bed reactor, Chem. Eng. Sci. 66 (2) (2011) 152–162, https://doi.org/10.1016/j.ces.2010.09.036.

[197] F. Manenti, S. Cieri, M. Restelli, G. Bozzano, Dynamic modeling of the methanol synthesis fixed-bed reactor, Comput. Chem. Eng. 48 (2013) 325–334, https://doi.org/10.1016/j.compchemeng.2012.09.013.

[198] F.W. Dittus, L.M.K. Boelter, Heat Transfer in Automobile Radiators of the Tubular Type, University of California Press, Berkeley, Calif, 1930.

[199] C.V.S. Murty, M.V. Krishna Murthy, Modeling and simulation of a top-fired reformed, Ind. Eng. Chem. Res. 27 (10) (1988) 1832–1840, https://doi.org/10.1021/ie00082a016.

[200] J.K. Rajesh, S.K. Gupta, G.P. Rangaiah, A.K. Ray, Multiobjective optimization of cyclone separators using genetic algorithm, Ind. Eng. Chem. Res. 39 (11) (2000) 4272–4286, https://doi.org/10.1021/ie990741c.

[201] C.L. Yaws, Chemical Properties Handbook: Physical, Thermodynamics, Environmental Transport, Safety and Health Related Properties for Organic, McGraw-Hill Education, 1999.

[202] B.E. Poling, J.M. Prausnitz, J.P. O'Connell, The Properties of Gases and Liquids, fifth ed., McGraw-Hill Education, 2001.

[203] A. Burcat, B. Ruscic, Third Millennium Ideal Gas and Condensed Phase Thermochemical Database for Combustion with Updates from Active Thermochemical Tables, 2005. Technical Report, ANL-05/20 (September), ANL-05/20 TAE 960.

[204] I.M. Alatiqi, A.M. Meziou, G.A. Gasmelseed, Modelling, simulation and sensitivity analysis of steam-methane reformers, Int. J. Hydrog. Energy 14 (4) (1989) 241–256, https://doi.org/10.1016/0360-3199(89)90061-X.

[205] U.I. Amran, A. Ahmad, M.R. Othman, Kinetic based simulation of methane steam reforming and water gas shift for hydrogen production using aspen plus, Chem. Eng. Trans. 56 (December) (2017) 1681–1686, https://doi.org/10.3303/CET1756281.

[206] F.J. Durán, F. Dorado, L. Sanchez-Silva, Exergetic and economic improvement for a steam methane-reforming industrial plant: simulation tool, Energies 13 (15) (2020) 2019, https://doi.org/10.3390/en13153807.

[207] A. Giwa, S.O. Giwa, Simulation, sensitivity analysis and optimization of hydrogen production by steam reforming of methane using Aspen plus, Int. J. Eng. Res. Technol. 2 (7) (2013) 1719–1729.

[208] E.M.A. Mokheimer, M.I. Hussain, S. Ahmed, M.A. Habib, A.A. Al-Qutub, On the modeling of steam methane reforming, J. Energy Resour. Technol. Trans. ASME 137 (1) (2015) 1–12, https://doi.org/10.1115/1.4027962.

[209] J. Pulgarin-León, J. Saavedra-Rueda, A. Molina, Sensitivity analysis of a methane steam reformer, in: XXV Congreso Colombiano de Ingeniería Química, September 2009, 2009.

[210] M.T. Sadeghi, M. Molaei, CFD simulation of a methane steam reforming reactor, Int. J. Chem. React. Eng. 6 (2008), https://doi.org/10.2202/1542-6580.1593.

[211] E. Shagdar, B.G. Lougou, Y. Shuai, E. Ganbold, O.P. Chinonso, H. Tan, Process analysis of solar steam reforming of methane for producing low-carbon hydrogen, RSC Adv. 10 (21) (2020) 12582–12597, https://doi.org/10.1039/c9ra09835f.

[212] M. Sinaei Nobandegani, M.R. Sardashti Birjandi, T. Darbandi, M.M. Khalilipour, F. Shahraki, D. Mohebbi-Kalhori, An industrial steam methane reformer optimization using response surface methodology, J. Nat. Gas Sci. Eng. 36 (2016) 540–549, https://doi.org/10.1016/j.jngse.2016.10.031.

[213] F.F. Tabrizi, S.A.H.S. Mousavi, H. Atashi, Thermodynamic analysis of steam reforming of methane with statistical approaches, Energy Convers. Manag. 103 (2015) 1065–1077, https://doi.org/10.1016/j.enconman.2015.07.005.

[214] B.V. Ayodele, S.I. Mustapa, T.A.R. Tuan Abdullah, Bin, & Salleh, S. F., A Mini-review on hydrogen-rich syngas production by thermo-catalytic and bioconversion of biomass and its environmental implications, Front. Energy Res. 7 (October) (2019) 1–6, https://doi.org/10.3389/fenrg.2019.00118.

[215] X.J. Lee, H.C. Ong, Y.Y. Gan, W.H. Chen, T.M.I. Mahlia, State of art review on conventional and advanced pyrolysis of macroalgae and microalgae for biochar, bio-oil and bio-syngas production, Energy Convers. Manag. 210 (March) (2020), https://doi.org/10.1016/j.enconman.2020.112707, 112707.

[216] J.D. Martínez, K. Mahkamov, R.V. Andrade, E.E. Silva Lora, Syngas production in downdraft biomass gasifiers and its application using internal combustion engines, Renew. Energy 38 (1) (2012) 1–9, https://doi.org/10.1016/j.renene.2011.07.035.

[217] A. Molino, V. Larocca, S. Chianese, D. Musmarra, Biofuels production by biomass gasification: a review, Energies 11 (4) (2018) 1–31, https://doi.org/10.3390/en11040811.

[218] R.G. dos Santos, A.C. Alencar, Biomass-derived syngas production via gasification process and its catalytic conversion into fuels by Fischer Tropsch synthesis: a review, Int. J. Hydrog. Energy 45 (36) (2020) 18114–18132, https://doi.org/10.1016/j.ijhydene.2019.07.133.

[219] N.A. Slavinskaya, U. Riedel, V.E. Messerle, A.B. Ustimenko, Chemical kinetic modeling in coal gasification processes: an overview, Eurasian Chem. Technol. J. 15 (1) (2013) 1–18, https://doi.org/10.18321/ectj134.

[220] M. Wang, Y. Wan, Q. Guo, Y. Bai, G. Yu, Y. Liu, H. Zhang, S. Zhang, J. Wei, Brief review on petroleum coke and biomass/coal co-gasification: syngas production, reactivity characteristics, and synergy behavior, Fuel 304 (April) (2021), https://doi.org/10.1016/j.fuel.2021.121517, 121517.

[221] X. Zhuang, Y. Song, H. Zhan, X. Yin, C. Wu, Gasification performance of biowaste-derived hydrochar: the properties of products and the conversion processes, Fuel 260 (June 2019) (2020), https://doi.org/10.1016/j.fuel.2019.116320.

[222] R.A. El-Nagar, A.A. Ghanem, Syngas production, properties, and its importance, in: C. Ghenai, A. Inayat (Eds.), Sustainable Alternative Syngas Fuel, IntechOpen, 2016. https://www.intechopen.com/books/advanced-biometric-technologies/liveness-detection-in-biometrics.

[223] M.M. Hasan, M.G. Rasul, M.M.K. Khan, N. Ashwath, M.I. Jahirul, Energy recovery from municipal solid waste using pyrolysis technology: a review on current status and developments, Renew. Sust. Energ. Rev. 145 (April) (2021), https://doi.org/10.1016/j.rser.2021.111073, 111073.

[224] M. Inayat, S.A. Sulaiman, J.C. Kurnia, M. Shahbaz, Effect of various blended fuels on syngas quality and performance in catalytic co-gasification: a review, Renew. Sust. Energ. Rev. 105 (July 2018) (2019) 252–267, https://doi.org/10.1016/j.rser.2019.01.059.

[225] J. Ren, J.P. Cao, X.Y. Zhao, F.L. Yang, X.Y. Wei, Recent advances in syngas production from biomass catalytic gasification: a critical review on reactors, catalysts, catalytic mechanisms and mathematical models, Renew. Sust. Energ. Rev. 116 (August) (2019), https://doi.org/10.1016/j.rser.2019.109426, 109426.

[226] E. Ranzi, T. Faravelli, F. Manenti, Pyrolysis, gasification, and combustion of solid fuels, in: Advances in Chemical Engineering, first ed., vol. 49, Elsevier Inc., 2016, https://doi.org/10.1016/bs.ache.2016.09.001.

[227] E. Ranzi, M. Corbetta, F. Manenti, S. Pierucci, Kinetic modeling of the thermal degradation and combustion of biomass, Chem. Eng. Sci. 110 (2014) 2–12, https://doi.org/10.1016/j.ces.2013.08.014.

[228] A.F. Amaral, G. Bozzano, C. Pirola, A.G. Goryunov, A.V. Chistyakov, F. Manenti, Self-sustainable bio-methanol & bio-char coproduction from 2nd generation biomass gasification, Chem. Eng. Trans. 57 (2017) 1045–1050, https://doi.org/10.3303/CET1757175.

[229] A. Bassani, C. Frau, E. Maggio, A. Pettinau, G. Calì, E. Ranzi, F. Manenti, Devolatilization of organo-sulfur compounds in coal gasification, Chem. Eng. Trans. 57 (2017) 505–510, https://doi.org/10.3303/CET1757085.

[230] M. Corbetta, A. Frassoldati, H. Bennadji, K. Smith, M.J. Serapiglia, G. Gauthier, T. Melkior, E. Ranzi, E.M. Fisher, Pyrolysis of centimeter-scale woody biomass particles: kinetic modeling and experimental validation, Energy Fuel 28 (6) (2014) 3884–3898, https://doi.org/10.1021/ef500525v.

[231] Z. Ravaghi-Ardebili, F. Manenti, M. Corbetta, C. Pirola, E. Ranzi, Biomass gasification using low-temperature solar-driven steam supply, Renew. Energy 74 (2015) 671–680, https://doi.org/10.1016/j.renene.2014.07.021.

[232] C. Lange, P. Barthelmäs, T. Rosnitschek, S. Tremmel, F. Rieg, Impact of hpc and automated cfd simulation processes on virtual product development—a case study, Appl. Sci. 11 (14) (2021), https://doi.org/10.3390/app11146552.

[233] B. Wu, G. Chen, J. Moreland, D. Huang, D. Zheng, C.Q. Zhou, Industrial application of CFD simulation and VR visualization, in: ASME 2010 World Conference on Innovative Virtual Reality, 2010, pp. 51–59, https://doi.org/10.1115/WINVR2010-3734.

[234] L. Xi, D.W. Yin, J.S. Park, Special issue "cfd modeling of complex chemical processes: multiscale and multiphysics challenges", Processes 9 (5) (2021) 10–12, https://doi.org/10.3390/pr9050775.

[235] F. Bezzo, S. Macchietto, C.C. Pantelides, A general framework for the integration of computational dynamics and process simulation, Comput. Chem. Eng. 1354 (2000). Computers & Chemical Engineering.

[236] F.R. Menter, J. Schütze, K.A. Kurbatskii, M. Gritskevich, A. Garbaruk, Scale-resolving simulation techniques in industrial CFD, in: 6th AIAA Theoretical Fluid Mechanics Conference, June, 2011, pp. 1–12, https://doi.org/10.2514/6.2011-3474.

[237] M. Kuroki, S. Ookawara, K. Ogawa, A high-fidelity CFD model of methane steam reforming in a packed bed reactor, J. Chem. Eng. Jpn 42 (SUPPL. 1) (2009) 73–78, https://doi.org/10.1252/jcej.08we256.

[238] M. Gür, E.D. Canbaz, Analysis of syngas production and reaction zones in hydrogen oriented underground coal gasification, Fuel 269 (65) (2020), https://doi.org/10.1016/j.fuel.2020.117331, 117331.

[239] N. Hanchate, V.S. Korpale, C.S. Mathpati, S.P. Deshmukh, V.H. Dalvi, Computational fluid dynamics of dual fluidized bed gasifiers for syngas production: cold flow studies, J. Taiwan Inst. Chem. Eng. 117 (2020) 156–163, https://doi.org/10.1016/j.jtice.2020.12.014.

[240] G.D. Wehinger, M. Kraume, V. Berg, O. Korup, K. Mette, R. Schlögl, M. Behrens, R. Horn, Investigating dry reforming of methane with spatial reactor profiles and particle-resolved CFD simulations, AICHE J. 62 (12) (2016) 4436–4452, https://doi.org/10.1002/aic.15520.

[241] D. Pashchenko, R. Mustafin, A. Mustafina, Steam methane reforming in a microchannel reformer: experiment, CFD-modelling and numerical study, Energy 237 (2021), https://doi.org/10.1016/j.energy.2021.121624.

[242] M. Behnam, A.G. Dixon, P.M. Wright, M. Nijemeisland, E.H. Stitt, Comparison of CFD simulations to experiment under methane steam reforming reacting conditions, Chem. Eng. J. 207–208 (2012) 690–700, https://doi.org/10.1016/j.cej.2012.07.038.

[243] H.O. Gómez, M.C. Calleja, L.A. Fernández, A. Kiedrzyńska, R. Lewtak, Application of the CFD simulation to the evaluation of natural gas replacement by syngas in burners of the ceramic sector, Energy 185 (2019) 15–27, https://doi.org/10.1016/j.energy.2019.06.064.

[244] M.E. Taskin, A.G. Dixon, M. Nijemeisland, E. Hugh Stitt, CFD study of the influence of catalyst particle design on steam reforming reaction heat effects in narrow packed tubes, Ind. Eng. Chem. Res. 47 (16) (2008) 5966–5975, https://doi.org/10.1021/ie800315d.

[245] A. Abdelkader, H. Daly, Y. Saih, K. Morgan, M.A. Mohamed, S.A. Halawy, C. Hardacre, Steam reforming of ethanol over Co3O4-Fe 2O3 mixed oxides, Int. J. Hydrog. Energy 38 (20) (2013) 8263–8275, https://doi.org/10.1016/j.ijhydene.2013.04.009.

[246] D.C. Dayton, B. Turk, R. Gupta, Syngas cleanup, conditioning, and utilization, in: Thermochemical Processing of Biomass, 2019, pp. 125–174, https://doi.org/10.1002/9781119417637.ch5.

[247] D. Sheldon, Methanol production—a technical history, Johnson Matthey Technol. Rev. 61 (3) (2017) 172–182, https://doi.org/10.1595/205651317X695622.

[248] E. Ryckebosch, M. Drouillon, H. Vervaeren, Techniques for transformation of biogas to biomethane, Biomass Bioenergy 35 (5) (2011) 1633–1645, https://doi.org/10.1016/j.biombioe.2011.02.033.

CHAPTER 3

Computational fluid dynamics simulation of natural gas reformers

Mohammad Hadi Sedaghat[a] and Mohammad Reza Rahimpour[b]
[a]Department of Mechanical Engineering, Technical and Vocational University (TVU), Tehran, Iran
[b]Department of Chemical Engineering, Shiraz University, Shiraz, Iran

1. Introduction

Over the years, significant advances have been made in the areas of reformers modeling. McGreavy and Newmann [1] seemingly did the first mathematical modeling of the complete reformer. The advances in numerical modeling approaches, the evolution of the computing power, a detailed understanding of physical and chemical phenomena in power plants as well as advances in numerical modeling of reaction kinetics and reactor models lead to simulate reformers more accurately. Computational fluid dynamics (CFD) is a computer-based analysis, which uses numerical methods to calculate velocity profile and temperature of the fluid associated with the phenomena such as chemical reactions. CFD has been regarded as a powerful tool to provide valuable insights into reformer modeling by using more comprehensive and accurate mathematical utilities including precise radiation mechanisms, turbulent model, combined physical and chemical models, etc. [2]. Discretization is an integral part of CFD modeling. CFD discretization is categorized as geometry, space, and time discretization.

I. Geometry discretization or mesh generation has been used to divide the geometry to a number of cells. Various types of meshes have been reported in the literature including triangle and quadrilateral for two-dimensional and tetrahedron, pyramid, triangular prism, hexahedron, and polyhedron for three-dimensional modeling. The connectivity of cells is another important characteristic of the grids described by unstructured, structured, or hybrid meshes. The grid size, cell number, and cell quality have been regarded as the most important parameters in this regard.

II. Space discretization has been employed to study spatial derivatives of the governing partial differential equations numerically. Finite difference, finite element, and finite volume methods have been regarded as three important approaches that have been used in CFD simulations to perform space discretization. Finite different method [3] has been

Advances in Synthesis Gas: Methods, Technologies and Applications. https://doi.org/10.1016/B978-0-323-91879-4.00013-8

regarded as the simplest method of discretization. Although finite difference method has the advantage of simple algorithm and efficient calculation speed, this method has some restricted requirements for grid type and mesh quality, and only structured grids with fine meshes can be used in this method [4]. Finite element method [5] is another method, which has been used for simulation of solid domains and non-Newtonian viscoelastic fluid. Although this method has the advantage of the better accuracy especially on poor grid quality, memory occupation and the high computational costs are the most important demerits of that. Furthermore, this method is not suitable for turbulent flow. Finite volume method [6,7] is the most commonly used model in CFD solvers. In this method, the governing equations are solved by integrating the equations of each control volume [8].

III. Time discretization has been used for unsteady problems, which calculate time-averaged values in momentum equations. Explicit and implicit schemes have been regarded as the two main methods to solve time-dependent equations. The explicit method has been known as a low-cost and efficient form of time discretization method, which requires a small time step for solving unsteady problems. Implicit scheme, which is more common in convection and diffusion processes, is more stable for large time steps.

The governing equations, which should be solved by CFD solvers to study reformers including continuity (Eq. 1a), momentum (Eq. 1b), energy (Eq. 1c), and species material (Eq. 1d) equations, are defined as follows [9]:

$$\frac{\partial}{\partial t}(\rho) + \nabla \cdot \left(\rho \vec{v}\right) = 0 \tag{1a}$$

$$\frac{\partial}{\partial t}\left(\rho \vec{v}\right) + \nabla \cdot \left(\rho \vec{v} \vec{v}\right) = -\nabla P + \nabla \cdot \left(\bar{\bar{\tau}}\right) + \rho \vec{g} \tag{1b}$$

$$\frac{\partial}{\partial t}(\rho E) + \nabla \cdot \left(\vec{v}(\rho E + P)\right) = \nabla \cdot \left[k_{eff} \nabla T - \left(\sum_i h_i J_i\right) + \left(\bar{\bar{\tau}} \cdot \vec{v}\right) + \vec{q}_{rad} \right] + S_h \tag{1c}$$

$$\frac{\partial}{\partial t}(\rho Y_i) + \nabla \cdot \left(\rho \vec{v} Y_i\right) = -\nabla \cdot \vec{J}_i + R_i \tag{1d}$$

Where ρ is the fluid density, $\vec{v}$ is the velocity vector, P is the static pressure, $\bar{\bar{\tau}}$ is the stress tensor, $\vec{g}$ is the gravitational acceleration, E is the specific internal energy, k_{eff} is the effective thermal conductivity, T is the temperature, h is the specific sensible enthalpy, and Y_i is the mass fraction. In addition, in Eq. (1c) $\nabla \cdot \left[k_{eff} \nabla T - \left(\sum_i h_i J_i\right) + \left(\bar{\bar{\tau}} \cdot \vec{v}\right) + \vec{q}_{rad} \right]$ represents four main mechanisms of heat transfer, i.e., conduction, species diffusion, viscous dissipation, and radiation in reformer, respectively. Other parameters in Eq. (1) can be defined as:

$$\vec{J}_i = \left(\rho D_i + \frac{\mu^t}{Sc^t}\right) \nabla Y_i \tag{2a}$$

$$\bar{\bar{\tau}} = \mu^l \left[\left(\nabla \vec{v} + \nabla \vec{v}^T \right) - \frac{2}{3} \nabla \cdot \vec{v} I \right] \tag{2b}$$

$$E = h + \frac{v^2}{2} - \frac{P}{\rho} \tag{2c}$$

$$h = \sum_i Y_i h_i \tag{2d}$$

$$h_i(T) = \int_{T_{ref}}^{T} C_{p,i} dT \text{ with } T_{ref} = 298.15 \, \text{K} \tag{2e}$$

$$S_h = -\sum_i \frac{h_i}{M_i} R_i \tag{2f}$$

$$R_i = \sum_{k=1}^{3} v_i^k R_i^k \tag{2g}$$

$$k_{eff} = k^l + k^t \tag{2h}$$

Where D_i is the mass diffusion coefficient, μ^t and μ^l are the fluid turbulent and laminar viscosity, respectively, I is the unit tensor, R is the volumetric consumption/formation rate, k^t and k^l are the effective turbulent and laminar thermal conductivity, respectively.

Recent CFD solvers have also the advantage of solving detailed three-dimensional velocity, temperature, and species distributions that are essential to study complicated phenomena [10] as well as having the powerful visualization utilities to study various geometries and boundary conditions [11]. CFD technology has been successfully used in simulating industrial furnaces [12–15] as well as tube reactors [16–19].

There are a different types of reforming technologies to produce syngas summarized by Korobitsyn et al. [20]. Steam methane reforming (SMR) has been perceived as the commonly used and economical method to produce syngas [21]. However, some relative demerits have also been reported in the literature regarding SMR:

I. The excess produced hydrogen should be omitted before using in downstream synthesis reactions [22–24].
II. Production of methanol [22–24], dimethyl ether and acetic acid [25], oxo-alcohols, and isocyanates [26].
III. Using water-gas shift reaction in SMR produces significant amounts of carbon dioxide in the presence of carbon monoxide [27–32].
IV. The requirement of the superheated steam as well as strongly endothermic reactions in SMR requires further input energy. For this reason, SMR plants are often coupled with the exothermic plants [25].

Partial oxidation reforming (POR), which proceeds exothermically, is regarded as the second most common technology in syngas production. In this type of reformers, partial oxidation happens, which leads to an incomplete combustion, and the products consist of a mixture of carbon monoxide, hydrogen, carbon dioxide, water, as well as unreacted fuel. The reaction rate of POR is faster than that of SMR, which leads to decrease in residence times [33]. In addition, in comparison with SMR, the selectivity to syngas products is higher in POR reformer [34], which can be made more compact [21,35]. Reducing NOx by using low-temperature flames is another merit of these reformers [36]. On the other hand, using pure oxygen feed, compression of outlet products to 20 bar, which is used in the industrial processes, and the diluted syngas are the most demerits of POR reformers [37]. Although partial oxidation has been used in process engineering applications, e.g. emission reduction in iron production [38], acetylene production [39], molybdenum reduction [40], as well as it has also been employed at a commercial scale for the production of syngas by Shell and Texaco [41], it is not widely used in recent reformers as compared with SMR.

Syngas with a low H_2/CO ratio can also be obtained by carbon dioxide reforming, which is called dry methane reforming (DMR) and considered as the less common technology to produce syngas. DMR has been perceived as a desirable form of reforming, especially when coupled with SMR, POR, and the water-gas shift to enhance the production of syngas [42]. This type of reforming also uses two strong greenhouse gases, which results in global warming reduction [37]. Due to high energy requirements as well as the lack of effective and economic catalysts [37], the dry reforming technology is at the research step, and few commercial plants used this method of reforming [43]. The main disadvantage of this method is that reaction takes place heterogeneously, which leads to soot deposition and sintering of the metal catalyst [44].

In this chapter, we begin with by presenting the governing equations, as well as kinetic relations, of the simulation of aforementioned reforming technologies. In the following, this chapter reviews the pertinent literature in CFD simulation of these technologies by identifying the areas meriting further exploration.

2. Steam methane reforming

As remarked earlier, SMR is considered as the most common commercial method of syngas production. The reactions in SMR occur in catalyst-filled tubes, which consist of two main processes: tube and furnace side.

In tube side process, methane reacts with superheated steam at high-pressure and high-temperature condition in an endothermic reaction and converts into hydrogen, carbon oxide, and carbon monoxide in the presence of a nickel-based catalyst [45,46]. The schematic diagram of industrial-scale SMR process is shown in Fig. 1 [47].

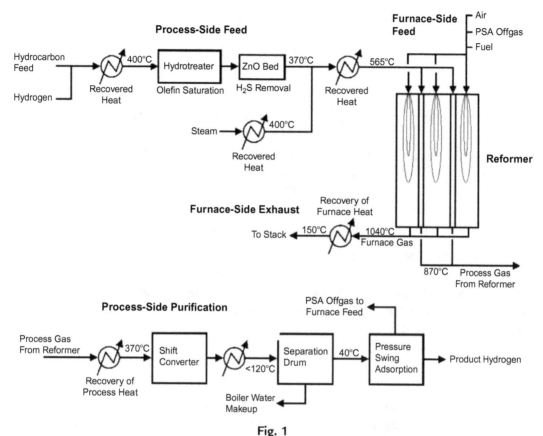

Fig. 1

The schematic diagram of industrial-scale SMR process [47]. *From L. Lao, A. Aguirre, A. Tran, Z. Wu, H. Durand, P.D. Christofides, CFD modeling and control of a steam methane reforming reactor, Chem. Eng. Sci. 148 (2016) 78–92, with permission from Elsevier.*

2.1 Reaction kinetic of SMR

The following intrinsic reaction kinetic model reported by Xu and Froment [48] has been implemented for the SMR reactions in most of the previous investigations:

$$CH_4 + H_2O \Leftrightarrow CO + 3H_2, \Delta H = 206 \text{ kJ/mol} \tag{3a}$$

$$r_1 = \frac{\frac{k_1}{P_{H_2}^{2.5}}\left(P_{CH_4}P_{H_2O} - \frac{P_{CO}P_{H_2}^3}{K_1}\right)}{DEN^2} \tag{3b}$$

$$CO + H_2O \Leftrightarrow CO_2 + H_2, \Delta H = -41 \text{ kJ/mol} \tag{4a}$$

$$r_2 = \frac{\frac{k_2}{P_{H_2}}\left(P_{CO}P_{H_2O} - \frac{P_{CO_2}P_{H_2}}{K_2}\right)}{DEN^2} \tag{4b}$$

$$CH_4 + 2H_2O \Leftrightarrow CO_2 + 4H_2, \Delta H = 165 \text{ kJ/mol} \tag{5a}$$

$$r_3 = \frac{\frac{k_3}{P_{H_2}^{3.5}}\left(P_{CH_4}P_{H_2O}^2 - \frac{P_{CO_2}P_{H_2}^4}{K_3}\right)}{DEN^2} \tag{5b}$$

$$DEN = 1 + \frac{P_{H_2O}K_{H_2O}}{P_{H_2}} + K_{CO}P_{CO} + K_{H_2}P_{H_2} + K_{CH_4}P_{CH_4} \tag{6}$$

where P_{H_2}, P_{CH_4}, P_{H_2O}, P_{CO}, and P_{CO_2} are the partial pressures of H_2, CH_4, H_2O, CO, and CO_2, respectively. In addition, K_{H_2}, K_{CH_4}, and K_{CO} are adsorption constants for H_2, CH_4, and CO, respectively; K_1, K_2, and K_3 are equilibrium constants of the reactions; and k_1, k_2, and k_3 are forward kinetic constant coefficients of the reactions in Eqs. (1a), (2a), and (3a), respectively, and DEN is a dimensionless parameter.

In the furnace side, open-flame furnaces have been used to supply the required heat, which comes from the combustion of mixture of methane, hydrogen, carbon dioxide, carbon monoxide, and air via radiative heat exchange. SMR reformers based on the locations of burners are summarized into four types: side-fired, terrace wall-fired, bottom-fired, and top-fired reformers [49].

2.2 CFD simulation of SMR

As remarked earlier, due to complex nature of the SMR, CFD simulations have been employed as a powerful tool to study this geometry more accurately in the literature. More detailed discussions and their distinguishing features of CFD simulation including simulation of various types of catalyst are available in the literature. Some studies have used more realistic characteristics of catalyst particles and reaction processes [10,50]. These studies used individual particles and bed segments [10] or full beds of spheres [51] catalyst model in a single tube. Dixon [18] used CFD simulation to study the effect of angle of the flow of a one-hole catalyst particle on the catalytic performance of SMR. Their results showed that with incidence angle, the flow through the particle hole reduces. Flat ends and the inside curved surface show decrease in heat and mass fluxes while they enhance with the outside curved surface. Their results also indicated that incident flow, which results in changes in particle surface, showed the maximum value of both reactant consumption and energy uptake. Taskin et al. [52] studied a three-dimensional CFD simulation of SMR with cylindrical catalyst particles to show the effect of the external flow field on the intraparticle processes. Their results [52] showed the steep temperature gradients took place at the tube wall where particles approach each other closely. Behnam et al. [16] performed a experimentally validated CFD model of SMR reaction conditions. Their results showed a reasonable agreement of CFD and experimental data for temperature distribution of catalyst particles for various amounts of inlet gas temperatures as well as the temperature reduction from outside to inside the pellets, which results from the reaction heat sink.

The types and methods of simulations have also been a matter of debate among the researchers. Dixon [53] in CFD-modeling study presented a comparison of the 1D, 2D, and 3D SMR with Comsol Multiphysics software, which used finite element method to solve the governing equations. Numerical results showed good agreement with the experiments for the 3D and 2D model. Ku et al. [54] used mixed Eulerian-Lagrangian method to simulate gas phase (Eulerian approach) and particle phase (Lagrangian approach), to study CFD modeling of the SMR with open-source software OpenFOAM. Pashchenko [55] studied the compression of the 1D, 2D, and 3D CFD simulation of SMR. The numerical simulation was devoted to study the wide range of the residence time as well as the radius-to-length ratios. Results showed that 1D model could be accurate for the large values of residence time (more than 8 $kg_{cat}s/mol_{CH_4}$) as well as the relative length (more than 15). For small values of the residence time and the relative length, the 3D and 2D models were reported to be more applicable. Pashchenko [56] in another investigation used ANSYS Fluent solver to study a 2D CFD model of preheated Ni-based catalyst SMR to obtain the transport phenomena into reaction space of the reformer. Lao et al. [47] used a 2D axisymmetric CFD simulation of an industrial-scale SMR reactor tube to simulate the transport and chemical reaction phenomena in the presence of catalyst packing (Fig. 2). To study the effect open-flame furnaces heat on the tube, they [47] applied a fourth-order polynomial temperature distribution on the tube side by using a least squares linear regression method.

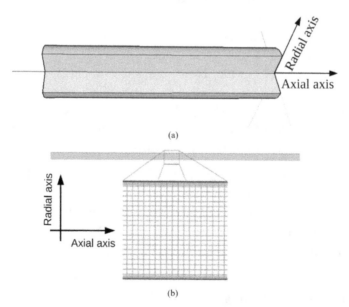

Fig. 2

The schematic geometry of 2D SMR tube used in study of Lao et al. [47] (A) 2D axisymmetric SMR reforming tube and (B) generated mesh of this geometry. *From L. Lao, A. Aguirre, A. Tran, Z. Wu, H. Durand, P.D. Christofides, CFD modeling and control of a steam methane reforming reactor, Chem. Eng. Sci. 148 (2016) 78–92, with permission from Elsevier.*

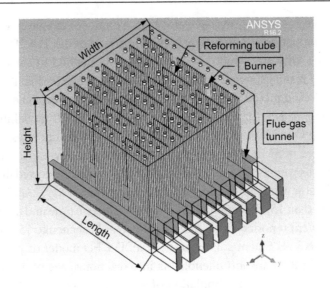

Fig. 3

The industrial-scale top-fired and cocurrent SMR reformer used in study of Tran et al. [9]. *From A. Tran, A. Aguirre, H. Durand, M. Crose, P.D. Christofides, CFD modeling of a industrial-scale steam methane reforming furnace, Chem. Eng. Sci. 171 (2017) 576–598, with permission from Elsevier.*

Wu et al. [57] extended this study in a closed-loop system equipped with a model predictive controller to determine the optimal values of outer reforming tube wall temperature. Tran et al. [9] in a 3D simulation studied an industrial scaled CFD model of SMR that consists of 96 burners, 336 reforming tubes, and 8 flue gas tunnels designed by Selas Fluid Processing Corporation, which is about 16 m wide, 16 m long, and 13 m tall (Fig. 3). They also used an experimental relation to estimate the radiative properties of a homogeneous gas in the furnace side of their model more accurately. The predictions and numerical results were in good agreement with prior results of SMR industrial reactors. Thereby, their data could be used in industrial operating conditions with sufficient accuracy.

These authors [49,58] in another 3D numerical investigations introduced a new method called furnace-balancing scheme based on the least squares regression approach to generate an optimal furnace-side feed distribution to enhance the thermal efficiency of a reformer. Some other investigations have also used CFD methods to study 2D [59–61] or 3D simulation [62–65] as well as optimization [66–68] of SMR reactors.

3. Partial oxidation reforming

In this type of reforming, a complete conversion of methane is obtained above 750°C, leading to H_2/CO ratio of 2 as follows:

$$CH_4 + \frac{1}{2}O_2 \Leftrightarrow CO + 2H_2, \Delta H = -44 \text{ kJ/mol} \tag{7}$$

This reforming method has some merits including a H_2:CO ratio up to 2:1 desirable for downstream use. Additionally, due to exothermic nature of this reaction, no fuel or less intensive energy is required, which can be suitable in compact reactors [27]. The pressures up to 75 bar and temperatures up to 1400°C are also needed for this type of reforming [20]. Noncatalytic oxidation (which uses thermal pathway) [21,35], as well as a lower-temperature catalytic pathway [21,24,33,36], has been employed to reform methane by POR. The catalytic partial oxidation has a higher flexibility and is less susceptible to soot formation and requires lower operating temperatures [69]. However, the carbon and soot deposition over the catalyst, which leads to inactivation [22], as well as the poisoning the catalyst by sulfur, which is often present in natural gas supplies [21], is the most demerit of this method.

Although noncatalytic oxidation has some disadvantages such as requiring high operating temperatures (1150–1315°C) [21] as compared with the steam methane reforming (700–800°C) [24], lower efficiencies than catalytic pathway [70], and more soot production under the rich conditions [35], it denotes an idealization of several industrially important applications, for it is less sensitive to fuel variation [35], and also has longer lower costs as well as operational lifetimes [70].

3.1 CFD simulation of POR

In fast reactions such as POR, heat and mass transfer play a vital role in simulating methane reforming reactors [71]. Due to the small size of these types of reactors, numerical simulations have been regarded as an appropriate choice of estimation of transport phenomena, velocity, and temperature distribution more accurately as well as to better understand the effective factors on producing syngas [72]. Detailed modeling characteristics are essential to estimate interplay between transport and kinetics of the catalytic POR of methane in numerical simulations [73–75]. Chen et al. [72] used two-dimensional CFD model to study catalytic POR in high-temperature conditions. The schematic geometry of this study is shown in Fig. 4. They studied the effect of homogeneous and heterogeneous reaction on the reaction products. The influence of some other factors such as reactor dimension, feed composition, pressure, and preheating temperature on the characteristics of the process for both air-feed and oxygen-feed systems was studied numerically. Their results showed the significant contribution of operating conditions on the homogeneous and heterogeneous reaction pathways. Their results also indicated that high preheating temperatures and pressures in large reactors would be useful for gas-phase chemistry, while surface chemistry works at low preheating temperatures and pressures in small reactors and leads to producing larger values of syngas.

Yu et al. [76] used a three-dimensional steady-state CFD simulation with ANSYS FLUENT solver to study the mixing behavior of the jets in cross-flow stream to produce cooled syngas from methane in the partial oxidation process. They studied the effects of various parameters such as the ratio of jet velocity to mainstream velocity, jet incident angle as well as Reynolds number on the process. Their results showed the angle of 145 degrees for the optimum jet

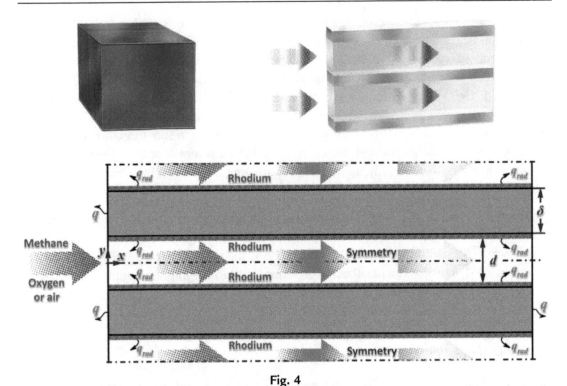

Fig. 4

The schematic geometry of POR CFD model used in study of Chen et al. [72]. *From J. Chen, W. Song, D. Xu, Catalytic partial oxidation of methane for the production of syngas using microreaction technology: a computational fluid dynamics study, Int. J. Hydrogen Energy 43(31) (2018) 14059–14077, with permission from Elsevier.*

incident angle. The velocity ratio in the range of 1–3 and 1.8–2.4 has been reported as the optimized ratio with jet incident angle smaller and larger than 130 degrees, respectively. The results also showed that Reynolds number has no great influence on the spatial distribution of the mixture. These authors in similar 3D CFD investigation [77] studied the residence time, mainstream temperature, and loss ratio of acetylene. Their results showed that compared with the jet-in-cross-flow, the impinging flow has a more uniform mixing and lower residence time. Additionally, the temperature decreasing rate of the mainstream is slower in the impinging flow compared with the jet-in-cross-flow. The loss ratio of acetylene in the quenching process is 2.89% and 1.45% for the jet-in-cross-flow and the impinging flow, respectively, which showed that the impinging flow configuration can be more efficient of POR of methane. Chen et al. [78] in another 3D CFD simulation studied the water reforming process for POR to acetylene by using the same jet-in-cross-flow configuration. Their results showed that the gas quenching process is more efficient due to the fact that the heat can be effectively recovered after the quenching at a loss of acetylene. Hettel et al. [79] proposed the CFD simulation of catalytic

POR by using the combination of OpenFOAM and DETCHEM solvers. Some other CFD numerical investigations have also used 2D [27,80] or 3D [81–83] model to simulate methane reformer by using POR reactors.

4. Dry methane reforming

Dry methane reforming (DMR) or CO_2 reforming is the third common form of reforming process. In this process, syngas with a H_2/CO ratio of 1:1 is produced according to the following reaction:

$$CH_4 + CO_2 \Leftrightarrow 2CO + 2H_2, \Delta H = 247 \text{ kJ/mol} \tag{8}$$

This reaction was originally proposed by Fischer and Tropsch as an alternative reaction to SMR [84]. Due to the absence of steam in this reaction, this process is also called "dry methane reforming (DMR)." DMR can convert both CH_4 and CO_2 into syngas that can be processed into liquid fuel, chemicals, or fertilizer agents and also indicates its essential potential for the nuclear and solar power plants due to its strong endothermicity ($\Delta H = 247 \text{kJ/mol}$). On the other hand, this highly endothermic process done at temperatures of 700–1000°C as well as coke formation, which leads to a deactivation of the catalyst, has been regarded as one of the largest obstacles for the industrial use [85–87].The DRM reaction for CO_2 reforming can proceed in the presence of Nickel and Ni/SiO_2 catalysts [88], transient metal catalysts such as Ni/Al_2O_3 [89], rhodium, or iridium, which are less susceptible to carbon formation [90] and other sources such as iron, cobalt, and nonmetallic catalysts [91]. In these reformers, the diameter of tubular fixed-bed reactors does not exceed 80 mm due to the thermal effect of the reaction [92].

4.1 Reaction kinetic of DMR

Several reaction kinetic models were reported in the literature for the DMR process. The power law, the Eley Rideal, and the Langmuir Hinshelwood models have been regarded as the most important ones. The power-law model shows the reaction rate of the DRM as follows [93]:

$$r = kP_{CH_4}^m P_{CO_2}^n \tag{9}$$

where k is the rate constant, P_{CH_4} and P_{CO_2} are the partial pressures of CH_4 and CO_2 respectively, m and n are the rate coefficients. This model has been rarely used due to some accuracy limitations.

The Eley Rideal model based on the reference composition was represented as the following equations [94]:

$$r_1 = \frac{kK_{CH_4}P_{CH_4}P_{CO_2}}{1 + K_{CH_4}P_{CH_4}} \tag{10a}$$

$$r_2 = \frac{kK_{CO_2}P_{CH_4}P_{CO_2}}{1 + K_{CO_2}P_{CO_2}} \qquad (10b)$$

where K_{CH_4} and K_{CO_2} are the equilibrium constants.

The Langmuir Hinshelwood model has been regarded as the most commonly used model employing the reaction rate of the DRM as [95]:

$$r = \frac{kK_{CH_4}K_{CO_2}P_{CH_4}P_{CO_2}}{\left(1 + K_{CO_2}P_{CO_2} + K_{CH_4}P_{CH_4}\right)^2} \qquad (11)$$

Based on the reported experimental and numerical investigations, this model is the best model in prediction of the reaction rates for various catalysts.

4.2 CFD simulation of DMR

Due to some complexities of the geometry as well as reaction mechanism of these reformers, some researchers used CFD tools to study the possibility of detailed quantitative comparisons between the predictions and observations. Benguerba et al. [96] in CFD simulation studied the nickel-based catalyst of DMR in the lab-scale reactor. The numerical results including concentrations and molar fluxes validated with the experimental data. Their results showed that in DMR, catalyst deactivation was not a critical issue in a catalytic fixed-bed membrane reactor. Wehinger et al. [97] in a 3D CFD simulation studied fixed-bed reactor for the DMR over rhodium (Fig. 5).

Their results [97] showed the significance interaction between chemical kinetics and heat and mass transfer in the reactor. They also showed the catalyst deactivation regions with carbon deposition in their simulation. In other CFD simulations [98–103], the same authors studied the effect of packing particle shapes, contact methods of particles, pore processes, particle dimensions, reactor configurations, and the micro kinetics of the reactions, for the simulation of DMR based on fixed-bed reactors. Zhang and Smith [104] in 3D CFD simulation studied the influence of the input radiative power and gas flow rate as well as the feedstock composition on the temperature distribution and the methane conversion of the nickel-based DRM reactors. Lee and Lim [105] in CFD modeling studied DMR by using molten salt as the heat source. They concluded that difference temperature of the heat source between the inlet and outlet of the reactor was not significant, which leads to a proportional methane conversion to the length when the catalyst volume is fixed. Their results also showed that molten salt heat resource can be more efficient compared with the gas heat source. They also concluded that the smaller radius of the input feed, the larger the methane conversion. They also observed the decrease in methane conversion results from enhancement of reformer bed. In another CFD investigation, [106], they studied the effect of high-temperature molten salt heat sources on the hydrogen production to find the best efficiency of the reactor for the

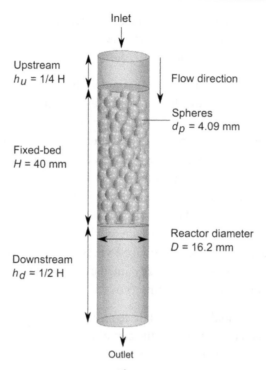

Fig. 5

The schematic geometry of DMR CFD model used in study of Wehinger et al. [97]. *From G.D. Wehinger, T. Eppinger, M. Kraume, Detailed numerical simulations of catalytic fixed-bed reactors: heterogeneous dry reforming of methane, Chem. Eng. Sci. 122 (2015) 197–209, With permission from Elsevier.*

methane conversion and hydrogen flow rate. Zheng et al. [107] recently used a CFD modeling to study the radiation absorption as well as chemical reactions in a solar DMR reactor. Their results showed that structural parameters, such as inclined angle, length of transition section, cut-off ratio, and acceptance half-angle of compound parabolic concentrator, could play a vital role in radiation heat flux density distributions, which leads to important variation of the chemical reaction.

5. Conclusion and future outlook

With the dramatic increase of computing power, CFD modeling has been regarded as a powerful tool to provide valuable insights into physical phenomena occurring in industrial applications such as reformers to study detailed representations of the reformer geometry as well as physical and chemical models. CFD modeling can estimate fluid flow, heat transfer, and associated phenomena such as chemical reactions by using numerical methods to solve mathematical models via software solvers. Additionally, present CFD solvers also have the advantage of powerful visualization capabilities to deal with various geometry characteristics and boundary conditions.

The literature currently available on the CFD simulation of the fluid flow and heat transfer characteristics as well as reaction mechanism associated with the various types of reforming process has been reviewed critically and thoroughly. Undoubtedly, due to large industrial applications of SMR, the simulation of these reformers has been studied most extensively. Hence, reliable values of various parameters such as catalyst characteristics, reactor configurations, residence time, types and methods of simulations, and so on are available for studying these reformers numerically. Some previous investigations developed a CFD model that simulated chemical reactions and transport phenomena of SMR in different computational domains while producing results consistent with available pilot-scale and full industrial-scale plants. Although most syngas has been produced via SMR in industry, it has some demerits such as being bulky, slow to start, and energy-intensive. Thereby, some alternate reforming methods have been reported in the literature to overcome these limitations.

POR of methane has the potential to address these issues as it is quick to start and exothermic. Besides, the thermal method in POX avoids the complications of deactivation, poisoning, and the expense of catalysts. Noncatalytic oxidation (which uses thermal pathway) and a lower-temperature catalytic pathway are used as two main methods in POX reforming. Due to the small size of these reactors, CFD modeling has been perceived as a powerful tool to study transport phenomena, velocity, and temperature distribution more accurately as well as to better understand the effective factors on producing syngas. Two- and three-dimensional CFD simulations showed the ability of this kind of modeling for parametric studies of a POR. The effects of reaction pathways, preheating temperature, pressure, feed composition, and reactor dimension, jet incident angle, the ratio of jet velocity to mainstream velocity, Reynolds number on the mixing behavior are some of the parameters that were performed in previous CFD simulations.

DMR has been perceived as another desirable form of reforming, especially when coupled with other types of reforming such as SMR, POR, and the water-gas shift as a way to improve the yield of syngas. Although DMR has some merits such as consuming two strong greenhouse gases, which results in global warming reduction, due to high energy requirements as well as the lack of effective and economic catalysts, the DMR technology is at the research step, and few commercial plants have used this method of reforming. Thereby, generally, only a small fraction of the currently available literature concerns CFD simulation of DMR compared with SMR and POR. Axisymmetric 2D as well as 3D CFD modeling has also been employed to study the effect of various parameters such as mean resident time, packing particle shapes, contact methods of particles, pore processes, particle dimensions, reactor configurations, the micro kinetics of the reactions inlet gas flow rate, the feedstock composition, and the input radiative power on the flow and temperature distribution as well as the methane conversion.

However, the aforementioned enhancement in CFD simulation of methane reformer is realized only at studding various parameters on reformers, and therefore, much work is required to arrive at optimal operating conditions, which would maximize the heat transfer as well as syngas production. Also, most of the available studies have ignored the variation of thermophysical properties of the fluid (heat capacity, density, thermal conductivity, viscosity) with temperature. Depending upon the type of the fluid flow characteristics (laminar or turbulent) and heat transfer model, consideration of temperature-dependent properties not only adds to the level of mathematical complexity, but it may also influence the rate of heat transfer. Also, the bulk of the literature pertains to the case of 2D simulations, whereas in most practical situations (such as industrial plants), the flow occurs in 3D geometrical configurations. Unfortunately, little is known about the radiation heat transfer in reactors. It also needs to be emphasized here that most of the currently available information is based on numerical solutions of the governing differential equations and the experimental results are indeed scarce in this field. It is hoped that the numerical results summarized in this chapter will act to stimulate experimental activity in the area, which is needed not only to substantiate and/or refute the numerical predictions, but also to span the range of conditions (which are currently beyond the reach of numerical simulations or vice versa.

References

[1] C. McGreavy, M. Newmann, Development of a mathematical model of a steam methane reformer, in: Conference on the Industrial Applications of Dynamic Modelling, Institution of Electrical Engineering, 1969.
[2] A. Aguirre, Computational Fluid Dynamics Modeling and Simulation of Steam Methane Reforming Reactors and Furnaces, UCLA, 2017.
[3] L.F. Richardson, IX. The approximate arithmetical solution by finite differences of physical problems involving differential equations, with an application to the stresses in a masonry dam, Philos. Trans. R. Soc. Lond. Ser. A 210 (459–470) (1911) 307–357.
[4] R.W. Jeppson, Limitations of some Finite Difference Methods in Solving the Strongly Nonlinear Equation of Unsaturated Flow in Soils, Utah Water Research Laboratory, Utah State University, Logan, UT, 1972. Report No. PRWG-59-c-8.
[5] R. Courant, Variational methods for the solution of problems of equilibrium and vibrations, Bull. Am. Math. Soc. 49 (1) (1943) 1–23.
[6] M.W. Evans, F.H. Harlow, E. Bromberg, The Particle-in-Cell Method for Hydrodynamic Calculations, Los Alamos National Lab NM, 1957.
[7] R.A. Gentry, R.E. Martin, B.J. Daly, An Eulerian differencing method for unsteady compressible flow problems, J. Comput. Phys. 1 (1) (1966) 87–118.
[8] A. Bakker, Lecture 5-Solution Methods Applied Computational Fluid Dynamics, Lectures of Fluent Instructor, Fluent Inc., New York, NY, 2002, pp. 41–43.
[9] A. Tran, et al., CFD modeling of a industrial-scale steam methane reforming furnace, Chem. Eng. Sci. 171 (2017) 576–598.
[10] A.G. Dixon, M. Nijemeisland, E.H. Stitt, Packed tubular reactor modeling and catalyst design using computational fluid dynamics, Adv. Chem. Eng. 31 (2006) 307–389.
[11] I. Uriz, et al., Computational fluid dynamics as a tool for designing hydrogen energy technologies, in: Renewable Hydrogen Technologies: Production, Purification, Storage, Applications and Safety, Elsevier Oxford, United Kingdom, 2013, pp. 401–435.

[12] M. Baburic, et al., Application of the conservative discrete transfer radiation method to a furnace with complex geometry, in: CHT-04-Advances in Computational Heat Transfer III. Proceedings of the Third International Symposium, Begel House Inc., 2004.

[13] Y. Han, R. Xiao, M. Zhang, Combustion and pyrolysis reactions in a naphtha cracking furnace, Chem. Eng. Technol. Ind. Chem. Plant Equip. Process Eng. Biotechnol. 30 (1) (2007) 112–120.

[14] M. Noor, A.P. Wandel, T. Yusaf, Detail guide for CFD on the simulation of biogas combustion in bluff-body mild burner, in: Proceedings of the 2nd International Conference of Mechanical Engineering Research (ICMER 2013), Universiti Malaysia Pahang, 2013.

[15] G.D. Stefanidis, et al., CFD simulations of steam cracking furnaces using detailed combustion mechanisms, Comput. Chem. Eng. 30 (4) (2006) 635–649.

[16] M. Behnam, et al., Comparison of CFD simulations to experiment under methane steam reforming reacting conditions, Chem. Eng. J. 207 (2012) 690–700.

[17] H. Calis, et al., CFD modelling and experimental validation of pressure drop and flow profile in a novel structured catalytic reactor packing, Chem. Eng. Sci. 56 (4) (2001) 1713–1720.

[18] A.G. Dixon, CFD study of effect of inclination angle on transport and reaction in hollow cylinder catalysts, Chem. Eng. Res. Des. 92 (7) (2014) 1279–1295.

[19] A. Guardo, et al., CFD flow and heat transfer in nonregular packings for fixed bed equipment design, Ind. Eng. Chem. Res. 43 (22) (2004) 7049–7056.

[20] M. Korobitsyn, et al., SOFC as a Gas Separator, Final Report NOVEM contract, 2000 (219.401–0012).

[21] R. Chaubey, et al., A review on development of industrial processes and emerging techniques for production of hydrogen from renewable and sustainable sources, Renew. Sustain. Energy Rev. 23 (2013) 443–462.

[22] M. Pen, J. Gomez, J.L.G. Fierro, New catalytic routes for syngas and hydrogen production, Appl. Catal. A Gen. 144 (1–2) (1996) 7–57.

[23] Y. Xu, et al., Numerical simulation of natural gas non-catalytic partial oxidation reformer, Int. J. Hydrogen Energy 39 (17) (2014) 9149–9157.

[24] P.D. Vernon, et al., Partial oxidation of methane to synthesis gas, Catal. Lett. 6 (2) (1990) 181–186.

[25] A.P. York, et al., Methane oxyforming for synthesis gas production, Catal. Rev. 49 (4) (2007) 511–560.

[26] M. Vascellari, et al., Flamelet/progress variable modeling of partial oxidation systems: from laboratory flames to pilot-scale reactors, Chem. Eng. Sci. 134 (2015) 694–707.

[27] Y. Bornstein, Numerical and Experimental Investigation of Partial Oxidation of Methane Over a Multiport Burner (Ph.D. thesis), University of Sydney, 2017.

[28] S. Dehghanpoor, et al., A feasibility study of the conversion of petrochemical off-gas streams to methanol over $CuO/ZnO/Al_2O_3$ catalyst, Top. Catal. (2022) 1–11.

[29] A. Mirvakili, A. Bakhtyari, M.R. Rahimpour, A CFD modeling to investigate the impact of flow mal-distribution on the performance of industrial methanol synthesis reactor, Appl. Therm. Eng. 128 (2018) 64–78.

[30] A. Mirvakili, M. Rahimpour, Mal-distribution of temperature in an industrial dual-bed reactor for conversion of CO2 to methanol, Appl. Therm. Eng. 91 (2015) 1059–1070.

[31] F. Samimi, N. Hamedi, M.R. Rahimpour, Green methanol production process from indirect CO_2 conversion: RWGS reactor versus RWGS membrane reactor, J. Environ. Chem. Eng. 7 (1) (2019) 102813.

[32] F. Samimi, D. Karimipourfard, M.R. Rahimpour, Green methanol synthesis process from carbon dioxide via reverse water gas shift reaction in a membrane reactor, Chem. Eng. Res. Des. 140 (2018) 44–67.

[33] L. De Rogatis, et al., Methane partial oxidation on NiCu-based catalysts, Catal. Today 145 (1–2) (2009) 176–185.

[34] A. Ashcroft, et al., Partial oxidation of methane to synthesis gas using carbon dioxide, Nature 352 (6332) (1991) 225–226.

[35] Z. Al-Hamamre, A. Al-Zoubi, The use of inert porous media based reactors for hydrogen production, Int. J. Hydrogen Energy 35 (5) (2010) 1971–1986.

[36] M. Lyubovsky, S. Roychoudhury, R. LaPierre, Catalytic partial "oxidation of methane to syngas" at elevated pressures, Catal. Lett. 99 (3) (2005) 113–117.

[37] Y.H. Hu, E. Ruckenstein, Catalytic conversion of methane to synthesis gas by partial oxidation and CO_2 reforming, Adv. Catal. 48 (1) (2004) 297–345.

[38] W.-H. Chen, C.-L. Hsu, S.-W. Du, Thermodynamic analysis of the partial oxidation of coke oven gas for indirect reduction of iron oxides in a blast furnace, Energy 86 (2015) 758–771.

[39] Q. Zhang, et al., Simulations of methane partial oxidation by CFD coupled with detailed chemistry at industrial operating conditions, Chem. Eng. Sci. 142 (2016) 126–136.

[40] M. Toledo, et al., Hydrogen production in ultrarich combustion of hydrocarbon fuels in porous media, Int. J. Hydrogen Energy 34 (4) (2009) 1818–1827.

[41] D. Eastman, Synthesis gas by partial oxidation, Ind. Eng. Chem. 48 (7) (1956) 1118–1122.

[42] C. Chen, et al., A "Swiss-Roll" fuel reformer: experiments and modeling, in: 9th US National Combustion Meeting, 2015.

[43] R. Kelling, G. Eigenberger, U. Nieken, Ceramic counterflow reactor for autothermal dry reforming at high temperatures, Catal. Today 273 (2016) 196–204.

[44] J. Gao, et al., Dry (CO_2) reforming, in: Fuel Cells: Technologies for Fuel Processing, Elsevier, 2011, pp. 191–221.

[45] J. Kroschwitz, in: J.I. Kroschwitz, M. Howe-Grant (Eds.), Glass, John Wiley & Sons, New York, 1994.

[46] M. Farsi, H. Shahhosseini, A modified membrane SMR reactor to produce large-scale syngas: modeling and multi objective optimization, Chem. Eng. Process. Process Intensif. 97 (2015) 169–179.

[47] L. Lao, et al., CFD modeling and control of a steam methane reforming reactor, Chem. Eng. Sci. 148 (2016) 78–92.

[48] J. Xu, G.F. Froment, Methane steam reforming, methanation and water-gas shift: I. Intrinsic kinetics, AICHE J. 35 (1) (1989) 88–96.

[49] A. Tran, et al., Temperature balancing in steam methane reforming furnace via an integrated CFD/data-based optimization approach, Comput. Chem. Eng. 104 (2017) 185–200.

[50] H. Freund, et al., Numerical simulations of single phase reacting flows in randomly packed fixed-bed reactors and experimental validation, Chem. Eng. Sci. 58 (3–6) (2003) 903–910.

[51] M. Kuroki, S. Ookawara, K. Ogawa, A high-fidelity CFD model of methane steam reforming in a packed bed reactor, J. Chem. Eng. Jpn. 42 (Supplement) (2009) s73–s78.

[52] M.E. Taskin, et al., Flow, transport, and reaction interactions for cylindrical particles with strongly endothermic reactions, Ind. Eng. Chem. Res. 49 (19) (2010) 9026–9037.

[53] A.G. Dixon, Local transport and reaction rates in a fixed bed reactor tube: endothermic steam methane reforming, Chem. Eng. Sci. 168 (2017) 156–177.

[54] X. Ku, T. Li, T. Løvås, CFD–DEM simulation of biomass gasification with steam in a fluidized bed reactor, Chem. Eng. Sci. 122 (2015) 270–283.

[55] D. Pashchenko, Effect of the geometric dimensionality of computational domain on the results of CFD-modeling of steam methane reforming, Int. J. Hydrogen Energy 43 (18) (2018) 8662–8673.

[56] D. Pashchenko, Numerical study of steam methane reforming over a pre-heated Ni-based catalyst with detailed fluid dynamics, Fuel 236 (2019) 686–694.

[57] Z. Wu, et al., Model predictive control of a steam methane reforming reactor described by a computational fluid dynamics model, Ind. Eng. Chem. Res. 56 (20) (2017) 6002–6011.

[58] A. Tran, et al., Steam methane reforming furnace temperature balancing via CFD model-based optimization, in: 2017 American Control Conference (ACC), IEEE, 2017.

[59] B.M. Cruz, J.D. da Silva, A two-dimensional mathematical model for the catalytic steam reforming of methane in both conventional fixed-bed and fixed-bed membrane reactors for the production of hydrogen, Int. J. Hydrogen Energy 42 (37) (2017) 23670–23690.

[60] J. Chen, et al., Retracted article: computational fluid dynamics modeling of the millisecond methane steam reforming in microchannel reactors for hydrogen production, RSC Adv. 8 (44) (2018) 25183–25200.

[61] D. Pashchenko, A. Eremin, Heat flow inside a catalyst particle for steam methane reforming: CFD-modeling and analytical solution, Int. J. Heat Mass Transf. 165 (2021) 120617.

[62] A.G. Beule, Physiology and pathophysiology of respiratory mucosa of the nose and the paranasal sinuses, GMS Curr. Top. Otorhinolaryngol. Head Neck Surg. 9 (2010).

[63] M. Upadhyay, et al., CFD simulation of methane steam reforming in a membrane reactor: performance characteristics over range of operating window, Int. J. Hydrogen Energy 46 (59) (2021) 30402–30411.

[64] S.I. Ngo, et al., Computational fluid dynamics model on a compact-type steam methane reformer for highly-efficient hydrogen production from natural gas, in: Computer Aided Chemical Engineering, Elsevier, 2018, pp. 307–312.

[65] D.D. Nguyen, et al., Computational fluid dynamics (CFD) modelling and optimum gap size of a compact steam methane reforming (SMR) reactor, in: Computer Aided Chemical Engineering, Elsevier, 2018, pp. 331–336.

[66] N.D. Vo, et al., Combined approach using mathematical modelling and artificial neural network for chemical industries: steam methane reformer, Appl. Energy 255 (2019) 113809.

[67] F. Rossi, M. Rovaglio, F. Manenti, Model predictive control and dynamic real-time optimization of steam cracking units, in: Computer Aided Chemical Engineering, Elsevier, 2019, pp. 873–897.

[68] M. Tutar, et al., Optimized CFD modelling and validation of radiation section of an industrial top-fired steam methane reforming furnace, Comput. Chem. Eng. 155 (2021) 107504.

[69] B.C. Enger, R. Lødeng, A. Holmen, A review of catalytic partial oxidation of methane to synthesis gas with emphasis on reaction mechanisms over transition metal catalysts, Appl. Catal. A Gen. 346 (1–2) (2008) 1–27.

[70] A. Loukou, et al., Experimental study of hydrogen production and soot particulate matter emissions from methane rich-combustion in inert porous media, Int. J. Hydrogen Energy 37 (21) (2012) 16686–16696.

[71] R. Quiceno, et al., Modeling the high-temperature catalytic partial oxidation of methane over platinum gauze: detailed gas-phase and surface chemistries coupled with 3D flow field simulations, Appl. Catal. A Gen. 303 (2) (2006) 166–176.

[72] J. Chen, W. Song, D. Xu, Catalytic partial oxidation of methane for the production of syngas using microreaction technology: a computational fluid dynamics study, Int. J. Hydrogen Energy 43 (31) (2018) 14059–14077.

[73] R. Schwiedernoch, et al., Experimental and numerical study on the transient behavior of partial oxidation of methane in a catalytic monolith, Chem. Eng. Sci. 58 (3–6) (2003) 633–642.

[74] P. Kraus, R. Lindstedt, Microkinetic mechanisms for partial oxidation of methane over platinum and rhodium, J. Phys. Chem. C 121 (17) (2017) 9442–9453.

[75] N. Vernikovskaya, et al., Transient behavior of the methane partial oxidation in a short contact time reactor: modeling on the base of catalyst detailed chemistry, Chem. Eng. J. 134 (1–3) (2007) 180–189.

[76] X. Yu, et al., Mixing behaviors of jets in cross-flow for heat recovery of partial oxidation process, Int. J. Chem. React. Eng. 15 (1) (2017).

[77] X. Yu, et al., CFD simulations of quenching process for partial oxidation of methane: comparison of jet-in-cross-flow and impinging flow configurations, Chin. J. Chem. Eng. 26 (5) (2018) 903–913.

[78] T. Chen, et al., Simulation of industrial-scale gas quenching process for partial oxidation of nature gas to acetylene, Chem. Eng. J. 329 (2017) 238–249.

[79] M. Hettel, et al., Numerical simulation of a structured catalytic methane reformer by DUO: the new computational interface for OpenFOAM® and DETCHEM™, Catal. Today 258 (2015) 230–240.

[80] J. Chen, W. Song, D. Xu, Computational fluid dynamics modeling of the catalytic partial oxidation of methane in microchannel reactors for synthesis gas production, Processes 6 (7) (2018) 83.

[81] K. Bawornruttanaboonya, et al., A computational fluid dynamic evaluation of a new microreactor design for catalytic partial oxidation of methane, Int. J. Heat Mass Transf. 115 (2017) 174–185.

[82] Q. Zhang, J. Wang, T. Wang, Effect of ethane and propane addition on acetylene production in the partial oxidation process of methane, Ind. Eng. Chem. Res. 56 (18) (2017) 5174–5184.

[83] Y. Liu, Q. Zhang, T. Wang, Detailed chemistry modeling of partial combustion of natural gas for coproducing acetylene and syngas, Combust. Sci. Technol. 189 (5) (2017) 908–922.

[84] S. Wang, G. Lu, G.J. Millar, Carbon dioxide reforming of methane to produce synthesis gas over metal-supported catalysts: state of the art, Energy Fuel 10 (4) (1996) 896–904.

[85] D. Chen, et al., Deactivation during carbon dioxide reforming of methane over Ni catalyst: microkinetic analysis, Chem. Eng. Sci. 56 (4) (2001) 1371–1379.

[86] J.M. Ginsburg, et al., Coke formation over a nickel catalyst under methane dry reforming conditions: thermodynamic and kinetic models, Ind. Eng. Chem. Res. 44 (14) (2005) 4846–4854.

[87] J. Guo, H. Lou, X. Zheng, The deposition of coke from methane on a $Ni/MgAl_2O_4$ catalyst, Carbon 45 (6) (2007) 1314–1321.

[88] A. Smitz, T. Yoshida, Minimizing carbon deposition on Ni/SiO_2 catalyst in CH_4-CO_2 reforming, in: 4th International Conference on Greenhouse Gases Control Technologies, Interlaken, Switzerland, 1998.

[89] G. Jones, et al., First principles calculations and experimental insight into methane steam reforming over transition metal catalysts, J. Catal. 259 (1) (2008) 147–160.

[90] J.R. Rostrup-Nielsen, Production of synthesis gas, Catal. Today 18 (4) (1993) 305–324.

[91] V.S. Arutyunov, O.V. Krylov, Oxidative conversion of methane, Russ. Chem. Rev. 74 (12) (2005) 1111.

[92] J.R. Rostrup-Nielsen, J. Sehested, J.K. Nørskov, Hydrogen and synthesis gas by steam-and CO_2 reforming, Adv. Catal. 47 (2002) 65–139.

[93] Y. Kathiraser, et al., Kinetic and mechanistic aspects for CO_2 reforming of methane over Ni based catalysts, Chem. Eng. J. 278 (2015) 62–78.

[94] N. Gokon, et al., Kinetics of CO_2 reforming of methane by catalytically activated metallic foam absorber for solar receiver-reactors, Int. J. Hydrogen Energy 34 (4) (2009) 1787–1800.

[95] B. Steinhauer, et al., Development of Ni-Pd bimetallic catalysts for the utilization of carbon dioxide and methane by dry reforming, Appl. Catal. A Gen. 366 (2) (2009) 333–341.

[96] Y. Benguerba, et al., Computational fluid dynamics study of the dry reforming of methane over Ni/Al_2O_3 catalyst in a membrane reactor. Coke deposition, Kinet. Catal. 58 (3) (2017) 328–338.

[97] G.D. Wehinger, T. Eppinger, M. Kraume, Detailed numerical simulations of catalytic fixed-bed reactors: heterogeneous dry reforming of methane, Chem. Eng. Sci. 122 (2015) 197–209.

[98] G.D. Wehinger, C. Fütterer, M. Kraume, Contact modifications for CFD simulations of fixed-bed reactors: cylindrical particles, Ind. Eng. Chem. Res. 56 (1) (2017) 87–99.

[99] G.D. Wehinger, F. Klippel, M. Kraume, Modeling pore processes for particle-resolved CFD simulations of catalytic fixed-bed reactors, Comput. Chem. Eng. 101 (2017) 11–22.

[100] T. Eppinger, et al., A numerical optimization study on the catalytic dry reforming of methane in a spatially resolved fixed-bed reactor, Chem. Eng. Res. Des. 115 (2016) 374–381.

[101] G.D. Wehinger, et al., Investigating dry reforming of methane with spatial reactor profiles and particle-resolved CFD simulations, AICHE J. 62 (12) (2016) 4436–4452.

[102] N. Jurtz, M. Kraume, G.D. Wehinger, Advances in fixed-bed reactor modeling using particle-resolved computational fluid dynamics (CFD), Rev. Chem. Eng. 35 (2) (2019) 139–190.

[103] G.D. Wehinger, T. Eppinger, M. Kraume, Evaluating catalytic fixed-bed reactors for dry reforming of methane with detailed CFD, Chem. Ing. Tech. 87 (6) (2015) 734–745.

[104] H. Zhang, J.D. Smith, Simulation of structural effects of monolith catalyst on methane dry reforming in a solar thermochemical reactor, in: 2019 Spring Meeting and 15th Global Congress on Process Safety, AIChE, 2019.

[105] S. Lee, H. Lim, The power of molten salt in methane dry reforming: conceptual design with a CFD study, Chem. Eng. Process. Process Intensif. 159 (2021) 108230.

[106] S. Lee, H. Lim, Variation of the number of heat sources in methane dry reforming: a computational fluid dynamics study, Int. J. Chem. Eng. 2021 (2021).

[107] H.-Y. Zheng, et al., Analysis of structure-induced performance in photothermal methane dry reforming reactor with coupled optics-CFD modeling, Chem. Eng. J. 428 (2022) 131441.

CHAPTER 4

Reforming process design and modeling: Steam, dry, and autothermal reforming

Mohammad Farsi

Department of Chemical Engineering, School of Chemical and Petroleum Engineering, Shiraz University, Shiraz, Iran

1. Introduction

The synthesis gas as a mixture of hydrogen, carbon monoxide species is a key intermediate feedstock in the chemical processes and is used to produce different products such as methanol, ammonia, dimethyl ether, ethylene glycol, synthetic fuels, methyl methacrylate, and various aldehydes and alcohols [1]. Commercial methods for producing syngas include steam reforming, partial oxidation, and autothermal reforming of hydrocarbons, as well as coal and biomass gasification. Although different hydrocarbon feedstocks are used in the reforming processes, methane is more attractive due to its high hydrogen-to-carbon ratio, low cost, availability, and low environmental impacts [2].

Steam reforming is an endothermic equilibrium limited reaction that light hydrocarbons and steam react over the catalyst at temperature range of 600–900°C and pressure range of 3–25 bar, depending on the applied technology [3]. Steam reforming as a mature technology is the most common and cost-effective catalytic route for the production of hydrogen-rich syngas. The first patents on steam methane reforming were published by BASF company in 1926s, and the reforming process was commercialized in the 1930s [4]. Currently, the ABB Lummus, Exxon Mobil, Halder Topsoe, Howe-Baker, ICI, KTI, Linde, Lurgi, M.W. Kellogg, Toyo, and Uhde are the main licensors of steam reforming technology in the world. Globally, steam reforming contributes up to 50% of the syngas production units in the world [5]. Despite the high quality of produced syngas and mature technology, the main bottlenecks in the steam reforming processes are low catalyst lifetime, impossibility to scale down, excessive energy input, and high capital cost. In the partial oxidation process, a substoichiometric mixture of methane and oxygen reacts to produce syngas with hydrogen-to-carbon-monoxide ratio of 2 [6]. The noncatalytic partial oxidation process is designed at temperature and pressure ranges of 1150–1500°C and 25–80 bar. Applying a catalyst in partial oxidation could moderate the operating temperature to 700–1000°C [7]. Partial oxidation has a lower cost and complexity compared with steam

Advances in Synthesis Gas: Methods, Technologies and Applications. https://doi.org/10.1016/B978-0-323-91879-4.00003-5

reforming. The disadvantages of partial oxidation are poor process controllability, the need to plan oxygen units, and high explosion risk. Autothermal reforming as the integration of steam reforming and noncatalytic partial oxidation is a heat-efficient route to produce high-quality syngas [8]. It was developed by Haldor Topsøe and Société Belge de l'Azote in the 1950s [9]. Currently, the main licensors of autothermal reforming technology are Air Liquide, Haldor Topsøe, Lurgi, ICI, and Foster Wheeler. Autothermal reforming has a higher hydrogen yield compared with partial oxidation and is more flexible in the terms of feed, heat, and operating condition compared with steam reforming [10]. Based on the economic analysis, the capital costs of an autothermal reforming plant are lower than that of a steam reforming plant up to 25% [11]. Dry reforming is a catalytic route to produce synthesis gas with hydrogen-to-carbon-monoxide ratio 1 at the temperature range of 1000–1200°C [12]. Although the technologies for autothermal and steam reforming are already mature, the dry reforming technology is at the research step, and few commercial plants are designed in Germany and Denmark [13]. The Calcor process licensed by Caloric and Sulfur-passivated reforming process commercialized by Haldor-Topsoe are the only available dry reforming units in the world that are used to produce carbon-monoxide-rich syngas [14]. Table 1 presents a comprehensive comparison between operating conditions and key design parameters of steam, autothermal, and dry reforming processes.

Globally, the main bottlenecks of reforming processes are the deactivation of catalyst by oxidation, sintering, poisoning, and coking [16,17]. Since methane molecules react on the metal surface, the reduced active phase could be oxidized by an oxidizing agent [18]. The nickel-based catalysts are susceptible to oxidation, and the oxidation rate is enhanced at high steam-to-carbon ratios. Sintering phenomenon is the agglomeration of dispersed active phase metals or crystallites to larger ones [19]. High temperature and steam-to-carbon ratio could enhance the rate of catalyst sintering in the reforming process [20]. Poison molecules are

Table 1 Comparison between steam, autothermal, and dry reforming processes [15].

	Steam reforming	Autothermal reforming	Dry reforming
Reactor type	Direct Fired Heater	Conventional Tubular	Conventional Tubular
Catalyst	Nickel-based	Nickel-based	Nickel-based
Auxilary facility	Desulfurization unit	Oxygen unit	Desulfurization unit
Temperature (°C)	600–900	850–1300	1000–1200
Pressure (bar)	3 – 25	20 – 70	10
H_2 to CO ratio	3	2.5	1
CH_4 conversion (%)	Up to 95	Up to 100	Up to 90
Steam consumption	High	Medium	No
Oxygen consumption	No	Low	No
Plant capacity	Medium	Large	Unknown
CO_2 emission	High[a]	Low	Low

[a]Steam reforming is the reference point.

chemically deposited on the catalyst surface, reducing the catalytic active site in the poisoning phenomenon. The poisoning rate of reforming catalyst is increased at low temperature and high-pressure conditions. In general, decreasing the operating temperature from 800°C to 500°C reduces the concentration of sulfur required for the poisoning of reforming catalyst up to 500 times. [21,22]. The coking is carbon condensation and deposition on the catalyst surface and results in catalyst deactivation, bed clogging, and pressure drop. The coke deposition on the catalyst is promoted when O/C and H/C ratios in the feed stream are low. In the dry reforming with stoichiometry feed, O/C and H/C ratios are 1 and 2, respectively. In the steam reforming of methane, O/C and H/C ratios are 1 and 6, and for partial oxidation approach 1 and 4, respectively. It concludes that the dry reforming process is more ready for coking against the steam and autothermal reforming. Compared with the commercial nickel-based catalyst, the noble metals such as Rh, Ru, Pt, Pd, and Ir are much more stable and robust against coking due to the difficulties of dissolving carbon in them [23]. However, the cost and availability have limited the application of nobel catalysts in the syngas production units [24]. In addition, applying Ni-based bimetallic catalysts led to lower coke deposition, better metal dispersion, smaller particle size, and the synergic interaction between Ni and promoter.

2. Reaction kinetic

2.1 Steam reforming

The steam reforming includes three reversible catalytic reactions including steam reforming of methane and water gas shift reactions as [25]:

$$CH_4 + H_2O \leftrightarrow CO + 3H_2 \tag{1}$$

$$CH_4 + 2H_2O \leftrightarrow CO_2 + 4H_2 \tag{2}$$

$$CO + H_2O \leftrightarrow CO_2 + H_2 \tag{3}$$

Based on the information offered in the literature, the proposed kinetic model by Xu and Froment could predict the rate of reforming reactions on commercial nickel-supported Al_2O_3 catalysts accurately [26]. In the temperature range of 557–700°C, the important side reactions in the reforming process are methane decomposition and Boudouard reactions.

$$CH_4 \leftrightarrow C + 2H_2 \tag{4}$$

$$2CO \leftrightarrow CO_2 + C \tag{5}$$

The most common catalyst used in steam reforming is nickel supported on α-alumina, magnesia, calcium aluminate, or magnesium aluminate [27,28]. In addition, the Group VIII non-noble metals are active and could progress the steam reforming reactions. However, iron is quickly oxidized, cobalt can't withstand steam, and precious metals such as rhodium, ruthenium, platinum, and palladium are too expensive. The presence of additives such as potassium could enhance the stability of nickel-based catalysts by reducing the coking rate [29].

2.2 Dry reforming

In the dry reforming process, carbon dioxide is used as an oxidizing agent to produce syngas from methane in the temperature range of 1000–1200°C. The dry reforming and reverse water gas shift reactions are:

$$CH_4 + CO_2 \leftrightarrow 2CO + 2H_2 \tag{6}$$

$$CO_2 + H_2 \leftrightarrow CO + H_2O \tag{7}$$

Although applied catalysts in the dry reforming process are primarily Rh, Ru, Pt, Co, and Ni [30], the nickel supported on metal oxides such as Al_2O_3, MgO, CeO_2, or La_2O_3 is more attractive due to availability and cost [31]. Promoting Ni catalysts with noble metals such as Rh, Pt, Pd, or Ru increases the activity and stability of dry reforming catalyst [32]. In addition, the use of basic compounds such as K_2O, MgO, CaO, La_2O_3 could decrease the carbon build-up [33]. The basic promoters enhance the adsorption and dissociation of carbon dioxide on the catalyst and result in lower carbon formation [34].

2.3 Autothermal reforming

Autothermal reforming of methane is a combination of adiabatic steam reforming and noncatalytic partial oxidation in series. The partial oxidation of methane as an exothermic reaction is the substoichiometric oxidation of methane.

$$CH_4 + \frac{1}{2}O_2 \leftrightarrow CO + 2\,H_2 \tag{8}$$

$$CH_4 + 2O_2 \leftrightarrow CO_2 + 2\,H_2 \tag{9}$$

$$H_2 + \frac{1}{2}O_2 \leftrightarrow H_2O \tag{10}$$

$$CO + \frac{1}{2}O_2 \leftrightarrow CO_2 \tag{11}$$

In general, the applied catalysts in steam reforming are used in the catalytic section of the autothermal reforming process [35].

3. Process design

3.1 Steam reforming

The steam reformer is a fired reactor consisting of a catalyst-filled coil tube hanged in a firebox. Although the conventional packed-bed coil reactors are used in the industrial reforming processes, the fluidized bed [36], sorption-enhanced [37], hydrogen permselective membrane [25,38] thermally coupled [39], and chemical looping configurations [40] have been proposed to apply in new designs. Traditional furnaces include a radiant part with burners and reforming tubes, as well as a convection section that recovers flue gas waste heat. The steam generation and feed preheating by flue gas stream in the economizer enhance the thermal efficiency of the process up to 85% [15]. Since the convective heat resistance in the coil limits the heat transfer from the skin toward the catalyst, applying the structured catalyst, tubes with small diameters, and passive flow baffles could enhance the rate of heat transfer [41]. New structures are focused on simplification of reformers and improving the heat transfer to achieve a simpler mechanical layout, easier maintenance and scale-up, and lower capital cost. The tube geometry, length, diameter and number of tubes, and burner location are key design factors in the reformers [42]. From a complexity viewpoint, increasing the tube length is more economical and effective against the number of tubes. However, tube length is limited by the risk of bending, vibration, and pressure drop. Since the reforming reactors operate under high-pressure conditions, increasing the pipe diameter leads to the use of thicker pipes. On the other hand, when the tubes with large diameters are applied in the reformer, up to $100°Ccm^{-1}$ temperature gradients could be developed near the wall in poor designs. Thus, as well as the higher capital cost of large diameter tubes, the located catalyst at the wall might be deactivated by sintering. Applying small-radius tubes could enhance the rate of heat transfer from the tube skin to centerline and decrease the radial temperature gradient in the reformer tube. The uniform and maximum heat flux could be achieved by optimal burner location and tube geometry. In general, a nonuniform temperature distribution in the furnace decreases firebox efficiency and energy loss [43]. There are four types of reformers based on the burner configuration including bottom-fired, terrace-wall fired, top-fired, and side-wall fired. Fig. 1 shows the typical furnace configurations in the steam reforming process. The bottom-fired type has an almost constant heat flux along the tube. The terrace-wall fired reformer is a modification of the bottom design, having a slightly lower tube wall temperature. The top-fired reformers have the highest heat flux and skin temperature. The side-wall fired design allows better regulation of the skin temperature and is more flexible in design and operation.

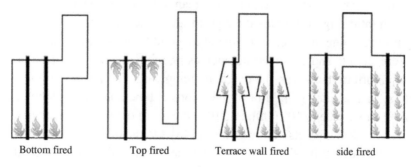

Bottom fired   Top fired   Terrace wall fired   side fired

Fig. 1
Typical furnace configurations in the steam reforming process.

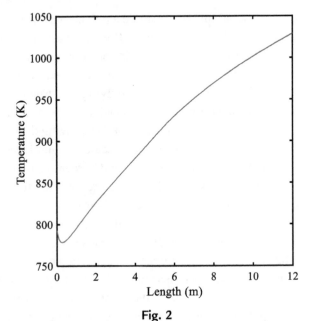

Fig. 2
Temperature profile along the coil tube in an industrial reformer [44].

Fig. 2 depicts the temperature profile along the coil tube in an industrial reforming reactor equipped with bottom and side burners [44]. The reforming reactions are endothermic and produced heat in the firebox supplies the heat of reactions. The rate of heat consumption dominates the rate of heat transmission from the firebox in the early stages of the proposed reformer, and the temperature of the gas stream steadily falls. Indeed, the transferred heat from the firebox cannot supply the heat of reforming reactions at the initial part of the reactor and temperature decreases gradually. Decreasing the reactant concentration reduces the rate of reforming reactions at the second part of the reformer and transferred heat from the firebox could supply the heat of reactions and temperature increases along the coil tube.

The main operating variables that influence the steam reforming are temperature, pressure, and steam-to-carbon ratio. Fig. 3 presents the equilibrium concentration of components in the methane steam reforming at different temperature conditions [45]. Because the reforming processes are reversible and endothermic, increasing the temperature improves the methane conversion equilibrium. However, the operating temperature of steam reformers is limited by various factors, such as the creep limit of the reformer tube material. Although the design lifetime of a reformer tube is 100 000h, a 20K increase in the tube wall temperature over the design temperature decreases its life by one-half [46]. On the other hand, the high temperature could degrade the catalyst by sintering. The lifetime of reforming tubes depends on the operating temperatures; so an increase in the skin temperature up to 20K can decrease the tube

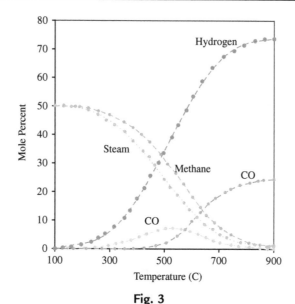

Fig. 3

Equilibrium concentration of components in the methane steam reforming [45].

lifetime by half [47]. According to Le Chatelier's principle, increasing steam-to-methane ratio and operating at low-pressure conditions increase the equilibrium conversion of methane. Since the steam reforming reactions are accompanied by a volume expansion, high-pressure conditions reduce the equilibrium conversion of methane. The steam-to-methane ratio affects fuel consumption, process efficiency, product quality, and catalyst lifetimeIn practice, the molar steam-to-methane ratio in the feed stream is managed in the 2.5–4.5 range to minimize coke formation on the catalyst surface and to save operating expenses [48,49].

3.2 Dry reforming

Fig. 4 shows the process flowsheet of the dry reforming unit with CO_2 recycling. The presented dry reforming process consists of two main sections including methane conversion and syngas purification stages. In the syngas purification stage, the unreacted CO_2 and produced water are separated and CO_2 is recycled to the conversion section. A suitable conversion configuration is the substitution of the fired heater and catalytic tubular reactor in the presented process with a fired reactor.

The main operating variables that influence the dry reforming of methane are temperature, pressure, and the molar ratio of carbon dioxide to methane in the feed stream [45]. Fig. 5 presents the equilibrium concentration of components in the dry reforming at different temperatures. Since the dry reforming is a reversible and endothermic reaction, an increase in temperature enhances the equilibrium conversion of methane and shifts the reforming reaction

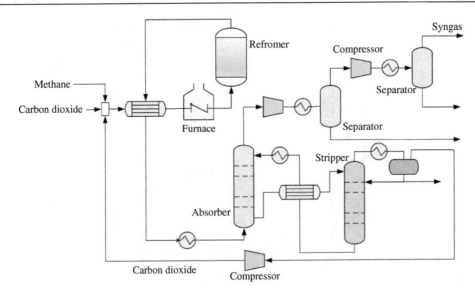

Fig. 4

The flowsheet of dry reforming with CO_2 recycle [50].

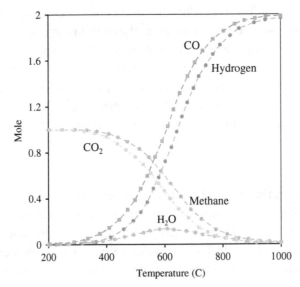

Fig. 5

Thermodynamic equilibrium plots for dry reforming at 1 atm [45].

toward hydrogen and carbon monoxide production. Although the dry reforming process is often run above 1000°C to minimize the coke formation on the catalyst, high temperature is one of the main drawbacks of dry reforming and leads to catalyst deactivation by sintering. According to Le Chatelier's principle, increasing carbon-dioxide-to-methane ratio and operating at low pressure increase the equilibrium conversion of methane.

3.3 Authothermal reforming

Autothermal reforming is a combination of steam reforming and partial oxidation in a single catalytic reactor or two reformers in series. The reformer consists of three zones including the burner, combustion zone, and catalytic zone [51]. Partial oxidation supplies the heat required for the endothermic reforming reactions in the catalytic section. In the catalytic section, steam and methane react and syngas is produced. It should be noted that reactant mixing is an important factor in autothermal reformers and affects the catalyst stability and quality of produced syngas. Incomplete mixing can lead to hot spots in the catalyst bed or on the refractory walls of the reformer. Possible mixing schemes include applying the bluff bodies, swirling, and acoustic enhancement. The risk of soot formation in an autothermal reformer depends on feed composition, temperature, pressure, and especially burner design. In the current technologies, setting the molar ratio of steam to carbon above 1.5 could minimize the soot formation.

The main operating variables that influence the autothermal reforming process are stream and oxygen to feed ratios in the feed stream, operating pressure, and temperature. Similar to steam reforming, the reforming in autothermal reformer is also favored by low pressures. Although partial oxidation supplies the required heat for steam reforming, applying an oxygen-rich feed stream increases methane oxidation and could reduce the H_2 to CO_2 ratio in the product. In addition, decreasing oxygen-to-hydrocarbon ratio leads to the incomplete conversion of methane.

4. Process modeling

4.1 Equilibrium approach

To compute the equilibrium concentration of species in a chemical system, two major thermodynamic frameworks are created. Stoichiometric analysis, which employs the equilibrium constants of preset reactions, is one method. If the system has a high number of reactions, this approach would be complicated. A nonstoichiometric approach, which is subject to the mass balance equations, minimizes the Gibbs free energy in the system. In the nonstoichiometric approach, the equilibrium concentration of species is determined by minimizing the total Gibbs free energy of the system at a specified pressure and temperature [52]. The total Gibbs free energy is given by the sum of the chemical potential of all chemical species as:

$$G_t = \sum_{i=1}^{N} n_i \mu_i \qquad (12)$$

The chemical potential of i species could be calculated by

$$\mu_i = G_i^o + RTln\left(\frac{f_i}{f_i^o}\right) \qquad (13)$$

The total Gibbs free energy can be expressed as:

$$G_t = \sum_{i=1}^{N} n_i G_i^o + \sum_{i=1}^{N} n_i RT \ ln \left(\frac{y_i P \emptyset_i}{P^o} \right) \tag{14}$$

According to the mass conservation law, the number of atoms at the initial and final states remains constant.

$$\sum_{i}^{N} n_i a_{i,j} - N_j = 0 \tag{15}$$

Where $a_{i,j}$ is the number of atoms of element j in the species of i, and N_i is the total number of atoms of element j. The equilibrium distribution of components is calculated by minimizing the Gibbs free energy of the system considering the mass balance equation as a constraint. The objective and constraints could be combined by applying the Penalty method as:

$$j = \sum_{i=1}^{N} n_i G_i^o + \sum_{i=1}^{N} n_i RT ln \left(\frac{y_i P \emptyset_i}{P^o} \right) + \left(\sum_{i}^{N} n_i a_{i,j} - N_j \right)^2 \tag{16}$$

The penalty approach is a good way to turn a constraint issue into an unconstrained problem. To find the optimal operating condition, the differential of the penalty function with respect to n_i is set to be zero. In general, the equilibrium approach is the simplest method to simulate the partial oxidation section in the autothermal reforming process [53].

4.2 Kinetic approach

Although the equilibrium model is simple, it could not predict the intermediate products and transient conditions. The results of the kinetic model approach equilibrium at the infinite time and could predict the process condition and intermediate species at a specified time. Thus, the kinetic models are more useful to simulate the reformers. The models used to characterize reformers may be divided into two categories: homogeneous models and heterogeneous models. [54]. The homogeneous models ignore the difference between catalyst and gas phases, while the phases are modeled separately considering the heat and mass transfer between them in the heterogeneous models. If the radial and axial heat and mass gradients have been accounted for in the models, the model becomes two-dimensional.

In general, external and internal heat and mass diffusions and chemical kinetics are the major limiting mechanisms in commercial reformers. Reactant diffusion from bulk onto the catalyst surface is known as the external diffusion, while internal diffusion is the mass transfer from the catalyst surface inside the catalyst. Since considering all of the transfer limitations and two-dimensional modeling makes the developed model complex and difficult to solve, the importance of resistances is evaluated by dimensionless numbers as:

- Nonadiabatic reactors can result in significant radial-temperature gradients near the wall due to heat transfer. In addition, the radial gradient of temperature and velocity could produce radial-concentration gradients. In general, the radial gradients are considerable when the catalyst-to-reactor diameter is small, the flow regime is laminar, and hot spots occur [55].
- Although characterization of flow regime in packed-bed by Reynolds number is not well defined due to the complex flow pattern in the packed-beds, the turbulent flow could be assumed when $Re > 10$. In most traditional reformers, the flow regime around the catalyst is turbulent.
- Mass and heat are transported from the bulk fluid to the porous catalyst's exterior surface during catalytic reactions [56]. The external heat and mass transfer resistances can be neglected when $N_{MR} < 0.15$.
- The reactants then diffuse into the catalyst from the outside, and the reaction occurs over the pores. The Weisz-Prater criterion is used to determine the importance of internal diffusion. In general, there is no concentration gradient inside the catalyst pellet when $N_{WP} \gg 1$. In most traditional reforming catalysts, the concentration gradient inside the catalyst limits the reaction and should be applied in the model [44].
- In the catalytic reactors, the reactions occur on the active site of the catalyst. The importance of temperature gradient inside the catalyst could be verified by the Biot number. When Bi number is sufficiently small, $Bi \ll 1$, temperature gradient in the catalyst could be ignored [57]. In most traditional reforming catalysts, inside the catalyst, the temperature gradient is small. [44].
- The axial diffusion of mass and heat could be ignored compared with the convection mechanism, when the heat and mass Péclet number are sufficiently large, $Pe \gg 1$, in the axial flow reformers [58]. In most traditional reformers, the axial diffusion of mass and heat along the fluid direction is negligible [44].

In the axial flow reformers, to create a mathematical model, an element in the reactor's axial direction is chosen, and mass and heat balance equations for the gas and solid phases are created. Considering the validity of one-dimensional and heterogeneous modeling with plug flow regime and lumped body assumptions, the steady-state mass and energy balance equations in the gas phase are:

$$-\frac{\partial(u_s C_i)}{\partial z} - k_{g_i} a_v (C_i - C_i^s) = 0 \tag{17}$$

$$-m^o C_P \frac{\partial T}{\partial z} - h_g a_v (T - T_s) + UA(T - T_f) = 0 \tag{18}$$

When the reformer is adiabatic such as the dry and catalytic section of autothermal reformers, the third part in the heat balance equation is neglected. The solid-phase mass and energy balance equations are as follows:

$$a_v k_{g_i} \left(C_i - C_i^s \right) - \rho_B a_t \sum_{j=1}^{N} r_{i,j} \eta_j = 0 \tag{19}$$

$$a_v h_g (T - T_s) - \rho_B a_t \sum_{j=1}^{N} r_j \eta_j \left(-\Delta H_j \right) = 0 \tag{20}$$

The total mole balance could be explained as

$$-\frac{\partial (u_s C_t)}{\partial z} - \sum_{i=1}^{N} k_{g_i} a_v \left(C_i - C_i^s \right) = 0 \tag{21}$$

When the Reynolds number of particle is in the range of 1–1000, the Ergun equation could be used to compute the pressure drop in the reactor [59]:

$$\frac{dP}{dz} = -\frac{150\,\mu_s (1 - \varepsilon_b)^2}{d_p^2 \varepsilon_b^3} u_s - \frac{1.75 (1 - \varepsilon_b)\rho_g}{d_p \varepsilon_b^3} u_s^2 \tag{22}$$

The effectiveness factor is used in the model to account for the influence of internal mass transfer resistance inside the catalyst. The effectiveness factor is defined as the ratio of reaction rate considering mass transfer resistance in the particle to the reaction rate applying the surface concentration in the particle. The effectiveness factor of catalyst is calculated as:

$$\eta_j = \frac{D_e A_C \left. \frac{dC_{i,j}}{dr} \right|_{r=R}}{r_{i,j} \big|_{C_s} \left(\frac{1}{6} \pi d_p^3 \right)} \tag{23}$$

The concentration gradient on the catalyst surface is determined by applying a mass balance inside the catalyst. The mass balance equation for the spherical catalyst is:

$$\frac{d^2 C_i}{dr^2} + \frac{2}{r} \frac{dC_i}{dr} - \frac{\rho_c}{D_e} \sum_{j=1}^{N} r_{i,j} = 0 \tag{24}$$

The boundary conditions of the mass balance equation for the spherical catalyst are

$$\left. \frac{dC_i}{dr} \right|_{r=0} = 0 \tag{25}$$

$$D_e \left. \frac{dC_i}{dr} \right|_{r=R} = k_g \left(C_i^s - C_i^g \right) \tag{26}$$

In general, modeling the combustion chamber in the steam and partial oxidation reformers is a very difficult task due to the complexity of the combustion mechanism and reactions, and the participation of gas, and equipment in the emission and absorption of heat. The well-stirred combustion chamber, long furnace, zone method, Monte Carlo, and Computational Fluid Dynamic models are common approaches to simulate the firebox [60]. In general, applying a large number of in the firebox and adequate mixing in the furnace result in the uniform temperature in the furnace.

4.2.1 Numerical solution

Although the conventional adiabatic tubular reactors are applied in the autothermal and dry reforming processes, the furnace reactors are used in the steam reforming of methane. The heat created in the firebox is transported to the outside wall of the coil tubes by radiation and convection processes, conducted down the tube thickness, and delivered to the feed via convection mechanisms in the steam reforming of methane. The mathematical model of tubular reactors can be solved straightforwardly by common numerical methods such as explicit and implicit Euler and Runge-Kutta, while the model of furnace reactor needs a try and error procedure. The below procedure could be used to solve the furnace reactor model.

- Initialization and set skin temperature and solution error
- Repeat until stopping criteria
- Determine the heat transfer rate between the chamber and the coil, Q_O
- Use a numerical approach to solve the balancing equations in the coil tube
- Determine the heat transfer rate between the gas and the coil, Q_I
- Calculate the solution error as the absolute difference of Q_O and Q_I
- If solution error is acceptable go to the next element, else correct skin temperature

4.3 Computational fluid dynamics simulation

In general, considering radiation and convection in the firebox, the geometry of and position of burners and coil, flow hydrodynamic, reaction kinetics, and heat and mass transfer resistances led to developing a comprehensive complex model for reformers [47]. Solving these models is computationally expensive and time-consuming. Computational fluid dynamics is a powerful discipline to model complex problems and solve the developed comprehensive models. In the first step, the geometric domain of the problem is determined and a suitable mesh structure is generated. The mesh quality is the most critical issue for accurate and successful CFD simulation. Due to the axisymmetric geometry of the reforming tube, a two-dimensional axisymmetric geometry and its corresponding mesh structure could be employed [61]. Then, the governing equations are selected and the developed model is solved numerically. The mass and momentum transport through the porous region is described by the continuity and Navier-Stokes equations for Newtonian fluids [62]. To estimate the pressure drop across the catalytic bed, the Ergun moment equation is useful. The reactions are incorporated into the system model with a kinetic model approach. The CFD simulation presents detailed results about the velocity, pressure, temperature, and concentration in the reformers and helps the designer to investigate the possibility of hotspot formation, fouling, and phase change. Based on the literature data, computational fluid dynamics is successful to simulate the industrial reformers [63,64].

5. Process optimization

5.1 Steady-state optimization

In general, process optimization is a specific discipline to adjust the operating condition of a process at the desired level based on a predefined objective without violating constraints and bounds. The main object in the reforming process is enhancing the syngas production capacity by manipulation of available inputs. The temperature of feed and furnace and the steam-to-methane ratio are the main manipulated variables in the steam reforming process, while the feed temperature, pressure, and carbon-dioxide-to-methane ratio are manipulated variables in the dry reforming. In autothermal reforming, the feed temperature, pressure, steam-to-methane ratio, and oxygen-to-methane ratio are the main manipulated variables. The steam-to-methane ratio is adjusted in the range of 2.5–4.5 to reduce coke accumulation on the catalyst surface, lowering operating limitations and costs [49]. In addition, the oxygen-to-methane ratio in commercial autothermal reformers is set in the range of 0.25–0.35. Due to safety and material limitations, the upper bound of skin temperature is limited to 1300K in conventional reformers [65]. Since the produced syngas is used in different plants, the lower and upper bounds of hydrogen-to-carbon-monoxide ratio in the products are determined based on the final products. The programmed optimization problem could be handled by heuristic optimization methods such as genetic algorithm and differential evolution.

5.2 Dynamic optimization

Due to the deactivation of reforming catalyst and decreasing the syngas production rate during the process run time, the dynamic modeling of process considering a detailed activity model and real-time optimization of process condition to obtain the optimal trajectory of inputs is a practical solution to maintain the production capacity at the desired level. Due to the very slow rate of catalyst deactivation, the pseudo steady-state modeling is appropriate and accurate to simulate the commercial reforming reactors [66]. In the dynamic optimization problem, the dynamic trajectory of manipulated variables is determined to achieve maximum process lifetime, production capacity, and product quality. The optimum control is a dynamic optimization approach devised to discover the ideal dynamic trajectory of choice variables as an extension of the calculus of variations. Based on the literature, applying the optimal trajectory of firebox temperature and steam-to-methane ratio in an industrial steam reformer could increase hydrogen production up to 11.6%.

6. Conclusion and future outlook

Generally, the process modeling and simulation are a useful tool in design, analysis, and optimization of chemical processes. Since the steam and autothermal reforming technologies are already mature, the process modeling is used to solve the process bottlenecks and optimize the process condition. However, the dry reforming is at the research and development stage and the modeling could be used in the design stage. In general, equilibrium and kinetic-based models are two different approaches to simulate the reforming process. The kinetic approach incorporates the reaction kinetics and transfer resistances in the model and could predict the process condition and intermediate products at a specified time, while the equilibrium stage could predict the final equilibrium condition. Due to the possibility of considering detailed transfer mechanism, geometry, fluid hydrodynamic, and reaction kinetics, computational fluid dynamics is a useful discipline to simulate the commercial reformers. In general, the modification of reformer structure to enhance the thermal efficiency, simplify the mechanical layout, maintenance, and scale-up, and decrease the capital cost is an active research area.

Abbreviations and symbols

A cross section (m^2)
C_i concentration of species (mol m^{-3})
C_p special heat capacity (kJ $kmol^{-1}K^{-1}$)
C^s concentration of species on surface catalyst (mol m^{-3})
ΔH_j heat of reaction j (kj $kmol^{-1}$)
d_p catalyst particle diameter (m)
ε_b packed-bed porosity
η_j effectiveness factor of reaction j
h_g heat transfer coefficient (J $s^{-1}m^{-2}K^{-1}$)
$k_{g,i}$ mass transfer coefficient for species (m s^{-1})
μ_g gas viscosity (pa s^{-1})
P total gas pressure (bar)
r_i rate of species i (kmol $kg^{-1}h^{-1}$)
ρ density (kg m^{-3})
T_s catalyst temperature (K)
U overall heat transfer coefficient (J $s^{-1}m^{-2}K^{-1}$)
u_s superficial gas mixture velocity (m s^{-1})
z axial direction (m)

References

[1] V.H.C. Manaças, Advanced Steady-State Modelling and Optimisation of Natural Gas Reforming Reactors, University of Lisbon, Portugal, 2013.

[2] H.-S. Roh, I.-H. Eum, D.-W. Jeong, Low temperature steam reforming of methane over Ni-Ce $(1 - x)$ Zr (x) O 2 catalysts under severe conditions, Renew. Energy 42 (2012) 212–216.

[3] B. Zhang, L. Wang, R. Li, Bioconversion and chemical conversion of biogas for fuel production, in: Advanced Bioprocessing for Alternative Fuels, Biobased Chemicals, and Bioproducts, 2019, pp. 187–205.

[4] P.L. Spath, D.C. Dayton, Preliminary Screening—Technical and Economic Assessment of Synthesis Gas to Fuels and Chemicals with Emphasis on the Potential for Biomass-Derived Syngas, National Renewable Energy Lab, Golden, CO.(US), 2003.

[5] A. Basile, S. Liguori, A. Iulianelli, Membrane reactors for methane steam reforming (MSR), in: Membrane Reactors for Energy Applications and Basic Chemical Production, Elsevier, 2015, pp. 31–59.

[6] J. Zhu, D. Zhang, K. King, Reforming of CH 4 by partial oxidation: thermodynamic and kinetic analyses, Fuel 80 (2001) 899–905.

[7] K. Hohn, L. Schmidt, Partial oxidation of methane to syngas at high space velocities over Rh-coated spheres, Appl. Catal. A Gen. 211 (2001) 53–68.

[8] S. Rowshanzamir, M. Eikani, Autothermal reforming of methane to synthesis gas: modeling and simulation, Int. J. Hydrogen Energy 34 (2009) 1292–1300.

[9] T. Christensen, I. Primdahl, Improve syngas production using autothermal reforming, Hydrocarb. Process. 73 (1994) (United States).

[10] G. Voitic, B. Pichler, A. Basile, A. Iulianelli, K. Malli, S. Bock, V. Hacker, Hydrogen production, in: Fuel Cells and Hydrogen, Elsevier, 2018, pp. 215–241.

[11] I. Dybkjaer, S. Winter Madsen, H. Topsøe, Advanced reforming technologies for hydrogen production, Hydrocarb. Eng. 3 (1997) 56–65.

[12] Y.T. Shah, T.H. Gardner, Dry reforming of hydrocarbon feedstocks, Catal. Rev. 56 (2014) 476–536.

[13] R. Kelling, G. Eigenberger, U. Nieken, Ceramic counterflow reactor for autothermal dry reforming at high temperatures, Catal. Today 273 (2016) 196–204.

[14] R. Carapellucci, L. Giordano, Steam, dry and autothermal methane reforming for hydrogen production: a thermodynamic equilibrium analysis, J. Power Sources 469 (2020) 228391.

[15] M. Korobitsyn, F.P. Van Berkel, G. Christie, SOFC as a Gas Separator, Final report, 2000.

[16] M.D. Argyle, C.H. Bartholomew, Heterogeneous catalyst deactivation and regeneration: a review, Catalysts 5 (2015) 145–269.

[17] D. Li, Y. Nakagawa, K. Tomishige, Methane reforming to synthesis gas over Ni catalysts modified with noble metals, Appl. Catal. A Gen. 408 (2011) 1–24.

[18] J.A. Moulijn, A. Van Diepen, F. Kapteijn, Catalyst deactivation: is it predictable?: what to do? Appl. Catal. A Gen. 212 (2001) 3–16.

[19] S. Aouad, M. Labaki, S. Ojala, P. Seelam, E. Turpeinen, C. Gennequin, J. Estephane, E. Abi Aad, A review on the dry reforming processes for hydrogen production: catalytic materials and technologies, Catal. Mater. Hydrog. Prod. Electro Oxid. React. Front. Ceram. Sci 2 (2018) 60–128.

[20] J. Sehested, J.A.P. Gelten, I.N. Remediakis, H. Bengaard, J.K. Nørskov, Sintering of nickel steam-reforming catalysts: effects of temperature and steam and hydrogen pressures, J. Catal. 223 (2004) 432–443.

[21] J. Rostrup-Nielsen, L.J. Christiansen, Concepts in Syngas Manufacture, World Scientific, 2011.

[22] P. Van Beurden, On the Catalytic Aspects of Steam-Methane Reforming, Energy Research Centre of the Netherlands (ECN), 2004. technical report I-04-003.

[23] Y. Song, E. Ozdemir, S. Ramesh, A. Adishev, S. Subramanian, A. Harale, M. Albuali, B.A. Fadhel, A. Jamal, D. Moon, Dry reforming of methane by stable Ni-Mo nanocatalysts on single-crystalline MgO, Science 367 (2020) 777–781.

[24] K.T. de Campos Roseno, R.M.d.B. Alves, R. Giudici, M. Schmal, Syngas production using natural gas from the environmental point of view, in: Biofuels-State of Development, 2018, pp. 273–290.

[25] M. Farsi, H. Shahhosseini, A modified membrane SMR reactor to produce large-scale syngas: modeling and multi objective optimization, Chem. Eng. Process. Process Intensif. 97 (2015) 169–179.

[26] J. Xu, G.F. Froment, Methane steam reforming, methanation and water-gas shift: I. Intrinsic kinetics, AIChE J. 35 (1989) 88–96.

[27] E. Meloni, M. Martino, V. Palma, A short review on Ni based catalysts and related engineering issues for methane steam reforming, Catalysts 10 (2020) 352.

[28] C.H. Bartholomew, R.J. Farrauto, Fundamentals of Industrial Catalytic Processes, John Wiley & Sons, 2011.

[29] G. Garbarino, F. Pugliese, T. Cavattoni, P. Costamagna, A study on CO2 Methanation and steam methane reforming over commercial Ni/calcium aluminate catalysts, Energies 13 (2020) 2792.

[30] R. Maurya, S.R. Tirkey, S. Rajapitamahuni, A. Ghosh, S. Mishra, Recent advances and future prospective of biogas production, in: Advances in Feedstock Conversion Technologies for Alternative Fuels and Bioproducts, Elsevier, 2019, pp. 159–178.

[31] D. Hu, C. Liu, L. Li, K.-L. Lv, Y.-H. Zhang, J.-L. Li, Carbon dioxide reforming of methane over nickel catalysts supported on TiO2 (001) nanosheets, Int. J. Hydrogen Energy 43 (2018) 21345–21354.

[32] M. García-Diéguez, I. Pieta, M. Herrera, M. Larrubia, L. Alemany, RhNi nanocatalysts for the CO2 and CO2+ H2O reforming of methane, Catal. Today 172 (2011) 136–142.

[33] Ş. Özkara-Aydınoğlu, A.E. Aksoylu, Carbon dioxide reforming of methane over Co-X/ZrO2 catalysts (X= La, Ce, Mn, Mg, K), Catal. Commun. 11 (2010) 1165–1170.

[34] A.F. Lucrédio, J.M. Assaf, E.M. Assaf, Reforming of a model biogas on Ni and Rh-Ni catalysts: effect of adding La, Fuel Process. Technol. 102 (2012) 124–131.

[35] Y. Ma, Y. Ma, G. Long, J. Li, X. Hu, Z. Ye, Z. Wang, C. Buckley, D. Dong, Synergistic promotion effect of MgO and CeO2 on nanofibrous Ni/Al2O3 catalysts for methane partial oxidation, Fuel 258 (2019) 116103.

[36] S. Hajizadeh, R. Habibi, G.R. Sotoudeh, N. Mostoufi, Modeling of a Fluidized-Bed Methane Steam Reformer, 2008.

[37] K. Tzanetis, C. Martavaltzi, A. Lemonidou, Comparative exergy analysis of sorption enhanced and conventional methane steam reforming, Int. J. Hydrogen Energy 37 (2012) 16308–16320.

[38] H.R. Shahhosseini, M. Farsi, S. Eini, Multi-objective optimization of industrial membrane SMR to produce syngas for Fischer-Tropsch production using NSGA-II and decision makings, J. Nat. Gas Sci. Eng. 32 (2016) 222–238.

[39] K.S. Patel, A.K. Sunol, Modeling and simulation of methane steam reforming in a thermally coupled membrane reactor, Int. J. Hydrogen Energy 32 (2007) 2344–2358.

[40] J. Adanez, A. Abad, F. Garcia-Labiano, P. Gayan, F. Luis, Progress in chemical-looping combustion and reforming technologies, Prog. Energy Combust. Sci. 38 (2012) 215–282.

[41] P. Erickson, J. Feinstein, M. Ralston, D. Davieau, I. Sit, Testing of methane steam reforming in a novel structured catalytic reactor providing flow impingement heat transfer, in: Proceedings of the AIChe 2010 Annual Meeting, Salt Lake City, Utah, 2010.

[42] P. Ferreira-Aparicio, M. Benito, J. Sanz, New trends in reforming technologies: from hydrogen industrial plants to multifuel microreformers, Catal. Rev. 47 (2005) 491–588.

[43] A. Kumar, M. Baldea, T.F. Edgar, O.A. Ezekoye, Smart manufacturing approach for efficient operation of industrial steam-methane reformers, Ind. Eng. Chem. Res. 54 (2015) 4360–4370.

[44] M. Sarkarzadeh, M. Farsi, M. Rahimpour, Modeling and optimization of an industrial hydrogen unit in a crude oil refinery, Int. J. Hydrogen Energy 44 (2019) 10415–10426.

[45] L. García, Hydrogen production by steam reforming of natural gas and other nonrenewable feedstocks, in: Compendium of Hydrogen Energy, Elsevier, 2015, pp. 83–107.

[46] D.A. Latham, K.B. McAuley, B.A. Peppley, T.M. Raybold, Mathematical modeling of an industrial steam-methane reformer for on-line deployment, Fuel Process. Technol. 92 (2011) 1574–1586.

[47] D. Latham, Mathematical Modelling of an Industrial Steam Methane Reformer, Queen's University Kingston, ON, 2008.

[48] J.R. Rostrup-Nielsen, J. Sehested, J.K. Nørskov, Hydrogen and synthesis gas by steam-and C02 reforming, ChemInform 34 (2002).

[49] M. Pen, J. Gomez, J.G. Fierro, New catalytic routes for syngas and hydrogen production, Appl. Catal. A Gen. 144 (1996) 7–57.

[50] W.L. Luyben, Design and control of the dry methane reforming process, Ind. Eng. Chem. Res. 53 (2014) 14423–14439.

[51] S. Ahmed, M. Krumpelt, Hydrogen from hydrocarbon fuels for fuel cells, Int. J. Hydrogen Energy 26 (2001) 291–301.

[52] H. Atashi, J. Gholizadeh, F.F. Tabrizi, J. Tayebi, S.A.H.S. Mousavi, Thermodynamic analysis of carbon dioxide reforming of methane to syngas with statistical methods, Int. J. Hydrogen Energy 42 (2017) 5464–5471.

[53] A. Tran, A. Aguirre, H. Durand, M. Crose, P.D. Christofides, CFD modeling of a industrial-scale steam methane reforming furnace, Chem. Eng. Sci. 171 (2017) 576–598.

[54] G. Froment, Analysis and Design of Fixed Bed Catalytic Reactors, in, ACS Publications, 1972.

[55] K.G. Gudekar, Modeling, Control, and Optimization of Fixed Bed Reactors, Texas Tech University, 2002.

[56] H.S. Fogler, Elements of Chemical Reaction Engineering, fourth ed., Pearson Education Limited, Edinburgh, 2014.

[57] J. Holman, Heat Transfer, ninth ed., 2002. New York.

[58] I. Alam, D.H. West, V. Balakotaiah, Transport effects on pattern formation and maximum temperature in homogeneous-heterogeneous combustion, Chem. Eng. J. 288 (2016) 99–115.

[59] E. Alpay, L. Kershenbaum, N. Kirkby, Pressure correction in the interpretation of microreactor data, Chem. Eng. Sci. 50 (1995) 1063–1067.

[60] A. Zamaniyan, Z.A. Taghi, Software development for simulation of reformer furnace, Iran. J. Chem. Chem. Eng. 25 (2006).

[61] M. Son, Y. Woo, G. Kwak, Y.-J. Lee, M.-J. Park, CFD modeling of a compact reactor for methanol synthesis: maximizing productivity with increased thermal controllability, Int. J. Heat Mass Transf. 145 (2019) 118776.

[62] S. Rowshanzamir, S. Safdarnejad, M. Eikani, A CFD model for methane autothermal reforming on Ru/γ-Al2O3 catalyst, Procedia Eng. 42 (2012) 2–24.

[63] J.G. Rebordinos, C. Herce, A. Gonzalez-Espinosa, M. Gil, C. Cortes, F. Brunet, L. Ferre, A. Arias, Evaluation of retrofitting of an industrial steam cracking furnace by means of CFD simulations, Appl. Therm. Eng. 162 (2019) 114206.

[64] A. Cherif, R. Nebbali, L. Nasseri, CFD study of ATR reaction over dual Pt-Ni catalytic bed, in: Advances in Renewable Hydrogen and Other Sustainable Energy Carriers, Springer, 2021, pp. 137–143.

[65] J. Rajesh, S.K. Gupta, G.T. Rangaiah, A.K. Ray, Multiobjective optimization of steam reformer performance using genetic algorithm, Ind. Eng. Chem. Res. 39 (2000) 706–717.

[66] M. Taji, M. Farsi, P. Keshavarz, Real time optimization of steam reforming of methane in an industrial hydrogen plant, Int. J. Hydrogen Energy 43 (2018) 13110–13121.

CHAPTER 5

Microreactor modeling and simulation for syngas production

Maryam Delshah[a], Shabnam Yousefi[a], Mohammad Amin Makarem[b], Hamid Reza Rahimpour[a], and Mohammad Reza Rahimpour[a]

[a]Department of Chemical Engineering, Shiraz University, Shiraz, Iran [b]Methanol Institute, Shiraz University, Shiraz, Iran

1. Introduction

Clean energy and alternative energy have become the main areas of investigation worldwide to develop sustainable energy. Significant areas of research and development include the production and purification of hydrogen, synthesis gas, and fuel processing for fuel cells [1]. Syngas is the short name of synthesis gas and includes H_2 and CO mixture. Syngas is a product of gasification, reforming, etc., processes using natural gas, biomass, biofuels, and feedstock. The initial use of syngas is the fuel's production, which means methanol and diesel fuel. Large amounts of waste gas are produced in some industrial processes with these specifications [2].

One of the applications of syngas is to feed hydrogen fuel cells straightly [3]. Hydrogen is easily extracted from gas and refined in fuel cells [2]. Some different routes have been proposed to convert syngas to transportation fuels. These routes are segmented as follows [4]:

I. Fischer-Tropsch synthesis;
II. Methanol;
III. Methanol-to-gasoline (MTG) and distillates (MOGD).

Fischer-Tropsch is the process by which diesel fuel production from synthesis gas relies on this process, a set of chemical reactions that convert CO and H_2 to liquid hydrocarbons [5]. Methane from landfills can be used as a primary feed for diesel fuel production. Because it is not derived from fossil fuels, it is technically considered biodiesel [6].

There are some trends in the application of synthetic gas. One is that a large portion of the hydrogen requirement has been provided from catalytic reforming units, but probably less existing due to limitations in the aromatic content of diesel and gasoline [7]. So, there is an

Advances in Synthesis Gas: Methods, Technologies and Applications. https://doi.org/10.1016/B978-0-323-91879-4.00002-3

increasing requirement for hydrogen plants to operate on hydrocarbon flows such as off-gases, natural gas, and naphtha. Another trend is converting natural gas into transportation fuels such as synthetic gasoline, methanol, DME, and diesel. Also, the choice of technology for syngas production depends on the scale of operation [8].

Syngas is obtained by different methods; for example, syngas could be produced from solid carbonaceous fuel and natural gas (>80% CH_4) using reforming reactions of methane [9]. However, the reforming reaction is not a gas supply but a molecular rearrangement. Partial oxidation of methane or natural gas is another way to produce syngas. Unlike the reforming reaction, partial oxidation is exothermal, which is also used to produce syngas from fuels [6].

Syngas production by steam reforming plays a central role in the chemical industry. The yield is high, but it is associated with significant investments [10]. Steam reforming of hydrocarbons is the leading industrial process used to produce hydrogen and synthesis gas [4]. The hydrocarbons conversion to syngas and hydrogen will play a great impress in the 21st century, from significant gases to hydrogen and liquid plants for refineries to small hydrogen supply units for fuel cells [8].

The development of research and technology in the production and refining of syngas and hydrogen and fuel processing for fuel cells has great potential to address three main challenges in the energy field [11]: (a) providing more clean fuels to meet the growing demand for liquid and gaseous fuels and electricity, (b) increasing energy yield for fuel and electricity generation, and (c) eliminating polluters and decoupling the link between energy greenhouse gas emissions and usage in end-use systems [1]. The schematic of syngas production is shown in Fig. 1.

Miniatured devices have attracted much attention due to their advantages such as reliability, good mixing characteristics, better control, good mass and heat transfer performance, etc., compared with conventional ones. Previous research has investigated the simulation and modeling of microreactors for syngas as well as hydrogen production. In the following, microreactor structure, advantages/disadvantages in addition to its modeling and simulation in different studies are discussed in detail [13].

2. Microreactor

To use CSR technology for small-scale portable and mobile applications, where compactness, simple fuel processor performance, and reasonable temperature control are essential, new designs of reactors such as microreactors, plate modifiers, and membrane reactors developed [14]. Microreactors are devices in which chemical reactions are performed in a chamber with regular sidelong dimensions of less than 1mm [7]. Microchannels are the most common type of this restriction [15]. Microreactors have a channel slit on the order of microscale, usually <1000 μm or mesoscale, and it is usually a continuous-flow reactor [16].

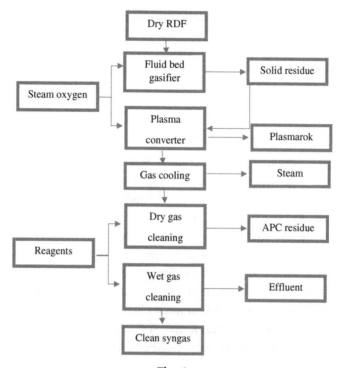

Fig. 1
Syngas production schematic [12].

Microreactors compared with conventional systems are more suitable for distributed hydrogen production because of the countless benefits listed below [1]:

- High-performance mass and heat transfer for extraction and multiphase reactions
- A powerful tool for process intensification and microscale processing [17]
- Good mixing properties
- The less holding volume of reagents
- Better control compared with traditional batch reactors [18]
- Hold great promise for new chemical pathways such as thermal inertia
- Allow for straightforward control of the reaction temperature as a parameter is very important
- The possibility of creating substantial surface-to-volume ratios for heterogeneous catalyst gas-phase reactions [19]
- Extensive advances in energy efficiency
- Reaction speed and efficiency
- Safety
- Reliability
- Scalability
- Much more accurate degree of process control [20]

At the same time, microreactors have several disadvantages compared with conventional devices:

- Clog or sediment in microchannels
- Permeation between channels
- The high expense of reactor construction [21]

There are many different ways to make microreactors, which are briefly mentioned below:

Developing microfluidic devices, the construction of microchannels is an important aspect. The number of production techniques has been expanded and used for various applications and materials over the years [22]. Essentially, microchannels of polymers, glass, silicon, and metal substrates are created. At the same time, most polymer and glass layers in chemical and biomedical devices, silicon, and metal layers for mechanical engineering and electronic applications are used [23]. Except for circular cross-sectional microchannels, other microchannels are built on the surface of the basic material. Circular microchannels are commonly made intranet the material via the bulk of it. A few microns thickness naturally remains on top of the circular cross-section microchannels along the length for optical and slightly transparency [24]. Various kinds of materials for various microchannel-based devices have been used. These materials can be segregated into the following three classifications [22]:

(I)   glass and polymeric layers: The polymer materials for microfluidic devices include polymethyl methacrylate (PMMA) [25] and poly dimethyl siloxane (PDMS) [26]. PMMA is completely transparent and can be used as an alternative to glass. It mostly possesses high biocompatibility, implants, and medication delivery system material used [27].

(II) Metallic layers: Microtechnology-based energy and chemical systems (MECS) increment the rate of mass and heat transfer in heat exchangers using microscale properties within the systems they are used to better surface-to-volume ratio. The temperature can also reach 650°C, and it is favorable to carry out many exothermic and endothermic reactions in such metal layers. [28]. Therefore, ordinary metals such as stainless steel are less preferred for MECS systems [29].

(III) Semiconductors, ceramics, and composites: Silicone is not only used for microelectronic systems but can also be used for applications such as fuel cells [30]. Except for silicon, quartz is also widely used as a layer for multiple devices based on microchannel. Quartz is somewhen more appropriate than silicon because (I) it is chemically inert and stable, (II) it is optically transparent, (III) it is cheaper than silicon, and (IV) it is a good electrical insulator [31].

Over the years, microchannels with various cross sections have been created, most of which include half-circular microchannels [32], U-shape, circular microchannels [24], rectangular microchannels, and square microchannels [32].

3. Modeling and simulation of microreactors

Research studies in engineering sciences are done in both experimental and modeling. This chapter focuses on modeling and simulation of microreactors for syngas production. The simulation is useful in these cases: (I) The real system is regularized, and components of the system and also their interaction can be defined. (II) When you need a tool that everyone involved can agree on a set of assumptions and then observe the results and effects of those assumptions. Also, (III) the animation from the simulation is a great training and educational tool for managers, supervisors, engineers, and laborers [33].

Here are some of the advantages and disadvantages of modeling and simulation. One of the advantages of simulation is the identification of problems, bottlenecks, and design shortfalls before building or modifying a system. Also, a reasonable simulation model provides numerical measurements of system performance and insights into system performance [34]. The computational cost is lower in the simulation, but they are time-consuming. One of the essential modeling processes is the production of syngas. Part of the microreactor modeling is done for the following reasons:

One of which is to study the effects of operating conditions on the performance of reactors [35]. Another to investigate the processes of heat transfer and reaction at the pore scale in a posted microreactor [36] and the process of hydrogen production in microreactors by the methanol-steam reforming reaction was also investigated [37] and for simulation of propane catalytic oxidation in a microreactor used a steady-state COMSOL multiphysics model [38]. Also, the influence of porosity, steam/DME feed ratio of DME SR, and wall temperature on the microreactor performance was also investigated [39].

Modeling is done thermodynamically, kinetically, and numerically. Among the software used in modeling and simulation are Ansys-Fluent, Open Foam, Aspen HYSYS, COMSOL, Ansys-CFX, etc., software; MATLAB software is used to perform numerical calculations. Ansys-Fluent, Open Foam, COMSOL, Ansys-CFX software falls within the scope of this chapter.

The following studies on modeling and simulation of microreactors in two-dimensional and three-dimensional sections are presented.

3.1 Two-dimensional models

This section discusses the modeling and simulation of microreactors to produce synthetic gases. Some authors have simulated two-dimensional microreactors, which are mentioned below:

Various studies have been performed on two-dimensional simulations of microreactors to produce hydrogen. One focuses on accurate modeling and simulation of an integrated

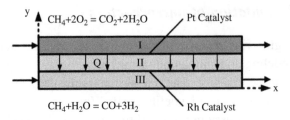

Fig. 2

Geometric structure of SRM microreactor. *From X. Zhai, S. Ding, Y. Cheng, Y. Jin, Y. Cheng, CFD simulation with detailed chemistry of steam reforming of methane for hydrogen production in an integrated micro-reactor, Int. J. Hydrogen Energy 35(11) (2010) 5383–5392, with permission.*

microreactor for the SRM reaction to specify the importance of the reactor ability's thermal conductivity and the interaction between endothermic and exothermic reactions for quick reactions in both channels. The equations that describe fluid dynamics, mass and heat transfer, and reaction behaviors in the gas phase are solved with FLUENT 6.3 software. Also, user-defined functions give concentration and temperature information in the flow field to the solver of stiff equations. The results showed that for rapid reactions in milliseconds, the thermal conductivity of the reactor wall is very important in the reactor design. In particular, a metallic wall is better than a ceramic wall in terms of thermal conductivity. Also, the characteristic width of 0.5mm as an appropriate channel size to balance the thoughts was considered. The following figures show the geometric structure of the microreactor and the temperature profile inside the reactor [40] (Figs 2 and 3).

The other modeling for producing hydrogen is the decomposition of ammonia in the complex geometry of the microreactor. The steady-state solutions of this problem were solved with

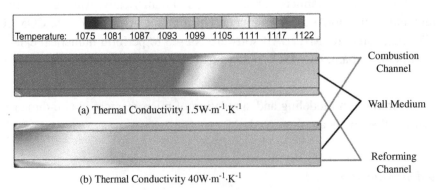

Fig. 3

The contour of the temperature within the reactors with ceramic walls and metallic. *From X. Zhai, S. Ding, Y. Cheng, Y. Jin, Y. Cheng, CFD simulation with detailed chemistry of steam reforming of methane for hydrogen production in an integrated micro-reactor, Int. J. Hydrogen Energy 35(11) (2010) 5383–5392, with permission.*

FLUENT (version 6.0) software and then the simulation results with experimental data as temperature and flow rate functions were compared. The results obtained from the simulations are, in addition to the increased surface area, these posts create an impressive transverse mixing that leads to better conversions, compared with the results of a CSTR and a tubular reactor, at least pressure drops [41]. Also, another study presents to: (1) develop a coupled LB algorithm and framework that can simulate several thermal and physicochemical operations at pore scale; (2) investigate several physicochemical and thermal phenomena at the pore scale and the coupling mechanism of microreactors; (3) determining the effect of operating conditions and geometry on the reactor performance and betterment of the conversion yield of the chemical reaction. Two assumptions are discussed in this study. First, because Reynolds number is small in microreactors, the fluid flow is considered laminar and incompressible; and second, mass and heat transfer has limited impacts on fluid flow. Therefore, the connection between the flow and heat/mass transfer is one-sided. Some results that came out from this simulation are that the size of the post greatly influences the yield of the reaction in the microreactor and reduces the size of the post to better the performance of the reactor because it can increment the surface area to volume ratio, and the microreactor performance under high temperature is favorable because it causes to high H_2 flow rate and high NH_3 conversion. Rising NH_3 flow rate reduces the NH_3 conversion and temperature distribution uniformity, but it can cause a high H_2 flow rate [36]. Another numerical simulation for hydrogen production is the steam reforming of dimethyl ether in a microreactor. In this study, the boundary condition of the inlet was velocity inlet and pressure outlet modeled for the outlet. Numerical and experimental results showed that the conversion of DME in the microreactor is more of a fixed-bed reactor [42]. Also, numerical results showed that the temperature and velocity distributions in the microreactor are more uniform and give more porosity DME conversion and higher CO concentration. Figs. 4 and 5 show the geometric models of fixed-bed reactors, Ni microfoam reactors, temperature contour with velocity vector in microreactors, and fixed-bed reactors [39].

One study investigated the effect of several operating parameters such as the reactant inlet temperature, the reactor channels width, the catalyst layers' thickness, and the methane concentration in the reactant combustion gases; then, the differences between the 2D and 3D simulations were analyzed and concentrated on the comparison between countercurrent (CTCF) and concurrent (CNCF) flow patterns of endothermic and exothermic reactant streams. The reaction system was mathematically modeled in two-dimensional and three-dimensional steady state, and the model solutions were performed by an FEM (COMSOL multiphysics) software. In both cases, the heat flux required to heat reaction mixtures and generate the exothermic process provides catalytic combustion [43]. Also, another two-dimensional simulation investigated the behavior of cold flow, airflow and, using the COMSOL multiphysics software, determined the great suitable geometry and configuration of the reactor. COMSOL multiphysics is a powerful interactive environment for modeling and solving various PDE-based scientific and engineering problems. COMSOL multiphysics uses FEM when

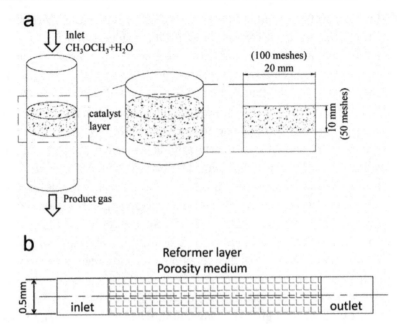

Fig. 4

The geometrical model of (A) the fixed-bed reactor and (B) the Ni foam microreactor. *From C.-F. Yan, W. Ye, C.-Q. Guo, S.-L. Huang, W.-B. Li, W.-M. Luo, Numerical simulation and experimental study of hydrogen production from dimethyl ether steam reforming in a micro-reactor, Int. J. Hydrogen Energy 39(32) (2014) 18642–18649, with permission.*

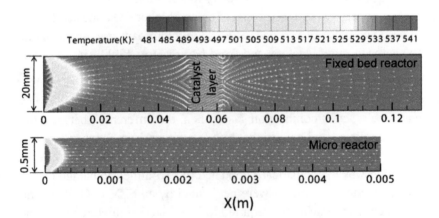

Fig. 5

The temperature gradient and the velocity vector field of the fixed-bed reactor and the microreactor. *From C.-F. Yan, W. Ye, C.-Q. Guo, S.-L. Huang, W.-B. Li, W.-M. Luo, Numerical simulation and experimental study of hydrogen production from dimethyl ether steam reforming in a micro-reactor, Int. J. Hydrogen Energy 39(32) (2014) 18642–18649, with permission.*

solving the PDEs. The simulation results of two flow regimes are shown for four geometries depending on the inlet velocity. At low flow rates (below 10m/s) for all geometries, the flow distribution is entirely defined by the wall friction.

The flow distribution is independent of the flow rate for operation below this velocity. At flow rates higher than a transitional velocity, inertial effects begin to impact the flow distribution with a constant relative standard deviation [38]. The other research has examined some microreactor configurations through mathematical modeling.

to increase overall performance. They analyzed a microreactor in the form of a square cross-section channel of parallel, perforated pin, wavy, oblique fin, serpentine, twisted, coiled with serpentine, and coiled with double serpentine channel geometries by using computational fluid dynamics. This study is a single-phase catalytic reaction between methane in a wide range of Reynolds numbers achieved. Some of the assumptions of fluid flow are laminar, steady, and Newtonian that follow the ideal gas condition; the catalytic reaction to be occurred, the wall is covered with a platinum catalyst. It was observed that the reactor design based on the coils creates much higher conversion in all Reynolds numbers than the rectilinear designs, and coil-based designs are favorable for achieving high conversions for applications where pumping power is not a topic [44]. To show the effect of geometric design parameters of fluid sections, such as arrangements and shapes, on reactor performance, namely mixing speed and product efficiency: the ratio of reaction rate to diffusion rate and the ratio of the dimensions of the average diffusion length in the two-dimensional paths in the reactor section use two-dimensional numbers. Researchers mixing by molecular diffusion and the reaction in the reactor with different designs and shapes of fluid components were simulated. They used CFD code for simulated by using Fluent software that the equations of momentum, energy, and mass conservation solved with this code and the equation of the mass conservation for the components A and B as reactants and R as the product is given below:

$$-v_Z \frac{\partial C_A}{\partial Z} + D_A \left(\frac{\partial C_A^2}{\partial x^2} + \frac{\partial C_A^2}{\partial y^2} \right) - k_1 C_A C_B = 0 \tag{1}$$

$$-v_Z \frac{\partial C_B}{\partial Z} + D_B \left(\frac{\partial C_B^2}{\partial x^2} + \frac{\partial C_B^2}{\partial y^2} \right) - (k_1 C_A C_B + k_2 C_B C_R) = 0 \tag{2}$$

That k_i is the reaction rate, C_j is the molar concentration of the component, D_j is the diffusion coefficient of the component, and v_Z is the velocity of the reactant fluid flows.

Also, for solving these equations, the second-order upwind scheme was used. Finally, they obtained that the first dimensionless number used to specify whether reactions proceed under reaction-controlled conditions even when the cross-sectional shape of fluid parts is changed. The second dimensionless number can display the impacts of designs and aspect ratios of cross-sectional shapes of fluid segments on the reactor performance [45]. Yang et al. [46]

simulated transport and reactions in microscale multiphase reactors using Open FOAM software. They couple hydrodynamics, reaction, and mass transfer modules to show that the simulations determine the temporal and spatial details of the mass transfer performance and record the increment of mass transfer in the presence of reactions. The CFD simulations are based on the VOF method. The VOF method is used to simulate microscale multiphase flow. They found that the peclet number to obtain Henry's law constant must be less than 0.5. Simulations of the transient concentration profiles showed that the multiphase transport in the segmented flow system is a two-regime process that is first affected by convection and then diffusion. The VOF method is used to model multiphase flow in structured microreactors. The volume-of-fluid method simulated the multiphase flows in the Eulerian reference frame, and the mesh was generated for calculating using Pointwise V17.1R4 software. Finally, a good agreement was reached between LIF visualization and CFD simulation [47]. The other researchers modeled membrane catalytic microreactors for yield manufacture of DME from synthesis gas with CO_2 gases. A two-dimensional reactor model has been developed to measure mass transfer and momentum transport in fluid phases, reaction transfer in the catalyst layer, and cross-mass membrane transfer. The mathematical model to consider the impacts of syngas composition, pressure, and temperature and the kind of dehydration catalyst on membrane helped carbon oxide conversions, and DME yield is used. Some conclusions obtained from this study are as follows:

- Integration SOD membrane layer enhancement of the CO and CO_2 conversion and DME performance.
- Rising temperature and pressure increase the benefits of membrane assistance in CO and CO_2 valuation and increase DME power.
- Reducing the infiltration channel pressure to increase the steam concentration gradient between the channels does not improve the reactor performance because a significant H_2 flow inhibits the rate of hydrogenation reactions.
- The syngas' rising inlet velocity to the infiltration channel causes cross-membrane steam flow and H_2 influx from/to the reaction channel [48].

3.2 Three-dimensional models

This section also models and simulates the studies performed on microreactors in three dimensions according to the previous section.

Sabziani et al. investigated modeling the process of producing hydrogen in microreactors with methanol-steam modification reaction. The simulation results show that Stephen Maxwell's method is more consistent with the experimental results. The difference is more at lower feed flow rates since the flow rate increases, the mechanism of mass transfer from diffusion to convection changes, reducing the difference [37]. The other three-dimensional simulation using the equations of continuity, momentum conservation, energy conservation, and chemical species conservation that are the governing equations for the flow field in the reactor and investigated the influences of main operating conditions on the performance of the reactor and achieved both hydrogen yield and high methanol conversion also low CO mole fraction [35].

These equations are shown in Eqs. (3)–(7)

$$\rho_{mix} \nabla \cdot \vec{V} = 0 \tag{3}$$

$$\rho_{mix} \vec{V} \cdot \nabla \vec{V} = -\nabla P + \nabla \cdot \left(\mu_{eff} \nabla \vec{V} \right) \tag{4}$$

$$\rho_{mix} \nabla \cdot \left(C_{P,eff} \vec{V} T \right) = \nabla \cdot \left(k_{eff} \nabla T \right) + \sum_{j=1}^{4} \left(\Delta H_j r_j \right) \tag{5}$$

$$\rho_{mix} \nabla \cdot \left(\vec{V} Y_i \right) = \rho_{mix} \nabla \cdot \left(D_{eff,i} \nabla Y_i \right) + \omega_i MW_i \tag{6}$$

$$\omega_i = \sum_{j=1}^{4} \nu_{ij} r_j \tag{7}$$

where ω_i is the production rate or consumption of species i, ν_{ij} is the stoichiometric coefficient of the i in reaction j, and r_j is reaction rate. Fig. 6 shows the geometry of the reactor with its specifications, and Figs. 7 and 8 show the velocity, pressure, and temperature in the reactor [35].

Another simulation was performed with the commercial ANSYS CFX software to produce hydrogen from the integration of methanol steam reforming and combustion in a microchannels reactor [49]. The model for the integration the kinetics of chemical reactions and to investigate

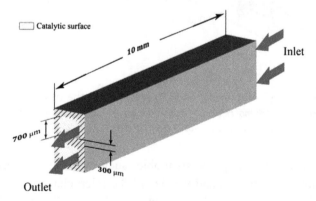

Fig. 6
Reactor specifications for the baseline case. *From M.A. Tadbir, M. Akbari, Integrated methanol reforming and oxidation in wash-coated microreactors: a three-dimensional simulation, Int. J. Hydrogen Energy 37(3) (2012) 2287–2297, with permission.*

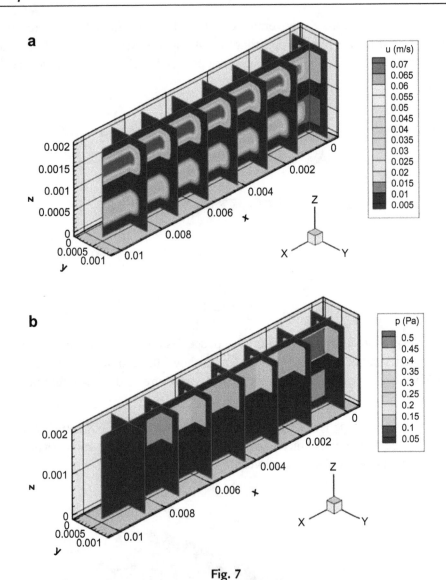

Fig. 7
The profile of (A) x-velocity component, and (B) pressure for the baseline conditions. *From M.A. Tadbir, M. Akbari, Integrated methanol reforming and oxidation in wash-coated microreactors: a three-dimensional simulation, Int. J. Hydrogen Energy 37(3) (2012) 2287–2297, with permission.*

the effect of some of the original process variables such as the velocity of the space of the reforming and combustible streams and the size of the microchannels is used [50].

A heterogeneous 3D steady-state model was completed using COMSOL Multiphysics and described concentration, temperature, velocity, and distributions within the surface-bed reactor system with structured catalysts. The results that were carried out from this simulation were

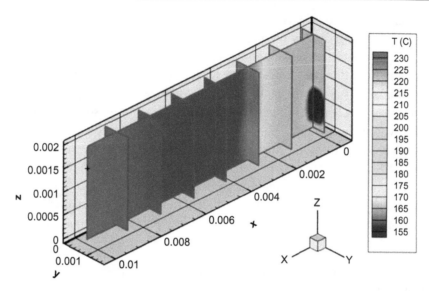

Fig. 8

The temperature profile in the reactor for the baseline case. *From M.A. Tadbir, M. Akbari, Integrated methanol reforming and oxidation in wash-coated microreactors: a three-dimensional simulation, Int. J. Hydrogen Energy 37(3) (2012) 2287–2297, with permission.*

between the reforming and combustion catalysts; the temperature difference is very close together [43].

The other researchers studied the reaction flow by evaluating the oxidation reaction of propane in a microreactor with the proposed two-dimensional geometry, based on an available kinetic model by using the CFD package of COMSOL Multiphysics. They developed a three-dimensional model to describe the gas flow and reaction in the microreactor and for an ideal gas and laminar flow inside the microchannels solved. The simulation results showed that the conversion of the output propane increases with increasing temperature [38].

Another study used ANSYS FLUENT 14.5 software to model the microreactor and analyzed reaction conversion, heat generation, temperature profiles, and selectivity in a reaction channel. One of the results of this study comparing a decoupled model with that of reaction and heat transfer shows that there is a difference of less than 1°C in temperature along the channel [51]. Three-dimensional computational fluid dynamics simulations analyze a helical microreactor for hydrogen combustion in a Pt catalyst with precise mechanisms for both catalytic and gas-phase reactions. A helical microreactor consists of a single channel accumulated in a spiral, starting at the center and twisting outward. The researchers compared the spiral with the conventional reactor channels. The spiral geometry was created with three rotations in FLUENT, and the meshing was organized into three-dimensional hexahedrons across the computational field. The steady-state equations of mass, energy, momentum, and conservation of species in three dimensions by ANSYS FLUENT 17.2 with SIMPLE algorithm were solved. The results showed the geometry of single spiral can potentially be an alternative design to improve thermal management and sustainability of catalytic microreactors, and combustion characteristics showed that the preheating of reactants and protection of reaction zone in the spiral reactor make its superior over the direct channel reactor [52]. One of the simulations performed by the researchers examined the heat transfer and combustion properties of catalytic microreactors with hydrogen fuel. They used the Open-FOAM code for steady numerical simulations. They found from the simulations that the catalytic conversion of the hydrogen is not limited to transport, and the ratio of wall-to-center hydrogen mass fractions is 15% higher at axial positions. This indicates the importance of precision surface reaction mechanisms in numerical simulations and calculated radiation yield; it increases with increasing inlet velocity and equivalence ratio depending on the difference of enthalpy between products and reactants [53].

Another paper simulated the adsorption estimation rate and apparent kinetic rate constants for optical decomposition of 4-nitrophenol in a rotary and single-pass photocatalytic microreactor. Adsorption and kinetic rate constants were evaluated by fitting a three-dimensional model with experimental data. Experimental data prediction was performed by creating a dynamic three-dimensional model in COMSOL Multiphysics. COMSOL multiphysics for solving partial differential equations can use the finite element method. The simulation result showed that increasing Co increases the initial degradation rate. Also, reducing the flow rate (Q), increasing microchannels length (L), increasing the overall conversion of 4-nitrophenol in solution, and (h) decreasing microchannels height [54].

Some researchers have modeled and simulated three-dimensional membrane microreactors, examples of which are given below:

In one study, the researchers with a mathematical model described the steady isothermal flow of microfluidic in a membrane microreactor. In this model, the Navier-Stokes equation with suitable boundary conditions is used for fluid penetration through the membrane and velocity slip in the walls to calculate the high Knudsen number. The equations of this model were solved analytically using finite Fourier transforms. Finally, they evaluated the influence of fluid penetration through the Knudsen number and the membrane on the velocity and pressure drop profile. Application of this modeling is used for apart hydrogen from the other sewerages in a microreformer [55]. The other study used membrane microreactors for Knoevenagel condensation reaction between ethyl cyano-acetate and benzaldehyde to make ethyl 2-cyano-3-phenyl acrylate. In this study, two miniature membrane reactors were PBMR and CMR. Zeolites as catalysts for the reaction and membrane separation have been incorporated. Using batch reaction kinetic data, they simulated the reaction in multichannel membrane microreactors, associated data from membrane separation, and published transport data.

The assumptions from this model are the steady-state and isothermal operation conditions. The researchers evaluated the effects of reactor geometry, membrane separation, and catalyst attributes, and the conclusions were compared with experimental data. Finally, they found that the thickness of the membrane is determined by the growth specification of the zeolite film. The thicker film is required to obtain the ZSM-5 membrane due to the crystalline shape of the zeolite column [56]. Chasanis et al. [57] have studied the influence of miniaturization and inlet velocity changes on the membrane performance microreactor numerically. From 3D CFD model was used to capture hydrodynamics and mass transfer in a micromembrane reactor consisting of a reaction channel and a permeate channel. The assumptions were considered isothermal, steady-state conditions, and ideal gas behavior. Also, because the Reynolds numbers were low, the flows in both microchannels were laminar. In this modeling, the equations of mass, momentum, and species were solved. They observed that by reducing the dimensions of the reactor, the reactor performance is significantly improved as a result of increasing the catalyst and the membrane surface per volume.

4. Conclusion

Miniatured devices, compared with conventional devices, have attracted much attention and some of their benefits were mentioned in this chapter. Due to the importance of modeling mentioned in this chapter, modeling is a method that has a low computational cost but is time-consuming, and one of the most important processes performed by modeling is the production of syngas gas. Many researchers have performed two-dimensional and three-dimensional simulations of microreactors to produce syngas. In this study, several examples of these simulations were discussed. Microreactors simulation can be used for studying the effects of operating conditions on the performance of reactors, investigating the processes of heat transfer and reaction at the pore scale in a posted microreactor, investigating modeling the process of producing hydrogen in microreactors with methanol-steam reforming reaction, studying the reaction flow by evaluating the oxidation reaction of propane in a microreactor, and examining some microreactor configurations through mathematical modeling that may increase overall performance. The results that researchers obtained from the microreactor simulations include that for rapid reactions in milliseconds, the ability of thermal conductivity of the reactor wall is very important in the reactor's design. The temperature and velocity distributions in the microreactor are more uniform and give more porosity DME conversion and higher CO concentration. Another result from the simulation of two flow regimes is shown that at low flow rates (below 10m/s) for all geometries, the flow distribution is completely determined by the wall friction, and at flow rates higher than a transitional velocity, the effects of inertial begin to impact on the flow distribution with a constant relative standard deviation. The Peclet number to obtain Henry's law constant must be less than 0.5. Stephen Maxwell's method is more consistent with experimental results. The difference between the two is more minor at higher flow rates because the mass transfer mechanism changes from diffusion to convection. In the other study, the results showed that the temperature difference is very close between the reforming and combustion catalysts, and the conversion of the output propane increases with increasing temperature. As Co increases, it increases the initial degradation rate and reduces the flow rate, and they obtained that by reducing the dimensions of the reactor, the reactor performance is significantly improved as a result of increasing the catalyst and the membrane surface per volume.

Abbreviations and symbols

CMR	catalytic membrane reactor
FEM	finite element method
MECS	microtechnology-based energy and chemical systems
MOGD	mobil's olefin-to-gasoline/distillate
MTG	methanol to gasoline
PBMR	packed-bed membrane reactor
PDEs	partial differential equations
PDMS	poly dimethyl siloxane
PMMA	poly methyl methacrylate
VOF	volume-of-fluid

References

[1] K. Liu, C. Song, V. Subramani, Hydrogen and Syngas Production and Purification Technologies, John Wiley & Sons, 2010.

[2] A.G. Capodaglio, S. Bolognesi, Ecofuel feedstocks and their prospects, in: Advances in Eco-Fuels for a Sustainable Environment, Elsevier, 2019, pp. 15–51.

[3] Z. Moravvej, et al., Thermochemical routes for hydrogen production from biomass, in: Advances in Bioenergy and Microfluidic Applications, Elsevier, 2021, pp. 193–208.

[4] A.P. York, et al., Methane oxyforming for synthesis gas production, Catal. Rev. 49 (4) (2007) 511–560.

[5] S. Saeidi, et al., Recent advances in reactors for low-temperature Fischer-Tropsch synthesis: process intensification perspective, Rev. Chem. Eng. 31 (3) (2015) 209–238.

[6] P. Basu, Biomass Gasification, Pyrolysis and Torrefaction: Practical Design and Theory, Academic Press, 2018.

[7] M.R. Kiani, et al., Catalytic membrane micro-reactors for fuel and biofuel processing: a mini review, Top. Catal. (2021) 1–20.

[8] J.R. Rostrup-Nielsen, New aspects of syngas production and use, Catal. Today 63 (2–4) (2000) 159–164.

[9] M. Parhoudeh, F. Farshchi Tabrizi, M.R. Rahimpour, Auto-thermal chemical looping reforming process in a network of catalytic packed-bed reactors for large-scale syngas production: a comprehensive dynamic modeling and multi-objective optimization, Top. Catal. (2022) 1–28.

[10] J.R. Rostrup-Nielsen, Production of synthesis gas, Catal. Today 18 (4) (1993) 305–324.

[11] M. Khanipour, et al., Enhancement of synthesis gas and methanol production by flare gas recovery utilizing a membrane based separation process, Fuel Process. Technol. 166 (2017) 186–201.

[12] M. Materazzi, R. Taylor, The GoGreenGas case in the UK, in: Substitute Natural Gas from Waste, Elsevier, 2019, pp. 475–495.

[13] D. Karimipourfard, et al., Mathematical modeling and optimization of syngas production process: a novel axial flow spherical packed bed tri-reformer, Chem. Prod. Process. Model. 13 (2) (2018).

[14] M. Makarem, Applications of minichannels in gas absorption: a review, Mod. App. Ocean Petrochem. Sci. 2 (5) (2019).

[15] P. Watts, C. Wiles, Recent advances in synthetic micro reaction technology, Chem. Commun. 5 (2007) 443–467.

[16] A.S. Bhangale, K.L. Beers, R.A. Gross, Enzyme-catalyzed polymerization of end-functionalized polymers in a microreactor, Macromolecules 45 (17) (2012) 7000–7008.

[17] X. Yao, et al., Review of the applications of microreactors, Renew. Sustain. Energ. Rev. 47 (2015) 519–539.

[18] M. Veeramani, S. Narasimhan, N. Bhatt, Identification of reaction systems using spectroscopic measurements and micro-reactors, in: Computer Aided Chemical Engineering, Elsevier, 2018, pp. 931–936.

[19] G. Veser, et al., A simple and flexible micro reactor for investigations of heterogeneous catalytic gas phase reactions, in: Studies in Surface Science and Catalysis, Elsevier, 1999, pp. 237–245.

[20] Wikipedia, Microreactor, 2021, Available from: https://www.wikipedia.com.

[21] A. Helmi, et al., Highly pure hydrogen production in a micro-channel membrane reactor for fuel cell applications-modeling and experimental work, in: 第五届世界氢能技术大会 (WHTC2013), 2013, pp. 1–2.

[22] S. Prakash, S. Kumar, Fabrication of microchannels: a review, Proc. Inst. Mech. Eng. B J. Eng. Manuf. 229 (8) (2015) 1273–1288.

[23] S. Prakash, et al., An experimental investigation on Nd: YAG laser microchanneling on polymethyl methacrylate submerged in water, Proc. Inst. Mech. Eng. B J. Eng. Manuf. 227 (4) (2013) 508–519.

[24] M. Abdelgawad, et al., A fast and simple method to fabricate circular microchannels in polydimethylsiloxane (PDMS), Lab Chip 11 (3) (2011) 545–551.

[25] A. Muck, et al., Fabrication of poly (methyl methacrylate) microfluidic chips by atmospheric molding, Anal. Chem. 76 (8) (2004) 2290–2297.

[26] N. Bao, et al., Fabrication of poly (dimethylsiloxane) microfluidic system based on masters directly printed with an office laser printer, J. Chromatogr. A 1089 (1–2) (2005) 270–275.

[27] J. Fernández-Pradas, et al., Femtosecond laser ablation of polymethyl-methacrylate with high focusing control, Appl. Surf. Sci. 278 (2013) 185–189.

[28] K. Kanlayasiri, B. Paul, A nickel aluminide microchannel array heat exchanger for high-temperature applications, J. Manuf. Process. 6 (1) (2004) 72–80.

[29] Y. Mishin, A. Lozovoi, A. Alavi, Evaluation of diffusion mechanisms in NiAl by embedded-atom and first-principles calculations, Phys. Rev. B 67 (1) (2003), 014201.

[30] A. Kamitani, et al., Microfabricated microfluidic fuel cells, Sens. Actuators B Chem. 154 (2) (2011) 174–180.

[31] S.-J. Qin, W.J. Li, Micromachining of complex channel systems in 3D quartz substrates using Q-switched Nd: YAG laser, Appl. Phys. A 74 (6) (2002) 773–777.

[32] H. Becker, L.E. Locascio, Polymer microfluidic devices, Talanta 56 (2) (2002) 267–287.

[33] M. Farniaei, et al., Syngas production in a novel methane dry reformer by utilizing of tri-reforming process for energy supplying: modeling and simulation, J. Nat. Gas Sci. Eng. 20 (2014) 132–146.

[34] J.S. Carson, Introduction to modeling and simulation, in: Proceedings of the Winter Simulation Conference, 2005, IEEE, 2005.

[35] M.A. Tadbir, M. Akbari, Integrated methanol reforming and oxidation in wash-coated microreactors: a three-dimensional simulation, Int. J. Hydrogen Energy 37 (3) (2012) 2287–2297.

[36] L. Chen, et al., Pore-scale simulation of coupled multiple physicochemical thermal processes in micro reactor for hydrogen production using lattice Boltzmann method, Int. J. Hydrogen Energy 37 (19) (2012) 13943–13957.

[37] J. Sabziani, A. Sari, A CFD simulation of hydrogen production in microreactors, Iran. J. Oil Gas Sci. Technol. 4 (1) (2015) 35–48.

[38] S. Odiba, et al., Computational fluid dynamics in microreactors analysis and design: application to catalytic oxidation of volatile organic compounds, J. Chem. Eng. Process Technol. 7 (2016).

[39] C.-F. Yan, et al., Numerical simulation and experimental study of hydrogen production from dimethyl ether steam reforming in a micro-reactor, Int. J. Hydrogen Energy 39 (32) (2014) 18642–18649.

[40] X. Zhai, et al., CFD simulation with detailed chemistry of steam reforming of methane for hydrogen production in an integrated micro-reactor, Int. J. Hydrogen Energy 35 (11) (2010) 5383–5392.

[41] S. Deshmukh, A. Mhadeshwar, D. Vlachos, Microreactor modeling for hydrogen production from ammonia decomposition on ruthenium, Ind. Eng. Chem. Res. 43 (12) (2004) 2986–2999.

[42] M.R. Kiani, et al., Novel gas-liquid contactors for CO_2 capture: mini-and micro-channels, and rotating packed beds, in: Advances in Carbon Capture, Elsevier, 2020, pp. 151–170.

[43] S. Vaccaro, P. Ciabelli, Results of modeling of a catalytic micro-reactor, in: 31st Meeting on Combustion, 2012.

[44] H. An, et al., Computational fluid dynamics (CFD) analysis of micro-reactor performance: effect of various configurations, Chem. Eng. Sci. 75 (2012) 85–95.

[45] N. Aoki, S. Hasebe, K. Mae, Geometric design of fluid segments in microreactors using dimensionless numbers, AICHE J. 52 (4) (2006) 1502–1515.

[46] L. Yang, M.J. Nieves-Remacha, K.F. Jensen, Simulations and analysis of multiphase transport and reaction in segmented flow microreactors, Chem. Eng. Sci. 169 (2017) 106–116.

[47] L. Yang, et al., Characterization and modeling of multiphase flow in structured microreactors: a post microreactor case study, Lab Chip 15 (15) (2015) 3232–3241.

[48] H.H. Koybasi, A.K. Avci, Modeling of a membrane integrated catalytic microreactor for efficient DME production from syngas with CO_2, Catal. Today 383 (2020) 133–145.

[49] M. Makarem, M. Farsi, M. Rahimpour, CFD simulation of CO_2 removal from hydrogen rich stream in a microchannel, Int. J. Hydrogen Energy 46 (37) (2021) 19749–19757.

[50] G. Arzamendi, et al., Integration of methanol steam reforming and combustion in a microchannel reactor for H2 production: a CFD simulation study, Catal. Today 143 (1–2) (2009) 25–31.

[51] C. Han, K.S. Kshetrimayum, Modeling, Simulation, and Design Procedure Development of Micro-Channel FT Reactor Using Computational Fluid Dynamics, 서울대교 대원, 2017.

[52] N. Yedala, A.K. Raghu, N.S. Kaisare, A 3D CFD study of homogeneous-catalytic combustion of hydrogen in a spiral microreactor, Combust. Flame 206 (2019) 441–450.

[53] R. Sui, et al., An experimental and numerical investigation of the combustion and heat transfer characteristics of hydrogen-fueled catalytic microreactors, Chem. Eng. Sci. 141 (2016) 214–230.

[54] A. Yusuf, G. Palmisano, Three-dimensional CFD modelling of a photocatalytic parallel-channel microreactor, Chem. Eng. Sci. 229 (2021), 116051.

[55] K.A. Alfadhel, M.V. Kothare, Microfluidic modeling and simulation of flow in membrane microreactors, Chem. Eng. Sci. 60 (11) (2005) 2911–2926.

[56] K.L. Yeung, et al., Experiments and modeling of membrane microreactors, Catal. Today 110 (1–2) (2005) 26–37.

[57] P. Chasanis, et al., Modelling and simulation of a membrane microreactor using computational fluid dynamics, in: Computer Aided Chemical Engineering, Elsevier, 2008, pp. 751–756.

CHAPTER 6

Simulation of biomass to syngas: Pyrolysis and gasification processes

José Antonio Mayoral Chavando[a], Valter Silva[a,b], M. Puig-gamero[c], João Sousa Cardoso[a,d], Luís A.C. Tarelho[b], and Daniela Eusébio[a]

[a]Polytechnic Institute of Portalegre, Portalegre, Portugal [b]Department of Environment and Planning, Centre for Environmental and Marine Studies (CESAM), University of Aveiro, Aveiro, Portugal [c]University of Castilla—La Mancha, Ciudad Real, Spain [d]Instituto Superior Técnico, Universidade de Lisboa, Lisboa, Portugal

1. Introduction

Plastic production has risen dramatically during the previous few decades. In 1950, the world produced 1.5 million tons of plastic. In 2015, it grew to more than 350 million tons (see Fig. 1) [1,2]. By 2015, cumulative plastic production had reached approximately 8 billion tons, with approximately 79% of that plastic ending up in landfills, dumps, or the natural environment, and only 9% and 12% of that plastic being recycled and incinerated, respectively, which is highly alarming given that the majority of plastic items never fully degrade eventually ending up in our oceans. By 2050, if current trends continue, our oceans may contain more plastic than fish [3].

Numerous measures can be taken to minimize the problem, ranging from public awareness campaigns to the use of sophisticated plastic recycling technologies such as pyrolysis and gasification. The resulting products of these technologies can be transformed into chemicals and biofuels, providing environmental benefits linked with the use of plastic wastes [4,5].

The economic feasibility of gasification and pyrolysis of residues is dependent on several parameters, including the availability of feedstock, operating conditions, and the capacity to predict the final products without conducting costly experimental testing [4]. Therefore, process simulation is a precious tool in the chemical industry for forecasting the final products and the behavior of a process under specified operating parameters. It also assists in identifying operational issues, namely variation sources and optimal operating conditions. Furthermore, process simulation provides technical knowledge to enable decision-makers to scale up a process idea or pilot-scale process.

Advances in Synthesis Gas: Methods, Technologies and Applications. https://doi.org/10.1016/B978-0-323-91879-4.00015-1

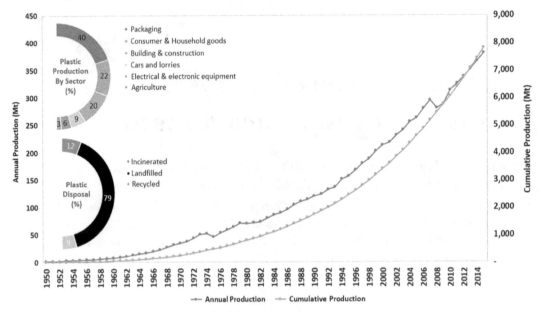

Fig. 1
Global plastic production and disposal overview [1,2].

Numerous simulation approaches exist. Chemical equilibrium, kinetics, and computational fluid dynamics (CFD) are often used [6–8]. The pros and cons of each are summarized in Table 1.

All simulation methodologies have advantages and disadvantages, and the models can become as complex as different evaluation parameters are integrated. Table 2 summarizes various works that have already simulated biomass gasification. However, most of them do not consider efficiency parameters. That is why this work also simulates the cogasification of high-density polyethylene (HDPE) mixed with wood using both a chemical equilibrium model and a kinetic model in Aspen Plus, considering efficiency parameters to get deeper into the analysis.

Utilizing HDPE blends results in gases with higher lower heating values (LHV), adding value to a product that has no other use and must be disposed of in landfills or open dumps. In addition, the simulation of the cogasification process will provide critical information for future research into the technique's practicality on a pilot scale at the University of Aveiro.

2. Problem domain identification of gasification process

This section defines the cogasification process's problem and introduces essential operational concepts, starting by defining cogasification. It is a thermal conversion process that combines at least two feedstocks to transform organic matter into valuable products such as syngas [15]. Cogasification, like gasification, is a process of partial oxidation or incomplete combustion [16]. Cogasification is also a combination of pyrolysis and combustion processes, where

Table 1 Simulation methodologies [6,7].

Methodology	Advantages	Disadvantages
Chemical equilibrium	− Easy to implement − Excellent convergence − Fair matching to the actual process − The employed software gives kinetics	− It does not consider hydrodynamic phenomena − The kinetic data of the software can not apply to a specific process
Kinetic	− More accurate than chemical equilibrium models − It can better predict the composition of the product	− The user must introduce all chemical reactions − The user must introduce all kinetic data − Kinetic data might vary from source to source − Some software does not consider hydro-dynamic phenomena
CFD	− Highly precise − It considers the kinetic − It considers hydrodynamic phenomena	− Expensive − Very complex compared to the other two methodologies − The user must supply reactions and kinetic data

Table 2 Gasification simulation works.

Ref.	Methodology	Kinetics	Hydrodynamics	Feedstock
[9]	Equilibrium with stoichiometric	Unknown	Unknown	Wood
[9]	Equilibrium without stoichiometric	Unknown	Unknown	Wood
[10]	Kinetics	Known	Unknown	Wood
[11]	Kinetics	Known	Partially known	Coal
[12]	CFD	Known	Known	Wood
[13]	CFD	Known	Known	Wood
[14]	CFD	Known	Known	Wood

pyrolysis, oxidation, and reduction reactions occur in different reactor zones depending on the type of reactor. For example, Fig. 2 presents these zones in a circulating fluidized bed (CFB) reactor.

According to the literature, specific reactions occur in each reactor zone. However, gasification is a complex and dynamic process, and what occurs in each reactor zone is unknown. Additionally, many reactions must operate concurrently and compete with one another. Nonetheless, kinetic models have been developed that incorporate the significant reactions, producing consistent results with the experimental results.

$$\text{Biomass} + \text{Air} \xrightarrow{Heat} N_2 + CO_2 + CO + H_2$$

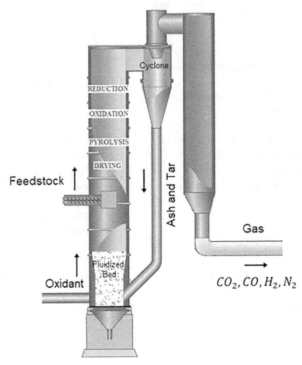

Fig. 2
Reaction zones in a circulating fluidized bed reactor.

2.1 Equivalence ratio

The equivalency ratio (ER) is defined as the mass ratio of fuel to oxidizer under stoichiometry circumstances [17]. According to several authors, ER is the most notable measure for improving the gas quality [18]. Indeed, its influence could be stronger than the reforming and cracking reaction rates [19]. The following equation mathematically depicts the concept:

$$ER = \frac{m_{air}}{m_{(air)st}} \tag{1}$$

where m_{air} is the mass of air corresponding to a particular ER, $m_{(air)st}$ is the mass of the air that stoichiometrically reacts with the mass of carbon and hydrogen in the feedstock, and it is calculated as follows:

$$m_{(air)st} = \left[\left(C_m * \frac{MW_{O_2}}{MW_C} \right) + \left(H_{2m} * \frac{0.5 MW_{O_2}}{MW_{H_2}} \right) \right] * \frac{0.79}{0.21} \tag{2}$$

where C_m is the mass of carbon in the feedstock, MW_C is the molecular weight of carbon, H_{2m} is the mass of hydrogen in the feedstock, MW_{H_2} is the molecular weight of hydrogen, and MW_{O_2} is the molecular weight of oxygen.

2.2 Gasification performance parameters

Several parameters can be used to determine the gasification performance, including gas LHV, cold gas efficiency (CGE), hot gas efficiency (HGE), and gas production, each of which is described in the following sections.

2.2.1 Gas LHV

The gas's lower heating value is determined by the contribution of each of its constituents, as expressed in the following equation:

$$LHV_{Gas} = \frac{\sum y_{iG}*m_i*LHV_i}{m_G} \tag{3}$$

where LHV_{Gas} denotes the produced gas' lower heating value, LHV_i refers to the produced gas' lower heating value of a component, m_i means the mass of a component of the produced gas, m_{Gas} represents the produced gas' mass, and y_{iGas} is the mass fraction of a component of the produced gas.

2.2.2 Cold gas efficiency (CGE)

It is the chemical energy of the gas by the chemical energy of the feedstock [20]. In other words, it is the output energy by input energy [21]. The following equation portrays the concept mathematically:

$$CGE = \frac{LHV_G*m_G}{LHV_F*m_F}*100\% \tag{4}$$

where CGE is the cold gas efficiency, LHV_{Feed} denotes the feed stream's LHV, LHV_{Gas} represents the produced gas' LHV, m_{Gas} indicates the mass of the produced gas, and m_{Feed} refers to the feed stream mass.

2.2.3 Hot gas efficiency (HGE)

The HGE, like the CGE, is the output energy divided by the input energy, but it also takes into account the sensible heat of the feedstock, as the following equation shows:

$$HGE = \frac{(LHV_{Gas}*m_{Gas}) + (Cp_{Gas}*\Delta T*m_G)}{LHV_{Feed}*m_{Feed}}*100\% \tag{5}$$

where Cp_{Gas} denotes the produced gas' specific heat, HGE indicates the hot gas efficiency, LHV_{Feed} represents the feed stream's lower heating value, LHV_{Gas} symbolizes the produced gas' lower heating value, m_{Gas} denotes the mass of the produced gas, m_F represents the mass of the feed stream, and ΔT is the temperature gradient.

2.2.4 Gas yield (Y_{gas})

Y_{gas} is the volume of the generated gas divided by the mass of feedstock, as expressed in the following equation:

$$Y_{gas} = \frac{V_{Gas}}{m_{Feed}} \qquad (6)$$

where m_{Feed} is the mass of the feed stream, m_{Feed} represents the volume of produced gas, and Y_{gas} denotes the amount of produced gas.

The efficiency parameters stated previously are discussed in the results and analysis section. The following part describes the gasification process using kinetic and equilibrium model simulations.

3. Process simulation

The simulation of a gasification process has different objectives, such as estimating the composition of the resulting gas, assessing the yields of gas, char, and tars, and evaluating the process's general efficiency. All this occurs without conducting any experiments in the laboratory or a pilot plant.

To conduct a process simulation, at least two critical elements must be considered: The first element is the simulation's characteristics, including the specifications of the components involved in the processes and the thermodynamic approach to be employed. The second element is the simulation model, either kinetic or equilibrium or CFD.

3.1 Simulation properties

The first step to simulate is to define the properties of our system. One of the properties is the component that will participate in the process.

3.1.1 Components

In Aspen, the components are generally classified as conventional and nonconventional components. Conventional components are the elements or molecules in the Aspen Plus default database. In a gasification process, the conventional components are N_2, H_2, O_2, H_2O, CO, CO_2, and CH_4.

On the other hand, nonconventional components are not in the Aspen Plus default database and must be specified. The gasification process usually involves solid feedstocks, such as agroforestry residues, biomass, HDPE, and ashes. They are defined only by enthalpy and density models and do not participate in phase and chemical equilibrium calculations. However, they can be further characterized in the stream input by providing proximate, ultimate, and sulfur feedstock analysis. That is why they must be decomposed into their primary elements, namely C, H, O, N, and S.

3.1.2 Thermodynamic method

The thermodynamic method is a set of models to calculate thermodynamic, kinetic, and transport properties. In other words, it is a subset of the property method that can be classified as an activity coefficient-based method or an equation of state method. Aspen Plus has several thermodynamical models. Some of them are presented in the following Table 3.

3.2 Equilibrium model

Equilibrium models can estimate the composition of the produced gas, the maximum yield, and the ideal parameters for energy efficiency and syngas heating value for each reactor based on the features of the biomass feedstock [22]. Equilibrium models are classified as nonstoichiometric or stoichiometric. Nonstoichiometric models are based on the Gibbs free energy minimization model, and stoichiometric consider a set of reactions [9,23].

The equilibrium method considers the direct minimization of Gibbs free energy with phase splitting to calculate equilibrium, equilibrium ratios, and the spontaneity of a reaction with acceptable accuracy. The general relationship between the Gibbs free energy and the equilibrium constant is given by the following equations:

$$\Delta G = \Delta G^0 + RT\ln Q \tag{7}$$

Table 3 Thermodynamic method.

Model	Type
ASME steam tables	Fundamental
Ideal Gas	Ideal
Lee–Kesler	Virial
Peng–Robinson	Cubic
Redlich–Kwong	Cubic
Redlich–Kwong–Aspen	Cubic
Standard Redlich–Kwong–Soave	Cubic
Schwartzentruber–Renon	Cubic
SRK-ML	Cubic

From Aspen Plus Software.

$$Q = K = \frac{[C]^c[D]^d}{[A]^a[B]^b} \tag{8}$$

where ΔG is the difference between the free energy of products and reactants at any moment, ΔG^0 is the difference between the free energy of products and reactants at standard conditions, R is gas constant, T is temperature, and Q is the reaction quotient. In the equilibrium $\Delta G = 0$. Thus Eq. (1) can be rearranged as follows:

$$\Delta G^0 = -RT\ln K \tag{9}$$

3.3 Kinetic model

Kinetic models use mathematics to describe the reaction kinetics of the primary reactions and, in some cases, the hydrodynamic interactions between phases within the gasifier [24]. To use kinetic models effectively, it is necessary to have information on the reaction rate, residence time, and in some cases, hydrodynamic parameters, such as gasifier geometry, minimum fluidization velocity, pressure drop, among others.

The kinetic model incorporates kinetics by considering the reaction rate term proportional to the fugacity difference. The reaction rate constant for Aspen plus environment can be defined in two distinct methods. These two alternative configurations are seen in the following equations. One variant includes a reference temperature. The remaining form does not. Naturally, these two types can be equivalent through a simple mathematical rearrangement.

$$\text{With regard to unspecified } r = AT^n e^{-E/RT}\prod_i C_i^{\alpha_i} \tag{10}$$

$$\text{With regard to specified: } r = A\left(\frac{T}{T_0}\right)^n e^{-\left(\frac{E}{R}\right)\left(\frac{1}{T}-\frac{1}{T_0}\right)}\prod_i C_i^{\alpha_i} \tag{11}$$

where r is the rate of reaction, A is the preexponential factor, T is temperature, T_0 is the reference temperature, n is the temperature exponent, E is the activation energy, R is the universal gas law constant, C is component concentration, and i is component index.

On the other hand, the rate constant can be calculated by the Arrhenius equation as shown:

$$k = Ae^{\left(\frac{-E}{RT}\right)} \tag{12}$$

where E is activation energy (J/kmol), R is gas constant (J/kmolK), T is temperature (K), and A is a preexponential factor. It is also called the frequency factor and describes how often two molecules collide. (Units of A and k depend on overall reaction order). $A = p \times Z$, where p is a steric factor, and Z is the collision frequency.

Table 4 Representative gasification reaction [25–28].

Reaction	ΔHR	Reaction #
Dry Biomass $\rightarrow v_{vol}$Volatils + v_{char}Char	−0.52 kJ/kg of biomass	(13)
$C + \frac{1}{2}O_2 \rightarrow CO$	−9250 kJ/kg of C	(14)
$C + CO_2 \rightarrow 2CO$	14,250 kJ/kg of C	(15)
$C + H_2O \rightarrow CO + H_2$	10,833 kJ/kg of C	(16)
$C + 2H_2 \rightarrow CH_4$	−6230 kJ/kg of C	(17)
$CO + \frac{1}{2}O_2 \rightarrow CO_2$	−10,105 kJ/kg of CO	(18)
$CH_4 + 1.5O_2 \rightarrow 2H_2O + CO$	−32,375 kJ/kg of CH_4	(19)
$H_2 + \frac{1}{2}O_2 \rightarrow H_2O$	−241,000 kJ/kg of H_2	(20)
$CO + H_2O \leftrightarrow H_2 + CO_2$	−1470 kJ/kg of CO	(21)
$CO_2 \rightarrow \frac{1}{2}O_2 + CO$	6410 kJ/kg of CO_2	(22)

Table 5 Kinetic constants of heterogeneous reactions.

Reaction	Rate (mol/m³ s)	A	E (J/kmol)	Ref.	#
(10)	$r_P = kC_{Biomass}$	14,400 [s⁻¹]	8.86×10^7	[26]	(23)
(11)	$r = 2\left(\frac{M_{Char}}{M_{O_2}}\right)v_p\left(\frac{kh_m}{k_{C1}+h_m}\right)[O_2]$	$1.7T_S$ [ms⁻¹]	7.48×10^7	[29]	(24)
(12)	$r = \left(\frac{M_{Char}}{M_{CO_2}}\right)v_p\left(\frac{kh_m}{k+h_m}\right)[CO_2]$	$3.42T_S$ [ms⁻¹]	1.30×10^8	[29]	(25)
(13)	$r = \left(\frac{M_{Char}}{M_{CO_2}}\right)v_p\left(\frac{kh_m}{k+h_m}\right)[CO_2]$	$5.71T_S$ [ms⁻¹]	1.30×10^8	[29]	(26)
(14)	$r = \left(\frac{M_{Char}}{M_{(H_2O)v}}\right)v_p\left(\frac{kh_m}{k+h_m}\right)[H_2O]$	$3.42 \times 10^{-3}T_S$ [ms⁻¹]	1.30×10^8	[29]	(27)

Table 6 Kinetic constants of homogeneous reactions.

Reaction	Rate (kmol/m³ s)	A	E (J/kmol)	Ref.	#
(15)	$r = Ae^{\left(\frac{-E}{RT}\right)}[CO][CO_2]^{0.25}[H_2O]^{0.5}$	2.24×10^{12}	1.70×10^8	[28]	(28)
(16)	$r = Ae^{\left(\frac{-E}{RT}\right)}[CH_4]^{0.7}[O_2]^{0.8}$	5.01×10^{11}	2.00×10^8	[28]	(29)
(17)	$r = Ae^{\left(\frac{-E}{RT}\right)}[H_2][O_2]$	9.87×10^8	3.10×10^7	[28]	(30)
(18)	$r = Ae^{\left(\frac{-E}{RT}\right)}[O_2]$	5×10^8	1.70×10^8	[28]	(31)
(19)	$r = A\left([CO][H_2O] - \frac{[CO_2][H_2]}{k_{WG,e}}\right)$	2780	1.26×10^7	[28]	(32)

Many chemical reactions occur during the gasification process, so it is necessary to simplify, considering the most representative reactions. Table 4 presents some of these reactions.

The previous reactions can be classified as homogeneous and heterogeneous reactions, and the kinetic constants are found in Tables 5 and 6, respectively.where v_p is solid particle volume (m^3), $h_{m, j}$ is mass transfer coefficient of the reaction, M_i is molecular weight of the species i (kg/kmol), k_j is kinetic constant of the reaction.

It is worth noting that kinetic models vary as a function of the number of equations considered and the reaction rate values, ranging from author to author.

3.4 Computational fluid dynamics model

CFD modeling is one of the most sophisticated techniques for analyzing the gasification process [30]. CFD models provide temperature profiles for the gasifier's solid, liquid, and gas phases, as well as a chemical and fluid mechanical characterization of the reactor bed [31]. In addition, CFD analysis may optimize gasifier design and investigate gasification process factors such as operating parameters, kinetics, and feedstock physical qualities [32,33].

The hydrodynamics of the gasification process significantly affect the production and composition of syngas since hydrodynamics parameters influence the heat and mass transfer of the gasification process [25] and the residence time distribution [34]. The typical hydrodynamic parameters are the minimum fluidization velocity, bed void fraction, and pressure drop [35].

Minimum fluidization velocity: It is defined as the lowest surface velocity at which the weight of the particles balances the drag force and the fluid's upward buoyant force [36]. In other words, it is the velocity necessary to start fluidization at which the weight of particles' gravitational force matches the drag on the particles from rising gas [37]. Aspen Plus determines minimum fluidization velocity as follow:

- Archimedes number (A_r)

$$A_r = \frac{d_p^3 \rho_g \left(\rho_p - \rho_g\right)}{\eta_g^2} \tag{33}$$

- Parameters a and b

$$a = \frac{1.75}{\phi \varepsilon_{gmf}^2} \tag{34}$$

$$b = \frac{150 \left(1 - \varepsilon_{gmf}\right)}{\phi^2 \varepsilon_{gmf}} \tag{35}$$

- Reynolds number at minimum fluidization (Re_{mf})

$$Re_{mf} = \frac{-b + \sqrt{b^2 + 4aA_r}}{2a} \tag{36}$$

- Minimum fluidizing velocity (U_{mf})

$$U_{mf} = \frac{\eta_g \, Re_{mf}}{d_p \rho_g}$$ (37)

where ρ_p is density of particles in the bed (kgm-3), d_p is mean size (diameter) of particles in the bed (m), ρ_g is density of fluidizing gas (kgm-3), η_g is viscosity of fluidizing gas (Pas), and ϕ is particle shape factor for bed material, which can be calculated as follows:

$$\phi = \frac{\text{Surface area of the sphere of same volume}}{\text{Surface area of the particle}}$$ (38)

A more accurate method of calculating U_{mf} uses the Ergun (1952) equation and assumes ε_{gmf}. Bed voidage at minimum fluidization and ϕ particle shape factor for bed material are known.

Aspen also utilizes the Wen and Yu correlation, which is defined by the following equations

$$U_{mf} = \left((1140 + 0.041 A_r)^{\frac{1}{2}} - 34 \right) \left(\frac{\mu}{\rho_g} d_p \right)$$ (39)

$$U_{mf} = 7.9 \times 10^{-3} d_p^{1.82} \Delta P^{0.94} \mu^{-0.88}$$ (40)

where all variables are in SI units and $Re_{mf} < -10$.

Shape factor and voidage: The proportion of the total volume available for fluid flow, so the bed's fractional volume filled with solid material is $(1 - e)$ [38].

Pressure drop (ΔP): It is the pressure change across a fluidized bed, equal to its effective weight per unit area. The pressure drop in a spouted fluidized bed increases with increasing bed height, spout diameter, and density of the particles. In contrast, the pressure drop decreases with increasing particle diameter and spout gas velocity [39].

$$\Delta P = \frac{M}{\rho_p S_b} \left(\rho_p - \rho_g \right) g_n$$ (41)

where M is mass of particles in the bed (kg), S_b cross-sectional area of the bed (m^2), and g_n is the acceleration due to gravity 9.81 (ms^{-2}).

4. Simulation results of the gasification process

4.1 Equilibrium model

María Gonzalez et al. [9] utilized Aspen plus to simulate the gasification of wood with air in a bubbling fluidized bed reactor, utilizing an equilibrium model, so there is no division of the reaction zones as in kinetic models. In addition, a calculator block was utilized to estimate the yields based on the wood component attributes. The thermodynamic method was utilized in the Peng–Robinson equation of state with Boston–Mathias modification.

The first step is to establish the components that will participate in the process and the parameters of nonconventional components such as wood. The parameters used in this simulation's wood are listed in Table 7.

This model utilized the following equations, where the components that will participate in the process can be noticed.

$$C + O_2 \rightarrow CO_2 \tag{42}$$

$$C + \frac{1}{2}O_2 \rightarrow CO \tag{14}$$

$$C + CO_2 \rightarrow 2CO \tag{15}$$

$$C + H_2O \rightarrow CO + H_2 \tag{16}$$

$$C + 2H_2O \rightarrow CO_2 + 2H_2 \tag{43}$$

$$C + 2H_2 \rightarrow CH_4 \tag{17}$$

$$CO + H_2O \rightarrow H_2 + CO_2 \tag{44}$$

Table 7 Biomass characterization equilibrium model [9].

Proxanal (wt.%, db)	
FC	19.78
VM	78.41
Ash	1.81
Ultanal (wt.%, db)	
Ash	1.81
Carbon	52.32
Hydrogen	6.0.21
Nitrogen	0.62
Chlorine	0
Sulfur	0.05
Oxygen	38.99
Sulfanal	
Pyritic	0
Sulfate	0.55
Organic	0

db; dry basis.

$$H_2O + CH_4 \rightarrow CO + 3H_2 \tag{45}$$

The thermodynamic chemical equilibrium model employs direct Gibbs free energy minimization with phase splitting to calculate equilibrium, equilibrium ratios, and the spontaneity of a reaction. Thus, it is advantageous when the reactions occurring are unknown or in large numbers.

A reaction is highly spontaneous when it does not need an external energy source to continue reacting, such as oxidation reactions. The variation of Gibbs free energy between the products and reactants (ΔG) measures the spontaneity of response. If $\Delta G < 0$, then it is a spontaneous reaction. While if $\Delta G > 0$, then it is a nonspontaneous reaction. The variation of Gibbs free energy can be mathematically expressed as follows:

$$\Delta G = \Delta H - T\Delta S \tag{46}$$

The variation of enthalpy between the products and reactants (ΔH) is heat released or absorbed at constant pressure. If $\Delta H < 0$, then the reaction is exothermic, releasing heat. While, if $\Delta H > 0$, then the reaction is endothermic, so it needs heat. Therefore, exothermic reactions favor spontaneity. The variation of entropy between the products and reactants (ΔS) is a measure of disorder. If $\Delta S < 0$ ($-$), then the disorder decreases. While, if $\Delta S > 0$ ($+$), then disorder increases. For example, when solid carbon reacts, forming gases, the disorder increases because molecules are more disordered than in a solid. Thus, ΔS will be higher than 0. Reactions that favor disorder ($\Delta S > 0$, or positive) favor spontaneity. In this regard, this model will favor oxidation reactions.

The gasification process is depicted in Fig. 3, where the RDF and wood streams are decomposed into their elemental forms (C, H, O, N, and S) in the DECOMP block. The decomposition is based on their proximate and ultimate analysis. Next, these elements react with a stream of air in the gasifier producing CO, CO_2, CH_4, H_2, and other components. Finally, the gas and by-products are separated.

The following figure (Fig. 4) compares experimental findings for wood gasification at ER=0.2 and 700°C to simulation results for wood gasification at the same operating parameters using a nonstoichiometric equilibrium model (Rgibbs) and a stoichiometric equilibrium model (Rstoic). As can be seen, the simulation values differ significantly from the experimental values. These findings may be explained because oxidation reactions are exothermic and very spontaneous, requiring no extra energy to continue reacting. This situation is the case with CH_4 combustion, which seems to be strongly favored in the simulation, and hence, CH_4 is not reported in the Gibbs reactor's generated gas. In contrast, a stoichiometric reactor produces a minimal volume compared with the experimental data. Consequently, the quantity of hydrogen stated in the experimental findings is substantially lower than the amount in the simulation. This situation may be connected to the oxidation of CH_4 that may create CO_2 and H_2.

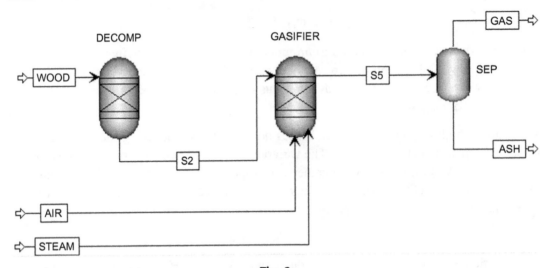

Fig. 3
Nonstochiometric equilibrium model [9].

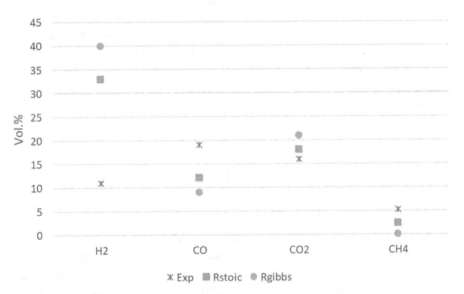

Fig. 4
Nonstochiometric equilibrium model comparison between experimental results with simulation results [9].

Li et al. [40] said that while the pure equilibrium predictions are distant from experimental data, they offer the thermodynamic upper and lower bounds for CO and CO_2. They claimed that experimental scatter is most likely due to system pressure and temperature changes between 155kPa and 1100K. The authors also utilized a kinetically modified model to demonstrate that the H_2 predicted closely fits the experimental data. However, the forecasted CH_4 concentration is one to three orders of magnitude lower than the experimental value. According to Coates, CH_4 high concentration represents a nonequilibrium species formed during pyrolysis and short residence time [41]. Most equilibrium models assume a residence time sufficient to achieve thermodynamic equilibrium [42,43]. However, this assumption might not be valid, resulting in gas composition quite different from experimental data. Furthermore, most simulations do not consider efficiency parameters to evaluate the model's accuracy.

The following section will present a simulation using an equilibrium model and HDPE–wood blends as feedstock.

4.1.1 HDPE–wood blends equilibrium simulations

The equilibrium model decomposes raw materials into their elemental form in a reactor (RYield block). Then these two streams are blended in a mixer (Mixer Block) and enter the gasifier (RGibbs Block). Finally, Aspen determines the yields based on the potential specified products and the direct minimization of Gibbs free energy with phase splitting. Fig. 5 depicts the process.

To validate the equilibrium model, wood was gasified at two ratios, producing gas with different LHV. Experimental values are compared with the simulation findings. Table 8 summarizes and contrasts the experimental with the simulation results.

When pine chips are gasified at 0.23 ER, experimental data indicate that the gas generated has an LHV of $6.2\,MJ/m^3$, but modeling results imply a value of $5.41\,MJ/m^3$. For 0.31 ER, experimental findings suggest that the gas generated has an LHV of $4.65\,MJ/m^3$, whereas experimental results indicate an LHV of $5.165\,MJ/m^3$. These figures indicate a relative

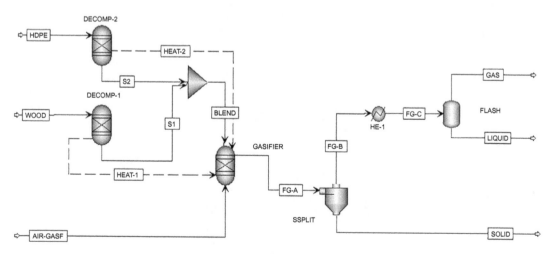

Fig. 5
HDPE–wood blends equilibrium simulations.

Table 8 LHV Comparison of experimental results and simulation results.

	Operating conditions			Volume fraction							LHV
	Mass flow (kg/h)	ER	Air (LPM)	CO_2	CO	H_2	N_2	CH_4	C_2H_4	H_2O	LHV (MJ/m^3)
EXP	14	0.23	200	0.1565	0.1827	0.0638	0.5204	0.052	0.0213	0	6.222
EXP	10	0.31	200	0.1637	0.1348	0.0562	0.5902	0.0388	0.0152	0	4.656
ASPEN	14	0.23	200	0.1712	0.1140	0.2567	0.3975	0.0287	0	0.0316	5.411
ASPEN	10	0.31	200	0.1367	0.1553	0.2309	0.4299	0.0153	0	0.0316	5.165

difference of 13.03% and −10.93% between the experimental and Aspen findings, respectively. These results provide an excellent estimate of the LHV of the resultant gas. In this manner, we will determine the LHV of several wood–HDPE blends.

The percentage corresponding to CO is significantly greater than in the simulation in the experimental data. In addition, while in the simulation findings, the H_2 percentage is more significant than the experimental results.

Three streams with different compositions of HDPE and wood were simulated, maintaining a constant ER of 0.25. Table 9, increasing the amount of HDPE increases the amount of carbon and hydrogen in the blend while decreasing the amount of oxygen, water, and ash. As a result, more oxygen will be required to attain the same ER for blends containing more HDPE, which will undoubtedly affect the gas composition, LHV, and efficiency parameters. As seen in the table, as the amount of HDPE in the blend increases, so does the LHV of the generated gas. This behavior is consistent with Daniel Pio's findings, which suggest that plastics enhance short-chain hydrocarbon synthesis, such as CH_4 and C_2H_4, resulting in higher LHV [44].

The composition of the blends and the ER affect the produced gas composition. Table 10 illustrates the influence of blends' composition on the composition of the produced gases. Take note that the ER of each blend is the same. As can be observed, the addition of HDPE increases CO and H_2 production. However, a diluting effect produced by N_2 is seen since mixtures containing more HDPE require more air to attain the same ER, diluting the produced gas.

Finally, Table 11 presents the efficiency parameters obtained in the equilibrium model. Considering the efficiency parameters ranges obtained experimentally, namely CGE and HGE, the simulation results fit very well in the range since the LHVs obtained are very similar to the experimental ones. However, the Y_{gas} results are very different from those typically obtained experimentally (1.52–2.10). This discrepancy is because the simulation yields more gas than the experimental results since the model does not adequately account for the production of tars, resulting in a gas mass increase and consequently to a rise of the Y_{gas}.

4.2 Kinetic model

Maria Puig et al. [10] utilized Aspen plus to simulate the gasification of wood with air in a bubbling fluidized bed reactor, where three zones mimic the reactor: the pyrolysis, the combustion, and the reduction zone. An independent Aspen block represented each zone. Furthermore, an external MS-Excel subroutine was used to forecast the yield and composition of char, gas, and tar. The kinetic model was calibrated and then verified using experimental results from the pilot-scale bubbling fluidized bed reactor operating at a range of equivalency ratios (from 0.17 to 0.35) and temperatures (from 709°C to 859°C).

Table 9 HDPE—wood blend LHV.

	Feedstock (kg/h)		Feedstock composition (kg/h)							Air			LHV	
Run	HDPE	Pine chips	C	H	O	N	S	H₂O	ASH	Air st (kg/h)	ER	Air for ER (kg/h)	LHV (MJ/m³)	LHV (MJ/kg)
1	0	10	4.1296	0.5874	4.1385	0.0178	0	1.1	0.0267	58.9269	0.25	14.7317	5.19	4.87
2	2.5	7.5	5.17945	0.7888	3.1646	0.0184	0.0018	0.825	0.022	75.4624	0.25	18.8656	5.48	5.49
3	5	5	6.2293	0.9902	2.1908	0.0189	0.0035	0.55	0.0174	91.9978	0.25	22.9995	5.65	5.94

Table 10 Equilibrium gas composition of wood–HDPE blends.

| Run | HDPE (%) | Gas mass fractions | | | | | | | Gas properties | | | |
		H_2O	CO_2	CO	H_2	N_2	CH_4	C_2H_4	m_G (kg/hr)	V_G (m^3/h)	CPMX-M (KJ/kg-K)	Mass density (kg/m^3)
1	0	0.0241	0.2630	0.1773	0.0200	0.5043	0.0113	0.0	23.1083	24.2176	1.2871	0.9542
2	25	0.0256	0.1847	0.1912	0.0236	0.5629	0.0118	0.0	26.5111	29.5639	1.3515	0.8967
3	50	0.0269	0.1294	0.1976	0.0266	0.6075	0.0118	0.0	29.9350	34.9641	1.4032	0.8562

Table 11 Efficiency parameters using the equilibrium model.

Run	HDPE (%)	m_F (kg/h)	LHV MJ/kg	T_G (K)	m_G (kg/h)	V_G (m³/h)	CP_G (MJ/kg K)	ΔT (K)	CGE (%)	HGE (%)	Y_{gas} (kg$_F$/V$_G$)
1	0	10.0	4.87	894.60	23.11	24.22	0.00129	596.45	59.89	69.32	2.422
2	25	10.0	5.49	913.04	26.51	29.56	0.00135	614.89	55.91	64.37	2.956
3	50	10.0	5.94	929.21	29.94	34.96	0.00140	631.06	53.42	61.39	3.496

Table 12 Kinetic model components [10].

Item	ID	Name	Type
1	CO	Carbon-monoxide	Conventional
2	CO_2	Carbon-dioxide	Conventional
3	CH_4	Methane	Conventional
4	C_2H_4	Ethylene	Conventional
5	H_2	Hydrogen	Conventional
6	N_2	Nitrogen	Conventional
7	O_2	Oxygen	Conventional
8	C_6H_6	Benzene	Conventional
9	NO	Nitric-oxide	Conventional
10	NO_2	Nitrogen-dioxide	Conventional
11	H_2S	Hydrogen-sulfide	Conventional
12	C	Carbon-graphite	Solid
13	S	Sulfur	Solid
14	BIOMASS	–	Nonconventional
15	CHAR	–	Nonconventional
16	ASH	–	Nonconventional
17	H_2O	Water	Conventional
18	C_6H_6O	Phenol	Conventional
19	$C_{10}H_8$	Naphthalene	Conventional

Following the previous procedure, the first step is to define the components that participated in the reaction. Table 12 presents the components

As shown in the component list, there are conventional and nonconventional components. They include biomass, char, and ashes and must be defined. On the other hand, tar components such as phenol, naphthalene, and benzene can be observed. However, it is worth noting that tars are composed of at least 34 components, as shown in the accompanying Table 13. Without a doubt considering all the components of tars will make a very complex simulation.

The nonconventional components are characterized by their enthalpy (enthalpy model), density, proximate, ultimate, and sulfur analysis. Table 14 presents the characterization of biomass. This simulation used a coal density model, deviating from the actual biomass density ($577 \, \text{kg/m}^3$ wet basis) and enthalpy values (18.8 MJ/kg) [44].

As shown in Table 15, five of the 10 main reactions determined in the previous section were used. In addition, this simulation uses nine more reactions that will undoubtedly affect the composition of the gas and the yields of each phase.

Table 13 Tar composition.

Item	ID	Name	ER 0.16 (wt%)	0.17	0.23	0.30
1	$C_6H_{14}O_2$	1,3-Pentanediol, 2-methyl	16.72	18.97	12.43	32.08
2	$C_2H_4O_2$	Acetic acid	29.78	19.27	35.51	16.47
3	$C_6H_8O_2$	2-Cyclopenten-1-one, 2-hydroxy-3-methyl-	1.23	–	1.75	1.48
4	$C_3H_6O_2$	2-Propanone, 1-hydroxy-	5.36	3.36	6.50	3.46
5	$C_8H_{18}O_2$	Propane, 2,2'-[ethylidenebis(oxy)]bis-	12.05	21.91	0.55	1.55
6	C_4H_8O	Cyclopropyl carbinol	0.88	–	1.54	1.84
7	$C_4H_6O_3$	Propanoic acid, 2-oxo-, methyl ester	1.29	–	2.15	–
8	$C_9H_{20}O$	Propane, 1,1-dipropoxy-	0.79	2.38	–	–
9	$C_5H_{10}N_2O$	2-Methyliminoperhydro-1,3-oxazine	1.67	–	–	–
10	$C_6H_{10}O_4$	1,2-Ethanediol, diacetate	–	–	1.30	–
11	C_5H_6O2	1,2-Cyclopentanedione	–	–	1.71	–
12	$C_8H_{17}NO$	Oxazolidine, 2,2-diethyl-3-methyl-	–	–	1.97	–
13	$C_{10}H_{22}O_3$	Propane, 2,2',2''-[methylidynetris(oxy)]tris-	–	6.34	–	–
14	C_7H_8O	Creosol	3.09	2.56	2.88	3.67
15	$C_9H_{10}O2$	2-Methoxy-4-vinylphenol	1.10	1.09	1.38	1.55
16	$C_8H_{10}O_3$	Phenol, 2,6-dimethoxy-	3.94	3.14	4.98	6.36
17	$C_{10}H_{12}O_2$	Phenol, 2-methoxy-4-(1-propenyl)-	1.36	3.48	1.38	4.03
18	$C_{11}H_{14}O_3$	Phenol, 2,6-dimethoxy-4-(2-propenyl)-	2.30	–	3.41	1.77
19	$C_9H_{12}O_2$	Phenol, 4-ethyl-2-methoxy-	0.91	–	1.32	1.34
20	$C_7H_8O_2$	Phenol, 2-methoxy-	–	–	1.67	1.77
21	C_7H_8O	Phenol, 3-methyl-	–	–	0.85	–
22	C_6H_6	Benzene	1.36	1.33	1.48	2.97
23	C_7H_8	Toluene	2.49	–	–	3.53
24	C_8H_{10}	Ethylbenzene	–	–	–	1.70
25	C_8H_8	Styrene	1.39	3.06	–	8.62
26	$C_9H_{12}O_3$	1,2,4-Trimethoxybenzene	4.54	–	4.98	–
27	$C_{10}H_{14}O_4$	Benzene, 1,2,3-trimethoxy-5-methyl-	1.32	–	1.54	–
28	$C_{24}H_{38}O_4$	Di-n-octyl phthalate	1.92	1.15	–	–
29	$C_5H_4O_2$	Furfural	4.48	2.80	5.50	5.16
30	$C_9H_{10}O$	Benzyl methyl ketone	–	2.26	2.32	–
31	$C_9H_{14}O_5$	Propanedioic acid, oxo-, bis(1-methylethyl) ester	–	–	0.89	–
32	$C_{11}H_{14}O$	2-Butanone, 3-methyl-1-phenyl-	–	1.25	–	–
33	$C_8H_{16}O_2$	Acetic acid, 3,3-dimethylbut-2-yl ester	–	5.03	–	–
34	$C_{10}H_8$	Naphthalene	–	0.62	–	0.64

Adapted from J. Kluska, M. Ochnio, Characteristic of tar content and syngas composition during beech wood updraft gasification, Eco-Energetics, 2 no. February 2019, 2018.

Table 14 Biomass characterization kinetic model [10].

Proxanal	
MOISTURE	11
FC	16.8
VM	71.1
ASH	1.1
Ultanal	
ASH	1.1
CARBON	50.8
HYDROGEN	6.5
NITROGEN	0.25
CHLORINE	0
SULFUR	0
OXYGEN	41.35
Sulfanal	
PYRITIC	0
SULFATE	0
ORGANIC	0

Table 15 Reactions of the kinetic model [10].

Reaction	Reaction #
Biomass $\rightarrow H_2 + CO + CO_2 + CH_4 + C_2H_4 + C_6H_6 + C_{10}H_8 + C_6H_6O + H_2O + char$	(47)
$\alpha C + O_2 \rightarrow 2(\alpha - 1)CO + (2 - \alpha)CO_2$	(48)
$CO + 0.5O_2 \rightarrow CO_2$	(18)
$CH_4 + 0.5O_2 \rightarrow CO + 2H_2$	(19)
$H_2 + 0.5O_2 \rightarrow H_2O$	(20)
$H_2O + CO \leftrightarrow H_2 + CO_2$	(21)
$C_6H_6O + 4O_2 \rightarrow 3\ H_2O + 6CO$	(49)
$C_6H_6 + 4.5O_2 \rightarrow 3\ H_2O + 6CO$	(50)
$C_{10}H_8 \rightarrow 6.5C + 0.5\ C_6H_6 + 0.5CH_4 + 1.5H_2$	(51)
$CO + H_2O \rightarrow H_2 + CO_2$	(52)
$C + H_2O \rightarrow CO + H_2$	(16)
$H_2O + CH_4 \rightarrow CO + 3H_2$	(53)
$C_6H_6O \rightarrow CO + 0.4C_{10}H_8 + 0.15C_6H_6 + 0.1CH_4 + 0.75H_2$	(54)
$C_6H_6O + 3H_2O \rightarrow 4CO + 0.5C_2H_4 + CH_4 + 3H_2$	(55)

Table 16 Kinetics of the simulation [10,45,46].

Reaction	Reaction rate	A	n	E (kJ/kmol)	#
(44)	$r = AT^n e^{\left(\frac{-E}{RT}\right)} * \frac{6}{dp}[O_2]$ $\alpha = \frac{1+2f}{1+f}$ $f = 4.72 \times 10^{-3} e^{\left(\frac{37,787}{RT}\right)}$	2.98×10^{13}	1	1.49×10^5	(56)
(15)	$r = Ae^{\left(\frac{-E}{RT}\right)}[CO][CO_2]^{0.25}[H_2O]^{0.5}$	1.78×10^{10}	0	1.80×10^5	(57)
(16)	$r = Ae^{\left(\frac{-E}{RT}\right)}[CH_4]^{0.7}[O_2]^{0.8}$	1.58×10^{12}	0	2.02×10^5	(58)
(17)	$r = Ae^{\left(\frac{-E}{RT}\right)}[H_2][O_2]$	1.08×107	0	1.25×10^5	(59)
(18)[a]	$r = Ae^{\left(\frac{-E}{RT}\right)}[CO][H_2O]$	230	0	1.25×10^5	(60)
(45)	$r = AT^n e^{\left(\frac{-E}{RT}\right)}[C_6H_6O]^{0.5}[O_2]$	6.5×10^4	1	1.25×10^4	(61)
(46)	$r = Ae^{\left(\frac{-E}{RT}\right)}[C_6H_6]^{0.5}[O_2]$	2.40×10^{11}	0	1.25×10^5	(62)
(47)	$r = Ae^{\left(\frac{-E}{RT}\right)}[C_{10}H_8]^{1.6}[H_2]^{-0.5}$	1.00×10^{14}	0	3.5E5	(63)
(48)	$r = Ae^{\left(\frac{-E}{RT}\right)}[CO][H_2O] - \frac{[CO_2][H_2]}{k_{eq}}$ $k_{eq} = 0.022e^{\left(\frac{34,730}{RT}\right)}$	230	0	12,560	(64)
(13)	$r = Ae^{\left(\frac{-E}{RT}\right)}[C][H_2O]$	0.008	0	49,887	(65)
(49)	$r = Ae^{\left(\frac{-E}{RT}\right)}[CH_4][H_2O]$	3.00×10^{13}	0	125,000 1.25E5	(66)
(50)	$r = AT^n e^{\left(\frac{-E}{RT}\right)}[C_6H_6O]$	1.00×10^7	0	100,000	(67)
(51)	$r = AT^n e^{\left(\frac{-E}{RT}\right)}[C_6H_6O]$	1.00×10^7	0	100,000	(68)

[a]Reversible reaction $A = -3.54$, $B = 4004$, Temperature $= 500°C$.

Thermal breakdown causes significant physical and chemical changes in the feedstock particles, culminating in gases and liquids (condensable vapors). The number and composition of the products are determined by the feedstock used and the gasifier's operating conditions. The dynamics of this process in heat degradation are highly complex, and several models have been proposed in the literature. Table 16 presents the data used in this simulation process.

The kinetic parameters of the eight reactions equal to those described in the previous section are different. This fact also greatly affects the composition of the syngas and the yields of each phase.

After defining the previous parameters, a flowsheet is performed using the available Aspen blocks. This simulation depicted the process, as Fig. 6 shows.

The flowsheet begins with a reactor (PYROLYSIS—RYield Block), which will imitate the pyrolysis stage, producing char and volatiles. Next, the drying and pyrolysis stages were simulated using an existing model [47], which provides accurate data on the pyrolytic products char, gas, and tar. This model considers eight species: tar (represented by a blend of C_6H_6, $C_{10}H_8$, and C_6H_6O), H_2, H_2O, CO, CO_2, CH_4, and dry ash-free char [10,47].

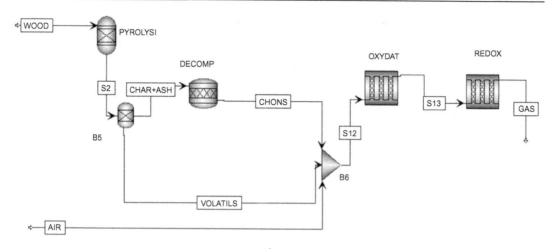

Fig. 6
Flowsheet of the kinetic model [10].

After the pyrolysis stage, products are separated into volatiles and char + ash (Sep Block). Next, the char is decomposed into its elemental form, C, H, O, N, and Ash. Then all elements are mixed in a single stream to go into the oxidation phase (RPlug Block). The previous section discussed the reactions that occur inside the reactor. These reactions generate heat, and the products of these processes also generate reduction reactions in another reactor (RPlug).

The simulation results are compared with experimental results in Fig. 7. The model predicted CO, CO_2, CH_4, and C_2H_4 with better accuracy than H_2 and N_2. These discrepancies between the simulation and experimental results can be attributed to a variety of factors, including the values of the kinetic parameters used, the number of established components, and, of course, the hydrodynamic effects that are not included in the simulation but have a significant effect on mass and heat transfer. Without a doubt, the results achieved with the gasification simulation employing a kinetic model are pretty close to those obtained experimentally. However, it should be noted that the quality of the results will rely on the number of reactions and kinetic parameters employed in the model.

Some researchers have used Aspen Plus to model biomass gasification in fluidized bed reactors. Marcantonio et al. [48] used Aspen Plus to model H2 generation from biomass gasification in fluidized bed reactors with various separation techniques. The authors determined that the quasi-equilibrium model utilized can make reasonable predictions about the gasification process. They study the effect of steam on gas composition. The kinetic data are not published, and efficiency parameters are not studied. Behesht et al. [35] used an approach similar to Maria Puig, as they first pyrolyzed the biomass in an RYield reactor based on empirical correlations. Similarly, it separates volatiles from solids (char and ashes) using a separation column and then gasifies the volatiles in an RCSTR type reactor, considering specific kinetic parameters (Different from those described in previous sections). It is worth mentioning that the solids are

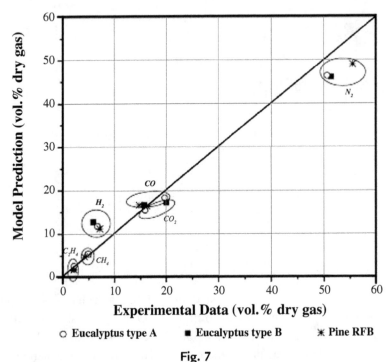

Fig. 7

Kinetic model, comparison between experimental results with simulation results.

bypassed and do not interact in the gasification reactor. On the other hand, Maria Puig decomposes the char into its primary elements (CHONS) to involve them in the gasification reactor. However, Behesht [49] employed different particle sizes, and the results indicated that the smaller diameters favor the production of H_2, CH_4, and CO, which is beneficial for the quality of the gas produced. This result may be due to better heat transfer in smaller particles. Furthermore, the study concludes that the influence of the ER is greater than the influence of other parameters, as other authors have suggested [50–52]. A steady-state model for biomass gasification was built using ASPEN PLUS, where the reaction kinetics and bed hydrodynamics were sent through a FORTRAN-coded library. Two sets of experimental data from pilot-scale bubbling fluidized bed gasification systems were used to confirm the simulation findings. The model can forecast gasifier performance under various operating variables, including temperature, steam, and equivalency. The model predictions agreed well with the observed values. The results also indicated that increasing the temperature enhanced the gasification process, boosting the total gas output by increasing hydrogen generation and decreasing the amount of tar in the gas. The influence of steam ratio on gas output was insignificant within the specified range of 0.5–1, while its effect on tar conversion was straightforward and advantageous at higher gasifier temperatures. However, some values may be inaccurate due to model constraints such as simplified bed hydrodynamics, omitted fluid mixing and mass transfer, and so on. The need to overcome these constraints will be the subject of future effort since in most simulations, that is why Cheolyong et al. [53] said that beyond comprehending the included phenomena, it is crucial to develop a new reaction engineering process, such as reversely analyzing it and upgrading the procedures toward a rational reactor designing tool. Using numerical models that assist in identifying the ideal reactor design and operating conditions for optimizing efficiency while decreasing unwanted components such as tar.

4.2.1 HDPE–wood blends equilibrium simulations

The kinetic model of the gasification process was constructed using the blocks and streams schematically depicted in Fig. 8.

The kinetic model begins with a mixer (MIXER—Mixer Block), which combines wood and HDPE, depending on the desired blend. Next, this blend will be added to a reactor (PYROLYSIS—RYield Block), which will imitate the pyrolysis stage, producing char and volatiles. To define this block, it is necessary to specify the product yields as Table 6 describes. Next, the drying and pyrolysis stages were simulated using an existing model [47], which provides accurate data on the pyrolytic products char, gas, and tar. This model considers eight species: tar (represented by a blend of C_6H_6, $C_{10}H_8$, and C_6H_6O), H2, H_2O, CO, CO_2, CH_4, and dry ash-free char. The model is based on the following empirical equations used in the Aspen mode through an external Excel subroutine. For more information, consult this reference [10]. To perform this simulation, it considered the reactions and kinetic parameters described in Tables 15 and 16, respectively.

The kinetic model is validated based on the gas composition produced at the Aveiro's facilities. The experimental findings were achieved by gasifying pine chips at the University of Aveiro's pilot-scale BFB reactor (80 kWth) [44]. The gasifier's biomass feed is fed by a screw, which controls the flow by changing the speed of the screw. On the other hand, the air is fed at a steady rate of 200LPM. Therefore, if it is required to use another ER, the biomass flow is decreased or raised, depending on the desired ER.

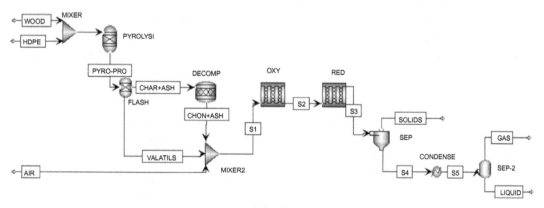

Fig. 8
HDPE–wood blends kinetic simulations.

The composition of the resultant gas in terms of CO_2, CO, CH_4, and C_2H_4 is remarkably comparable to the experimental data, as shown in Table 17. Therefore, the model is validated.

Kinetic models provide more precision when identifying the composition of a product since they employ the chemical kinetics of the critical reactions that occur in a process. Table 18 presents the kinetic model results for three wood–HDPE blends (0, 25, and 50% wt. HDPE) at a constant ER of 0.25. It can be seen that as the percent weight of HDPE grows, the concentrations of CO_2, H_2, CH_4, and C_2H_4 drop. On the other hand, the concentration of CO and N_2 rises. The increase in N2 is because blends with a higher proportion of HDPE contain more carbon and hydrogen, necessitating the addition of a greater volume of air to reach the required ER (0.25). This situation has a considerable diluting impact on the generated gas, lowering its LHV. However, the quantity of gas produced is larger in blends containing a larger share of HDPE.

The variables required to compute the efficiency parameters are listed in Table 19, along with the efficiency parameters for each blend. According to Daniel Pio et al. [44], the CGE ranges between 40% and 52% and the CGE between 63% and 84%. A noticeable difference in these parameters may be noticed on a bullish basis. This trend is because the LHVs of the gases produced in the simulation are considerably higher than those obtained experimentally. The numerator of Eqs. (4) and (5) contains the LHV of the produced gas. Thus, having a larger LHV than the experimental LHV results in higher efficiency parameters.

4.3 Gasification optimization

One of the most critical aspects to optimize is the mass ratio of oxidizing agent to solid fuel. To determine the optimum values of these parameters, various variables such as the amount and composition of the gas, reactor temperature, total carbon conversion, gas tar content, and gas heating value must be observed under various values of the oxygen to mass feedstock ratio and the steam to mass feedstock ratio. This method may be implemented in a sensitive analysis block by designating all specified variables as variables and selecting the mass flow of the OXYGEN and STEAM streams, respectively. Observable variables include the mole fraction of the primary gas component, feedstock mass flow, reactor temperature, tar yield, and producer gas mole flow.

Another critical point to evaluate is the reactions considered in the model since some responses will have a more significant effect than others. For example, in this model, it was deemed that the pyrolysis stage produces benzene, phenol, naphthalene, and water as the liquid phase of pyrolysis. These components are partially oxidized in a reactor, and then the produced gases pass to a reduction stage to deliver the final gas. If more products generated in the pyrolysis phase are specified, more reactions that oxidize these components formed in the gasification phase will have to be determined; these reactions can alter the composition of the resulting gas and the number of tars, char, and gas.

Table 17 Kinetic model validation.

	Operating conditions			Volume fraction								
	Mass flow (kg/h)	ER	Air (LPM)	CO_2	CO	H_2	N_2	CH_4	C_2H_4	C_2H_6	C_3H_8	H_2O
EXP	14	0.23	200	0.1565	0.1827	0.0638	0.5205	0.0520	0.0214	0.0027	0.0005	0
EXP	10	0.31	200	0.1637	0.1349	0.0562	0.5903	0.0388	0.0153	0.0008	0.0001	0
ASPEN	14	0.23	200	0.1508	0.1780	0.1188	0.4265	0.0538	0.0199	0.0000	0.0000	0.02696
ASPEN	10	0.31	200	0.148874	0.162028	0.098	0.48601	0.0424	0.0156	0.0000	0.0000	0.02695

Table 18 Kinetic gas composition of wood–HDPE blends.

Run	HDPE (%)	Gas mass fractions									m_G (kg/h)	V_G (m³/h)	Gas properties			
		H_2O	CO_2	CO	H_2	N_2	CH_4	C_2H_4	Benzene	$C_{10}H_8$			CPMX-M (kJ/kg K)	Mass Density (kg/m³)	LHV (MJ/m³)	LHV (MJ/kg)
1	0	0.0171	0.2253	0.1599	0.0068	0.4886	0.0223	0.0143	0.0311	0.0347	23.1767	20.1659	1.1368	1.1493	9.18	7.15
2	25	0.0172	0.2024	0.1640	0.0060	0.5381	0.0152	0.0093	0.0251	0.0227	26.9481	23.6065	1.1202	1.1416	7.28	5.71
3	50	0.0172	0.1836	0.1683	0.0054	0.5756	0.0099	0.0056	0.0206	0.0137	30.6935	27.0131	1.1072	1.1362	5.88	4.63

Table 19 Efficiency parameters using the kinetic model.

Run	HDPE (%)	m_F (kg/hr)	LHV MJ/kg	T_G (K)	m_G (kg/hr)	V_G (m³/hr)	CP_G (MJ/kg K)	ΔT (K)	CGE (%)	HGE (%)	Y_{gas} (kgF/VG)
1	0	10.0	7.15	950.15	23.18	20.17	0.00114	652.00	88.17	97.30	2.017
2	25	10.0	5.71	950.15	26.95	23.61	0.00112	652.00	59.10	66.66	2.361
3	50	10.0	4.63	950.15	30.69	27.01	0.00111	652.00	42.74	49.40	2.701

5. Simulation results of the pyrolysis process

Pyrolysis is simulated as a subprocess of the gasification step. However, the equilibrium model does not treat it as a distinct stage since the Gibbs reactor used to mimic the gasification reactor comprises all phases of the gasification process (drying, pyrolysis, oxidation, and reduction). Additionally, this reactor is based on direct Gibbs free energy minimization through phase splitting to determine equilibrium and equilibrium ratios. Thus, the user is limited to specifying potential products. By contrast, the kinetic model addresses these four steps individually to construct the gasification process, allowing for independent analysis of pyrolysis. Indeed, pyrolysis is the initial step of the gasification process in the examined simulation by Maria Puig et al. [10]. The pyrolysis zone was simulated by an existing model created by Neves et al. [47]. They consider three main products solid (char and ashes), liquids (bio-oil, tars, and water), and gases (short-chain hydrocarbons).

The solid phase can be calculated as follows:

$$Y_{char} = \frac{Y_S - Y_{a.S}}{1 - Y_{a.S}} \tag{69}$$

where Y_{char} is the yield of char without ashes, Y_S is the yield of the solid phase (dry fuel basis), $Y_{a.\,S}$ is the ash content of solid phase, which is assumed as equal to the ash content of the feedstock ($Y_{a.\,F}$).

The liquid phase can be calculated as follows:

$$Y_{liquid} = +Y_{tar} + Y_{H_2O} + Y_M \tag{70}$$

where Y_{liquid} is the yield of liquid, Y_{tar} is the yield of tar or bio-oil produced, Y_{H_2O}, is the water form during the pyrolysis, and Y_M is the moisture of the feedstock.

According to Neves, the gas phase is often reported as yields or Nm^3 gas/kg fuel. When represented as yields, the stoichiometry of biomass pyrolysis may be determined as follows:

$$(Y_F + Y_{a.F}) \frac{1}{1 - Y_M} = Y_{char} + Y_{tar} + Y_{H_2O} + Y_M + Y_{gas} + Y_{a.F} \tag{71}$$

where Y_F is the dry ash-free feedstock, $Y_{a.\,F}$ is ash content of the feedstock, Y_M is the moisture of the feedstock, Y_{char} is the yield of char without ashes, Y_{tar} is the yield of tar, Y_{H_2O} is the water form during the pyrolysis, Y_M is the moisture of the feedstock, and Y_{gas} is the gas yield, which can also be expressed as:

$$Y_{gas} = Y_{C_xH_x} + Y_{CH_4} + Y_{CO} + Y_{CO_2} + Y_{H_2} \tag{72}$$

Daniel Neves developed an empirical model based on experimental data to anticipate the general patterns in product distribution as a linear function of temperature.

Table 20 Pyrolysis results of wood [54].

ID	Simulation yield	Experimental yield
Char	0.20648514	0.2473
H_2O	0.27589955	58.16
C_6H_6O	0.32163098	
C_6H_6	0.05360516	
CO	0.09350016	0.0378
CO_2	0.04135178	0.1888
H_2	0.00035234	0.0021
CH_4	0.00717488	0.0011
C_2H_4	0	0.0033
Phases		
Char	0.20648514	0.2473
Tar	0.53605163	58.16
Gases	0.25746323	17.11

$$Y_{char} = 0.106 + 2.43 \bullet \exp\left(-0.66 \bullet 10^{-2} \bullet T\right) \text{ where } R^2 = 0.56 \tag{73}$$

The yields of volatile species (as and liquid phase) are forecasted using a set of linear equations, which define the overall elemental mass balance and the trends of selected empirical data as a function of temperature. For more information, consult [10,47].

The simulation results of wood are presented in Table 20, and it is contrasted with experimental results under the same operational conditions (500°C).

It can be seen that there is a significant difference between the experimental results and the results obtained in the simulation. This situation is because this pyrolysis model considers linear equations with R2 less than 0.80 in some cases, which is entirely understandable since the information used in this model is from a variety of feedstocks. Furthermore, the model considers only eight species forms, which is not valid. Nevertheless, despite these discrepancies, the model is helpful to forecast the phase yields and gas composition.

5.1 Pyrolysis optimization

This simulation model considers the effects of temperature. However, it only considers eight compounds as the pyrolytic products: tar (benzene, phenol, naphthalene, and water), H_2, H_2O, CO, CO_2, CH_4, and dry ash-free char. One way to optimize the simulation to obtain more realistic results would be to increase the number of reactions in the model. However, adding more reactions would make the model more complex and challenging to compile.

One way to optimize the simulation to obtain more realistic results would be to increase the number of reactions in the model. However, adding more reactions would make the model more complex and challenging to compile. That is why the formation of the most representative components should be chosen wisely. For example, Table 20 mentions that 1,3-Pentanediol, 2-methyl, and acetic acid are the most representative compounds accounting together 38.24%–48.55% depending on the ER used.

6. Conclusion

This book chapter evaluated some gasification and pyrolysis simulations developed in Aspen Plus, considering the equilibrium and kinetic models. Both models have advantages and disadvantages. For example, the equilibrium model can determine an LHV of the produced gas very close to the experimental results. However, the composition of the produced gas varies considerably from the experimental results. On the other hand, the kinetic model offers values of the composition of the produced gas significantly closer to the experimental results but an LHV that is further away from them. These variations will impact the efficiency parameters. For example, the equilibrium model assumes no tar formation, so Y_{gas} is far superior to what is reported in practical studies. In contrast, if the kinetic model reports higher LHV values than the experimental ones, the CGE and HGE parameters will vary significantly from the experimental efficiency parameters.

The success of the kinetic model is a function of the reactions and kinetic parameters considered in the simulation since these factors determine the gas composition. It is worth noting that these parameters vary from author to author. The considerations made in the process also greatly influence the results. For example, some authors consider pyrolysis and drying in the first stage of the gasification process. Therefore, the pyrolysis results will determine the reactions in the following stages, namely oxidation and reduction. It was observed that the model used for pyrolysis considers the decomposition of biomass into nine components, benzene, phenol, naphthalene, and water representing the liquid phase, H_2, CO, CO_2, CH_4 represents the gas phase, and char the solid phase. These results are far from the reality since it is reported that tar is composed of at least 32 elements, with 1,3-Pentanediol, 2-methyl, and acetic acid being the most predominant compounds, which were not considered in the pyrolysis model, and therefore the oxidation and reduction reactions that involve these compounds. Another essential factor that influences the results is the hydrodynamic parameters such as minimum fluidization velocity that determines the residence time of the reagents and, therefore, it will influence the results.

Despite these limitations presented by both models, their combination strengthens the analysis of the pyrolysis and gasification process, thus having a good idea of the behavior of these processes without the need to develop experimentation, which can be expensive and time-consuming. It also gives us the flexibility to evaluate many raw materials and even combinations of these, as was the case with HDPE–wood blends. Thus, it can visualize a possible use of problematic materials such as plastics, adding value to a product that has no other reason than to be disposed of in sanitary landfills in the best of cases.

Acknowledgments

The authors would also like to express their gratitude to the Fundação para a Ciência e a Tecnologia (FCT) for the grant SFRH/BD/146155/2019 and for the projects SAICTALT/39486/2018, PTDC/EME-REN/4124/2021, and CEECIND/2021/02603. Thanks are also due to FCT, Ministry of Science, Technology and Higher Education (MCTES) for the financial support to CESAM (UIDP/50017/2020 + UIDB/50017/2020 + LA/P/0094/2020), through national funds.

References

[1] European Parliament, Plastic Waste and Recycling in the EU: Facts and Figures, 2018, Dec. 19 https://www.europarl.europa.eu/news/en/headlines/priorities/fighting-plastic-pollution/20181212STO21610/plastic-waste-and-recycling-in-the-eu-facts-and-figures. (accessed May 28, 2021).

[2] M. Eriksen, et al., Plastic pollution in the World's oceans: more than 5 trillion plastic pieces weighing over 250,000 tons afloat at sea, PLoS One 9 (12) (2014), https://doi.org/10.1371/journal.pone.0111913.

[3] UNEP, Our Planet Is Drowning in Plastic Pollution—it's Time for Change, 2018. https://www.unep.org/interactive/beat-plastic-pollution/. (accessed May 31, 2021).

[4] J. Cardoso, V.B. Silva, D. Eusébio, Implementation guidelines for modelling gasification processes in computational fluid dynamics: a tutorial overview approach, in: Inamuddin, A. Asiri (Eds.), Nanotechnology in the Life Sciences, Springer Science and Business Media B.V, 2020, pp. 359–379.

[5] V. Ferraz de Campos, V. Silva, J.S. Cardoso, P. Brito, C. Tuna, J. Silveira, A review of waste management in Brazil and Portugal: waste-to-energy as pathway for sustainable development, Renew. Energy 178 (2021) 802–820.

[6] L. Moretti, F. Arpino, G. Cortellessa, S. Di Fraia, M. Di Palma, L. Vanoli, Reliability of equilibrium gasification models for selected biomass types and compositions: an overview, Energies 15 (2022) 61.

[7] V.B.R.E. Silva, J. Cardoso, How to approach a real CFD problem—a decision-making process for gasification, in: Comput. Fluid Dyn. Appl. to Waste-to-Energy Process., Jan. 2020, pp. 29–83, https://doi.org/10.1016/B978-0-12-817540-8.00002-9.

[8] J. Cardoso, V. Silva, D. Eusébio, P. Brito, R.M. Boloy, L. Tarelho, J. Silveira, Comparative 2D and 3D analysis on the hydrodynamics behaviour during biomass gasification in a pilot-scale fluidized bed reactor, Renew. Energy 131 (2019) 713–729.

[9] M. Pilar González-Vázquez, F. Rubiera, C. Pevida, D.T. Pio, L.A.C. Tarelho, Thermodynamic analysis of biomass gasification using Aspen plus: comparison of stoichiometric and non-stoichiometric models, Energies 14 (1) (2021) 189, https://doi.org/10.3390/en14010189.

[10] M. Puig-Gamero, D.T. Pio, L.A.C. Tarelho, P. Sánchez, L. Sanchez-Silva, Simulation of biomass gasification in bubbling fluidized bed reactor using aspen plus®, Energy Convers. Manag. 235 (2021), https://doi.org/10.1016/j.enconman.2021.113981, 113981.

[11] T.M. Ismail, M. Shi, J. Xu, X. Chen, F. Wang, M.A. El-Salam, Assessment of coal gasification in a pressurized fixed bed gasifier using an ASPEN plus and Euler-Euler model, Int. J. Coal Sci. Technol. 7 (3) (2020) 516–535, https://doi.org/10.1007/s40789-020-00361-w.

[12] V. Silva, J. Cardoso, Computational Fluid Dynamics Applied to Waste-to-Energy Processes, Elsevier, 2020.

[13] B. Pandey, Y.K. Prajapati, P.N. Sheth, CFD analysis of biomass gasification using downdraft gasifier, Mater. Today Proc. 44 (2020) 4107–4111, https://doi.org/10.1016/j.matpr.2020.10.451.

[14] U. Kumar, M.C. Paul, CFD modelling of biomass gasification with a volatile break-up approach, Chem. Eng. Sci. 195 (2019) 413–422, https://doi.org/10.1016/j.ces.2018.09.038.

[15] A.Y. Mutlu, Ö. Yücel, F. Elmaz, Evaluating the effect of blending ratio on the co-gasification of high ash coal and biomass in a fluidized bed gasifier using machine learning, Mugla J. Sci. Technol. 5 (1) (2019) 1–15, https://doi.org/10.22531/muglajsci.471538.

[16] S.-L. Chang, C.Q. Zhou, Combustion and thermochemistry, in: C.J. Cleveland (Ed.), Encyclopedia of Energy, Elsevier, 2004, pp. 595–603.

[17] P. Palies, Premixed swirling flame stabilization, in: P. Palies (Ed.), Stabilization and Dynamic of Premixed Swirling Flames, Elsevier, 2020, pp. 105–158.

[18] S.A. Salaudeen, P. Arku, A. Dutta, Gasification of plastic solid waste and competitive technologies, in: S.M. Al-Salem (Ed.), Plastics to Energy: Fuel, Chemicals, and Sustainability Implications, Elsevier, 2018, pp. 269–293.

[19] R. Xiao, B. Jin, H. Zhou, Z. Zhong, M. Zhang, Air gasification of polypropylene plastic waste in fluidized bed gasifier, Energy Convers. Manag. 48 (3) (2007) 778–786, https://doi.org/10.1016/j.enconman.2006.09.004.

[20] A.M. Shakorfow, Operating and performance gasification process parameters, Int. J. Sci. Res. 5 (6) (2016) 1768–1775.

[21] P. Ponangrong, A. Chinsuwan, An investigation of performance of a horizontal agitator gasification reactor, Energy Procedia 157 (2019) 683–690, https://doi.org/10.1016/j.egypro.2018.11.234.

[22] N. Mazaheri, A.H. Akbarzadeh, E. Madadian, M. Lefsrud, Systematic review of research guidelines for numerical simulation of biomass gasification for bioenergy production, Energy Convers. Manag. 183 (2019) 671–688, https://doi.org/10.1016/J.ENCONMAN.2018.12.097.

[23] P. Sharma, B. Gupta, M. Pandey, K. Singh Bisen, P. Baredar, Downdraft biomass gasification: a review on concepts, designs analysis, modelling and recent advances, Mater. Today Proc. 46 (2021) 5333–5341, https://doi.org/10.1016/J.MATPR.2020.08.789.

[24] B. Das, A. Bhattacharya, A. Datta, Kinetic modeling of biomass gasification and tar formation in a fluidized bed gasifier using equivalent reactor network (ERN), Fuel 280 (2020), https://doi.org/10.1016/J.FUEL.2020.118582, 118582.

[25] C. Di Blasi, Modeling wood gasification in a countercurrent fixed-bed reactor, AICHE J. 50 (9) (2004) 2306–2319, https://doi.org/10.1002/AIC.10189.

[26] F.V. Tinaut, A. Melgar, J.F. Pérez, A. Horrillo, Effect of biomass particle size and air superficial velocity on the gasification process in a downdraft fixed bed gasifier. An experimental and modelling study, Fuel Process. Technol. 89 (11) (2008) 1076–1089, https://doi.org/10.1016/J.FUPROC.2008.04.010. undefined.

[27] M.L. de Souza-Santos, Solid Fuels Combustion and Gasification : Modeling, Simulation, and Equipment Operation, Marcel Dekker, New York, 2004.

[28] A. Barreto, Modelación de la gasificación de biomasa en un reactor de lecho fijo en contracorriente, Universidad de los Andes, 2010.

[29] K.M. Bryden, K.W. Ragland, Numerical modeling of a deep, fixed bed combustor, Energy Fuels 10 (1996) 269–275.

[30] V.B.R.E. Silva, J. Cardoso, Introduction and overview of using computational fluid dynamics tools, in: Comput. Fluid Dyn. Appl. to Waste-to-Energy Process, Jan. 2020, pp. 3–28, https://doi.org/10.1016/B978-0-12-817540-8.00001-7.

[31] N. Couto, V. Silva, C. Bispo, A. Rouboa, From laboratorial to pilot fluidized bed reactors: analysis of the scale-up phenomenon, Energy Convers. Manag. 119 (2016) 177–186.

[32] C. Siripaiboon, P. Sarabhorn, C. Areeprasert, Two-dimensional CFD simulation and pilot-scale experimental verification of a downdraft gasifier: effect of reactor aspect ratios on temperature and syngas composition during gasification, Int. J. Coal Sci. Technol. 7 (3) (2020) 536–550, https://doi.org/10.1007/S40789-020-00355-8/FIGURES/9.

[33] J. Cardoso, V. Silva, D. Eusébio, P. Brito, M.J. Hall, L. Tarelho, Comparative scaling analysis of two different sized pilot-scale fluidized bed reactors operating with biomass substrates, Energy 151 (2018) 520–535.

[34] C.E. Agu, C. Pfeifer, M. Eikeland, L.A. Tokheim, B.M.E. Moldestad, Measurement and characterization of biomass mean residence time in an air-blown bubbling fluidized bed gasification reactor, Fuel 253 (2019) 1414–1423, https://doi.org/10.1016/J.FUEL.2019.05.103.

[35] S.M. Beheshti, H. Ghassemi, R. Shahsavan-Markadeh, Process simulation of biomass gasification in a bubbling fluidized bed reactor, Energy Convers. Manag. 94 (2015) 345–352, https://doi.org/10.1016/J.ENCONMAN.2015.01.060.

[36] P.D.S. de Vasconcelos, A.L.A. Mesquita, Minimum and full fluidization velocity for alumina used in the aluminum smelter, Int. J. Eng. Bus. Manage. 3 (4) (2011) 7–13, https://doi.org/10.5772/50943.

[37] J.A. Tom, E.O. Ehirim, Correlation of minimum fluidization, fluidization and terminal velocities in the design of fluidized bed electrochemical reactor, Int. J. Eng. Mod. Technol. 4 (3) (2018). Accessed: Dec. 20, 2021. [Online]. Available www.iiardpub.org.

[38] R.P. Chhabra, J.F. Richardson, Particulate systems, in: Non-Newtonian Flow Process Ind, Jan. 1999, pp. 206–259, https://doi.org/10.1016/B978-075063770-1/50006-3.

[39] E.R. Monazam, W. Ronald, J. Weber, Analysis of maximum pressure drop for a flat-base spouted fluid bed, Chem. Eng. Res. Des. 122 (2017) 43–51. Accessed: Dec. 20, 2021. [Online]. Available: https://www.osti.gov/pages/servlets/purl/1477173.

[40] X. Li, J.R. Grace, A.P. Watkinson, C.J. Lim, A. Ergüdenler, Equilibrium modeling of gasification: a free energy minimization approach and its application to a circulating fluidized bed coal gasifier, Fuel 80 (2) (2001) 195–207, https://doi.org/10.1016/S0016-2361(00)00074-0.

[41] R.L. Coates, C.L. Chen, B.J. Pope, Coal devolatilization in a low pressure, low residence time entrained flow reactor, Adv. Chem. Ser. 13 (1974) 92–107, https://doi.org/10.1021/BA-1974-0131.CH007.

[42] A. Gagliano, F. Nocera, M. Bruno, G. Cardillo, Development of an equilibrium-based model of gasification of biomass by Aspen plus, Energy Procedia 111 (2017) 1010–1019, https://doi.org/10.1016/J.EGYPRO.2017.03.264.

[43] A. Gambarotta, M. Morini, A. Zubani, A non-stoichiometric equilibrium model for the simulation of the biomass gasification process, Appl. Energy 227 (2018) 119–127, https://doi.org/10.1016/J.APENERGY.2017.07.135.

[44] D.T. Pio, L.A.C. Tarelho, A.M.A. Tavares, M.A.A. Matos, V. Silva, Co-gasification of refused derived fuel and biomass in a pilot-scale bubbling fluidized bed reactor, Energy Convers. Manag. 206 (2020), https://doi.org/10.1016/j.enconman.2020.112476, 112476.

[45] C.K. Westbrook, F.L. Dryer, Chemical kinetic modeling of hydrocarbon combustion, Prog. Energy Combust. Sci. 10 (1) (1984) 1–57, https://doi.org/10.1016/0360-1285(84)90118-7.

[46] W.M. Champion, C.D. Cooper, K.R. Mackie, P. Cairney, Development of a chemical kinetic model for a biosolids fluidized-bed gasifier and the effects of operating parameters on syngas quality, J. Air Waste Manag. Assoc. 64 (2) (2013) 160–174, https://doi.org/10.1080/10962247.2013.845619.

[47] D. Neves, H. Thunman, A. Matos, L. Tarelho, A. Gómez-Barea, Characterization and prediction of biomass pyrolysis products, Prog. Energy Combust. Sci. 37 (5) (2011) 611–630, https://doi.org/10.1016/J.PECS.2011.01.001.

[48] V. Marcantonio, M. De Falco, M. Capocelli, E. Bocci, A. Colantoni, M. Villarini, Process analysis of hydrogen production from biomass gasification in fluidized bed reactor with different separation systems, Int. J. Hydrogen Energy 44 (21) (2019) 10350–10360, https://doi.org/10.1016/J.IJHYDENE.2019.02.121.

[49] A. Erkiaga, G. Lopez, M. Amutio, J. Bilbao, M. Olazar, Influence of operating conditions on the steam gasification of biomass in a conical spouted bed reactor, Chem. Eng. J. 237 (2014) 259–267, https://doi.org/10.1016/J.CEJ.2013.10.018.

[50] U. Arena, Fluidized bed gasification, in: Fluid. Bed Technol. Near-Zero Emiss. Combust. Gasif, Jan. 2013, pp. 765–812, https://doi.org/10.1533/9780857098801.3.765.

[51] V.S. Sikarwar, M. Zhao, Biomass gasification, in: Encycl. Sustain. Technol, Jan. 2017, pp. 205–216, https://doi.org/10.1016/B978-0-12-409548-9.10533-0.

[52] P. Kaushal, R. Tyagi, Advanced simulation of biomass gasification in a fluidized bed reactor using ASPEN PLUS, Renew. Energy 101 (2017) 629–636, https://doi.org/10.1016/j.renene.2016.09.011.

[53] C. Choi, W. Zhang, K. Fukumoto, H. Machida, K. Norinaga, A review on detailed kinetic modeling and computational fluid dynamics of thermochemical processes of solid fuels, Energy Fuel 35 (7) (2021) 5479–5494, https://doi.org/10.1021/ACS.ENERGYFUELS.0C04052/SUPPL_FILE/EF0C04052_SI_001. PDF.

[54] J.A.M. Chavando, E.C.J. de Matos, V.B. Silva, L.A.C. Tarelho, J.S. Cardoso, Pyrolysis characteristics of RDF and HPDE blends with biomass, Int. J. Hydrogen Energy (2021), https://doi.org/10.1016/J. IJHYDENE.2021.11.062.

Modelling and simulation of syngas purification processes

Modeling and simulation of membrane-assisted separation of carbon dioxide and hydrogen from syngas

Nayef Ghasem

Department of Chemical and Petroleum Engineering, UAE University, Al-Ain, United Arab Emirates

1. Introduction

Syngas is mainly produced from natural gas sources. However, biomass sources have attracted attention. Syngas is released in the gasification processes, or the steam methane reforming mainly consists of hydrogen and carbon dioxide. The removal or the reduction of carbon dioxide emissions in the downstream process is obligatory. Many industries connected by carbon capture are characterized as hydrogen production systems. Hydrogen is utilized to generate electricity in fuel cells or as a heat source. It can also be used in the petroleum refinery and fertilizer industry [1]. Global concerns about the danger of the increased temperature of global heating have led industrial nations to decrease carbon emissions. In order to achieve this goal, researchers are trying to improve new solutions to reduce carbon dioxide emissions from current and different fossil fuel use. A promising method in power generation is the gasification integrated combined cycle, in which oil, gas, and coal are cracked down to produce a valued natural gas type of fuel called synthesis gas, where it is converted by shift reaction into a gas mixture of carbon dioxide and hydrogen. This gas contained around 60mol% hydrogens and 40 mol% carbon dioxide. The removal of carbon dioxide from this mixture is a crucial element. The resultant hydrogen stream is used in fuel cells and gas turbines to produce electricity [2]. Hydrogen is a promising clean energy source, a combustible gas with free pollutants and high heating value. Approximately one-third of carbon dioxide emissions are released from power plants, which use fossil fuels to generate electricity in power generation plants. Petroleum refineries, fertilizer, and cement plants release massive carbon dioxide into the atmosphere. The most significant amount of carbon dioxide emissions that arise from power plants is through coal combustion.

Generally, amine-based solvents capture carbon dioxide from the gas stream in packed-bed towers in petroleum refineries and the petrochemical industry through traditional absorption

Advances in Synthesis Gas: Methods, Technologies and Applications. https://doi.org/10.1016/B978-0-323-91879-4.00019-9

and adsorption methods. Compared with the traditional absorption column, membrane separation processes are good candidates because of their low energy consumption, simplicity, lower capital cost, compactness, and ease of scale-up [3–5]. Both hydrogen- and carbon-dioxide-selective membranes can be employed to separate hydrogen and carbon dioxide for syngas. The kinetic diameter of hydrogen (2.89 A°) is less than carbon dioxide (3.30A°) [6]. Therefore, membrane separation techniques attracted the attention of many researchers [4–6].

The separation and purification of hydrogen and carbon dioxide from syngas are essential to meet industrial application requirements [7–10]. Gas separation in membranes depends on membrane physical and chemical properties, the size, shape, and polarity of permeant species, and the interaction between permeant gas and membrane. The membrane properties and the nature of the permeant species controlled the membrane separation process [11–15]. Nanofiller and polymer matrix are suitable membrane selecting materials for hydrogen and carbon dioxide capture [16]. The nanofillers such as molecular sieve, zeolite, metal–organic frameworks (MOFs) allow gases to pass through the provided porous channels, and the polymer matrix offers good mechanical properties for the fabricated membrane [17]. The membrane separation processes are responsive to environment with affordable cost and designed effective. It has wide range of applications in environmental process dealing with energy saving and water purification, global warming, and climate changes [18].

Methods of successfully separating carbon dioxide must apply to industrially appropriate gas streams at representative pressure and temperatures. Several hydrogen-selective and carbon-dioxide-selective membranes are available for syngas purification at various gas pressures in the literature [19,20]. The raw material entering the gasifier reacts with carbon dioxide, water vapor, and oxygen. Therefore, the gas released from gasification is treated and cleaned enough to use. A purification process eliminates impurities such as tar, ash, carbon dioxide, sulfur compounds, methane, and water vapor.

The fundamental design membrane module is essential to develop an efficient membrane separation process [21–25]. Computational fluid dynamics (CFD) modeled the hydrogen removal from a gas mixture of hydrogen, CO inside a ceramic membrane module. The membrane is porous glass porous silica utilized steam-reforming process [26]. Expected CFD model predictions of separation factor were related to the results calculated from the plug flow model against membrane permeability and gas flow rate. CFD is a fluid mechanics knowledge to solve mathematical models of a three-dimensional coordinate that reflects even the detailed shape of the device. Before model implementation, model validation and hence a mesh study should be conducted to ensure that the model could predict the experimental data and the mesh size does not influence the results. This chapter presents modeling and simulation of the performance of the transportation of hydrogen/carbon dioxide of H_2-selective membranes for syngas purification.

2. Polymeric membrane separation

Polymeric membrane separation methods are technically and economically viable due to low-maintenance cost, compatible with large gas volumes, no phase change, and less energy-intensive. Membrane fabricated from polybenzimidazole (PBI) shows excellent chemical and thermal resistance, good mechanical properties, and high transition temperature (Fig. 1). In addition, the membrane can work at high pressure and temperature [27].

Aspen software simulated the PBI membrane to remove hydrogen and carbon dioxide from a feed composed of 55% hydrogen, 41% carbon dioxide, and the balance of carbon monoxide, hydrogen, N_2, CH_4, hydrogen sulfide at 250°C around 50 atm. Fig. 2 shows the simulated membrane performance for the capture of hydrogen and carbon dioxide. Results revealed that the purity of hydrogen increased from 91% to 95.6%. By contrast, the percent recovery decreased from 98% to 28%, with increased feed flux from 35 to 200 ft^3/ft^2.h. The case is opposite for carbon dioxide, the percent purity deceased and percent recovery increased with increased feed flux. As a result, hydrogen purity decreased from 87.7 to 47.2%, while the percent recovery increased from 90% to 99% with increased volumetric feed flux from 35 to 200 ft^3/ft^2.h. Modeling and sensitivity analysis is developed to study the two-stage membrane system to investigate the influence of temperature, feed pressure, selectivity, and membrane permeance on the overall cost [21].

3. Combined gasification and multistage membrane separation

Fig. 3 is a schematic of an integrated gasification combined cycle power plant to capture carbon emissions utilizing multistage membranes. Gasification took place at temperatures around 700°C and pressure above 50 bar. Syngas from coal gasifier is produced in the existence of steam and oxygen. O_2 is extracted from the flue gas using an oxygen separating unit. The syngas from the gasification process is treated through a quencher for tar and solid particulate removal before being sent to a reactor chain for water gas sifter reactions. Carbon monoxide reacted with steam to produce carbon dioxide and hydrogen. Hydrogen and carbon dioxide are separated and purified using multiple-staged membrane separation techniques. Multistage and single-stage

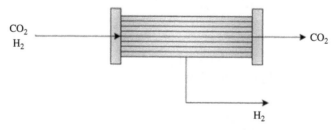

Fig. 1
Schematic of selective membrane separation method.

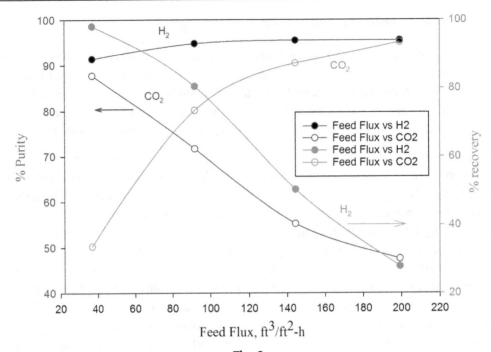

Fig. 2
Membrane simulated performance for the effect of feed flux on hydrogen and carbon dioxide percent purity and recovery [27].

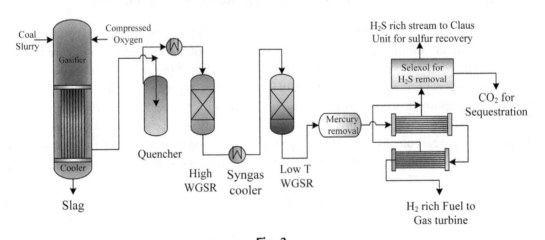

Fig. 3
Schematic of the integrated gasification combined cycle technology for carbon capture [21].

membranes were operated at 16 and 23 bar. Incorporation of membrane separation with shift reaction results in an energy penalty of 8–14% units, based on the staging and pressure. The membrane technology reduces the CO_2 emissions more than 50% relative to the integrated gasification combined cycle technology. In recent years, a progress has been made in utilizing membrane technology for gas separation. The membrane got the advantage of being small in size, easy to operate and maintain. The applicability of polymeric membrane for the recovery of CO_2 from flue gas, and easy to scale-up, no by-product pollution such as solids and spent solutions [28]. Hydrogen is a demanding industrial gas; high-purity hydrogen is required for many applications. Recent developments in both CO_2- and H_2-selecive membrane for hydrogen purification are studied [28]. With CO_2-selective membrane, hydrogen purification is attained [29].

Other methods such as solvent absorption and solid adsorption are widely known carbon dioxide removing processes. Carbon capture with integrated gasification cycle technology (IGCC) is used in power generation plants on a large scale. After removing carbon dioxide and hydrogen, the syngas from the power plant is utilized to generate electricity through a gas turbine. The polymeric membrane is a good candidate for syngas purification at a temperature of 107°C and a pressure up to 15 bar. Amines are carbon dioxide carriers dispersed in the membrane matrix [21].

4. Modeling and simulation of the two-stage separation

This section described the process of modeling and performing a sensitivity analysis for the operation cost of removing carbon dioxide from syngas for a power plant in a two-stage membrane. First, biomass or coal is utilized to produce synthetic gas (syngas) in a power plant by gasification. Then, the syngas is purified and employed a feed to manufacture value-added chemicals or electricity generation through the fuel cell gas or gas turbine. The power plant with an electrical energy gasifier was the base of a case study [21]. Fig. 4 sketched the model described process. The process separates hydrogen and carbon dioxide in a two-stage membrane. The model considered the following simplifying assumption in the model development [22]:

1. Steady-state operation.
2. Neglect the gas pressure drop in the axial direction of the feeding section [30].

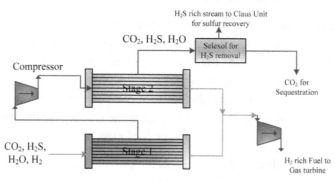

Fig. 4

Two-stage membrane process flow diagram for carbon dioxide and hydrogen removal [21].

3. We can neglect axial diffusion compared with axial convection.
4. Gas permeance is constant along the module length.
5. They assumed consistent selectivity for both stages.

The molar balance on each component is as follows:

The change in the molar rate in the feed side (n_f),

$$\frac{dn_{fi}}{dz} = -\pi d J_i \tag{1}$$

The component molar flow rate in the sweep side (n_s),

$$\frac{dn_{si}}{dz} = -\pi d J_i \tag{2}$$

where d is the hollow fiber diameter (cm), and J is the flux at steady state (mol/(m^2 s). The component flux across the membrane is permeance (P_i) times the difference in the partial pressure between feed (p_f) and sweep sides (p_s), determined using the mole fraction on the feed side (x) and the sweep side (permeate) mole fraction (y_i).

$$J_i = P_i\left(p_f \times x_i - p_s \times y_i\right) \tag{3}$$

Results revealed that the operation cost could be reduced if operating at high pressure through reducing energy consumption. The high selectivity of carbon dioxide/hydrogen reduces the operating cost if working at around 23 bars. The modeling, simulation, and cost-sensitivity study reveals the efficient performance of the carbon dioxide selectivity in a polymeric membrane for the purification of syngas in a two-stage scheme. Fig. 5 shows the plot of the influence of carbon dioxide on hydrogen selectivity on the selective polymeric membrane versus the cost of electricity of a two-stage membrane process (stage 1 operates at an optimum 23 bar and stage 2 at 3 bar and a temperature of 107°C). A large amount of hydrogen is produced in stage 1. The cost of electricity decreased when the membrane manufacturing cost decreased in the first stage with high selectivity for CO/hydrogen as it is associated with a decrease in the area required for the membrane in the second stage.

By contrast, in stage 2, the cost of electricity increased with increased carbon dioxide to the hydrogen-selective membrane. The membrane performance drops due to the decline in the carbon dioxide driving force and the high infusing of carbon dioxide through the stage 2 membrane. Accordingly, developing membranes that withstand high pressure is essential for carbon dioxide/hydrogen-selective membranes to reduce energy consumption. Membrane thickness harms membrane performance. The higher the membrane thickness increased the membrane resistance and hence increased the cost of electricity.

Across-linked membrane fabricated from sulfonated polybenzimidazole (SPBI), and polyvinyl alcohol (PVA) for the carbon dioxide selectivity facilitated transport membrane separation [31]. The fabrication and characterization of the membrane are for the process of purifying

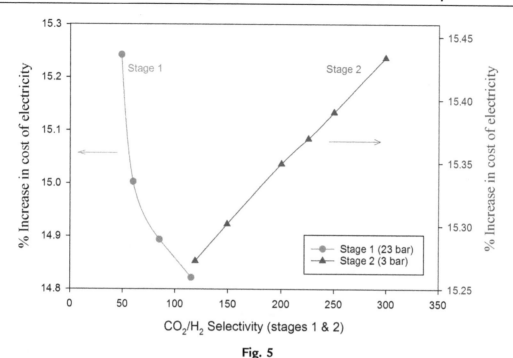

Fig. 5

Effect of carbon dioxide/hydrogen selectivity in a two-stage polymeric selective membrane of stage 1 (23 bar) and stage 2 (3 bar) on the cost of electricity cost at a temperature of 107°C [21].

high-pressure syngas. The cross-linked fabricated membranes have a high selectivity for carbon dioxide/hydrogen. The best carbon dioxide separation for the highest performance for the cross-linked fabricated PVA membrane was observed at a temperature of 106°C. By contrast, the highest version for the membrane fabricated from SPBI-1 copolymer membranes was at 100°C due to their low hydrophilicity [32]. For the two-component carbon dioxide and hydrogen represented by i and j, respectively. Assuming that carbon dioxide permeations higher than hydrogen, the flux (J_i) and membrane selectivity (α_{ij}) are defined for the component i over component j. The membrane characterized for the separation performance for two parameters. One is the separation factor (or selectivity), defined as

$$\alpha_{ij} = \frac{y_i/y_j}{x_i/y_j} = \frac{P_i}{P_j} \tag{4}$$

Which is the ratio of the retentate mol fraction (x) to permeate mole fraction (y). The permeability (P_i) defined as,

$$P_i = \frac{N_i}{\Delta p_i/l} \tag{5}$$

The permeability is the ratio of the membrane steady-state permeation flux (N) to thickness (l), the ($\Delta p/l$) is the difference of partial pressure between the feed and permeate sides (atm) to the

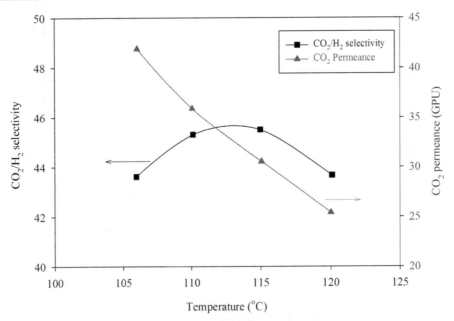

Fig. 6

Effects of temperature on the Syngas-PVA-1 membrane versus carbon dioxide/hydrogen selectivity and carbon dioxide permeance. The gas flow rate ($60\,cm^3$/min) being fed, the pressure of feed side (15atm), sweep side (1atm), gas sweep flow rate ($30cm^3$/min) [32]. 1 GPU$=3.35\times10^{-10}$mol/(m^2 s Pa) [32].

membrane thickness (l). The permeance (P_i/l) is the gas permeation unit (GPU) with its 10^{-6} cm^3 (STP)/(cm^2 s cmHg).

Fig. 6 shows the temperature influence on the membrane's selective carbon dioxide/hydrogen properties and carbon dioxide permeance. As the temperature increased, the carbon dioxide/hydrogen increased; this is due to carbon dioxide carrier having a high reaction and diffusion rates. By contrast, the temperature is more significant than 110°C, and the carbon dioxide/hydrogen selectivity decreased as the temperature increased due to the reduction of water retention in the membrane. Additionally, the carbon dioxide permeance had the opposite reaction, dropping as the applied temperature increased due to the decrease in the membrane water-retaining property. However, the carbon dioxide/hydrogen selectivity maintained its position at a relatively stable value at the applied temperature range [32].

5. Modeling and economic analysis

Modeling and sensitivity analyses of membrane process overall cost as a process for carbon capture and air as a sweeping fluid were performed to investigate the impact of operating parameters on membrane performance, permeance, and selectivity on the cost of electricity and capture cost. The study investigated the influence of the operating parameters and the membrane cost on the entire process cost. The process feed pressure is atmospheric and hence the low price of the membrane module. The study provides combined modeling with the sensitivity analysis of the cost. The study highlights membrane fabrication enhancement for better performance [33]. The process-achieved separation depends on the mass transfer driving force through the membrane along with the membrane properties.

5.1 Energy balance

The overall energy balance of the feed side,

$$\frac{d(\sum (n_{fi}H_{fi}))}{dz} = -\pi d\left(\Sigma(J_i H_{fi})\right] - \pi d U\left(T_f - T_s\right)\right) \tag{6}$$

The right-hand side (first term) represents the permeating components enthalpy from the feed side, the heat transfer between sweep and feed. The second term on the right-hand side includes conductive heat transfer across the membrane, and the left-hand side is the variation in the feed stream enthalpy. The energy balance on the sweep side

$$\frac{d(\sum (n_{si}H_{si}))}{dz} = -\pi d\left(\Sigma(J_i H_{fi})\right] - \pi d U\left(T_f - T_s\right)\right) \tag{7}$$

More details can be found elsewhere [33]. The approaches for the carbon dioxide removal from power plants are categorized as precombustion and postcombustion carbon capture. In precombustion carbon capture processes, the carbon dioxide removal involves treating the boiler exhaust gases immediately before entering the stack, while in postcombustion, carbon dioxide is eliminated from the fume exhaust. In postcombustion, the carbon dioxide capture process necessitates a significant design challenge because of the low partial pressure of the carbon dioxide in the flue gas. Gas separation membrane is a promising technology for the carbon dioxide capture from industrial flue gases, mainly from postcombustion power plant flue gases [34]. Hydrogen is a clean, efficient energy transporter to lessen the rising global energy and environmental crisis. Therefore, efficient hydrogen purification technologies are essential. The distinct advantages of gas separating membrane-based methodology for hydrogen purification compared with conventional separation methods have concerned significant courtesy [15–18]. A summary presents the recent progress on the essential carbon dioxide and hydrogen selective membrane processes in the following sections.

5.2 Hydrogen purification and carbon capture in precombustion process

The hot syngas (200°C) after the water gas shift reaction (WGS) contains around 30% carbon dioxide, 45% hydrogen, 20%–25% hydrogen vapor, and less than 1% CO and 0.5% hydrogen trace amount of N_2, Ar, NH_3, HCL, and CH_4. Therefore, the carbon dioxide removed from streams contains hydrogen in a precombustion process. Gasification takes place at a pressure of

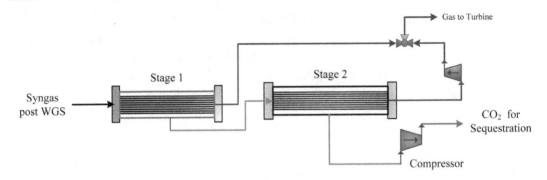

Fig. 7

Schematic of the precombustion membrane separation process [37].

around 50 bars. In the membrane separation process, high-pressure favored membrane separation. By contrast, polymeric carbon-dioxide-selective membrane is unsuitable for operation at high temperatures than the cross-linked matrix and facilitated transport member [20,35]. CFD technique is used to model membrane modules without the need to provide detailed descriptions of the models. A porous ceramic membrane was used to remove hydrogen from the hydrogen/CO gas mix [36].

The carbon-dioxide-selective membrane can separate undesired residuals (e.g., CO and N_2) in the syngas separate from carbon dioxide along with hydrogen. The process enables high carbon dioxide purity in the sequestered stream. Therefore, the amount of CO is higher in the fuel stream when low conversion in the WGS reactor for lower carbon dioxide recovery is required. Usually, an additional combustion unit is added after the capture of carbon dioxide for membranes with hydrogen selectivity [37]. The nitrogen has separation from carbon dioxide and has to be done in an extra step before it's captured because the hydrogen-selective membrane cannot separate the nitrogen. Therefore, this process favors the reduction in the total consumption of energy by the air separation process. In addition, the selectivity of carbon dioxide is enhanced utilizing the facilitated transport membrane, which is not restricted by the upper bound [33]. The water-gas shift (WGS) process permits the CO/hydrogen ratio modification by reacting the syngas with steam. The water detached into oxygen and hydrogen, and the oxygen reacts with CO in the syngas stream to produce carbon dioxide (Fig. 7).

5.3 Carbon capture in the postcombustion process

Removing carbon dioxide from a stream of flue gas streams is a challenging application process for most separation methods, attributed to the carbon dioxide concentration and the low pressure of the inlet gas, combined with the massive gas flow rate. The power plant's tremendous volumetric flow rate of the flue gas stream indicates the plants required extensive membrane areas. By contrast, producing membranes for this application is not a difficult job. The problem is in the significant, expensive compression, which requires high-energy

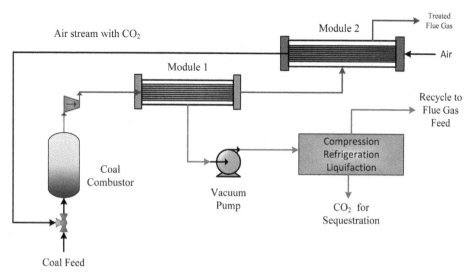

Fig. 8
Postcombustion with multistage membrane gas separation technique [37].

consuming equipment. After sulfur removal from flue gas in a typical power plant using coal is close to atmospheric pressure. The gas temperature ranges between 50°C and 60°C with 10%–15% carbon dioxide and 70% N_2. The partially low carbon dioxide pressure of 0.12–0.15 bar resulted in an intense driving force for carbon dioxide transport [37]. It is essential to identify the significant parameters telling the membrane performance: the flux and membrane selectivity. At a given selectivity, the carbon dioxide purity decreased at a lower pressure ratio.

Higher carbon dioxide recovery can decrease the driving power and expand the area of membrane; 90% recovery of carbon dioxide higher than 95% purity carbon dioxide is not possible with a single-stage membrane separation unit unless the recycling is at a pressure ratio of 5–10, which is more practical, the pressure is acquired by compression from the feed and permeate evacuation [38,39]. Permeate and retentate staging and recycling were used to improve simulation performance and study preferred purity and recovery objectives. Fig. 8 is a schematic diagram of a postcombustion process for the carbon dioxide recovery from the flow gas stream. Coal enters the coal combustion chamber combusted, and the flue gases are compressed and entered a two-stage membrane separation process. The carbon dioxide is recovered after being compressed, refrigerated, and liquefied, and the sweeping medium is air.

5.4 CFD modeling of the membrane module

5.4.1 Numerical model

The mass balances describing the transport and reactions in the shell side are given by diffusion-convection equation at steady state

$$\nabla(-D\nabla C_i) + v \cdot \nabla C_i = R \qquad (8)$$

Since the shape of the membrane is cylindrical, cylindrical coordinate is considered.

$$-D\left(\frac{\partial^2 C_i}{\partial r^2} + \frac{1}{r}\frac{\partial C_i}{\partial r} + \frac{1}{r^2}\frac{\partial^2 C_i}{\partial \theta^2} + \frac{\partial^2 C_i}{\partial z^2}\right) + \left(\frac{v_r \partial C_i}{\partial r} + \frac{v_\theta}{r}\frac{\partial C_i}{\partial \theta} + v_z\frac{\partial C_i}{\partial z}\right) = R \qquad (9)$$

where D_i is the inter diffusion coefficient of species i, C_i is the species concentration, and v is the superficial velocity. The R term corresponds to the reaction rate expression. Under the plug flow assumption, the molecular diffusion term is canceled from Eq. (2)

$$\left(\frac{v_r \partial C_i}{\partial r} + \frac{v_\theta}{r}\frac{\partial C_i}{\partial \theta} + v_z\frac{\partial C_i}{\partial z}\right) = R \qquad (10)$$

The rate of hydrogen permeation per unit area of the membrane is written in terms of Fick's first law as follows:

$$J_{H_2} = \frac{Q}{\Delta x}\left\{P^n_{H_2,ret} - P^n_{H_2,perm}\right\} \qquad (11)$$

where Δx is the thickness of the membrane, Q is the hydrogen permeation coefficient per unit area, $P_{H_2, ret}$ is the partial pressures of hydrogen in the retentate. $P_{H_2, perm}$ is the partial pressures of hydrogen in the permeate side. The rate of hydrogen permeation per unit length of catalyst bed (J_{H_2}) is linked with hydrogen partial pressure ($n = 0.76$) [40].

$$J_{H_2} = Pe\left(P^{0.76}_{H_2,lumen} - P^{0.76}_{H_2,shell}\right) \qquad (12)$$

where C is the concentration, h enthalpy, p pressure, ke effective thermal conductivity, J diffusion flux, E energy, D diffusion coefficient, v velocity vector, τ stress tensor, S_E energy source, S_i source of species, and Pe the hydrogen permeability. Simulations were performed using the described design and operating conditions (600K, 0.5MPa, 0.1m/s). The hydrogen permeability value of the Pd/Ag membrane of 1.0×10^{-4} (mol/m^2 sPa$^{0.5}$) was used [41]. In the CFD modeling for the hydrogen separation through a palladium-based membrane for hydrogen and carbon dioxide separation, the grid size was tested for the range of 0.05–1mm. The solution becomes independent of the cell size at the mesh length of 0.22mm [42]. The CFD dedicated program Comsol Multiphysics 5.6 is used to solve the process system model to perform the simulation in 2D space. The case is to separate 80% hydrogen and 20% carbon dioxide in a Pd-based membrane (length 1m, 0.02m diameter, 0.0007 m thickness). Fig. 9 represents the case study of hydrogen separation through a Pd-based membrane.

The CFD modeling result and the calculated results obtained from the plug flow model are in reasonable agreement, assuming a perfect low hydrogen permeability (lower than 10^{-7}mol/m^2 s Pa); dynamics flow is achieved.

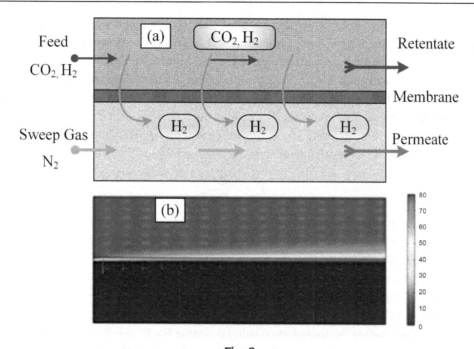

Fig. 9
(A) Mapped distribution mesh of the hydrogen separation system, (B) CFD simulation using Comsol Multiphysics 5.6.

Nevertheless, the CFD simulation prediction shows a decreased flux and separation factor for the hydrogen than the results obtained by the plug flow model for the condition where the expected results were high hydrogen recovery when using higher hydrogen permeability (greater than 10^{-6} mol/m^2 sPa). This shows the significance of bearing in mind that considering nonideal fluid dynamics in the design of a membrane module is more appropriate to the actual system imitation process. Most developed mathematical models assumed ideal gas flow, resulting in sometimes underestimating, or overestimating, the performance of membrane modules. This discrepancy is produced partly by the not-perfect flow regime that functions the geometry or the scale of the membrane module [43–45]. The CFD simulations of the hydrogen/carbon monoxide infusion through the membranes were performed using the tube geometry model [46].

The CFD model can give a membrane module design as well as predict selectivity. The tube geometry and two-channel models of membrane units test the gas permeation results of the CFD model simulations into the membrane unit of a gas mixture (hydrogen and CO) were tested. The CFD simulation results agreed with those from the plug flow model for

P less than 10^{-7} mol/(m^2 sPa), while there is an inconsistency when the used P is higher than 10^{-6} mol/(m$_2$ sPa), and θ is more significant than 0.1. The concentration of the polarization at the surface of the membrane is attributed to the gas pressure distribution. The permeation of hydrogen/CO passing the membranes simulated with the CFD program was validated with the experimental data of the tube geometry model. The concentration polarization depends on the feed volume flow rate (F), the hydrogen permeance (P), surface area (S), the total membrane pressure difference (p), hydrogen selectivity (α), and geometry module. The CFD is characterized result introduces the dimensionless parameter, θ, defined as:

$$\theta = \frac{PS\Delta P}{F} \tag{13}$$

The acquired dimensionless parameter is beneficial since the performance of the membrane is detailed only as a θ function with the ideal flow that is assumed in the module. Within the calculations, the geometry of the module, S, α, and ΔP are fixed constants. Consequently, when performing the concentration polarization analysis, a function depends on P and F. The two parameters being calculated, the fraction weight of the hydrogen in the exit of permeate side (Y_{H_2}), and the cut of hydrogen (R_{H_2}). The model evaluated the concentration polarization; assuming ideal flow within the model, then using the same θ will result in constant parameters. The Y_{H_2} and R_{H_2} are estimated at ideal flow conditions based on the plug flow model. Fig. 2 shows the calculated (Y_{H_2}) for membranes with different P versus θ. In the plug flow model (ideal flow), the Y_{H_2} slightly decreases as θ increases. The increase in θ resembles the decrease of F at constant P. The total flux of the hydrogen inside the membrane rises as the θ goes up. Given that condition, the difference in the partial pressure of hydrogen will decrease.

In a carbon dioxide capture from a postcombustion of a power station (1000MW capacity of the coal-fired plant). Zho et al. [38] conducted a sequence of parametric studies of gas separation processes using a single-stage membrane. The PRO/II software package is used as the process simulator, utilizing the binary gas separation membrane model built-in PRO/II. The model considered the developed polyimide membrane properties [47]. The study investigated the impact of membrane property and operating conditions on the performance of the investigated membrane. The simulation results revealed that on-stage membrane process could not achieve the desired high degree of separation and higher purity of carbon dioxide using the developed current membrane technology. Membrane gas separation with a multistage approach is the best to achieve the separation and purity targeted value, combining the advantages of the membrane with high selectivity and permeability. Hybrid coupling of membrane arrangement in multistage arrangement with carbon dioxide liquefaction process obtained the required carbon dioxide purity.

Carbon dioxide capture by hollow fiber membrane was simulated [48]. The achieved simulation was using developer software included in the Aspen Hysys software package.

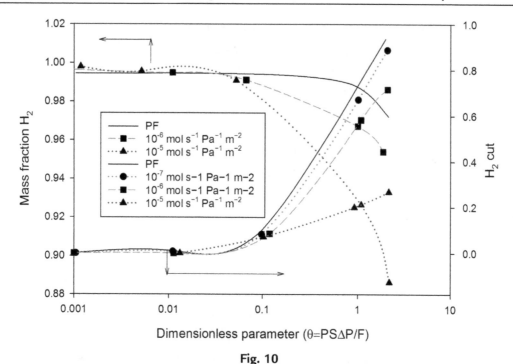

Fig. 10

Comparison of CFD model prediction and ideal flow model at different permeabilities, PF refers to the plug flow model, solid lines [36].

The membrane modules are in different configurations, cocurrent and countercurrent, and thoroughly mixed. The optimal alignment design is a single-stage membrane arrangement. The best membrane performance is obtained using the countercurrent model compared with the combined and entirely varied current. The configuration depends on the required space of the membrane and the energy demand (Fig. 10).

Based on economic evaluation, the three-stage membrane was optimized. The process design is based on a flue gas stream from a 400 MW power plant of burnt coal. The performance of hollow fiber membranes needs further improvement to reduce the capital cost of carbon dioxide capture from the industrial-scale process.

Table 1 lists the various types of membrane utilized in carbon dioxide and hydrogen separation. The polymeric membrane is usually fabricated from multiple polymers; cross-linked polyvinyl alcohol (PVA)-polysiloxane (POS)/fumed silica (FS), polyethylene glycol diacrylate (PEGDA), poly (ethylene glycol) diacrylate (PEGDA), polyacrylic acid (PAA), 2-aminoisobutyric acid potassium (AIBA-K), polyethyleneimine (PEI), sulfonated polybenzimidazole (SPBI), *N,N*-dimethylglycine lithium salt (DMG-Li), poly (1-trimethylsilyl-1-propyne) (PTMSP).

Table 1 Comparison of membrane performance from literature.

Membrane	Conditions T (°C) (atm)	Feed gas composition on dry basis	Selectivity (carbon dioxide/hydrogen)	Ref.
PVA-POS/23.5 wt% FS	107, 3	20% carbon dioxide, 80% hydrogen	87	[20]
PVA-POS/23.5 wt% FS	107, 0.95	20% carbon dioxide, 80% hydrogen	103	[20]
PEGDA/PEGMES-30	35, 13	Pure gas	13	[49]
PEGDA 700/10 wt% FS	23, 6.8	Pure gas	9.5	[50]
Crosslinked PVA (PAA, AIBA-K)	110, 0.56	20% carbon dioxide, 40% hydrogen, 40% N_2	170	[19]
SPBI (PEI, PAA, AIBA-K)	100, 0.41	20% carbon dioxide, 40% hydrogen, 40% N_2	65	[31]
Crosslinked PVA (DMG-Li, PEI)	80, 0.5	20% carbon dioxide, 40% hydrogen, 40% N_2	25	[51]
Arginine salt-chitosan	110, 0.5	10% carbon dioxide, 10% hydrogen, 80% N_2	20	[52]
PTMSP ([(CH$_3$)4N] F·4hydrogenO)	50, 0.9	25% carbon dioxide, 51% hydrogen, 24% CH_4	55	[53]

From the membrane configurations and structure listed in Table 1, the following conclusion could be extracted [20]:

- Due to the carrier saturation phenomenon, the carbon dioxide/hydrogen selectivity and carbon dioxide permeability will decrease as the pressure of the feed increased.
- The carbon dioxide/hydrogen selectivity and carbon dioxide permeability will be more excellent as the gas relative humidity increases.
- The concentration of amine in the membrane influences the separation performance; the performance of the membrane will improve as the amine increases. As the temperature increased, the membrane performance decreased [54].
- The addition of fuming silica to the Cross-linked Polyvinyl alcohol (PVA)-Polysiloxane (POS)/Fumed Silica (FS) shows a substantial effect on membrane separation performance.

6. Conclusion and future outlook

Modeling and simulations of membrane and processes for carbon dioxide and hydrogen separation from the syngas provide a significant and noteworthy improvement in energy efficiency possible at low-temperature absorbent-based Selexol process currently employed for the syngas treatment and cleanup. In terms of progress, hydrogen membranes fabricated from Pd-alloy composite approached implementation in gas capture processes. Additional work must increase the membrane performance and stability in impurities such as CO, sulfur, and Hg, existing in a coal gasification situation. Polymeric hydrogen membranes provide an excellent alternative for the fractional capture of carbon dioxide at a low capital cost. In a polymeric membrane for hydrogen separation, as the hydrogen and steam from the syngas are recovered at low pressure, these permeate the sides and need to be recompressed before entering the gas turbine, the process loses the advantage of having the gas turbine high-pressure steam diluent in the syngas.

References

[1] Y. Han, W.S.W. Ho, Facilitated transport membranes for H2 purification from coal-derived syngas: a techno-economic analysis, J. Membr. Sci. (2021), https://doi.org/10.1016/j.memsci.2021.119549, 119549.

[2] S.M. Klara, R.D. Srivastava, U.S., DOE integrated collaborative technology development program for CO2 separation and capture, Environ. Prog. 21 (2002) 247–253, https://doi.org/10.1002/ep.670210414.

[3] S.R. Reijerkerk, R. Jordana, K. Nijmeijer, M. Wessling, Highly hydrophilic, rubbery membranes for CO2 capture and dehydration of flue gas, Int. J. Greenhouse Gas Control 5 (2011) 26–36, https://doi.org/10.1016/j.ijggc.2010.06.014.

[4] A. Ghadimi, M. Amirilargani, T. Mohammadi, N. Kasiri, B. Sadatnia, Preparation of alloyed poly(ether block amide)/poly(ethylene glycol diacrylate) membranes for separation of CO2/H2 (syngas application), J. Membr. Sci. 458 (2014) 14–26, https://doi.org/10.1016/j.memsci.2014.01.048.

[5] M.T. Ravanchi, S. Sahebdelfar, F.T. Zangeneh, Carbon dioxide sequestration in petrochemical industries with the aim of reduction in greenhouse gas emissions, Front. Chem. Eng. China 5 (2011) 173–178, https://doi.org/10.1007/s11705-010-0562-1.

[6] N. Mehio, S. Dai, D. Jiang, Quantum mechanical basis for kinetic diameters of small gaseous molecules, J. Phys. Chem. A 118 (2014) 1150–1154, https://doi.org/10.1021/jp412588f.

[7] Y. Han, W.S.W. Ho, Polymeric membranes for CO2 separation and capture, J. Membr. Sci. 628 (2021), https://doi.org/10.1016/j.memsci.2021.119244, 119244.

[8] S.C. Rodrigues, J. Sousa, A. Mendes, Chapter 13—facilitated transport membranes for CO2/H2 separation, in: A. Basile, E.P. Favvas (Eds.), Current Trends and Future Developments on (Bio-) Membranes, Elsevier, 2018, pp. 359–384, https://doi.org/10.1016/B978-0-12-813645-4.00013-1.

[9] M. Klepić, K. Setničková, M. Lanč, M. Žák, P. Izák, M. Dendisová, A. Fuoco, J.C. Jansen, K. Friess, Permeation and sorption properties of CO2-selective blend membranes based on polyvinyl alcohol (PVA) and 1-ethyl-3-methylimidazolium dicyanamide ([EMIM][DCA]) ionic liquid for effective CO2/H2 separation, J. Membr. Sci. 597 (2020), https://doi.org/10.1016/j.memsci.2019.117623, 117623.

[10] Y. Zhao, B.T. Jung, L. Ansaloni, W.S.W. Ho, Multiwalled carbon nanotube mixed matrix membranes containing amines for high pressure CO2/H2 separation, J. Membr. Sci. 459 (2014) 233–243, https://doi.org/10.1016/j.memsci.2014.02.022.

[11] N. Ghasem, Modeling and simulation of the simultaneous absorption/stripping of CO2 with potassium Glycinate solution in membrane contactor, Membranes 10 (2020), https://doi.org/10.3390/membranes10040072.

[12] N. Ghasem, CFD simulation of CO2 absorption by water-based TiO2 nanoparticles in a high pressure stirred vessel, Sci. Rep. 11 (2021) 1984, https://doi.org/10.1038/s41598-021-81406-1.

[13] M.S. Yanan Zhao, Carbon Dioxide-Selective Membranes Containing Sterically Hindered Amines, The Ohio State University, 2013.

[14] N. Ghasem, M. Al-Marzouqi, A. Duidar, Effect of PVDF concentration on the morphology and performance of hollow fiber membrane employed as gas-liquid membrane contactor for CO_2 absorption, Sep. Purif. Technol. 98 (2012), https://doi.org/10.1016/j.seppur.2012.06.036.

[15] N. Ghasem, M. Al-Marzouqi, L. Zhu, Preparation and properties of polyethersulfone hollow fiber membranes with o-xylene as an additive used in membrane contactors for CO_2 absorption, Sep. Purif. Technol. 92 (2012), https://doi.org/10.1016/j.seppur.2012.03.005.

[16] P.S. Goh, K.C. Wong, L.T. Yogarathinam, A.F. Ismail, M.S. Abdullah, B.C. Ng, Surface modifications of nanofillers for carbon dioxide separation nanocomposite membrane, Symmetry (Basel) 12 (2020), https://doi.org/10.3390/sym12071102.

[17] Z. Qiao, Z. Wang, C. Zhang, S. Yuan, Y. Zhu, J. Wang, PVAm–PIP/PS composite membrane with high performance for CO_2/N_2 separation, AICHE J. 59 (2012) 215–228, https://doi.org/10.1002/aic.

[18] J. Poudel, J.H. Choi, S.C. Oh, Process design characteristics of syngas (CO/H2) separation using composite membrane, Sustainability 11 (2019), https://doi.org/10.3390/su11030703.

[19] J. Zou, W.S.W. Ho, CO2-selective polymeric membranes containing amines in crosslinked poly(vinyl alcohol), J. Membr. Sci. 286 (2006) 310–321, https://doi.org/10.1016/j.memsci.2006.10.013.

[20] R. Xing, W.S.W. Ho, Crosslinked polyvinylalcohol-polysiloxane/fumed silica mixed matrix membranes containing amines for CO2/H2 separation, J. Membr. Sci. 367 (2011) 91–102, https://doi.org/10.1016/j.memsci.2010.10.039.

[21] V. Vakharia, K. Ramasubramanian, W.S. Winston Ho, An experimental and modeling study of CO2-selective membranes for IGCC syngas purification, J. Membr. Sci. 488 (2015) 56–66, https://doi.org/10.1016/j.memsci.2015.04.007.

[22] K. Tahvildari, S.M.R. Razavi, H. Tavakoli, A. Mashayekhi, R. Golmohammadzadeh, Modeling and simulation of membrane separation process using computational fluid dynamics, Arab. J. Chem. 9 (2016) 72–78, https://doi.org/10.1016/j.arabjc.2015.02.022.

[23] M. Azizi, S.A. Mousavi, CO2/H2 separation using a highly permeable polyurethane membrane: molecular dynamics simulation, J. Mol. Struct. 1100 (2015) 401–414, https://doi.org/10.1016/j.molstruc.2015.07.029.

[24] L. Li, S. Fan, G. Yang, Q. Chen, J. Zhao, N. Wei, W. Meng, J. Fan, H. Yang, Continuous simulation of the separation process of CO2/H2 by forming hydrate, Chem. Eng. Sci. X. 7 (2020), https://doi.org/10.1016/j.cesx.2020.100067, 100067.

[25] G.O. Aksu, H. Daglar, C. Altintas, S. Keskin, Computational selection of high-performing covalent organic frameworks for adsorption and membrane-based CO2/H2 separation, J. Phys. Chem. C 124 (2020) 22577–22590, https://doi.org/10.1021/acs.jpcc.0c07062.

[26] C. Egawa, T. Nishida, S. Naito, K. Tamaru, Ammonia decomposition on (1110) and (001) surfaces of ruthenium, J. Chem. Soc. Faraday Trans. 1 Phys. Chem. Condens. Phases 80 (1984) 1595–1604, https://doi.org/10.1039/F19848001595.

[27] G. Krishnan, D. Steele, K. O'Brien, R. Callahan, K. Berchtold, J. Figueroa, Simulation of a process to capture CO2 from IGCC syngas using a high temperature PBI membrane, Energy Procedia 1 (2009) 4079–4088, https://doi.org/10.1016/j.egypro.2009.02.215.

[28] C.Y. Chuah, X. Jiang, K. Goh, R. Wang, Recent Progress in mixed-matrix membranes for hydrogen separation, Membranes 11 (2021), https://doi.org/10.3390/membranes11090666.

[29] L. Yu, M. Nobandegani, J. Hedlund, High performance fluoride MFI membranes for efficient CO2/H2 separation, J. Membr. Sci. 616 (2020), https://doi.org/10.1016/j.memsci.2020.118623, 118623.

[30] Y. Chu, A. Lindbråthen, L. Lei, X. He, M. Hillestad, Mathematical modeling and process parametric study of CO2 removal from natural gas by hollow fiber membranes, Chem. Eng. Res. Des. 148 (2019) 45–55, https://doi.org/10.1016/j.cherd.2019.05.054.

[31] H. Bai, W.S.W. Ho, New carbon dioxide-selective membranes based on sulfonated polybenzimidazole (SPBI) copolymer matrix for fuel cell applications, Ind. Eng. Chem. Res. 48 (2009) 2344–2354, https://doi.org/10.1021/ie800507r.

[32] H. Bai, W.S.W. Ho, Carbon dioxide-selective membranes for high-pressure synthesis gas purification, Ind. Eng. Chem. Res. 50 (2011) 12152–12161, https://doi.org/10.1021/ie2007592.

[33] K. Ramasubramanian, H. Verweij, W.S. Winston Ho, Membrane processes for carbon capture from coal-fired power plant flue gas: a modeling and cost study, J. Membr. Sci. 421–422 (2012) 299–310, https://doi.org/10.1016/j.memsci.2012.07.029.

[34] T.C. Merkel, H. Lin, X. Wei, R. Baker, Power plant post-combustion carbon dioxide capture: an opportunity for membranes, J. Membr. Sci. 359 (2010) 126–139, https://doi.org/10.1016/j.memsci.2009.10.041.

[35] S. Yousef, J. Šereika, A. Tonkonogovas, T. Hashem, A. Mohamed, CO2/CH4, CO2/N2 and CO2/H2 selectivity performance of PES membranes under high pressure and temperature for biogas upgrading systems, Environ. Technol. Innov. 21 (2021), https://doi.org/10.1016/j.eti.2020.101339.

[36] H. Takaba, S.I. Nakao, Computational fluid dynamics study on concentration polarization in H 2/CO separation membranes, J. Membr. Sci. 249 (2005) 83–88, https://doi.org/10.1016/j.memsci.2004.09.038.

[37] K. Ramasubramanian, Y. Zhao, W.S.W. Ho, CO2 capture and H2 purification: prospects for CO2-selective membrane processes, AICHE J. 59 (2013) 1033–1145, https://doi.org/10.1002/aic.

[38] L. Zhao, E. Riensche, R. Menzer, L. Blum, D. Stolten, A parametric study of CO2/N2 gas separation membrane processes for post-combustion capture, J. Membr. Sci. 325 (2008) 284–294, https://doi.org/10.1016/j.memsci.2008.07.058.

[39] R. Bounaceur, N. Lape, D. Roizard, C. Vallieres, E. Favre, Membrane processes for post-combustion carbon dioxide capture: a parametric study, Energy 31 (2006) 2556–2570, https://doi.org/10.1016/j.energy.2005.10.038.

[40] M. Gee, The water gas shift reaction assisted by palladium membrane reactor, Ind. Eng. Chem. Res. 1 (1991) 585–589, https://doi.org/10.1080/14790720408668193.

[41] G. Barbieri, A. Brunetti, G. Tricoli, E. Drioli, An innovative configuration of a Pd-based membrane reactor for the production of pure hydrogen: experimental analysis of water gas shift, J. Power Sources 182 (2008) 160–167, https://doi.org/10.1016/j.jpowsour.2008.03.086.

[42] R. Ben-Mansour, A. Abuelyamen, M.A. Habib, CFD modeling of hydrogen separation through Pd-based membrane, Int. J. Hydrogen Energy 45 (2020) 23006–23019, https://doi.org/10.1016/j.ijhydene.2020.06.141.

[43] C.T. Blaisdell, K. Kammermeyer, Counter-current and co-current gas separation, Chem. Eng. Sci. 28 (1973) 1249–1255, https://doi.org/10.1016/0009-2509(73)80077-6.

[44] W.P. Walawender, S.A. Stern, Analysis of membrane separation parameters. II. Counter-current and cocurrent flow in a single permeation stage, Sep. Sci. 7 (1972) 553–584, https://doi.org/10.1080/00372367208056054.

[45] S.A. Stern, S.C. Wang, Countercurrent and cocurrent gas separation in a permeation stage. Comparison of computation methods, J. Membr. Sci. 4 (1978) 141–148, https://doi.org/10.1016/S0376-7388(00)83290-1.

[46] K. Haraya, T. Hakuta, H. Yoshitome, S. Kimura, A study of concentration polarization phenomenon on the surface of a gas separation membrane, Sep. Sci. Technol. 22 (1987) 1425–1438, https://doi.org/10.1080/01496398708058408.

[47] S. Kazama, S. Morimoto, S. Tanaka, H. Mano, T. Yashima, K. Yamada, K. Haraya, K. Thambimuthu, Cardo polyimide membranes for CO2 capture from flue gases, in: E.S. Rubin, D.W. Keith, C.F. Gilboy, M. Wilson, T. Morris, J. Gale (Eds.), Greenhouse Gas Control Technologies 7, Elsevier Science Ltd., Oxford, 2005, pp. 75–82, https://doi.org/10.1016/B978-008044704-9/50009-4.

[48] X. He, J. Arvid Lie, E. Sheridan, M.B. Hägg, CO2 capture by hollow fibre carbon membranes: experiments and process simulations, Energy Procedia 1 (2009) 261–268, https://doi.org/10.1016/j.egypro.2009.01.037.

[49] H. Lin, E. Van Wagner, J.S. Swinnea, B.D. Freeman, S.J. Pas, A.J. Hill, S. Kalakkunnath, D.S. Kalika, Transport and structural characteristics of crosslinked poly (ethylene oxide) rubbers, J. Membr. Sci. 276 (2006) 145–161, https://doi.org/10.1016/j.memsci.2005.09.040.

[50] N.P. Patel, A.C. Miller, R.J. Spontak, Highly CO2-permeable and selective polymer nanocomposite membranes, Adv. Mater. 15 (2003) 729–733, https://doi.org/10.1002/adma.200304712.

[51] Y.-H. Tee, J. Zou, W.S.W. Ho, CO2-selective membranes containing dimethylglycine mobile carriers and polyethylenimine fixed carrier, J. Chin. Inst. Chem. Eng. 37 (2006) 37–47.

[52] S.J. Caldwell, B. Al-Duri, N. Sun, C. Sun, C.E. Snape, K. Li, J. Wood, Carbon dioxide separation from nitrogen/hydrogen mixtures over activated carbon beads: adsorption isotherms and breakthrough studies, Energy Fuel 29 (2015) 3796–3807, https://doi.org/10.1021/acs.energyfuels.5b00164.

[53] R. Quinn, J.B. Appleby, G.P. Pez, New facilitated transport membranes for the separation of carbon dioxide from hydrogen and methane, J. Membr. Sci. 104 (1995) 139–146, https://doi.org/10.1016/0376-7388(95)00021-4.

[54] Y. Zhao, W.S.W. Ho, CO2-selective membranes containing sterically hindered amines for CO2/H2 separation, Ind. Eng. Chem. Res. 52 (2013) 8774–8782, https://doi.org/10.1021/ie301397m.

Further reading

K. Prakash, Simulation and optimization of hot syngas separation processes in integrated gasification, Massachusetts Institute of Technology, United States, 2009 (Master thesis).

P. Li, Z. Wang, Z. Qiao, Y. Liu, X. Cao, W. Li, J. Wang, S. Wang, Recent developments in membranes for efficient hydrogen purification, J. Membr. Sci. 495 (2015) 130–168, https://doi.org/10.1016/j.memsci.2015.08.010.

S.A. Wassie, S. Cloete, V. Spallina, F. Gallucci, S. Amini, M. van Sint Annaland, Techno-economic assessment of membrane-assisted gas switching reforming for pure H2 production with CO2 capture, Int. J. Greenhouse Gas Control 72 (2018) 163–174, https://doi.org/10.1016/j.ijggc.2018.03.021.

CHAPTER 8

Simulation of cyclone separator for particulate removal from syngas

Minhaj Uddin Monir[a], Azrina Abd Aziz[b], Abu Yousuf[c], Jafar Hossain[a], Ahosan Habib[d], Kuaanan Techato[e], and Khamphe Phoungthong[f]

[a]Department of Petroleum and Mining Engineering, Jashore University of Science and Technology, Jashore, Bangladesh [b]Faculty of Civil Engineering Technology, University Malaysia Pahang, Gambang, Malaysia [c]Department of Chemical Engineering and Polymer Science, Shahjalal University of Science and Technology, Sylhet, Bangladesh [d]Geological Survey of Bangladesh, Dhaka, Bangladesh [e]Faculty of Environmental Management, Prince of Songkla University, Songkhla, Thailand [f]Environmental Assessment and Technology for Hazardous Waste Management Research Center, Faculty of Environmental Management, Prince of Songkla University, Songkhla, Thailand

1. Introduction

Global energy consumption is quickly expanding, and fossil-based fuels provide a significant portion of that need. However, reserves of fossil fuels are decreasing and will be diminished next few decades. For this reason, alternative and sustainable energy sources are being sought [1,2]. Syngas will be the appropriate solution for alternatives of fossil fuels [3,4] which is mainly composed of hydrogen (H_2) and carbon-monoxide (CO) mixtures. Gasification is a thermochemical conversion procedure for changing natural gas, coal, and biomass into syngas via interactions with oxygen and a variety of other sources [5,6]. In certain situations, this syngas is a fuel gas mixture composed of hydrogen, carbon monoxide, carbon dioxide, methane, and other gases [7]. This varies greatly depending on the raw material and gasification technology used; nevertheless, syngas is typically 30%–60% of carbon monoxide, 25%–30% of hydrogen, 0%–50% of methane, and 5%–15% of carbon dioxide, with a lower or greater amount of water vapor [8,9].

Moreover, gasification-based raw syngas includes gaseous components such as CO, H_2, CO_2, and H_2O [7]. As a residual of the gasifying agent, little quantities of Ar and O_2 remain in the syngas. The syngas composition is critical information for defining and calculating the best approach for measuring contaminants in the syngas. The composition of syngas's principal components changes according to the internal conversion from the water gas shift reactions (WGS). Syngas accounts for 2% of total basic energy consumption. The vast majority of syngas

Advances in Synthesis Gas: Methods, Technologies and Applications. https://doi.org/10.1016/B978-0-323-91879-4.00008-4

Table 1 Composition of syngas [5].

Character	Unit	Wood	Coal
H_2	Vol%	30–45	25–30
CO	Vol%	25–35	30–60
CO_2	Vol%	20–30	5–15
CH_4	Vol%	1–10	0–5
$C_2 + HC$	Vol%	1–2	N/A
N_2	Vol%	1–5	0.5–4
NH_3	ppm	500–1000	0–3000
H_2S	ppm	50–120	2000–10,000

Table 2 Properties of syngas [11].

Fuel properties	Unit
Volume flowrate (Q) (m^3/s)	0.1434
Gas velocity (m/s)	48.95
Density (kg/m^3)	0.1116
Dynamic viscosity (kg/ms)	2.612×10^{-5}
Heat transfer coefficient (W/mK)	0.2641
kinetic viscosity (m^2/s)	2.341×10^{-4}
Temperature (K)	330.4
Specific heat (Kj/mole kg C)	3.524

is used to produce fertilizer (55%); the second greatest portion is utilized in the oil refining process (22%); and lesser quantities are used to produce methanol. If the syngas is generated from biomass, it has a low carbon monoxide content but a high proportion of hydrocarbon. However, some hydrogen sulfide may be found when coal is used to produce syngas. To prevent corrosion, it is necessary to remove contaminants from syngas, particularly ammonia and hydrogen sulfide. Olalekan and Ogedengbe [10] reported an optimal purification technique for particle-rich synthetic gas (syngas) in a cyclone containing gasifier. They studied easily available biomass feedstocks such as wood, animal waste, and agricultural products. Tables 1 and 2 show the composition of syngas, as well as its features, for quality syngas produced from various sources.

2. Syngas impurities

Impurity levels are strongly influenced by raw material types and syngas production methods. The size of gasifier particle matter ranges from less than 1 m to more than 100 m, and its composition varies depending on the raw material and technique. Most syngas requires greater than 99% of particulate removal [12]. Removal of the particulate matter is necessary for syngas.

Table 3 Syngas impurities produced during biomass gasification.

Syngas impurities	Compounds	Amounts of impurities	References
Sulfur rich impurities	COS, H_2S	<0.0001 (% Vol.)	[13–16]
Hydrogen halide impurities	HF, HCl	<0.00001 (% Vol.)	[14,15,17]
Metal carbonyl impurities	Ni $(CO)_4$, Fe $(CO)_5$	–	[13–15]
Nitrogen rich impurities	NO, NH_3, N_2O HCN	<1 (% Vol.)	[13–15,17]
Alkali metals	K, Na	<0.001 (% Vol.)	[13–15,18]
Heavy metals	Cr, Zn, Mn, Hg, etc.	<0.0001 (% Vol.)	[13–15,17]
Tars	Naphthalene, Benzene	<0.2 (% Vol.)	[14,15,17]
Oxygen	O_2	<0.0001 (% Vol.)	[14,15,18]
Other elements	B, As, Se, etc.	–	[14,15]

To remove syngas particle matter, technologies such as cyclones, purification, and electrostatic precipitation have been used [5]. Syngas is sometimes blended with Ar, N_2, H_2S, COS, NH_3, and ash at a volume percentage of 0%–1%, and contaminants (Fe, Cl, Si species, metals, etc.) at a volume percentage of 100 ppm [13]. During gasification, several types of impurities are typically found in syngas (Table 3). The following are the most common unwanted syngas impurities:

2.1 Particle impurity

Dust is a fine-grained particle (organic or inorganic) generated by gasification of syngas. The most important factor is the quality of the biomass-based syngas [15]. The diameter of the particle is reduced during the thermal conversion of biomass by raising the gasification temperature. However, elutriation has a substantial impact on particle size reduction, particularly when feedstock is fragmented. Furthermore, heating alters its chemical characteristics, resulting in an increase in residual solid content. These variables include the kinds of gasifying feedstock, reactor types, and operating parameters such as pressure, temperature, and gasifying agents, as described by Rauch, Hrbek, and Hofbauer [19]. According to Nwokolo, Mamphweli, and Makaka [20], the cyclone separator is typically connected to the reactor to remove particles from produced syngas.

2.2 Tar compound impurity

Tar is a flammable, black, viscous liquid hydrocarbon produced by the gasification of fuel resources. It is the highly significant product of biomass gasification [21]. The amount of tar's undesirable by-product is affected by a number of variables. When these chemicals are employed directly, they may block transportation engines. This by-product is also utilized for road construction, coating, and wood preservation [22]. Temperature, equivalent ratios, and thermal cracking treatment all have an impact on the quality [23]. In this respect, temperature is critical in the removal of tar component from syngas. The concentration of tar dropped as the temperature rose. Another important parameter that increased the oxidation reactions during char volatilization is the equivalent ratio. As a result of raising the equivalent ratios, the concentration of tar rose, as reported by Adhikari, Abdoulmoumine, Nam, and Oyedeji [24].

2.3 Ammonia (NH$_3$) impurity

Ammonia (NH$_3$) impurity is a nitrogen and hydrogen chemical that is also produced during the gasification [15]. Existence of nitrogen in the biomass or coal caused the formation of this kind of contaminant. It is also influenced by the nitrogen concentration of various feedstock. Furthermore, the working parameters of the gasification have an impact on the formation of nitrogenous pollutants. NH$_3$ is produced as a result of the presence of steam during gasification.

2.4 H$_2$S and COS impurity

H$_2$S and COS are two toxic and flammable syngas contaminants that have an impact on syngas properties. As a consequence, during gasification, these gases are engaged in a variety of interactions with H$_2$S and other sulfur contaminants, affecting the quality of syngas [15].

2.5 Mercury impurity

Mercury is a hazardous contaminant found in syngas. Heavy metals including Hg, As, Se, Cu, Pd, Cd, and Zn are found in trace concentrations in producer syngas. Furthermore, as Sikarwar, Zhao, Fennell, Shah, and Anthony [25] observed, removing oxidized Hg in the form of HgS is more challenging.

3. Technologies for particles removal from syngas

3.1 Cyclone separator

Although a cyclone separator has a complicated flow pattern, it is the most basic and not expensive systems with relation to gas particulate substance dissociation. Because of its high efficiency, versatility, simple construction, and minimal maintenance, cyclone separators are frequently utilized. There are several kinds of cyclones used to separate solid particles [26]. The centrifugal separation type of cyclone separator is most often used to remove solid particles from gas. The cyclone's separation performance is determined by particle properties. Most of cyclone studies has been on micron-sized powder materials, with some studies focusing on granular materials, but mostly on heavy particles [27]. Pressure drop intake and outlet of cyclone separators are both significantly influenced by inlet velocity, and pressure drop constantly rises as input velocity rises. The maximum efficiency of inlet velocity (Eq. 1).

Table 4 Parameter used in simulation [29,30].

Solid phase		Gas phase	
Dimensions of the particles	**2 (mm)**	**Types of gases**	**Steam**
Solidity of the particles	2740 (kg/m^3)	Solidity	1.225 (kg/m^3)
Input acceleration of particles	3 (m/s)	Inlet acceleration of gas	20 (m/s)
The factor of viscous forces	0.3	Density	1.8×10^{-5} (kg/m/s)
Young modulus	1×10^7 (N/m2)	Time step	0.0001 (s)
Poisson's fraction	0.25	Column numbers	62,105
Factor of damping	0.3	Column types	Six plane faces
Factor of moving Rub	6×10^{-5}		

$$V_{MEV} = 231.6 \left(\frac{4g\mu_g \rho_g}{3p_g^2} \right) \left(\frac{b/D}{1 - b/D} \right) b^{0.2} \tag{1}$$

When the input velocity is within the proper gas velocity range, a higher separation efficiency with a lower pressure drop may be produced [28].

It is critical to obtain an accurate assessment of cyclone effectiveness and convection when designing a cyclonic separation system, which is dependent on cyclone design, geometry, particle characteristics, intake air velocity, and operating conditions. The height and diameter, cylinder elevation, cyclone detector diameter and depth, entire depth of the cyclone, and conical point dimension are the five key geometrical characteristics that impact cyclone performance. Finding a combination of cyclone efficiency and geometric design of cyclones is the key problem in employing it as a final separating device. As a result, the dimensions of the cyclone inlet have a major influence on dissociation actions and behaviors [26]. The parameters used in simulation are shown in Table 4.

Fig. 1 represents the standard cyclone separator geometry. The relevant criteria must be considered when calculating the optimum speed intake frequency of a gas-solid cyclone separator: ignoring the radial gas velocity; particle shape is spherical; wall-particle collisions; Gas moves with particles in the radial and vertical orientations; and disregarding particle movement prior to the initial particle-wall collisions. According to Saputro et al. [11], the thickness, roughness, and density of cyclone solid bodies are usually 2mm, 0.046, and 1643kg/m^3, respectively. Fig. 2 depicts the inlet velocity of the maximum-efficiency computation method [28].

3.2 Grade efficiency for cyclone separator

The grading efficiency controls cyclone separator performance, and separation efficiency is related to the particle diameter. The most often used grade efficiency model (Eq. 2) is shown below:

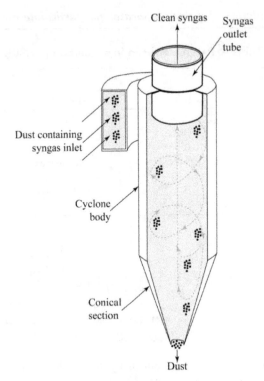

Fig. 1
Diagram of a cyclone separator [31].

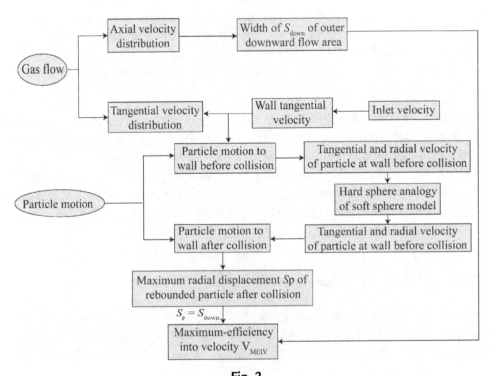

Fig. 2
The procedure for measuring the required inlet velocities in a step-by-step manner [28].

$$\eta_\delta = 1 - \exp\left(-\alpha\delta^\beta\right) \tag{2}$$

Where α and β are influenced by factors such as operation requirements, particle properties, and cyclone structures. The task at hand is to determine an initial set of components using dimensional analysis in order to accomplish dimension reduction and model scalability. The first demonstrates the shortcomings of the standard regression model and then employs the framework of the generalized linear model [32].

3.3 Pressure loss in cyclone

Pressure loss into a cyclone happens, as does the fiction with both the boundaries and irreversible drops inside this center of the machine, as well as total pressure loss (Eq. 3),

$$\Delta p = \Delta p_{body} + \Delta p_x \tag{3}$$

It is known as the Euler number in mathematical expression in Eq. (4):

$$E_u = \frac{1}{\frac{1}{2}\rho v_{in}^2}\left[\Delta p_{body} + \Delta p_x\right] \tag{4}$$

The cyclone cell's wall breakdown is specified by Eq. (5)

$$\Delta p_{body} = f\frac{A_R}{0.9Q}\frac{\rho}{2}\left(v_{\theta w}v_{\theta cs}\right)^{1.5} \tag{5}$$

Where v_{in} is the average input velocity area, Q is the gaseous volumetric flow ratio, A_R is the overall interior zone of cyclone friction wear, $v_{\theta w}$ is the frequency near the fence, and $v_{\theta cs}$ is the tangential acceleration of the gases just at central core diameter.

The pressure loss is characterized by a loss of pressure in the center with cyclone detector by Eq. (6).

$$\Delta p_x = \left[2 + \left(\frac{v_{\theta cs}}{v_x}\right)^2 + 3\left(\frac{v_{\theta cs}}{v_x}\right)^{4/3}\right]\frac{1}{2}\rho v_x^2 \tag{6}$$

Where average axial velocity is v_x [33].

Gravity concentration is a wide word that refers to a range of techniques for mineral particle separation that are essentially based on variations in specific gravity. Aside from specific gravity differences, there are several other elements that affect and influence the separation in a simultaneous and cumulative way. Some of the variables that determine the result of a separation include fluid qualities such as viscosity and buoyancy, the medium in which the separation is performed, the shape and size of the particles, and so on.

Dry scrubbers, also called as cyclone separators, are separation devices that remove particulate matter from exhaust gases by using the concept of inertia. Because it separates bigger particles of fine particulates, cyclone separators are just one of those air quality management devices called as precleaners. This decreases the future requirement for finer filtering systems to cope with larger, more abrasive particles. Furthermore, a multicyclone system may run several cyclone separators at the same time.

It is essential to consider that the size of cyclones may vary substantially. The amount of flue gas that must be filtered dictates the size of the cyclone; hence, larger cyclones are necessary for heavier operations. Multiple variations of the same cyclone type, for example, may exist, with diameters varying from 1.2 to 1.5m (approximately 4–5ft) with about 9m long (30ft)—roughly the height of a three-story skyscraper. Modabberifar, Nazaripoor, and Safikhani [34] reported that cyclones are used for a variety of functions, including air pollution management and manufacturing applications. Separating cyclones have been regarded as crucial particle separation equipment because of its advantageous qualities such as relative ease of manufacture, cheap operating costs, and excellent compatibility with hard conditions.

4. Simulation technology

The syngas produced by biomass gasification might be utilized to produce chemicals such as methanol. Before methanol synthesis, the syngas produced by biomass gasification has to be purified. The majority of adsorption processes using Mono-Ethanol Amine (MEA) technology are costly. However, the modeling process is critical for determining the impact of working situations on plant operation. Process simulation also saves money and time [35]. In simulation the particles did not interlude with one another [36]. There are many commercial simulators, with Aspen Plus being one of them. In order to improving syngas composition, the gasification step has been emphasized using Aspen Plus simulation software.

4.1 Aspen plus modeling

The gasification parameters are optimized using the Aspen Plus simulator. Prior to the gasification run, the experimental number, duration, and cost have to be decreased. The assumptions are used to run the simulation (Table 5). The simulation is done using the proximate and ultimate findings of raw materials that correspond to the needed temperatures for cellulose, hemicellulose, and lignin degradation. The experimental results are compared with the simulated results. In this aspect, the simulation seems to be a viable tool for valuing product syngas and predicting gasification behavior. The exact connection between temperature, pressure, and mole fractions is referred to in this simulation [39]. At the end of the simulation, optimized pressure and temperature are set and the experiments are run for gasification process.

Table 5 Description of ASPEN PLUS units [37,38].

Unit	Category
RSTOIC	Reduces the fuel's moisture content
	Simulates the generation of HCl, NH_3, and H_2S
RYIELD	Fuel element decomposition and product distribution
	It simulates the synthesis of toluene, naphthalene, and benzene
RGIBBS	By lowering Gibbs free energy, the atom balancing element is utilized to simulate single molecule equilibrium
RCSTR	A continuous-stirred container reaction model. When the kinetics of the reaction are known, the above method is helpful. When solids are included in the reactions, this model comes in handy
	Drying, pyrolysis, partial oxidation, and gasification are all simulated
CYCLONE	Distinguishes between gases and solids

For the simulation, the following assumptions have been assessed:

- Gasification procedure is isothermal and steady state
- Volatile matters are consisted of H_2, CO, CO_2, CH_4, and H_2O by drying and pyrolysis
- Mainly char contains 100% carbon
- All gases treat ideally
- The organic substances considered are toluene, naphthalene, and benzene.
- The inorganic substances are H_2S, NH_3, and HCl [37]

Aspen Plus is a system simulation software, which assists in the physical, chemical, and biological processes stated by Kaushal and Tyagi [38]. They reported to have employed mass and energy balance equations, as well as a phase equilibrium database, to simulate chemical systems involve material, fluid, and vapor streams under certain conditions. Because Aspen Plus's solution algorithms are more complicated than Aspen Hysys's, warnings are extremely likely (Table 6). Aspen Plus has simplified model development and upgrade throughout the years, and tiny parts of complicated and interconnected systems may be developed and tested as independent modules before they are merged [3]. The simulation results through Aspen Plus simulator are shown in Fig. 3.

In the existing libraries, the Aspen Plus simulator is used to optimize the gasification parameters. The gasification process must be modeled based on specific assumptions. In this regard, the experimental data are compared with the simulated outcomes. It also reduces the cost and duration of the experiment. As a result, the simulation seems to be a viable choice for estimating the value of product syngas and projecting gasification activities (Table 6). Despite the fact that the simulation and modeling of biomass gasification have been extensively studied in the literature, there has been little study on the critical parameter optimization for the gasification of diverse kinds of energy resources. The accuracy of the link between temperature, pressure, and mole fractions is a benefit of this simulation.

Table 6 Calculation model parameter [27].

	Term	Details	Exponent	Grade
CFD	Element	Liquid/Steam	Solidity(kg/m3)	1.225
			Viscosity (Kg/m/s)	$1.789e-5$
		Solid	Solidity(kg/m3)	1385
	Boundary condition	Inlet velocity	Gas velocity (m/s)	2,3.5,5,7.5,10,12.5
		Outflow/gas Outlet	Outflow flow Weighting	1
		Outflow/Solid Outlet	Outflow flow Weighting	0
		Wall	Wall movement	Stationary wall
			Shear state Roughness length (mm)	No slip
				0
			Constant Roughness	0.01
		Time step size	Fixed time step (s)	$6e-4$
DEM	Material	Particle	Poisson's ratio	0.206
			Shear modulus (Pa)	1e7
			Density (kg/m3)	40
	Interaction	Particle-Particle	Coefficient of restitution	0.307
			Coefficient of static friction	2
			Rolling friction coefficient	0.2
			Coefficient of restitution	0.047
			Coefficient of static friction	0.894
		Particle-wall	Rolling friction coefficient	0.2
	Simulator	Number of time scales	Constant step of time (s)	$6e-6$

4.2 Computational fluid dynamics

Computational fluid dynamics (CFD) may simulate continuous fluid flow by evaluating mass, energy, chemical processes, and combustion conservation equations while discretizing the domain into cells. CFD modeling is now an effective method for designing, simulating, and optimizing many fluidized systems. Many researchers used DEM-CFD numerical approach in order to overcome error and distinction among both separate component stage and liquid-level stage. Depending on the impacts of particle fluid interactions, DEM-CFD coupling may be configured as data transmission can be single or multiple way [26].

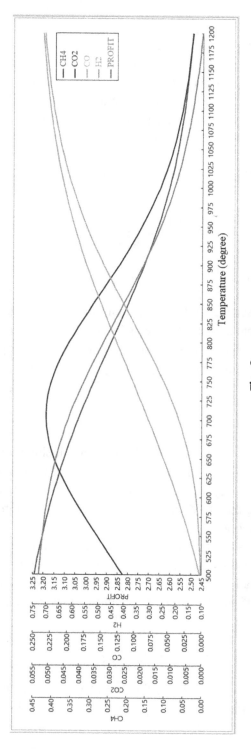

Fig. 3

Simulation curve having CO, H_2, CO_2 and CH_4 using Aspen Plus simulator [3,40].

4.3 CFD modeling

The gas flow pattern within the cyclone separator is designed to be significantly rigid; accurate turbulence models are needed to represent the flow field. At the moment, complicated turbulent fluid systems are created using CFD methods and fast and sophisticated computer processing equipment. For turbulent flow modeling, the standard k- epsilon model is well known and widely used. For successfully simulating the air flow field characteristics, the k-epsilon model is consulted and applied. Where k and epsilon signify turbulent kinetic energy and breakdown rate, accordingly, which are simulated in Fluent to use the different transport formulas generated out from Navier-Stokes formulas of equilibrium mass and energy preservation in Eqs. (7) and (8).

$$\frac{\partial}{\partial t}\left(\rho_f k\right) + \frac{\partial}{\partial x}\left(\rho_f k u_f\right) = \frac{\partial}{\partial x}\left[\alpha_k \mu_{f_{eff}} \frac{\partial k}{\partial x}\right] + G_k + G_b - \rho\varepsilon - Y_M + S_k \tag{7}$$

$$\frac{\partial}{\partial t}\left(\rho_f \varepsilon\right) + \frac{\partial}{\partial x}\left(\rho_f \varepsilon u_f\right) = \frac{\partial}{\partial x}\left[\alpha_\varepsilon \mu_{eff} \frac{\partial\varepsilon}{\partial x}\right] + C_1 \frac{\varepsilon}{k}(G_k + C_3 G_b) - C_2\rho_f \frac{\varepsilon^2}{k} - R_\varepsilon + S_\varepsilon \tag{8}$$

Where ρ_f is the liquid consistency, u_f is the element of flow rate, $\mu_{f_{eff}}$ is the unstable velocity of a flow, G_b is the creation of potential energy due to disturbance, G_k is the production of turbulent kinetic energy as a result of overall velocity profile, α_k and α_ε are the opposite feasible Prandtl numbers for k and ε, accordingly and C_1, C_2, and C_3 are constants [26].

4.4 DEM modeling

DEM is now widely utilized in many industrial operations, including particulate combining, dehydrating foods, drug companies, mining, crop processes, transfer chute design, pneumatic conveying methods, breaking, separating, handling, and fluid-particle fluxes [41]. Based on Newton's motion equations, the DEM can mimic particulate flow translations and rotating particle motion [42]. CFD research using ANSYS cyclone separator using Auxiliary cone capable of capturing particles ranging in size from 0.1 to 20 μm [43]. It may also calculate equation of motion by following each particle's path using the Lagrangian formulation approach, which is mathematically linked with time. Because the gas volume is substantially lower than for the particles size, forces such as the virtual mass force and the Basset force are frequently overlooked (Eq. 9).

$$m_p \frac{dv_p}{dT} = F_p^c + m_p g + F_p^{f_p} \tag{9}$$

Where m_p is the molecule weight, $F_p^{f_p}$ is the liquid contact pressure on the particles. Particle contact pressure is F_p^c, gravitational force is g, and particle translating velocity is v_p (Eqs. 10–11).

$$F_n^{(ij)} = \left(k_n^{(ij)} \Delta x_p^{(ij)} + c_n^{(ij)} \Delta v_n^{(ij)} \right) \tag{10}$$

$$k_t^{(ij)} = \left(k_t^{(ij)} \Delta x_p^{(ij)} + c_t^{(ij)} \Delta v_t^{(ij)} \right) \tag{11}$$

Where normal force is F_n, stiffness and dumping constant is $k_n^{(ij)}$ and $c_n^{(ij)}$, respectively, in normal constant and $k_t^{(ij)}$ and $c_t^{(ij)}$ in tangential constant, the overlap in between particles is denoted by x_p.

When there are a lot of particles in a granular flow, the particle form becomes very important. In general, various parameters influence simulation duration, including particle number in the domain sector, component roundness, particle's physical location, connection laws (linear or non-linear), moving borders, and additional pressures such as as cohesiveness and coupling via CFD (Eq. 12).

$$F = F_n^T + F_t^T \tag{12}$$

Where F represents the total force applied on the particles, F_n^T describes the average elastic-plastic impact energy at the present time (T), and F_t^T represents the tangential elastic-plastic contacts pressure at the present time (T). As is well known, in DEM codes, a collection of actual contact models is employed to explain genuine touch and interactions. To numerically characterize this structure as follows (Eqs. 13–15),

$$F_n^T = \min \left(F^{T-TD} + K_{nu}.ds_n.K_{nl}.s_n \text{ if } ds_n \geq 0 \tag{13}$$

$$F_n^T = \max \left(F_n^{T-TD} + K_{nu}.ds_n.0.001.K_{nl}.s_n \text{ if } ds_n < 0 \tag{14}$$

$$ds_n = S_n^T - S_n^{T-dT} \tag{15}$$

Where former time pressure is F_n^{T-TD} and times step is dT. S_n^{T-dT} and S_n^T are the standard overlapping values at the past and new periods, respectively, and K_{nl} and K_{nu} are the loaded and unloaded touch stiffnesses. The loading stiffness of the particle is discussed below Eqs. (16)–(19).

$$\text{For particles,} K_{nlp} = E_p.L \tag{16}$$

$$\text{For boundaries,} K_{nlb} = E_b.L \tag{17}$$

$$\text{For particle} - \text{particle connection,} K_{nl} = \frac{K_{nlp_i}.K_{nlp_j}}{\left(K_{nlp_i} + K_{nlp_j} \right)} \tag{18}$$

$$\text{For particle boundary connection,} K_{nl} = \frac{K_{nlp}.K_{nlb}}{\left(K_{nlp} + K_{nlb} \right)}$$

$$K_{nu} = \frac{K_{nl}}{\varepsilon^2} \tag{19}$$

where E_p is the bulk Young's modulus of the particle material, E_b is the border element Young's modulus, and L is the grain size. The restoration coefficient is ε. It should be noted that K_{nu} is restricted by a value of 0.001. $K_{nl} \cdot s_n$ in order for the applied load to be zero prior to the particles colliding. Because the normal contact occurs at two distinct values of K_{nl} and K_{nu}, the timestep in the hysteretic linear spring models may be determined as stated in below Eq. (20).

$$\Delta T = \min \left(\frac{\pi}{2N'_{\Delta T}} \right) \sqrt{\frac{m^*}{K_{nl}}} \frac{\pi}{8} \sqrt{\frac{m^*}{K_{nu}}} \tag{20}$$

where m^* is the contact's effective mass, and $N'_{\Delta T}$ is a user-input parameter that specifies the minimum number of timesteps each loading cycle [42].

4.5 CFD-DEM coupling

A one-way Rocky-fluent connection was performed to validate the true segregation and purification properties of the air circulation in the backward gas cyclone separator. The drag correlation model, together with pressure gradient forces, is important in this coupling for characterizing and deals with both the interface of surrounding fluids, spherical and nonspherical particulates, and particle diameter. Because it takes particle shape and alignment into account, the Ganser correlation model is employed for both spherical and nonspherical particles. The Ganser model equation has two censored design parameters in Eq. (21) where K_1 is Stokes form parameter and K_2 is the Newton's form parameter. The drag force of a solid sphere and an irregular form particle is defined by the first component formula in Eq. (22). The second factor equation (Eq. 23) explains the ratio between arbitrary particle drag coefficient and spherical particle drag coefficient.

$$\frac{C_d}{K_2} = \frac{24}{R_e K_1 K_2} \left[1 + 0.1118 (R_e K_1 K_2)^{0.6567} \right] + \frac{0.4305}{1 + \frac{3305}{R_e K_1 K_2}} \tag{21}$$

$$K_1 = \left(\frac{d_p}{3 d_v} + \frac{2}{3} \varnothing^{\frac{-1}{2}} \right)^{-1} - 2.25 \frac{d_v}{D} \tag{22}$$

$$K_2 = 10^{1.8148(-\log_{10} \varnothing)^{0.5743}} \tag{23}$$

Where C_d is drag coefficient, d_p is the diameter of spherical particles, R_e is the Reynold number; these are in the very same manner, $\varnothing$ is the sphericity of the molecule, d_v is a circular particle's dimension, and D is the domain diameter [44]. The CFD parameters for the simulation process are shown in Table 7.

4.6 Mathematical model

The distinct element technique is used to characterize the solid material, which acts as a distinct phase. The translational and rotational movements of a particle at any time according to the model, as Newton's motion law states (Eqs. 24–25).

Table 7 CFD parameters for simulation [11].

Setting specifications	
Model	Stress-based
Turbulence model: RANS	RSM, the typical wall purpose
Trajectory (LPT) Lagrangian	Each connection in a discrete element system
Particle track	Maximum quantity of laps 1e − 06
	Variable of stride length 5
Solution method	Connection of pressure gradient SIMPLE
Terms of convergence	10^3
Spatial discretization	Pressure momentum: second order upwind
	Turbulent kinetic energy, turbulent dissipation rate

$$m_i \frac{dv_i}{dt} = f_{p-f,i} + \sum_{j=1}^{k_i} \left(f_{c,ij} + f_{d,ij} \right) + m_i g \qquad (24)$$

And

$$I_i \frac{d\omega_i}{dt} = \sum_{j=1}^{k_i} \left(T_{ij} + M_{ij} \right) \qquad (25)$$

Where m_i, I_i, v_i, and ω_i are gradually the mass, of translational and rotational motions of the moment of inertia. Gas-solid interaction force f_{p-fi} acts on solids both particulate i and j that take into account the contact pressure, $f_{c,ij}$ and adhesive dumping pressure $f_{d,ij}$ and $m_i g$ is the gravitational forces. The rolling friction torque is M_{ij} and it is adversary to the rotation of the particle. The k_i particles in contact particle i are the interparticle forces summation. The tangential forces cause particle i to generate torques, T_{ij} to rotate because the contact point between particles i and j acts by the interparticle forces. The solids motion is modeled using DEM at the discrete particle contacts, while the fluid flow is modeled using CFD at the computation cellular levels. DEM will always provide data on the locations and motions of individual atoms. These measurements would be used in CFD to investigate the gas flow pattern, which produces the liquid force applied on the individual atoms [29].

To characterize the removal process of ultralight expanded graphite (EG) particles, a CFD-DEM model in an Euler-Lagrangian environment is created. Because the simulated EG elements are comparable to spherical particles of the same volume in terms of calculation time, the remaining mathematical equations are all associated with the belief. User-defined functions (UDF) are used to add interfacial changes to the liquid theory basis functions.

The current average Navier-Stokes formula is used to solve the continuous phase in the cyclone separator as shown below (Eqs. 26–27):

$$\frac{\partial}{\partial t} \left(\varepsilon_g \rho_g \right) + \nabla \cdot \left(\varepsilon_g \rho_g u \right) = 0 \qquad (26)$$

$$\frac{\partial}{\partial t}\left(\varepsilon_g \rho_g u\right) + \nabla \cdot \left(\varepsilon_g \rho_g uu\right) = -\nabla p + \nabla \cdot (\varepsilon_g T) + \varepsilon_g \rho_g g - S + \nabla\left(-\rho_g \overline{u'u'}\right) \tag{27}$$

Where ε_g, ρ_g, u, and u' are the spatial voidage portion, gaseous concentration, standard flow rate, and turbulence velocity profile, in that order; P denotes pressure and is the tensor of local tension. The compressive pressure between both the gas and the particle is denoted by S. To explain the intense whirling flow of gas in the separator, the Reynolds stress turbulence model (RSM) theory in the business program FLUENT is used (Eq. 28):

$$\frac{\partial}{\partial t}\left(\rho_g \overline{u_i'u_j'}\right) + \frac{\partial}{\partial x_k}\left(\rho_g u_k \overline{u_i'u_j'}\right) = D_{T,ij} + p_{ij} + \varnothing_{ij} + \varepsilon_{ij} \tag{28}$$

Local time differential and diffusion term are the two things on the left side of the center, while the four elements on the right side of the mean are turbulent diffusion period, stress generation term, pressures strained period, and evaporation period, in that order (Eqs. 29–32).

$$D_{T,ij} = -\frac{\partial}{\partial x_k}\left[\rho_g \overline{u_i'u_j'u_k'} + \overline{p'\left(\delta_{kj}u''\,l_i + \delta_{ik}u_j'\right)}\right] \tag{29}$$

$$p_{ij} = -\rho_g\left(\overline{u_i'u_k'}\frac{\partial u_j}{\partial x_k} + \overline{u_j'u_k'}\frac{\partial u_i}{\partial x_k}\right) \tag{30}$$

$$\varnothing_{ij} = \overline{p'\left(\frac{\partial u_i'}{\partial x_j} + \frac{\partial u_j'}{\partial x_i}\right)} \tag{31}$$

$$\varepsilon_{ij} = -2\mu\overline{\frac{\partial u_i'}{\partial x_k}\frac{\partial u_j'}{\partial x_k}} \tag{32}$$

Where i, j, k denote the three Cartesian coordinate system directions, and $u_i'u_j'u_k'$ denote speed variations [27].

4.7 Numerical model

The precise description of turbulent flow behavior in order to simulate turbulent flow in cyclones is important to CFD's efficacy. FLUENT contains a number of turbulence models that may be used to simulate turbulent flow in a cyclone separator. The RSM demanded that the formula for every Reynolds stress factor be solved. Assume a precise forecast of the swirl flow pattern, axial velocity, tangential velocity, and pressure drop for the cyclone separator. The finite elements technique is utilized to characterize the model's ordinary differential equations by using the SIMPLEC (semi-implicit method for pressure-linked equations consistent) technique for pressure velocity connection and the QUICK technique to insert the factors on the controlled volume's interface. The residence times of two cyclones are fairly close. With the initial grid point, multiple degrees of grid refinement are applied in different sections of the boundary layer. Elsayed and Lacor [33] perform these simulations on an eight-node CPU Opteron 64 Linux cluster using the commercial program FLUENT. The model parameters are shown in Table 6.

4.8 Comparative analysis using simulation software and applications

Several commercial software packages such as HSC stoichiometric calculation program, Aspen Plus huge basic chemical flow simulation software, response surface methodology (RSM), a high-performance language for technical computing software MATLAB, and CFD software Fluent are commonly used to simulate the gasification process [8,45,46]. Some benefits of computer-aided simulation include the ability to rapidly evaluate the effectiveness of synthesized process flowsheets and conduct process synthesis operations. It may be used in conjunction with process synthesis to create optimal integrated designs while minimizing experimental and scale-up efforts. RSM is a useful statistical technique that has used both statistical and mathematical approaches to create, enhance, and optimize the gasification process. It also examines certain significant reactions that are either embraced by some variables [47,48]. The Aspen Plus simulator is often used to avoid complicated processes and create the simplest feasible model that includes the primary gasification reactions and the reactor's gross physical features [49]. Additionally, Sadhwani, Adhikari, Eden, and Li [50] used the ASPEN PLUS simulator to develop a process model for predicting steady-state performance of biomass CO_2 gasification in a fluidized bed gasifier. This model is concerned with both hydrodynamic characteristics and reaction kinetics. Inayat, Ahmad, Mutalib, and Yusup [51] used a MATLAB simulation model to investigate the effect of temperature, steam-to-biomass ratio, and sorbent/biomass ratio on hydrogen (H_2) generation performance in a steam gasification. They discovered that when the temperature and steam-to-biomass ratio increased, the hydrogen concentration and yield increased, but the thermodynamic efficiency decreased. There are numerous differences between them when it comes to simulating the gasification process, such as the simulation concept, the simulation foundation, the input and output of data, and so on. The major changes between the kinetic model and the updated chemical kinetics utilized in the CFD model are the frequency factor and activation energy values [52]. The use of a CFD solver for 2D modeling and simulation of biomass gasification in a bubbling fluidized bed gasifier. According to Anil, Rupesh, Muraleedharan, and Arun [45], ANSYS FLUENT 15.0 was used to create a model for predicting the impacts of equivalency ratio, steam-to-biomass ratio, steam temperature, and air preheating on product gas composition.

5. Recent advances and application

For a long time, the CFD-DEM method has been used to model DPRS. The sophisticated models are mostly limited to 2D solutions because of the high collision frequency of a large number of particles and the modeling scheme of gas-particle interactions at the particle scale.

The computation has been extended to include 3D applications. DEM was used to simulate the solid phase, while LES has used to model the turbulent fluid flow. Chemical processes in the char including O_2, CO, CO_2, NO, and NO_2 heterogeneous reactions and CO, O_2, NO, and N_2O homogeneous reactions. The 3D simulations of biomass pyrolysis employing fluidized bed reactors and discrete particles are provided. However, the simulations and inert sand particles have been denoted as a continuum by two discrete and relative particles. The complex reactive model comprised heat up, drying, particle shrinkage, primary and secondary pyrolysis, gasification, and tar degradation processes [53].

6. Conclusion and future outlook

The presence of particle impurity in syngas might create problems and reduce syngas quality. For the removal of particles from syngas, cyclone separators are often used. Various simulators are often used to optimize the parameters, lowering the experimental cost and time. This chapter represents the cyclone separator simulation for removing particulate from the syngas. Although a cyclone separator has a complicated flow pattern, Aspen Plus and CFD-DEM have been used to investigate this simulation. The ASPEN PLUS simulator used to create a process model for forecasting steady-state CO_2 gasification performance in a fluidized bed gasifier. CFD modeled continuous fluid flow by solving conservation equations for mass, energy, chemical reactions, and combustion while discretizing the domain into cells. Mathematical modeling and CFD analysis were utilized to investigate the influence of the cyclones geometric factor on cyclone efficiency and an optimal cyclone geometric ratio. In terms of describing interparticle collisions, CFD-DEM outperforms other approaches and more realistic, practical, and interactive simulations will be necessary in the future. Even though the obtained separation performance is considered suitable for the cyclonic separation process under the investigated operation conditions, the computation technique leads to a better understanding of how to develop and optimize the geometry and operational conditions of the gas cyclone separator.

Over the past 15years, CFD modeling has been more popular. In most studies, in the simulation analysis, the mean particle size acquired from tests was used as the represented particle size. Nonetheless, particle size variation impacts the gas particle flow dynamics for probable solid segregation and response rates. The fluid temperature has an effect on particle surface area. Particle size reduces as the reaction progresses, allowing the particle phase to be converted to gas products. The particle phase is continually made up of freshly supplied particles, partially converted particles, and completely transformed particles. It is quite difficult to detect changes in particle size during simulation [53]. CFD fluid flow is at the computational cell level, while DEM solid flow modeling is at the individual particle level [29]. The calculation is too costly for such a large number of particles and chemical processes. In most CFD-DEM investigations, solid particles have a spherical form, which does not correspond to reality. Changes in particle size and form complicate touch detection and CPU memory [53].

CFD simulation is utilized to forecast the flow and reactive properties. For improved control and design of industrial reactors and full-loop system simulations, more efficient and less expensive CFD methodologies are necessary. CFD-DEM is superior to other methods in terms of accuracy in describing interparticle collisions. In the future, more realistic, practical, and interactive simulations will be required. The CFD-DEM approach provides hidden benefits for simulating nonspherical particles. As a result, more emphasis should be placed on simulating nonspherical components in reacting flows [53].

Pressure drop, which is one of the most critical operating features for gas cyclones, is the primary metric for validation. It is established in experiments that the velocity reduces with increased solid loading, and this can be accounted for by the current numerical model. The developed CFD-DEM model is capable of capturing the basic flow parameters in a gas cyclone. The advantages of the CFD-DEM model have the first principal gateway and extensive micro dynamic data on gas-solid flow at particulate matter. [29]. To validate the accepted results, the prediction must be compared with experimental data. The present simulations are based on excellent tangential and axial velocity profiles measured at an axial point. Despite the uncertainty of the turbulence whirling flows of cyclones, the correlation among models and observations is deemed to be rather adequate. [33].

A CFD-DEM simulation for ultralight particles in cyclone separators was developed and effectively used to predict the removal process at various intake speeds. The following are some of the properties of ultralight particulate:

- It is simple to separate from the input velocity, every molecule in the gas may be completely removed.
- The flow field in the separator is hardly replicated by ultralight particles. A gas-solid dual phase's axial and tangential velocities are nearly similar. The tangential velocity distribution at the cone separation zone wall varies as intake velocity increases.
- The turbulent diffusion of particles is higher at initially in the top region of the cyclone and eventually lessens as such input flow rate increases.
- The particle gravity is less than the gas-particle interaction and collision forces. The major cause of supplementary breakdown is particle interactions of varying sizes [27].

References

[1] A. Sunny, P. Solomon, K. Aparna, Syngas production from regasified liquefied natural gas and its simulation using Aspen HYSYS, J. Nat. Gas Sci. Eng. 30 (2016) 176–181, https://doi.org/10.1016/j.jngse.2016.02.013.

[2] M.Y. Hasan, M.U. Monir, M.T. Ahmed, A.A. Aziz, S.M. Shovon, F. Ahamed Akash, M.F. Hossain Khan, M.J. Faruque, M.S. Islam Rifat, M.J. Hossain, P. Kundu, R. Akter, S. Ali, Sustainable energy sources in Bangladesh: a review on present and future prospect, Renew. Sustain. Energy Rev. 155 (2022), https://doi.org/10.1016/j.rser.2021.111870, 111870.

[3] M.U. Monir, A.A. Aziz, R.A. Kristanti, A. Yousuf, Co-gasification of empty fruit bunch in a downdraft reactor: a pilot scale approach, Bioresour. Technol. Rep. 1 (2018) 39–49, https://doi.org/10.1016/j.biteb.2018.02.001.

[4] M.U. Monir, M.Y. Hasan, M.T. Ahmed, A.A. Aziz, M.A. Hossain, A.S.M. Woobaidullah, P.K. Biswas, M.N. Haque, Optimization of fuel properties in two different peat reserve areas using surface response methodology and square regression analysis, Biomass Convers. Biorefin. (2021), https://doi.org/10.1007/s13399-021-01656-x.

[5] L. Yang, X. Ge, Biogas and Syngas Upgrading, Advances in Bioenergy, Elsevier, 2016, pp. 125–188, https://doi.org/10.1016/bs.aibe.2016.09.003.

[6] M.U. Monir, A.A. Azrina, R.A. Kristanti, A. Yousuf, Gasification of lignocellulosic biomass to produce syngas in a 50 kW downdraft reactor, Biomass Bioenergy 119 (2018) 335–345, https://doi.org/10.1016/j.biombioe.2018.10.006.

[7] M.U. Monir, A. Abd Aziz, R.A. Kristanti, A. Yousuf, Syngas production from co-gasification of forest residue and charcoal in a pilot scale downdraft reactor, Waste Biomass Valoriz. 11 (2) (2020) 635–651, https://doi.org/10.1007/s12649-018-0513-5.

[8] F.Z. Mansur, C.K.M. Faizal, M.U. Monir, N.A.F.A. Samad, S.M. Atnaw, S.A. Sulaiman, Co-gasification between coal/sawdust and coal/wood pellet: a parametric study using response surface methodology, Int. J. Hydrogen Energy 45 (32) (2020) 15963–15976, https://doi.org/10.1016/j.ijhydene.2020.04.029.

[9] M.U. Monir, A. Yousuf, A.A. Aziz, S.M. Atnaw, Enhancing co-gasification of coconut Shell by reusing char, Indian J. Sci. Technol. 10 (6) (2017) 1–4, https://doi.org/10.17485/ijst/2017/v10i6/111217.

[10] O.O. Olalekan, E.O. Ogedengbe, Optimal technique for separation of particle-rich syngas in cyclone for efficient methanation, in: 14th International Energy Conversion Engineering Conference, 2016, p. 5018.

[11] H. Saputro, T. Firdani, R. Muslim, Y. Estriyanto, D. Wijayanto, S. Lasmini, The CFD simulation of cyclone separator without and with the counter-cone in the gasification process, in: IOP Conference Series: Materials Science and Engineering, IOP Publishing, 2018. p. 012142.

[12] P.J. Woolcock, R.C. Brown, A review of cleaning technologies for biomass-derived syngas, Biomass Bioenergy 52 (2013) 54–84, https://doi.org/10.1016/j.biombioe.2013.02.036.

[13] A.B. Mansfield, M.S. Wooldridge, The effect of impurities on syngas combustion, Combust. Flame 162 (5) (2015) 2286–2295, https://doi.org/10.1016/j.combustflame.2015.01.026.

[14] M. Kobayashi, Dry Syngas Purification Processes for Coal Gasification Systems, Elsevier, 2020.

[15] M.U. Monir, A. Yousuf, A.A. Aziz, Syngas fermentation to bioethanol, in: A. Yousuf, D. Pirozzi, F. Sannino (Eds.), Lignocellulosic Biomass to Liquid Biofuels, Academic Press, 2020, pp. 195–216, https://doi.org/10.1016/B978-0-12-815936-1.00006-X.

[16] D. Xu, D.R. Tree, R.S. Lewis, The effects of syngas impurities on syngas fermentation to liquid fuels, Biomass Bioenergy 35 (7) (2011) 2690–2696, https://doi.org/10.1016/j.biombioe.2011.03.005.

[17] G. Iaquaniello, A. Mangiapane, Integration of biomass gasification with MCFC, Int. J. Hydrogen Energy 31 (3) (2006) 399–404, https://doi.org/10.1016/j.ijhydene.2005.09.010.

[18] K. Göransson, U. Söderlind, J. He, W. Zhang, Review of syngas production via biomass DFBGs, Renew. Sustain. Energy Rev. 15 (1) (2011) 482–492.

[19] R. Rauch, J. Hrbek, H. Hofbauer, Biomass gasification for synthesis gas production and applications of the syngas, Wiley Interdiscip. Rev.: Energy Environ. 3 (4) (2014) 343–362.

[20] N. Nwokolo, S. Mamphweli, G. Makaka, An investigation into heat recovery from the surface of a cyclone dust collector attached to a downdraft biomass gasifier, Appl. Therm. Eng. 98 (Supplement C) (2016) 1158–1164, https://doi.org/10.1016/j.applthermaleng.2016.01.014.

[21] M.U. Monir, F. Khatun, U.R. Ramzilah, A.A. Aziz, Thermal effect on co-product tar produced with syngas through co-gasification of coconut Shell and charcoal, IOP Conf. Ser.: Mater. Sci. Eng. 736 (2020), https://doi.org/10.1088/1757-899x/736/2/022007, 022007.

[22] M.J. Spear, 11—Preservation, protection and modification of wood composites, in: M.P. Ansell (Ed.), Wood Composites, Woodhead Publishing, 2015, pp. 253–310, https://doi.org/10.1016/B978-1-78242-454-3.00011-1.

[23] M.U. Monir, F. Khatun, A. Abd Aziz, D.-V.N. Vo, Thermal treatment of tar generated during co-gasification of coconut shell and charcoal, J. Clean. Prod. 256 (2020) 1–9, https://doi.org/10.1016/j.jclepro.2020.120305.

[24] S. Adhikari, N. Abdoulmoumine, H. Nam, O. Oyedeji, 16—Biomass Gasification Producer Gas Cleanup, Bioenergy Systems for the Future, Woodhead Publishing, 2017, pp. 541–557, https://doi.org/10.1016/B978-0-08-101031-0.00016-8.

[25] V.S. Sikarwar, M. Zhao, P.S. Fennell, N. Shah, E.J. Anthony, Progress in biofuel production from gasification, Prog. Energy Combust. Sci. 61 (2017) 189–248, https://doi.org/10.1016/j.pecs.2017.04.001.

[26] M.A. El-Emam, W. Shi, L. Zhou, CFD-DEM simulation and optimization of gas-cyclone performance with realistic macroscopic particulate matter, Adv. Powder Technol. 30 (11) (2019) 2686–2702.

[27] H. Zhou, Z. Hu, Q. Zhang, Q. Wang, X. Lv, Numerical study on gas-solid flow characteristics of ultra-light particles in a cyclone separator, Powder Technol. 344 (2019) 784–796, https://doi.org/10.1016/j.powtec.2018.12.054.

[28] Q. Wei, G. Sun, J. Yang, A model for prediction of maximum-efficiency inlet velocity in a gas-solid cyclone separator, Chem. Eng. Sci. 204 (2019) 287–297.

[29] K. Chu, B. Wang, D. Xu, Y. Chen, A. Yu, CFD-DEM simulation of the gas-solid flow in a cyclone separator, Chem. Eng. Sci. 66 (5) (2011) 834–847, https://doi.org/10.1016/j.ces.2010.11.026.

[30] S. Wang, H. Li, R. Wang, X. Wang, R. Tian, Q. Sun, Effect of the inlet angle on the performance of a cyclone separator using CFD-DEM, Adv. Powder Technol. 30 (2) (2019) 227–239, https://doi.org/10.1016/j.apt.2018.10.027.

[31] L. Wang, Theoretical Study of Cyclone Design, Texas A&M University, 2004.

[32] J. Wang, Y. Guo, J. Pan, Grade-efficiency modeling and robust parameter design for gas-solids cyclone separators, Powder Technol. 345 (2019) 457–467.

[33] K. Elsayed, C. Lacor, Optimization of the cyclone separator geometry for minimum pressure drop using mathematical models and CFD simulations, Chem. Eng. Sci. 65 (22) (2010) 6048–6058, https://doi.org/10.1016/j.ces.2010.08.042.

[34] M. Modabberifar, H. Nazaripoor, H. Safikhani, Modeling and numerical simulation of flow field in three types of standard new design cyclone separators, Adv. Powder Technol. (2021), https://doi.org/10.1016/j.apt.2021.09.037.

[35] M. Puig-Gamero, J. Argudo-Santamaria, J. Valverde, P. Sánchez, L. Sanchez-Silva, Three integrated process simulation using aspen plus: pine gasification, syngas cleaning and methanol synthesis, Energ. Conver. Manage. 177 (2018) 416–427.

[36] J. Derksen, S. Sundaresan, H. Van den Akker, Simulation of mass-loading effects in gas-solid cyclone separators, Powder Technol. 163 (1–2) (2006) 59–68.

[37] V. Marcantonio, E. Bocci, J.P. Ouweltjes, L. Del Zotto, D. Monarca, Evaluation of sorbents for high temperature removal of tars, hydrogen sulphide, hydrogen chloride and ammonia from biomass-derived syngas by using Aspen Plus, Int. J. Hydrogen Energy 45 (11) (2020) 6651–6662, https://doi.org/10.1016/j.ijhydene.2019.12.142.

[38] P. Kaushal, R. Tyagi, Advanced simulation of biomass gasification in a fluidized bed reactor using ASPEN PLUS, Renew. Energy 101 (2017) 629–636, https://doi.org/10.1016/j.renene.2016.09.011.

[39] A.J. Keche, A.P.R. Gaddale, R.G.J.C.T. Tated, Simulation of biomass gasification in downdraft gasifier for different biomass fuels using ASPEN PLUS, Clean Techn. Environ. Policy 17 (2) (2015) 465–473.

[40] M.U. Monir, A.A. Aziz, A. Yousuf, Integrated technique to produce sustainable bioethanol from lignocellulosic biomass, Mater. Lett.: X 13 (2022), https://doi.org/10.1016/j.mlblux.2022.100127, 100127.

[41] H.P. Zhu, Z.Y. Zhou, R.Y. Yang, A.B. Yu, Discrete particle simulation of particulate systems: a review of major applications and findings, Chem. Eng. Sci. 63 (23) (2008) 5728–5770, https://doi.org/10.1016/j.ces.2008.08.006.

[42] M. Emam, L. Zhou, W. Shi, H. Chen, True shape modeling of bio-particulate matter flow in an aero-cyclone separator using CFD-DEM simulation, Comput. Part. Mech. 8 (2021) 1–17, https://doi.org/10.1007/s40571-020-00383-w.

[43] H. Saputro, S. Supriyadi, I. Muttaqin, V. Fadlullah, L. Fitriana, T. Firdani, R. Muslim, K. Khaniffudin, S. Lasmini, V. Sutrisno, The improvement of synthetic gas quality on the gasification process of palm starch waste through the applied of cyclone separator, IOP Conf. Ser.: Mater. Sci. Eng. 434 (2018), https://doi.org/10.1088/1757-899X/434/1/012181, 012181.

[44] S. Kuang, M. Zhou, A. Yu, CFD-DEM modelling and simulation of pneumatic conveying: a review, Powder Technol. 365 (2020) 186–207, https://doi.org/10.1016/j.powtec.2019.02.011.

[45] M. Anil, S. Rupesh, C. Muraleedharan, P. Arun, Performance evaluation of fluidised bed biomass gasifier using CFD, Energy Procedia 90 (2016) 154–162, https://doi.org/10.1016/j.egypro.2016.11.180.

[46] P. Kaushal, J. Abedi, N. Mahinpey, A comprehensive mathematical model for biomass gasification in a bubbling fluidized bed reactor, Fuel 89 (12) (2010) 3650–3661, https://doi.org/10.1016/j.fuel.2010.07.036.

[47] B. Morero, E.S. Groppelli, E.A. Campanella, Evaluation of biogas upgrading technologies using a response surface methodology for process simulation, J. Clean. Prod. 141 (2017) 978–988.

[48] A. Sarrai, S. Hanini, N. Merzouk, D. Tassalit, T. Szabó, K. Hernádi, L. Nagy, Using central composite experimental design to optimize the degradation of tylosin from aqueous solution by photo-Fenton reaction, Materials 9 (6) (2016) 428.

[49] S.M. Atnaw, S.A. Sulaiman, S. Yusup, A simulation study of downdraft gasification of oil-palm fronds using ASPEN PLUS, J. Appl. Sci. 11 (11) (2011) 1913–1920.

[50] N. Sadhwani, S. Adhikari, M.R. Eden, P. Li, Aspen plus simulation to predict steady state performance of biomass-CO2 gasification in a fluidized bed gasifier, Biofuels Bioprod. Biorefin. 12 (3) (2018) 379–389.

[51] A. Inayat, M.M. Ahmad, M.A. Mutalib, S. Yusup, Effect of process parameters on hydrogen production and efficiency in biomass gasification using modelling approach, J. Appl. Sci. (Faisalabad) 10 (24) (2010) 3183–3190.

[52] U. Kumar, M.C. Paul, CFD modelling of biomass gasification with a volatile break-up approach, Chem. Eng. Sci. 195 (2019) 413–422, https://doi.org/10.1016/j.ces.2018.09.038.

[53] W. Zhong, A. Yu, G. Zhou, J. Xie, H. Zhang, CFD simulation of dense particulate reaction system: approaches, recent advances and applications, Chem. Eng. Sci. 140 (2016) 16–43, https://doi.org/10.1016/j.ces.2015.09.035.

CHAPTER 9

Tar and inorganic contaminant removal from syngas: Modeling and simulation

Enrico Bocci[a], Vera Marcantonio[b], and Andrea Di Carlo[c]

[a]Marconi University, Rome, Italy [b]Unit of Process Engineering, Department of Engineering, University Campus Bio-Medico di Roma, Rome, Italy [c]L'Aquila University, L'Aquila, Italy

1. Introduction

Nowadays, struggles toward global warming and climate change united to the necessity for energy security and national independency pointed out the importance in finding out a trustable alternative to fossil fuels. Biomass is the largest source of renewable energy, and it is ranked at the fourth place among the energy sources only after oil, coal, and natural gas; moreover, residual biomass, e.g., from agro-industry to municipal wastes, is inherently inexpensive [1–3]. These are the reasons why biomass is actually considered the fuel of renewable origin most suitable to replace the fossil ones.

Researchers point out the biomass gasification as one of the most advantageous thermochemical processes, due to its capacity to maintain a high-rate fuel gas production along to lower investment costs [4,5]. This process is promoted at high temperature, typically 750–1000°C or higher, depending on the specific technology of use, by processing the biomass feedstock with an oxidizing agent (oxygen, air, steam, or a mix of them). The process leads to the production of a fuel gas commonly called syngas, or producer gas, made of hydrogen, carbon monoxide, carbon dioxide, methane, and steam, along with several undesired by-products [6]. These undesired by-products are organic tars and inorganic impurities (e.g., hydrogen sulfide, hydrogen chloride, alkali, and ammonia). Tars are typically aromatic hydrocarbons, and since they have toxic and carcinogenic characteristics are necessary to remove them during the syngas cleaning. Sulfur compounds are responsible for catalyst poison and for pollution. Ammonia compounds are a poison for catalyst and are responsible of NO_x production [7]. Chlorine compounds and alkali are responsible of corrosion of metal equipment, health problems, and environmental issues. Then, raw syngas must be purified of the contaminants before its use in downstream applications.

Advances in Synthesis Gas: Methods, Technologies and Applications. https://doi.org/10.1016/B978-0-323-91879-4.00009-6

Process and system simulation models have obtained great interest in the evaluation of the best option to achieve the lowest level of contaminants, since they allow to have a good description of both chemical and physical phenomena occurring into each unit with minimal time and costs [8].

In this chapter, a review of the most promising simulation models for organic and inorganic removal is introduced and described.

2. Tar

Tar is defined as hydrocarbons with a molecular weight higher than that of benzene [9]. The representative composition of biomass tar from gasification processes at temperatures around 800°C is reported in Fig. 1, and it varies depending on the biomass feedstock, the type of gasifier, the variety of gasifying agent, and operating conditions.

Particulates and tar are regarded as the most considerable contaminants created through the gasification of biomass (around 3–100g/Nm^3). As reported by Rakesh et al. [11], tar generation is made in two different paths:

(1)   The instantaneous release of aromatic structures in lignin devolatilization;
(2)   The operation of hydrogen removal and acetylene introduction.

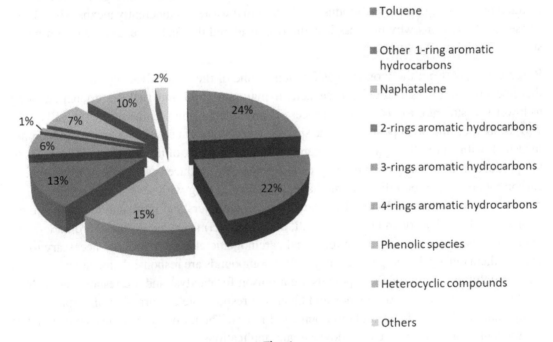

Fig. 1
Typical composition of biomass tar [10].

Tar yield and tar content reduction are inversely proportional to equivalence ratio, in fact bigger amount of oxygen allows to have a reaction with volatile compounds [12].

Moreover, tar yield is quite influenced by the residence time, rather than the tar composition, which is critically affected; actually the amounts of the compounds, which contain oxygen fall with the increasing of the residence time, determining a drop in the yield of one- and two-ring species (keeping out benzene and toluene) and an increase in the yield of three-ring and four-ring compounds [11].

The overall reactions (1–6), which are quite resctricted by kinetics, are essentials for the conversion of tar compounds [13–15].

$$\text{Steam Reforming} : C_nH_m + nH_2O \rightarrow C_{n'}H_{m'} + \left(n + \frac{m}{2}\right)H_2 + nCO \tag{1}$$

$$\text{Dry reforming} : C_nH_m + nCO_2 \rightarrow \left(\frac{m}{2}\right)H_2 + 2nCO \tag{2}$$

$$\text{Hydrocracking} : C_nH_m + H_2 \rightarrow C_{n'}H_{m'} \tag{3}$$

$$\text{Thermal cracking} : C_nH_m \rightarrow C_{n'}H_{m'} + H_2 \tag{4}$$

$$\text{Carbon formation} : C_nH_m \rightarrow nC + \left(\frac{m}{2}\right)H_2 \tag{5}$$

$$\text{Water-Gas shift} : CO + H_2 \leftrightarrow CO_2 + H_2O \tag{6}$$

It is common to declare kinetics for tar conversion by fixing an all-inclusive rate of tar retiral that combines the reactions [16].

Milne et al. [17] availed of molecular beam mass spectrometry (MBMS) to indicate that a methodical techinique to classify pyrolysis products as primary, secondary, and tertiary can be utilized in order to make an efficnet comparison of the different products that came out of every kind of gasifier. The authors selected four prevalent product classes:

(1) Primary products, which arecellulose-derived products, hemicellulose-derived products, and lignin-derived methoxyphenols;
(2) Secondary products, which are: phenolics and olefins;
(3) Alkyl tertiary products, which are: methyl derivatives of aromatics;
(4) Condensed tertiary products: benzene, naphthalene, acenaphthylene, anthracene/ phenanthrene, pyrene. Tar might decompose due to reforming reactions, cracking, and solid carbon formation, and this let to fouling and plugging. Then, tar is often dangerous due to their carcinogenic nature.

Fig. 2 shows the trend of the four "tar" component classes with respect to temperature at 300 ms (0.3 s) gas-phase residence time.

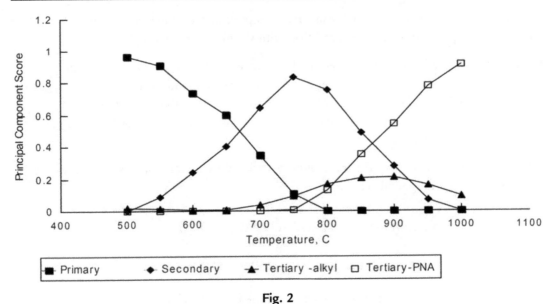

Fig. 2

The trend of the four "tar" component classes as a function of temperature at 300ms (0.3s) gas-phase residence time [17].

Table 1 Average amount of tar in biomass gasification and syngas common applications limits [18,19].

	Biomass gasification	Ammonia production	Gas turbine	SOFC	Methanol synthesis	ICE
Tars (mg/Nm^3)	3000–100,000	<0.1	<30	<100	<0.1	<30

So, it is important to reduce the tar amount at the exit of the gasifier, before the downstream applications. In Table 1, the average level of tar in biomass gasification and the limit for the most common industrial technologies are shown.

3. Tar removal

There are various applications that can be used in order to remove tar from raw syngas. Commonly, tar removal technologies are divided into physical methods (such as cyclone, scrubbers, or filters) and thermochemical methods (converting tar into syngas through catalyst). Thermochemical transformation technologies grant the growth of the global efficiency of the gasification process through the conversion of tar into applicable gas product, so this kind of application has conquered considerable interest [20]. And, according to the location in which tar is removed, the approach is distinguished into primary method, when it happens inside the

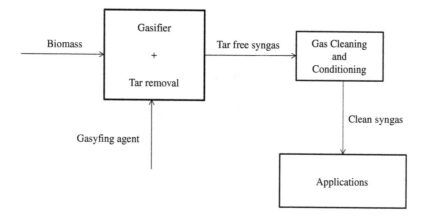

Fig. 3
Primary method—tar reduction prototype.

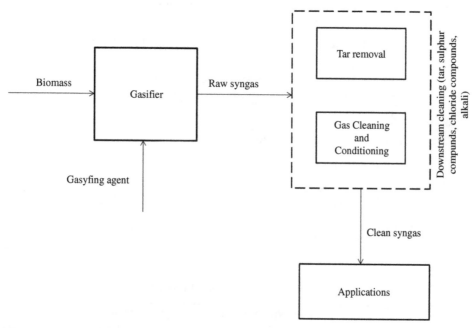

Fig. 4
Secondary method—tar reduction prototype.

gasifier, and secondary methods, when it happens outside of the gasifier. The prototype of primary and secondary methods is showed in Fig. 3 and Fig. 4, respectively.

Primary methods are based on thermochemical processes and reduce the formation of tar and/or convert tar into syngas into the gasifier reactor. Secondary methods are based on chemical or physical treatment that separate the tar or convert the tar downstream the gasifier.

3.1 Tar removal: Physical methods

Physical cleaning approaches are divided into:

- Dry cleaning approach, which are normally carried out both in the temperature range 200–500°C and 600–900°C [21,22];
- Wet cleaning approach, which is usually conducted after cooling, so around 20–60°C [23].

Table 2 and Table 3 show the most common dry and wet cleaning applications, respectively.

It is common to use cooling tower and scrubbing tower as the earliest wet scrubbing component following cyclones. The condensation of the heavy tar happens in there. Nevertheless, tar drops and gas/liquid hazes are dragged into the gas flow, and this makes this kind of tar removal quite ineffective. Venturi scrubbers are commonly the succeeding step.

Electrostatic precipitators (ESPs) are extensively utilized to take off fine solids and liquid drops from gas streams. Even if this technology is efficacious with liquid droplets, it is rather ineffective when tar is gas. This clearly indicates that, if the goal is tar elimination, it is better to work at mid-temperature. In that case, gas should be extinguished before using ESP. ESP

Table 2 Dry cleaning methods [24].

Technology	Tar removal efficiency (%)	Operational temperature (°C)	References
Cyclonic separator	30–70	100–900	[25,26]
Fabric filter	0–50	Up to 600	
Sand bed filter	50–90	20	
Ceramic quartz filter	75–95	650–770	[27]
Adsorbent activated carbon filter	80	20	[28]
Porous catalytic filter	77–99	800–900	[29,30]

Table 3 Wet cleaning methods [24].

Technology	Tar removal efficiency (%)	Operational temperature (°C)	References
Impinger	70	50	[31]
Three water impinger in series	>95	50	
Wet electrostatic precipitator (ESP)	40–70	20–30	[32]
Venturi scrubber	50–90	20–100	[25,33]
Packed bed scrubber	75	300	[34]
Bio-oil scrubber	60	50	[28]
Water scrubber	22	20–100	[25,35]

technology is determined by the transit of the gas across a high-voltage area, which has a negative charge. Particles that have become negative charged are then collected into an area in which are located opposite charge plates that remove the negative particles from the stream. ESP operation could be damaged by too much high conduciveness and also by too much low conductivities; thus, it is necessary to have a right balance in order to assure efficient operations. This suggests that tar may affect significantly the pattern of an ESP [23].

Wet scrubbing is well improved, and the efficiency of tar elimination lets us suppose that the final requirement level of tar contained in the syngas is already satisfied for downstream applications. Since results come from a wide variety of sources, it could be useful to have some trials on pilot plants with different types of feedstocks and gasification processes. Demisting is considered as wet scrubbing. The quantity of wastewater produced from this technology is not so relevant; on the contrary, the contamination level is very relevant due to the toxicity of particulates and tar; this makes the treatment expensive and not affordable in several cases. Nowadays, there exist many technologies to minimize the production of wastewater. Although, it is possible to indicate as inferior limit of wastewater production the amount of water that is already in the producer gas. There are several methods to clean wastewater, the most common involve wet oxidation, adsorption on active carbon and/or carbon-rich ashes from gasification, and biological treatment. Nevertheless, these treatment technologies are leaked of a full optimization. In the near future, if well optimized, they could provide a helpful possibility for commercial applications [17].

Anyway, at low temperature, tar condensation happens causing fouling, clogging, and plugging of the equipment and deactivation of bed catalyst. This is the reason why tar removal should be conducted at the temperature close to that of the gasification.

3.2 Tar removal: Thermochemical methods

- Thermal destruction approach is fundamentally to destroy the refractory aromatics, without a catalyst. It happens when temperature is over 1000°C and at reasonable residence times. Besides economics and materials issues, such thermal decomposition can form a sediment that can be more inconvenient than the aromatics depending on the processes. Among the light aromatic compounds, benzene is suggested as the least thermally reactive.
- Oxidative approach for tar conversion happens when oxygen is added selectively to different stages, so tar can be preferentially oxidized. The addition of oxygen or air to gasification process seems to create more refractory tar but at lower levels, while favoring the conversion of primaries.
- Catalytic approach for tar removal transformation happens mostly following two approaches: conversion inside the gasifier (primary methods), by adding catalyst material to the bed inventory, or conversion outside the gasifier (secondary methods), in a separated

reactor downstream. The second approach achieves a higher efficacy in tar removal, but it is not easy to set on small and medium applications [10,36,37].

In order to select the best catalyst, the following criteria must be considered [23,38–42]:

- The porosity of the material;
- The capacity to operate at elevated content of H_2, CO, CO_2, and H_2O at 650–950°C;
- The capacity to supply a pertinentH_2/CO ratio for the process;
- The ability to resist at the deactivation process (due to poisoning, fouling, etc.);
- The faculty of easy regeneration;
- The price, which should be the lower;
- The resistance to weakening.

Natural dolomite $CaMg(CO_3)_2$ is one of the most common catalysts used to remove tar. Natural dolomite has the advantage to be extremely cheap and to allow the achievement of 95% or more tar conversion [43–45]. It has been indicated as the most effective catalyst to eliminate tar in syngas [46]; furthermore, when dolomite is used, it is possible to reach the highest gas yield and heating value (HHV).

It has been experimentally demonstrated that the use of dolomite in the concept of secondary method gives higher efficiency for tar reduction instead of its using in primary method [47].

The conversion rates of tar when reacting with dolomite (which is a primary sorbent for NH_3, H_2S and HCl) are shown in Table 4.

Other common catalysts that can be used are the ones based on Ni, olivine $FeMg(SiO_4)_2$, and lime $Ca(OH)_2$.

- Thermal plasma approach is utilized in gasification reactor to improve the quality of syngas. It has been demonstrated that plasma operation in a two-stage fluidized bed gasifier to convert solid waste into syngas [54] is able to reduce the tar concentration. The plasma is able to decompose the condensable tar, and there is a great interest in the application of plasma since it is possible to obtain tar-free syngas to begin with. The minor size of equipment and the ability to deal with different kinds of dangerous and heterogeneous materials are benefits of this process. Besides, it was demonstrated as too complicated

Table 4 Conversion rate (X) of components that react with dolomite [45,48–53].

X_{C6H6} (%)	50
X_{C10H8} (%)	82
X_{C7H8} (%)	80
X_{H2S} (%)	85
X_{HCl} (%)	85
X_{NH3} (%)	95

system for the reduction of condensable tar. A considerable quantity of energy is necessary to clean the huge amount of gas, and this is the reason why this approach is uneconomical for cleaning purposes.

3.3 Tar removal and conversion: Modeling and simulation

A kinetic model for tar cracking has been suggested by Di Biasi [55], who collected various kinetic constants from literature sources. The limitation of the model is that the used are considered tar as a single lump component of $C_aH_bO_c$ type or do not specify its chemical formula. Also Boroson et al. [56] proposed a kinetic model for tar cracking based on resulting experimental data of wood gasification. They modeled tar as $C_xH_yO_z$, and after cracking, they considered it yields H_2, CO, CO_2, CH_4, and a secondary tar:$C_{x1}H_{y1}O_{z1}$. The limitation of the model is that primary and secondary tar components cannot be directly determined; it is a lumped component. Corella et al. [57] suggested two kinetic models for catalytic tar removal in biomass gasification assuming tar as a continuous mixture or being composed of six different lumps. The first approach developed gives details about the molecular weight of tar and the variation of tar molecular weight distribution. The second approach developed grants ranking of tar species depending on their reactivity. The proposed six-lump reaction network for catalytic tar elimination is shown in Fig. 5.

The suggested set of kinetic equations was:

$$r_1 = \frac{C_{1,0}dX_1}{d\tau} = k_{61}C_6 + k_{21}C_2 - k_1C_1 \tag{R1}$$

$$r_2 = \frac{C_{2,0}dX_2}{d\tau} = k_{42}C_4 + k_{32}C_3 - k_{26}C_2 - k_{21}C_2 \tag{R2}$$

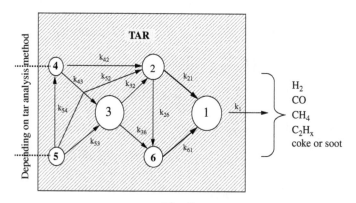

Fig. 5
Proposed six-lump reaction network for catalytic tar elimination [57].

$$r_3 = \frac{C_{3,0}dX_3}{d\tau} = k_{53}C_5 + k_{43}C_4 - (k_{32} + k_{36})C_3 \tag{R3}$$

$$r_4 = \frac{C_{4,0}dX_4}{d\tau} = k_{54}C_5 + k_{43}C_4 - k_{42}C_4 \tag{R4}$$

$$-r_5 = \frac{C_{5,0}dX_5}{d\tau} = (k_{53} + k_{54})C_5 \tag{R5}$$

$$r_6 = \frac{C_{6,0}dX_6}{d\tau} = k_{36}C_3 + k_{26}C_2 - k_{61}C_6 \tag{R6}$$

$C_{i,0}$ at the exit of the gasifier, that is also the inlet of the catalytic bed, had the following values:

$$C_{1,0} = 4400 \pm 400 \, mg/m_n{}^3$$

$$C_{2,0} = 2900 \pm 200 \, mg/m_n{}^3$$

$$C_{3,0} = 1320 \pm 120 \, mg/m_n{}^3$$

$$C_{4,0} = 900 \pm 120 \, mg/m_n{}^3$$

$$C_{5,0} = 530 \pm 300 \, mg/m_n{}^3$$

$$C_{6,0} = 200 \pm 100 \, mg/m_n{}^3$$

The following values of k_{ij} were obtained:

$$k_{54} = 19 \left[m^3 \left(T_{\text{catalytic bed}}, \text{wet} \right) \right] (\text{kg of catalyst})^{-1} \, h^{-1}$$

$$k_{53} = 34 \left[m^3 \left(T_{\text{catalytic bed}}, \text{wet} \right) \right] (\text{kg of catalyst})^{-1} \, h^{-1}$$

$$k_{52} = 3.5 \left[m^3 \left(T_{\text{catalytic bed}}, \text{wet} \right) \right] (\text{kg of catalyst})^{-1} \, h^{-1}$$

$$k_{43} = 8 \left[m^3 \left(T_{\text{catalytic bed}}, \text{wet} \right) \right] (\text{kg of catalyst})^{-1} \, h^{-1}$$

$$k_{42} = 32 \left[m^3 \left(T_{\text{catalytic bed}}, \text{wet} \right) \right] (\text{kg of catalyst})^{-1} \, h^{-1}$$

$$k_{32} = 0.1 \left[m^3 \left(T_{\text{catalytic bed}}, \text{wet} \right) \right] (\text{kg of catalyst})^{-1} \, h^{--1}$$

$$k_{36} = 33 \left[m^3 \left(T_{\text{catalytic bed}}, \text{wet} \right) \right] (\text{kg of catalyst})^{-1} \, h^{-1}$$

$$k_{26} = 24 \left[m^3 \left(T_{\text{catalytic bed}}, \text{wet} \right) \right] (\text{kg of catalyst})^{-1} \, h^{-1}$$

$$k_{21} = 50 \left[m^3 \left(T_{\text{catalytic bed}}, \text{wet} \right) \right] (\text{kg of catalyst})^{-1} \, h^{-1}$$

$$k_{61} = 296 \left[m^3 \left(T_{\text{catalytic bed}}, \text{wet} \right) \right] (\text{kg of catalyst})^{-1} \, h^{-1}$$

$$k_1 = 51 \left[m^3 \left(T_{\text{catalytic bed}}, \text{wet} \right) \right] (\text{kg of catalyst})^{-1} \, h^{-1}$$

CFD models on hot gas filtration can ameliorate the accuracy of multiphase phenomenon modeling, especially for tar removal. Savuto et al. [58] developed a 3D-CFD model by means of FLUENT software for tar removal through ceramic candles containing a layer of Ni-based catalyst allocated in the freeboard of the gasifier. The model parameters came from experimental results. The tar species considered are: benzene (C_6H_6), toluene (C_7H_8), and naphthalene ($C_{10}H_8$); the kinetic parameters inserted in the model, taken from literature, were relative to tar steam reforming reactions on Ni-based catalysts. Gasifier freeboard input compositions and operating conditions for simulation in dry molar fraction were: $H_2/CO/CO_2/CH_4/H_2O/C_6H_6/C_7H_8/C_{10}H_8 = 40.04/29.05/22.35/8.56/15.67/10.40/2.30/2.04$. The simulation output was $H_2/CO/CO_2/CH_4/H_2O/C_6H_6/C_7H_8/C_{10}H_8 = 48.42/29.11/29.34/3.13/3.31/0.69/0.56$. The simulation modeled showed that small injections of oxygen in the freeboard can increase tar conversion up to 77%.

1D-thermodynamic equilibrium models are also taking into account for the modeling of tar removal. Marcantonio et al. [7] developed a model by means of Aspen Plus software simulated the tar removal in the stage of catalytic filter candle filled by Ni in the freeboard of the gasifier and also in the bed of reactor made of calcined dolomite. The tars considered were benzene, toluene, and naphthalene. The catalytic filter was simulated through a stoichiometric reactor, and the reactions considered were:

$$C_6H_6 + 6\,H_2O \rightarrow 6\,CO + 9\,H_2 \tag{R7}$$

$$C_7H_8 + 7\,H_2O \rightarrow 7\,CO + 11\,H_2 \tag{R8}$$

$$C_{10}H_8 + 10\,H_2O \rightarrow 10\,CO + 14\,H_2 \tag{R9}$$

Data for reaction conversions came from lab-scale gasification tests [59], and they are: X_{C6H6} (%)/X_{C7H8} (%)/X_{C10H8} (%) = 95/92/90.

The in-bed tar removal was modeled trough a Gibbs reactor; the reactions of components with calcined dolomite were not taken into account, and the conversion rates assumed for tar came from lab-scale gasification tests, and they are: X_{C6H6} (%)/X_{C7H8} (%)/X_{C10H8} (%) = 50%80/82.

4. Sulfur and chlorine contaminants

Sulfur compounds (mainly hydrogen sulfide, H_2S) and chlorine compounds (mainly hydrogen chloride, HCl) are the inorganic contaminants contained in biomass-derived syngas. The amount of the contaminants in syngas composition depends on the chemical composition (C, O, H, N, S, Cl) of the biomass. In lignocellulosic biomass, the contents (expressed on a dry and ash free basis) of S and Cl are generally around 0.01%–0.2%. Lignocellulosic biomass has very low sulfur and chlorine content compared with coal and MSW. In Table 5, you can find the reported chemical composition of:

Table 5 Chemical composition of the most common biomass feedstock.

Biomass		Moisture (%)	Ash (%)	VM (wt%)	FC (wt%)	C (wt%)	H (wt%)	N (wt %)	S (wt%)	Cl (wt%)	O (wt%)	HHV (kJ/kg_dry)	References
Woods	Branches and leaves from pine tree	10–56	0.05–2.2			48.3–55.4	5.2–7.8	0.16–2.06	0.14–0.66		35.3–44.1	19,364–21,713	[60]
		31.8											[61]
		50–56	0.4–2.2			51.8	6.1	0.3	0.01	0.004	41.2	17,133	[62]
				79.4–85.7	14.0–18.7	46.5–48.6	6.7–6.9	0.13–0.74	0.03–0.86		42.1–46.4		[63]
	From fir	44.2	010	87.30	12.60	50.64	6.18	0.06	0.02		43.0	20,370	[64]
			2.86	85.1	14.9	49.4	5.5	0.16	0.03		44.8		[65]
	Branches and leaves from	24–60	0.62–1.14	80.9–83.8	15.6–18.1	49.3–50.5	6.0–6.1	0.33–0.77	0.01–0.03		41.8–43.5	19,640–19,960	[66]
	poplar tree	6.0–6.5	1.70–2.18	76.0–76.9	15.3–15.4	45.1–45.7	6.3	0.11–0.15	0.14–0.16		47.7–48.3	18,386–18,420	[67]
			1.43			48.5	5.9	0.47	0.01		43.7	19,380	[68]
			21	76.8	1.4	36.2	4.3	1.1	1.2		51.6		[69]
			2			48	6.2	0.4	0.03	<0.01	43	19,800	[70]
Straws	Beech	5.90	0.82	84	15.40	48.26	5.80	0.29	0.03		44.80	18,690	[71]
	Wheat straw	8.5	4.99	65.6	21.0	41.8	4.8	0.5	0		38	17,250	[72]
		6.4	8.1	75.6		47.1	5.7	0.75	0.07	0.07		17,000	[73]
		7.5	4.4	79	15	47.2	6.6	0.25	0.03		40.3	18,610	[74]
		7.5	6	72.69	18.09	42.89	5.81	0.98	0.05	0.12	40.93		[75]
		7.6	9.22	69.10	17.61	41.93	5.79	0.91	0.08	0.21	37.79	17,250	[76]
		7.6	13.29									16,660	[76]
	Oil Rape straw	10.5	5.6	80.2	3.7	42.9	5.8	0.87	0.58		49.9	17,700	[77]
		9	6.1	74.6	19.3	43.8	5.8	0.83	0.61		42.9	16,960	[78]
			3.5	69.6	17.9	37.8	4.6	0.76		0.04	44.3	18,100	[79]
			6.2			47.1	5.9	0.84	0.27	0.47	40		[80]
		7.6		95.5					0.134			17,900	[81]
	Maize stover	5.5	7.4	71.6	15.6							16,714	[82]
		8	5.7	76.8	17.5	48.8	6.2	0.5	0.1		44.4	18,450	[83]
		5	7.0	74.2	13.8	44.2	5.5	0.53	0.1	0	37.7	18,050	[84]
				90.6				1.21					[85]
		6.8	5.5	75.7	16.9	33.5	5.7	0.52	0.07		54.7	17,310	[86]

Agricultural wastes													
Sugarbeet leaves	9.4	2.3	71.9	16.5	40.8	5.9	1.3	0.19		45.7	17,005	[87]
	6.1	3	74.2	16.7	43.4	6.3	1.4	0.2		48.7	16,100	[88]
	12.5	9	79	16.9	45.7	6.8	1	0.51		46.7	18,700	[89]
	9.5	12.7–13.9	76	15	38.9	5.7	1.2	0.24	0.002	47.5	15,095	[90]
			72.3		41.0	5.0	0.97			40.2	15,900	[91]
Sunflower stalk		11.95			46.77	5.51	1.89	0.21	0.52	33.15	19,120	[92]
		11.14			42.74	5.03	0.99	0.11	0.78	39.21	16,320	[93]
		8.82			45.51	5.03	0.31	0.03	0.68	39.62	17,040	[94]
Olive tree pruning	3.8	2.9	72.6	21.9	51.2	6.7	1	0.15	0.02	39.3	19,830	[95]
		1.7	77.9	21.3	48.5	5.3	1.12	0.02	0.02	44.7	19,100	[96]
		0.8			44.5	5.8	0.7	<0.01	<0.01		17,320	[97]
		5.3					0.84	0.0185	0.0005		19,060	[98]
Vine pruning	14.9	1.55	80.8	17.66	49.57	5.96	0.10	0.006	0.05	42.80	18,950	[76]
	3.67	1.42	77.80	19.58	46.28	6.28	0.54	0.01	0.008	46.89		[99]
Almond shell	3.3	0.6	80.3	15.8	50.5	6.6	0.21	0.18			18,200	[100]
Hazelnut shell	12.9	5.3					0.22	0.11				[101]
Grape seeds	7.6	2.3	69.58	20.5	49.9	6.3	1.3			32.3	21,290	[102]
Olive pomace	10	3.9	77.6–78.0	18.6–20.1	51.3	5.9	0.95		0.39	36.3–37.3	21,700–22,300	[103]
	5.5–10	1.9–3.8	43.2–80.2	9.9–44.3	53.2–53.5	8.9–9.2	0.6–1.0	0.0–0.3	0.09–0.80	33.6–46.2	19,500–25,460	[104]
	6.9	2.6–9.0	71.9	16.1	46.8–54.9	5.7–8.2	0.7–2.2	0.1	0.80	32.0	20,520	[105]
	5.5–10	5.2	77.8–80.9	15.3–18.5	47.9	6.1	1.8	<0.1		34.0–40.1	20,610–23,390	[106]
		0.56–5.60			52.3–55.2	7.5–8.2	0.06–2.22					[107]
	36.3	5.95	73.01	21.04	51.84	7.14	2.79	0.06	0.08	32.14	21,350	[76]

Continued

Table 5 Chemical composition of the most common biomass feedstock—cont'd

Biomass		Moisture (%)	Ash (%)	VM (wt%)	FC (wt%)	C (wt%)	H (wt%)	N (wt %)	S (wt%)	Cl (wt%)	O (wt%)	HHV (kJ/kg_dry)	References
Residual woods	Wood sawdust	4.8	1.27	78.3	15.6	46.5	5.8	0.1	0.01	0.03	41.4	19,260	[108]
		16.4	0.37	81.0	18.6	48.6	5.8	2.39	3.56	0.4	43.2	15,100	[109]
			0.13	81.6	18.4							20,581	[110]
		11.2	0.41	81.8	17.8	49.40	5.84	0.43	<0.01	<0.01	43.92	20,160	[76]
		8.9	0.54	81.20	18.26	45.81	5.85	0.1	<0.01	<0.01	47.69	18,090	[76]
		6.6	0.43	84.66	14.91	47.07	6.15	0.1	<0.01	<0.01	46.24	18,480	[76]
		24.5	1.45	81.50	17.05	49.88	5.80	1.06	0.02	<0.01	41.79	19,140	[76]
Black liquor from paper mill		8.6	32.5	50.8	8.04	23.5	4.8	0.08	4.4	0.08	34.8	10,550	[111]
		3.81	32.4	45.7	18.9	31.07	2.96	0.55	0.93		29.09		[112]
						34–39	3–5	0.05–0.2	3–7	0.2–2	33–38		[113]
Municipal wastes	MSW		9.4			6.3	4.9						[114]
		20.6	48.28	43.54	8.18	33.27	3.87	0.15	0.74	0.12	13.57	12,080	[76]
		6.2	15.8	68.1	9.9	46.2	7.7	1.7	0.23		22.3	19,220	[115]
		3.3	9.1	79.7	7.2	63.6	8.1	0.4	0.11		27.1	15,978	[116]
		60.6	4.0	31.9	3.6	55.2	5.3	2.6	0.29	0.07	34.8	12,420	[117]
		4.6	5.3	77.9	16.7	14.6	2.1	0.4	0.09		7.6		[118]
		59.6	15.5	21.6	3.4	41.2–	5.4–	1.2–	0.18–	0.45–	28.9–		[119]
		25	18.5–22.6			42.3	5.5	1.3	0.23	0.71	32.3		[120]
	OFMSW	23.0	47.01	40.32	12.67	32.65	4.43	2.37	0.2	0.4	12.94	11,190	[76]
		10.8	7.7	76.9	15.4	52.3	6.5	2.7	0.3		38.3	19,900	[121]
		2.1	12.9	68.8	16.3	58.6	6.7	2.2	0.3		17.1		[122]
										1.2			[123]
Digestates			15.7–35.8	78.4–85.1	14.9–21.6	34.3–42.6	4.0–5.0	1.9–2.1	0.2–0.4		23.9–34.3	13,400–17,000	[124]
			12.4	67.0	20.6	42.5	5.9	1.8			49.8	19,740	[125]
		6.8	25.7	55.5	18.8	35.8	9.5	3.2	0.3		25.5		[126]
		2.3	54.8	33.0	9.9	21.8	2.6	1.3	0.54	0.34	18.6	8920	[127]
			16.0–55.5	36.2–70.2	2.1–13.8	24.1–44.1	1.7–5.1	1.5–3.2	0.2–1.5		16.4–31.3	14,900–17,800	[128]
		71.2	25.81	63.97	10.22	32.22	4.51	3.07	0.97	0.1	33.32	13,700	[76]

Manure of bovine, swine and poultry												
Bovine dry matter	84.3		77.0	11.0	25.9	6.8	0.7	0.58				[129]
	9.5		45.8	12.5	30.5–35.1	5.9	2.2–2.3					[130]
Swine manure		12.0	82.9	22.6	6.9	5.7	0.4	0.37		49.9	13,426	[131]
		32.2	64.8		41.1	5.1	2.7					[132]
	6.9–8.6	17.1	58.9–60.2	7.2–7.7	41.2	7.5–7.9	3.9	0.2		35.7	15,800	[133]
		12.6	69.5	9.3	44		2.9	1.7–1.9		26.8–31.5		[134]
		25.2–26.4	73.8	6.4	54.3–58.4		4.3–5.1					[135]
Poultry manure		21.2									19,500	[136]
	12.8–62.5	19.9	26.6–64.4		14.8–36.0	5.7–8.2	1.9–2.4	1.2	1.4		11,552–14,587	[137]
		10.6–15.4										[138]
		24.4			37.8	4.8	1.9	0.1		31	15,100	[139]
	18.1–70.9	12.0–44.4	50.4–69.9	5.2–18.1	21.9–42.9	2.5–5.6	1.7–5.5	0.28–0.68		29.2–34.1	8577–16,546	[140]
					44.7		24.3	0.1				[141]
							4.6	0.46		23.9		[142]
	90	32	67		41		2.8	1.18	0.71		11,552–14,587	[143]

- Primary residues from industrial forestry activities;
- Agricultural wastes;
- Secondary wastes from the woodworking industries;
- Secondary residues from industrial processing of agricultural products;
- Municipal waste;
- Wood waste;
- Digestate generated in biogas production;
- Livestock manures.

The lower the content of S and Cl in the inlet biomass, the lower the content of sulfur and chlorine compounds in the syngas.

H_2S is catalyst poison and provoke pollutions. HCl is responsible for corrosion of metal equipment, health problems, and environmental issues [50,144–146]. Those are the reasons why is fundamental to clean the raw syngas before the usages in downstream applications [147,148]. Table 6 indicates contaminant limits for the most common commercialized applications.

4.1 Hydrogen sulfide and hydrogen chlorine removal

It is possible to operate both at mid-temperature (in the range 400–600°C) and at high temperature (in the range of 600–850°C) regarding gas cleaning technologies. Higher temperatures are preferable since guarantee considerable economic advantageous since gas cleaning and conditioning happen at temperature near to the temperature of the gasification process [52,153]. There are some characteristics that make a desulfurization adsorbent the ideal one: high adsorption capacity (which means to lower the needed adsorbent quantity and also the equipment size), fast adsorption kinetics (literature experiments have demonstrated that the best kinetic of desulfurization should be first-order dependence in H_2S), high equilibrium constant, cheapness, fast and cheap regeneration capability, ideally regenerable [52,154].

The reaction (R10) is the solid-gas reaction happening when the metal oxide meets the contaminant:

$$MO_{solid} + H_2S_{gas} \leftrightarrow MS_{solid} + H_2O \qquad (R10)$$

Table 6 Typical H_2S and HCl concentration in biomass-waste-derived syngas and target-level-associated syngas applications [40,50,149–152].

Contaminants (ppm)	Biomass waste gasification	Ammonia production	Gas turbine	SOFC	Methanol synthesis
H_2S	20–500	<0.1	<20	<1	<0.5
HCl	30–1000	<1.5	<1.5	<1	<1

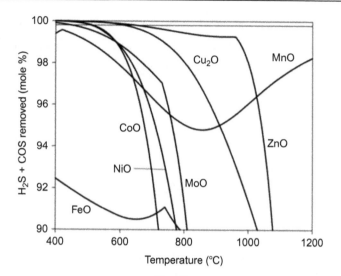

Fig. 6
Adsorption of metal oxides at their thermodynamic equilibrium [158].

The equilibrium constant of reaction (R10) is written in Eq. (R11):

$$K_{eq,H2S} = \frac{[H_2O][MS]}{[H_2S][MO]} = \frac{[H_2O]}{[H_2S]} \tag{R11}$$

Studies from literature indicate zinc-based sorbents as the best option for hydrogen sulfide removal [146,155,156], thanks to the capacity to catch almost the entire amount of H_2S at 400°C. As the temperature increases as the Zn devolatilizes, it results in lowering of the adsorption capacity. Besides zinc-based sorbents, also cerium- and copper-based sorbents can reduce H_2S below 1 ppm [19,146,156,157]. Generally, the H_2S cleaning capacity of different oxides is: Stannum < Nickel < Iron < Manganese < Molybdenum < Cobalt < Zinc < Copper and Cerium [19] and is shown in Fig. 6.

Table 7 reports the most significant properties of the most common metal oxides for H_2S removal.

Slimane and Abbasian [161] fabricated two types of CuO sorbents drugged with Cu and Mn content (the sorbent IGTSS-179, which works at 500–600°C and the sorbent IGTSS-326A, which works at 450–600°C), those kinds of sorbents can obtain a satisfying desulfurization efficiency ($H_2S < 1$ ppmv).

Zeng et al. [162] studied the ability of reduced cerium oxide in desulfurization. Their studies highlighted that CeO_n ($n<2$) has higher desulfurization capacity than CeO_2. The biggest benefit noticed was that elemental sulfur was created directly in the regeneration of the sulfide product, Ce_2O_2S, with SO_2.

Table 7 Properties of metal oxides for H_2S removal [48,159,160].

Sorbent	Chemical formula	Theoretical sorption capacity (gS/g sorbent)	Temperature range (°C)	Price (USD/kg)
Cerium oxide	Ce_2O_3	0.093	500–700	5.9
Copper oxide	Cu_2O	0.224	540–700	5.8
Zinc oxide	ZnO	0.395	450–650	2.6
Manganese oxide	MnO	0.400	400–900	4.2
Iron oxide	FeO	0.245	450–700	1.4

Ikenaga et al. [163] investigated carbon materials as a support of zinc ferrite ($ZnFe_2O_4$). Their experiments pointed out that the support of zinc ferrite increases the adsorption capacity with respect to the experiments carried out without support. At 500°C, the concentration of H_2S lowers from 4000ppmv under 1ppmv. It is possible to regenerate these kinds of sorbents adding 50vol% O_2/Air gas at 450°C for 30min with a restrict drop-in sulfur activity.

The removal of HCl is mostly based on alkali-based sorbents [149,151,164,165]. The temperature range for the experiments concerning CaO or $CaCO_3$ sorbents is above 500°C [166–168]. Verdone et al. [169] established that Na_2CO_3 sorbent is able to remove the most amount of HCl at 400–500°C. Ohtsuka et al. [170] conducted an experiment at 400°C and tested $NaAlO_2$ for hydrogen chloride removal; the experiment indicated that at the set temperature, the concentration of HCl drops from 200ppm in the inlet gas to less than 1ppm. Then, they pointed out $NaAlO_2$ has higher efficiency in HCl capture than Na_2CO_3. The presence of H_2S does not affect severely the ability of $NaAlO_2$ and Na_2CO_3 to capture HCl. Krishnan et al. [171] investigated nahcolite ($NaHCO_3$) as HCl sorbent removal, and they demonstrated it reduced HCl from 1750ppm to lower than 1ppm in the temperature range of 400–600°C. Dou et al. [49] indicated nahcolite as the favorable alkali-based sorbent, this sorbent is able to reduce the content of HCl below 1ppmv at 525–650°C.

When nahcolite is used, the reaction (R12) happens:

$$NaHCO_3(s) + HCl(g) \rightarrow NaCl(s) + H_2O(g) + CO_2(g) \tag{R12}$$

Over 550°C, water molecules are separated, following reaction (R11). Then, $NaHCO_3$ decomposes to $Na_2CO_3(s)$, which reacts with HCl to produce NaCl:

$$2NaHCO_3(s) \rightarrow Na_2CO_3(s) + H_2O(g) + CO_2(g) \tag{R13}$$

$$Na_2CO_3(s) + 2HCl(g) \rightarrow 2NaCl(s) + H_2O(g) + CO_2(g) \tag{R14}$$

Literature indicates 550°C as the best operative temperature of $NaHCO_3$ to reduce HCl under 1ppmv.

Ren et al. [164] studied an alkali-based potassium carbonate sorbent for HCl removal and obtained a reduction of 54% (considering a temperature of 500°C), 51% (considering a temperature of 300°C), and 32% (considering a temperature of 20°C). From the thermodynamic point of view, the adsorption of HCl by $Na_2CO_3(s)$ is favorite at mid-temperature; increasing temperature is favorable from the kinetic point of view until the limit of salt decomposition of alkali chloride.

Baek et al. [151] experimentally investigated the behavior of unprocessed and processed potassium-based CO_2 sorbents in HCl removal. The test was set through a micro fluidized-bed reactor (working at 300°C and 20bar) and a bench-scale bubbling fluidized-bed reactor (working at 540°C and 20bar). Tests indicated that HCl reduced from 150 to 900ppmv to 5ppmv and from 130 to 390ppmv to 1ppmv.

4.2 Hydrogen sulfide and hydrogen chlorine removal modeling

Marcantonio et al. [7] modeled a thermodynamic equilibrium system to simulate the hydrogen sulfide removal by means of Aspen Plus software, using an equilibrium reactor, which involves the solid-gas reaction between a metal oxide and the contaminants (R9) and (R10). The great advantage of the proposed model is that it works with every sorbent, taking into account its sorption capacity and working temperature range.

Shareefdeen et al. [172] modeled a kinetic study of H_2S removal for a novel biofilter media. They showed that a first-order model suits better the data:

$$C_{out} = C_{in}\ \exp\left(-k_1 EBRT\right) \tag{R15}$$

k_1 is the first-order reaction rate constant (s^{-1}) defined as:

$$k_1 = \frac{A_s}{m} \sqrt{\frac{X_V \cdot \mu^* \cdot f(X_V) \cdot D_{H_2S,W}}{K \cdot Y}}\ \tanh\left(\beta_2\right) \tag{R16}$$

$$\beta_2 = \delta \sqrt{\frac{X_{V\mu^*}}{K \cdot Y \cdot f(X_V) \cdot D_{H_2S,W}}} \tag{R17}$$

K is the Monod kinetics constant (kgm^{-3}), EBRT is the empty bed residence time, $f(X_V)$ is the ratio of diffusivity of a compound in the biofilm to that in water, $D_{H2S,\ w}$ is the diffusivity of H_2S in water $(m^2\ s^{-1})$, Y is the yield coefficient, which is equal to the amount of biomass produced/substrate consumed, μ^* is the specific growth rate (s^{-1}), A_S is the biofilm surface area per unit volume of biofilter (m^{-1}), δ is the biofilm depth (m), m is the dimensionless Henry's constant of the pollutant.

The kinetic constant was used to develop a correlation, which can be used in the prediction of outlet concentration and sizing a biofilter that was packed with the media. In order to evaluate the effect of dispersion, a basic Gaussian dispersion and US-EPA AER-MOD models were subsequently used "with" and "without biofilters."

Meng et al. [173] did a kinetic and thermodynamic simulation for sulfidation phenomena using the software package Factsage (GTT Technologies) equilibrium and phase stability diagram model by applying the free energy minimization method. Dolomite, limestone, lime, and several metal oxides, such as CuO, ZnO, FeO, and MnO, were chosen as desulfurization sorbent candidates. However, the reality of desulfurization process is more complex than just equilibrium reaction, and a pure-equilibrium approach offers strict limits.

Marcantonio et al. [7] modeled a thermodynamic equilibrium system to simulate the hydrogen chloride removal by means of Aspen Plus software, using a Gibbs reactor and considering nahcolite ($NaHCO_3$) as sorbent, since it is able to lower HCl below 1 ppm at 526–650°C at which reaction (R6) take place.

Bal et al. [174] used a Design Expert software version 7.0.0 to predict the removal efficiency of HCl by means of NaOH in a submerged self-priming Venturi scrubber and from the ANOVA test, the meaning of process variables was investigated. The removal efficiency was well predicted by a quadratic equation. Optimum conditions were observed at $55.18 \, \mathrm{ms}^{-1}$ of throat gas velocity, 405.10 ppm of inlet HCl concentration, and 0.0038 N of NaOH concentration in scrubbing liquid. The removal efficiency of hydrogen chloride was 90.80%. HCl removal efficiency ($Y\%$) was well modeled by the quadratic Eq. (R18).

$$Y(\%) = 85.84 + 1.73X_1 + 1.72X_2 + 0.72X_3 + 0.51X_1X_2 + 0.19X_1X_3 + 0.4X_2X_3 + 0.65X_1{}^2 - 0.088X_2{}^2 - 0.51X_3{}^2$$

(R18)

$X_1, X_2, \ldots X_n$ are the independent process variables.

Dou et al. [175] modeled the reaction between sorbent and HCl vapor as a first-order equation with respect to initial HCl (R19), where c_0 is the inlet HCl concentration, c is the HCl concentration in the bed, q represents chloride content of sorbent, which is kg chloride absorbed of per kg sorbent, q_0 corresponds to saturation chloride content of 0 sorbent, which represents the saturated absorbing condition of the sorbent, and x is the conversion of sorbent,

$$\frac{c}{c_0} = \frac{q}{q_0} = x$$

(R19)

The obtained rate of hydrogen chloride and sorbent reaction could be either in control of chemical steps with finite rates or of a mix of chemical reaction and mass transfer steps. The impact of external diffusion is very low against the other steps. The chemical kinetics is assumed to be of first order as function of hydrogen chloride concentration and zero order for sorbent. The particles of sorbent were considered spherical. The process was considered to be isothermal. If reaction is governed by the chemical rate of conversion at the grain surface, the data can be modeled through use of the shrinking core model expression.

$$1 - (1-x)^{1/3} = g(x) = t/\tau_g \tag{R20}$$

Where τ_g is the time required for complete conversion ($x=1$) and it can be obtained:

$$\frac{1}{\tau_g} = \frac{bk_s c_0 m}{\rho r} \tag{R21}$$

If reaction is controlled by reactant diffusion through the product layer, it will follow the expression:

$$1 - 3(1-x)^{2/3} + 2(1-x) = p(x) = t/\tau_p \tag{R22}$$

Where t indicates the time required for complete conversion ($x=1$) and can be got as:

$$\frac{1}{\tau_p} = \frac{6bD_e c_0 m}{\rho r^2} \tag{R23}$$

Commonly, reactions are controlled by the union of the two approaches described above. So, it is possible to conclude that the model follows the expression:

$$g(x) + \delta^2 p(x) = t/\tau_g \tag{R24}$$

$$\delta^2 = \tau_p/\tau_g \tag{R25}$$

5. Alkali compounds and how to remove them

Alkali are problematic contaminants that must be reduced under acceptable levels (for the most common downstream applicatios <0.1 mg/Nm^3). Based on the above results reported by Stemmler et al. [176], it is possible to represent with HCl all the halides, with H_2S all the sulfur compounds and with KCl and KOH all the alkali. In literature there is a very small number of research studies on alkali removal methodologies. Stemmler et al. [176] carried out a test on aluminosilicate as sorbents for KCl removal at high temperature. The gas stream was filled with 20 ppm of KCl. The tests indicated that aluminosilicates are eligible sorbents for KCl removal under 100 ppm in the temperature range of 800–900°C. However, kaolinite was the only sorbent that kept the 100 ppm limit longer than 50 h.

Tran et al. [177] studied the adsorption of KCl by means of kaoline, varying the kaolin bed temperature in the range of 750–950°C and keeping the inlet of KCl at 550°C. Diversifying the volumetric gas flow rate, it is possible to obtain different superficial gas velocities. The mass of kaolin pellets (0.7–1.0 mm) utilized for everyone experiment was varied correspondingly in order to keep the contact time between the gas and the solid unchanged (0.008 s). In the

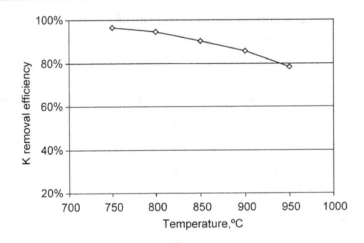

Fig. 7

The impact of bed temperature on the efficiency of KCl capture [177].

experiments, the KCl input concentration was set at the constant value of 1.1 ppm. KCl conversion was constant at 98.4%, and since KCl is independent from conversion on the superficial gas velocity, it is possible to conclude that external mass transfer was not the rate-limiting step. Generally, the conversion grew when the kaolin pellet size was reduced. The experimental activities were performed in air, and the flow rate was set at 300 mL/min. The amount of kaolin pellets for each experiment was 0.2 g, and the dynamic diameter was 0.7–1.0 mm. As indicated in Fig. 7, the increase of temperature determined the decrease of KCl conversion by kaolin pellets, but the conversion was above 78% in the whole range studied. In air, at 850°C, kaolin captures irreversibly KCl in gas phase to form water-insoluble products.

Wang et al. [178] investigated the condensation of alkali metals and showed that at 400°C, they are cubic crystals by homogeneous-monomolecular nucleation. The scheme is reported in Fig. 8.

So, choosing 400°C, it is ensured that alkali metals are condensed. Since the best temperature for hydrogen chloride removal, with nacholite, is in the range of 400–500°C, if this temperature is maintained, no sorbent will need to be included for the alkali removal, since they are already solid.

5.1 Alkali removal modeling

There are very few studies about alkali removal modeling. Moradi et al. [179] modeled the removal of alkali compounds, by means of Aspen Plus software, using a stoichiometric reactor, which simulated the kaolin bed to remove alkali compounds at the same temperature of the gasification process. It was assumed that kaolin bed can eliminate alkali with a conversion rate of 71% for both the NaOH and KOH at 800°C according to [177].

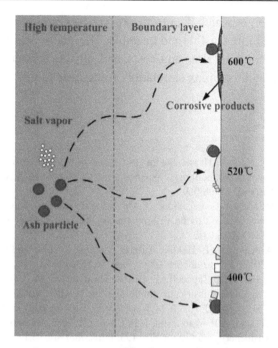

Fig. 8

Condensation mechanism of alkali vapors at different surface temperature.

Kerscher et al. [180] developed a 1-D kinetic simulation model of alkali removal by kaolin. The fixed bed of sorbent material is modeled as a one-dimensional, isothermal plug flow reactor. Considering a constant flow velocity along the length of the reactor, the differential equation of the mass transport can be derived from the mass balance of an infinitesimal element:

$$\frac{\delta c_{Alk}}{\delta t} = u_z \cdot \frac{\delta c_{Alk}}{\delta z} - \frac{1-\varepsilon}{\varepsilon} \cdot \frac{r_{sorption} \cdot S_s \cdot \rho_s}{M_{Alk}} \tag{R26}$$

Where c_{Alk} is the alkali concentration at the generic time t and axial position z, u_z is the axial flow velocity, ε is the porosity of the packed bed, $r_{Sorption}$ is the intrinsic reaction rate due to the alkali sorption, S_S is the specific surface area of the sorbent, ρ_S is the apparent density of the sorbent, and M_{Alk} is the molar mass of the alkali species. Diffusion and mixing effects were not taken into account. The equation system was solved numerically by the finite difference method.

6. Conclusion

Modeling the removal of organic and inorganic compounds has been demonstrated to be a powerful way to predict and evaluate the chemical and physical behavior of the real units in which the removal happens. The most promising approach is the synergy between experimental data and simulation model, in which data from experiments are used to implement simulation model, in order to make it more feasible and trustable. This pattern allows to optimize the whole configuration plant with minimal time and costs, and for this reason, it is necessary to keep modeling more accurate and complex simulation in the near future.

References

[1] L.P.R. Pala, Q. Wang, G. Kolb, V. Hessel, Steam gasification of biomass with subsequent syngas adjustment using shift reaction for syngas production: an Aspen plus model, Renew. Energy 101 (2017) 484–492, https://doi.org/10.1016/J.RENENE.2016.08.069.

[2] E. Bocci, M. Sisinni, M. Moneti, L. Vecchione, A. Di Carlo, M. Villarini, State of art of small scale biomass gasification power systems: a review of the different typologies, Energy Procedia 45 (2014) 247–256, https://doi.org/10.1016/J.EGYPRO.2014.01.027.

[3] S. Thapa, P.R. Bhoi, A. Kumar, R.L. Huhnke, Effects of syngas cooling and biomass filter medium on tar removal, Energies 10 (2017), https://doi.org/10.3390/en10030349.

[4] V. Marcantonio, M. De Falco, M. Capocelli, E. Bocci, A. Colantoni, M. Villarini, Process analysis of hydrogen production from biomass gasification in fluidized bed reactor with different separation systems, Int. J. Hydrog. Energy 44 (2019) 10350–10360, https://doi.org/10.1016/J.IJHYDENE.2019.02.121.

[5] X.T. Li, J.R. Grace, C.J. Lim, A.P. Watkinson, H.P. Chen, J.R. Kim, Biomass gasification in a circulating fluidized bed, Biomass Bioenergy 26 (2004) 171–193, https://doi.org/10.1016/S0961-9534(03)00084-9.

[6] T. Qi, T. Lei, B. Yan, G. Chen, Z. Li, H. Fatehi, et al., Biomass steam gasification in bubbling fluidized bed for higher-H2 syngas: CFD simulation with coarse grain model, Int. J. Hydrog. Energy 44 (2019) 6448–6460, https://doi.org/10.1016/J.IJHYDENE.2019.01.146.

[7] V. Marcantonio, E. Bocci, J.P. Ouweltjes, L. DelZotto, D. Monarca, Evaluation of sorbents for high temperature removal of tars, hydrogen sulphide, hydrogen chloride and ammonia from biomass-derived syngas by using Aspen plus, Int. J. Hydrog. Energy 45 (2020).

[8] V. Marcantonio, E. Bocci, D. Monarca, Development of a chemical quasi-equilibrium model of biomass waste gasification in a fluidized-bed reactor by using Aspen plus, Energies 13 (2019), https://doi.org/10.3390/en13010053.

[9] K. Maniatis, A.A.C.M. Beenackers, Tar protocols. IEA bioenergy gasification task: introduction, Biomass Bioenergy 18 (2000) 1–4, https://doi.org/10.1016/S0961-9534(99)00072-0.

[10] G. Guan, M. Kaewpanha, X. Hao, A. Abudula, Catalytic steam reforming of biomass tar: prospects and challenges, Renew. Sust. Energ. Rev. 58 (2016) 450–461, https://doi.org/10.1016/j.rser.2015.12.316.

[11] N. Rakesh, S. Dasappa, A critical assessment of tar generated during biomass gasification—formation, evaluation, issues and mitigation strategies, Renew. Sust. Energ. Rev. 91 (2018) 1045–1064, https://doi.org/10.1016/j.rser.2018.04.017.

[12] C.M. Kinoshita, Y. Wang, J. Zhou, Tar formation under different biomass gasification conditions, J. Anal. Appl. Pyrolysis 29 (1994) 169–181, https://doi.org/10.1016/0165-2370(94)00796-9.

[13] L. Devi, K.J. Ptasinski, F.J.J.G. Janssen, S.V.B. Van Paasen, P.C.A. Bergman, J.H.A. Kiel, Catalytic decomposition of biomass tars: use of dolomite and untreated olivine, Renew. Energy 30 (2005) 565–587, https://doi.org/10.1016/j.renene.2004.07.014.

[14] H.N.T. Nguyen, N. Berguerand, G.L. Schwebel, H. Thunman, Importance of decomposition reactions for catalytic conversion of tar and light hydrocarbons: an application with an ilmenite catalyst, Ind. Eng. Chem. Res. 55 (2016) 11900–11909, https://doi.org/10.1021/acs.iecr.6b03060.

[15] D.L. Trimm, The formation and removal of coke from nickel catalyst, Catal. Rev. 16 (1977) 155–189, https://doi.org/10.1080/03602457708079636.

[16] T. Herrmann, M. Dillig, M. Hauth, J. Karl, Conversion of tars on solid oxide fuel cell anodes and its impact on voltages and current densities, Energy Sci. Eng. 5 (2017) 194–207, https://doi.org/10.1002/ese3.166.

[17] T.A. Milne, R.J. Evans, Biomass Gasifier "Tars": Their Nature, Formation, and Conversion, Constraints, 1998, https://doi.org/10.2172/3726.

[18] W. Torres, S.S. Pansare, J.G. Goodwin, Hot gas removal of tars, ammonia, and hydrogen sulfide from biomass gasification gas, Catal. Rev. 49 (2007) 407–456, https://doi.org/10.1080/01614940701375134.

[19] P.V. Aravind, W. De Jong, Review evaluation of high temperature gas cleaning options for biomass gasification product gas for solid oxide fuel cells, Prog. Energy Combust. Sci. 38 (2012) 737–764, https://doi.org/10.1016/j.pecs.2012.03.006.

[20] H. Noichi, A. Uddin, E. Sasaoka, Steam reforming of naphthalene as model biomass tar over iron-aluminum and iron-zirconium oxide catalyst catalysts, Fuel Process. Technol. 91 (2010) 1609–1616, https://doi.org/10.1016/j.fuproc.2010.06.009.

[21] E. Simeone, M. Siedlecki, M. Nacken, S. Heidenreich, W. de Jong, High temperature gas filtration with ceramic candles and ashes characterisation during steam-oxygen blown gasification of biomass, Fuel 108 (2013) 99–111, https://doi.org/10.1016/j.fuel.2011.10.030.

[22] S. Tuomi, E. Kurkela, P. Simell, M. Reinikainen, Behaviour of tars on the filter in high temperature filtration of biomass-based gasification gas, Fuel 139 (2015) 220–231, https://doi.org/10.1016/j.fuel.2014.08.051.

[23] S. Anis, Z.A. Zainal, Tar reduction in biomass producer gas via mechanical, catalytic and thermal methods: a review, Renew. Sust. Energ. Rev. 15 (2011) 2355–2377, https://doi.org/10.1016/j.rser.2011.02.018.

[24] M.L. Valderrama Rios, A.M. González, E.E.S. Lora, O.A. Almazán del Olmo, Reduction of tar generated during biomass gasification: a review, Biomass Bioenergy 108 (2018) 345–370, https://doi.org/10.1016/j.biombioe.2017.12.002.

[25] K.M.K. Prabhansu, P. Chandra, P.K. Chatterjee, A review on the fuel gas cleaning technologies in gasification process, J. Environ. Chem. Eng. 3 (2015) 689–702, https://doi.org/10.1016/J.JECE.2015.02.011.

[26] P. Hasler, T. Nussbaumer, Gas cleaning for IC engine applications from fixed bed biomass gasification, Biomass Bioenergy 16 (1999) 385–395, https://doi.org/10.1016/S0961-9534(99)00018-5.

[27] W. De Jong, Ö. Ünal, J. Andries, K.R.G. Hein, H. Spliethoff, Biomass and fossil fuel conversion by pressurised fluidised bed gasification using hot gas ceramic filters as gas cleaning, Biomass Bioenergy 25 (2003) 59–83, https://doi.org/10.1016/S0961-9534(02)00186-1.

[28] S. Nakamura, S. Kitano, K. Yoshikawa, Biomass gasification process with the tar removal technologies utilizing bio-oil scrubber and char bed, Appl. Energy 170 (2016) 186–192, https://doi.org/10.1016/j.apenergy.2016.02.113.

[29] L. Ma, H. Verelst, G.V. Baron, Integrated high temperature gas cleaning: tar removal in biomass gasification with a catalytic filter, Catal. Today 105 (2005) 729–734, https://doi.org/10.1016/j.cattod.2005.06.022. Elsevier.

[30] K. Engelen, Y. Zhang, D.J. Draelants, G.V. Baron, A novel catalytic filter for tar removal from biomass gasification gas: improvement of the catalytic activity in presence of H2S, Chem. Eng. Sci. 58 (2003) 665–670, https://doi.org/10.1016/S0009-2509(02)00593-6.

[31] D. Khummongkol, C. Tangsathitkulchai, A model for tar-removal efficiency from biomass-produced gas impinging on a water surface, Energy 14 (1989) 113–121, https://doi.org/10.1016/0360-5442(89)90054-6.

[32] S.V.B. van Paasen, L.P.L.M. Rabou, R. Bär, Tar removal with a wet electrostatic precipitator (ESP); A parametric study, in: Exhib. Present. "The 2nd World Conf. Technol. Biomass Energy, Ind. Clim. Prot. Rome, Italy, 10–14 May 2004, 2004.

[33] J. Han, H. Kim, The reduction and control technology of tar during biomass gasification/pyrolysis: an overview, Renew. Sust. Energ. Rev. 12 (2008) 397–416, https://doi.org/10.1016/j.rser.2006.07.015.

[34] A.G. Bhave, D.K. Vyas, J.B. Patel, A wet packed bed scrubber-based producer gas cooling-cleaning system, Renew. Energy 33 (2008) 1716–1720, https://doi.org/10.1016/j.renene.2007.08.014.

[35] N.A. Ahmad, Z.A. Zainal, Performance and chemical composition of waste palm cooking oil as scrubbing medium for tar removal from biomass producer gas, J. Nat. Gas Sci. Eng. 32 (2016) 256–261, https://doi.org/10.1016/j.jngse.2016.03.015.

[36] S. Nilsson, A. Gómez-Barea, D. Fuentes-Cano, P. Ollero, Gasification of biomass and waste in a staged fluidized bed gasifier: modeling and comparison with one-stage units, Fuel 97 (2012) 730–740, https://doi.org/10.1016/j.fuel.2012.02.044.

[37] B.S. Huang, H.Y. Chen, J.H. Kuo, C.H. Chang, M.Y. Wey, Catalytic upgrading of syngas from fluidized bed air gasification of sawdust, Bioresour. Technol. 110 (2012) 670–675, https://doi.org/10.1016/j.biortech.2012.01.098.

[38] F. Pinto, H. Lopes, R.N. André, I. Gulyurtlu, I. Cabrita, Effect of catalysts in the quality of syngas and by-products obtained by co-gasification of coal and wastes. 1. Tars and nitrogen compounds abatement, Fuel 86 (2007) 2052–2063, https://doi.org/10.1016/j.fuel.2007.01.019.

[39] M. Asadullah, Biomass gasification gas cleaning for downstream applications: a comparative critical review, Renew. Sust. Energ. Rev. 40 (2014) 118–132, https://doi.org/10.1016/j.rser.2014.07.132.

[40] N. Abdoulmoumine, S. Adhikari, A. Kulkarni, S. Chattanathan, A review on biomass gasification syngas cleanup, Appl. Energy 155 (2015) 294–307, https://doi.org/10.1016/j.apenergy.2015.05.095.

[41] Y. Richardson, J. Blin, A. Julbe, A short overview on purification and conditioning of syngas produced by biomass gasification: catalytic strategies, process intensification and new concepts, Prog. Energy Combust. Sci. 38 (2012) 765–781, https://doi.org/10.1016/j.pecs.2011.12.001.

[42] D. Buentello-Montoya, X. Zhang, J. Li, V. Ranade, S. Marques, M. Geron, Performance of biochar as a catalyst for tar steam reforming: effect of the porous structure, Appl. Energy 259 (2020), https://doi.org/10.1016/j.apenergy.2019.114176, 114176.

[43] A. Orío, J. Corella, I. Narváez, Performance of different dolomites on hot raw gas cleaning from biomass gasification with air, Ind. Eng. Chem. Res. 36 (1997) 3800–3808, https://doi.org/10.1021/ie960810c.

[44] J. Delgado, M.P. Aznar, J. Corella, Biomass gasification with steam in fluidized bed: effectiveness of CaO, MgO, and CaO-MgO for hot raw gas cleaning, Ind. Eng. Chem. Res. 36 (1997) 1535–1543, https://doi.org/10.1021/ie960273w.

[45] V. Pallozzi, A. Di Carlo, E. Bocci, M. Carlini, Combined gas conditioning and cleaning for reduction of tars in biomass gasification, Biomass Bioenergy 109 (2018) 85–90, https://doi.org/10.1016/j.biombioe.2017.12.023.

[46] F. Pinto, R.N. André, C. Carolino, M. Miranda, P. Abelha, D. Direito, et al., Effects of experimental conditions and of addition of natural minerals on syngas production from lignin by oxy-gasification: comparison of bench- and pilot scale gasification, Fuel 140 (2015) 62–72, https://doi.org/10.1016/j.fuel.2014.09.045.

[47] J. Corella, M.P. Aznar, J. Gil, M.A. Caballero, Biomass gasification in fluidized bed: where to locate the dolomite to improve gasification? Energy Fuels 13 (1999) 1122–1127, https://doi.org/10.1021/ef990019r.

[48] X. Meng, W. De Jong, R. Pal, A.H.M. Verkooijen, In bed and downstream hot gas desulphurization during solid fuel gasification: a review, Fuel Process. Technol. 91 (2010) 964–981, https://doi.org/10.1016/j.fuproc.2010.02.005.

[49] B. Dou, C. Wang, H. Chen, Y. Song, B. Xie, Y. Xu, et al., Research progress of hot gas filtration, desulphurization and HCl removal in coal-derived fuel gas: a review, Chem. Eng. Res. Des. 90 (2012) 1901–1917, https://doi.org/10.1016/j.cherd.2012.04.009.

[50] C.(.C.). Xu, J. Donald, E. Byambajav, Y. Ohtsuka, Recent advances in catalysts for hot-gas removal of tar and NH3 from biomass gasification, Fuel 89 (2010) 1784–1795, https://doi.org/10.1016/J.FUEL.2010.02.014.

[51] E. Bjoerkman, K. Sjoestroem, Decomposition of ammonia over dolomite and related compounds, Energy Fuel 5 (1991) 753–760, https://doi.org/10.1021/ef00029a023.

[52] S. Cheah, D.L. Carpenter, K.A. Magrini-Bair, Review of mid- to high-temperature sulfur sorbents for desulfurization of biomass- and coal-derived syngas, Energy Fuels 23 (2009) 5291–5307, https://doi.org/10.1021/ef900714q.

[53] Z.U. Din, Z.A. Zainal, Biomass integrated gasification-SOFC systems: technology overview, Renew. Sust. Energ. Rev. 53 (2016) 1356–1376.

[54] M. Materazzi, P. Lettieri, R. Taylor, C. Chapman, Performance analysis of RDF gasification in a two stage fluidized bed-plasma process, Waste Manag. 47 (2016) 256–266, https://doi.org/10.1016/J.WASMAN.2015.06.016.

[55] C. Di Blasi, Modeling chemical and physical processes of wood and biomass pyrolysis, Prog. Energy Combust. Sci. 34 (2008) 47–90, https://doi.org/10.1016/J.PECS.2006.12.001.

[56] M.L. Boroson, J.B. Howard, J.P. Longwell, W.A. Peters, Product yields and kinetics from the vapor phase cracking of wood pyrolysis tars, AICHE J. 35 (1989) 120–128, https://doi.org/10.1002/AIC.690350113.

[57] J. Corella, M.A. Caballero, M.-P. Aznar, C. Brage, Two advanced models for the kinetics of the variation of the tar composition in its catalytic elimination in biomass gasification, Ind. Eng. Chem. Res. 42 (2003) 3001–3011, https://doi.org/10.1021/IE020401I.

[58] E. Savuto, A. Di Carlo, K. Gallucci, S. Stendardo, S. Rapagnà, 3D-CFD simulation of catalytic filter candles for particulate abatement and tar and methane steam reforming inside the freeboard of a gasifier, Chem. Eng. J. 377 (2019), https://doi.org/10.1016/J.CEJ.2018.10.227.

[59] E. Savuto, A. Di Carlo, A. Steele, S. Heidenreich, K. Gallucci, S. Rapagnà, Syngas Conditioning by Ceramic Filter Candles Filled With Catalyst Pellets, Placed Inside the Freeboard of a Fluidized Bed Steam Gasifier, To Be Submitt, 2018.

[60] L. Núñez-Regueira, J. Rodríguez-Añón, J. Proupín, A. Romero-García, Energy evaluation of forest residues originated from pine in Galicia, Bioresour. Technol. 88 (2003) 121–130, https://doi.org/10.1016/S0960-8524(02)00275-4.

[61] M.J. Feria, A. Rivera, F. Ruiz, E. Grandal, J.C. García Domínguez, A. Pérez, et al., Energetic characterization of lignocellulosic biomass from Southwest Spain, Int. J. Green Energy 8 (2011) 631–642, https://doi.org/10.1080/15435075.2011.600378.

[62] E. Alakangas, Properties of wood fuels used in Finland—BIOSOUTH -project, in: Bulk Solids Handling, 2005.

[63] H.F.S. dos Viana, A.M. Rodrigues, R. Godina, J.C.O. de Matias, L.J.R. Nunes, Evaluation of the physical, chemical and thermal properties of Portuguese maritime pine biomass, Sustainability 10 (2018) 2877, https://doi.org/10.3390/su10082877.

[64] O. Kitani, C.W. Hall, Biomass Handbook, Gordon and Breach Science, 1989.

[65] M. Skrizovska, H. Veznikova, P. Roupcova, Inclination to self-ignition and analysis of gaseous products of wood chips heating, Acta Chim. Slov. 13 (2020) 88–97, https://doi.org/10.2478/acs-2020-0013.

[66] J.M. Gómez-Martín, M. Castaño-Díaz, A. Cámara-Obregón, P. Álvarez-Álvarez, M.B. Folgueras-Díaz, M.A. Diez, On the chemical composition and pyrolytic behavior of hybrid poplar energy crops from northern Spain, Energy Rep. 6 (2020) 764–769, https://doi.org/10.1016/j.egyr.2019.09.065. Elsevier Ltd.

[67] P. Álvarez-Álvarez, C. Pizarro, M. Barrio-Anta, A. Cámara-Obregón, J. Bueno, A. Álvarez, et al., Evaluation of tree species for biomass energy production in Northwest Spain, Forests 9 (2018) 160, https://doi.org/10.3390/f9040160.

[68] P. Sannigrahi, A.J. Ragauskas, G.A. Tuskan, Poplar as a feedstock for biofuels: a review of compositional characteristics, Biofuels Bioprod. Biorefin. 4 (2010) 209–226, https://doi.org/10.1002/bbb.206.

[69] T. Wu, J.F. Yan, K.Q. Shi, H.T. Zhao, Y.D. Wang, Transformation of minerals in coal and biomass upon heating and their impacts on co-firing, Adv. Mater. Res. 742 (2013) 249–253, https://doi.org/10.4028/www.scientific.net/AMR.742.249. Trans Tech Publications Ltd.

[70] V. Bert, J. Allemon, P. Sajet, S. Dieu, A. Papin, S. Collet, et al., Torrefaction and pyrolysis of metal-enriched poplars from phytotechnologies: effect of temperature and biomass chlorine content on metal distribution in end-products and valorization options, Biomass Bioenergy 96 (2017) 1–11, https://doi.org/10.1016/j.biombioe.2016.11.003.

[71] C. Storm, H. Spliethoff, K.R.G. Hein, Generation of a gaseous fuel by gasification or pyrolysis of biomass for use as reburn gas in coal-fired boilers, in: Proc. 5th Eur. Conf. Ind. Furn. Boil, 2000, pp. 689–699.

[72] M. Danish, M. Naqvi, U. Farooq, S. Naqvi, Characterization of South Asian agricultural residues for potential utilization in future "energy mix.", Energy Procedia 75 (2015) 2974–2980, https://doi.org/10.1016/j.egypro.2015.07.604. Elsevier Ltd.

[73] A. Demirbas, Effects of moisture and hydrogen content on the heating value of fuels, Energy Sources Part A 29 (2007) 649–655, https://doi.org/10.1080/009083190957801.

[74] S. Paniagua, A.I. García-Pérez, L.F. Calvo, Biofuel consisting of wheat straw-poplar wood blends: thermogravimetric studies and combustion characteristic indexes estimation, Biomass Convers. Biorefin. 9 (2019) 433–443, https://doi.org/10.1007/S13399-018-0351-5.

[75] M. Mierzwa-Hersztek, K. Gondek, M. Jewiarz, K. Dziedzic, Assessment of energy parameters of biomass and biochars, leachability of heavy metals and phytotoxicity of their ashes, J. Mater. Cycles Waste Manage. 21 (2019) 786–800, https://doi.org/10.1007/s10163-019-00832-6.

[76] L. Del Zotto, E. Bocci, F. Fontana, M. Martini, S. Meola, C. Amoruso, et al., DELIVERABLE D2.1 Biomass Selection and Characterization for Small-to-Medium Scale Gasification-SOFC CHP Plants, 2019.

[77] L. Xu, Y. Jiang, L. Wang, Thermal decomposition of rape straw: pyrolysis modeling and kinetic study via particle swarm optimization, Energy Convers. Manag. 146 (2017) 124–133, https://doi.org/10.1016/j.enconman.2017.05.020.

[78] Q. Ma, L. Han, G. Huang, Potential of water-washing of rape straw on thermal properties and interactions during co-combustion with bituminous coal, Bioresour. Technol. 234 (2017) 53–60, https://doi.org/10.1016/j.biortech.2017.03.018.

[79] Phyllis2—rape (#539).

[80] Feedstock Selection, Characterization and Preparation.

[81] T. Cástková, Š. Hýsek, A. Sikora, O. Schönfelder, M. Böhm, Chemical and physical parameters of different modifications of rape straw (*Brassica napus* L.), Bioresources 13 (2018) 104–114, https://doi.org/10.15376/biores.13.1.104-114.

[82] X. Fang, L. Jia, Experimental study on ash fusion characteristics of biomass, Bioresour. Technol. 104 (2012) 769–774, https://doi.org/10.1016/j.biortech.2011.11.055.

[83] R. López, C. Fernández, J. Fierro, J. Cara, O. Martínez, M.E. Sánchez, Oxy-combustion of corn, sunflower, rape and microalgae bioresidues and their blends from the perspective of thermogravimetric analysis, Energy 74 (2014) 845–854, https://doi.org/10.1016/j.energy.2014.07.058.

[84] Phyllis2—corn stover (#889).

[85] Y. Tian, H. Zhang, L. Zheng, S. Li, H. Hao, H. Huang, Effect of Zn addition on the cd-containing anaerobic fermentation process: biodegradation and microbial communities, Int. J. Environ. Res. Public Health 16 (2019) 2998, https://doi.org/10.3390/ijerph16162998.

[86] J.S. Tumuluru, C.T. Wright, R.D. Boardman, T. Kremer, Proximate and Ultimate Compositional Changes in Corn Stover During Torrrefaction Using Thermogravimetric Analyzer and Microwaves, vol. 2, American Society of Agricultural and Biological Engineers, St. Joseph, MI, 2012, pp. 1348–1365, https://doi.org/10.13031/2013.41777. Dallas, Texas, July 29–August 1, 2012.

[87] X. Liu, M. Chen, D. Yu, Oxygen enriched co-combustion characteristics of herbaceous biomass and bituminous coal, Thermochim. Acta 569 (2013) 17–24, https://doi.org/10.1016/j.tca.2013.06.037.

[88] M. Yilgin, N. Deveci Duranay, D. Pehlivan, Co-pyrolysis of lignite and sugar beet pulp, Energy Convers. Manag. 51 (2010) 1060–1064, https://doi.org/10.1016/j.enconman.2009.12.010.

[89] P. Brachi, E. Riianova, M. Miccio, F. Miccio, G. Ruoppolo, R. Chirone, 40th Meeting of the Italian Section of the Combustion Institute, 2017.

[90] R. García, C. Pizarro, A. Álvarez, A.G. Lavín, J.L. Bueno, Study of biomass combustion wastes, Fuel 148 (2015) 152–159, https://doi.org/10.1016/j.fuel.2015.01.079.

[91] M.K. Fine, C.M. Nyberg, Final Report Summary For "Promoting Standardization of Combustion Characteristics for Biofuels" Industrial Commission of North Dakota Contract No. R-009-021, 2012.

[92] Phyllis2—sunflower (#554).

[93] Phyllis2—sunflower (#555).

[94] Phyllis2—sunflower (#556).

[95] M. Barbanera, E. Lascaro, V. Stanzione, A. Esposito, R. Altieri, M. Bufacchi, Characterization of pellets from mixing olive pomace and olive tree pruning, Renew. Energy 88 (2016) 185–191, https://doi.org/10.1016/j.renene.2015.11.037.

[96] I. Iáñez-Rodríguez, M.Á. Martín-Lara, G. Blázquez, Ó. Osegueda, M. Calero, Thermal analysis of olive tree pruning and the by-products obtained by its gasification and pyrolysis: the effect of some heavy metals on

their devolatilization behavior, J. Energy Chem. 32 (2019) 105–117, https://doi.org/10.1016/j.jechem.2018.07.002.

[97] D. Vamvuka, V. Tsamourgeli, M. Galetakis, Study on catalytic combustion of biomass mixtures with poor coals, Combust. Sci. Technol. 186 (2014) 68–82, https://doi.org/10.1080/00102202.2013.846331.

[98] D.J. Vega-Nieva, L. Ortiz Torres, J.L. Míguez Tabares, J. Morán, Measuring and predicting the slagging of woody and herbaceous Mediterranean biomass fuels on a domestic pellet boiler, Energy Fuels 30 (2016) 1085–1095, https://doi.org/10.1021/acs.energyfuels.5b02495.

[99] L.J.R. Nunes, A.M. Rodrigues, J.C.O. Matias, A.I. Ferraz, A.C. Rodrigues, Production of biochar from vine pruning: waste recovery in the wine industry, Agriculture 11 (2021) 489, https://doi.org/10.3390/agriculture11060489.

[100] P. Chen, Y. Cheng, S. Deng, X. Lin, G. Huang, R. Ruan, et al., Utilization of almond residues, Int. J. Agric. Biol. Eng. 3 (2010) 1–18, https://doi.org/10.3965/j.issn.1934-6344.2010.04.0-0.

[101] M. Valverde, R. Madrid, A.L. García, F.M. del Amor, L. Rincón, Use of almond shell and almond hull as substrates for sweet pepper cultivation. Effects on fruit yield and mineral content, Spanish J. Agric. Res. 11 (2013) 164–172, https://doi.org/10.5424/sjar/2013111-3566.

[102] N. Petrov, T. Budinova, M. Razvigorova, V. Minkova, Preparation of activated carbon from cherry stones, apricot stones and grape seeds for removal of metal ions from water, in: 2nd Olle Lindström Symp. Renew. Energy-Bioenergy, 1999.

[103] M. Sert, D. Selvi Gökkaya, N. Cengiz, L. Ballice, M. Yüksel, M. Sağlam, Hydrogen production from olive-pomace by catalytic hydrothermal gasification, J. Taiwan Inst. Chem. Eng. 83 (2018) 90–98, https://doi.org/10.1016/j.jtice.2017.11.026.

[104] C. Buratti, S. Mousavi, M. Barbanera, E. Lascaro, F. Cotana, M. Bufacchi, Bioresource technology thermal behaviour and kinetic study of the olive oil production chain residues and their mixtures during co-combustion, Bioresour. Technol. 214 (2016) 266–275, https://doi.org/10.1016/j.biortech.2016.04.097.

[105] M. Bennini, A. Koukouch, I. Bakhattar, M. Asbik, T. Boushaki, B. Sarh, et al., Characterization and combustion of olive pomace in a fixed bed boiler: effects of particle sizes, Int. J. Heat Technol. 37 (2019) 229–238, https://doi.org/10.18280/ijht.370128.

[106] Phyllis2—olive pomace pellets (#2893).

[107] T. Miranda, A. Esteban, S. Rojas, I. Montero, A. Ruiz, Combustion analysis of different olive residues, Int. J. Mol. Sci. 9 (2008) 512–525, https://doi.org/10.3390/IJMS9040512.

[108] Phyllis2—Sawdust (BM4) (#3057).

[109] Z. Chaula, M. Said, G. John, Thermal characterization of pine sawdust as energy source feedstock, J. Energy. Technol. Policy 4 (2014) 57–64.

[110] C. Ulloa, X. García, Burnout synergic or inhibiting effects in combustion of coal-sawdust blends, Ing. Investig. 34 (2014) 29–32, https://doi.org/10.15446/ing.investig.v34n2.39473.

[111] Phyllis2—black liquor (#1396).

[112] C. Cao, L. Xu, Y. He, L. Guo, H. Jin, Z. Huo, High-efficiency gasification of wheat straw black liquor in supercritical water at high temperatures for hydrogen production, Energy Fuels 31 (2017) 3970–3978, https://doi.org/10.1021/acs.energyfuels.6b03002.

[113] P. Bajpai, Properties, composition, and analysis of black liquor, Pulp Pap. Ind. (2017) 25–38, https://doi.org/10.1016/b978-0-12-811103-1.00002-4. Elsevier.

[114] R. Wang, R. Deplazes, F. Vogel, D. Baudouin, Continuous extraction of black liquor salts under hydrothermal conditions, Ind. Eng. Chem. Res. 60 (2021) 4072–4085, https://doi.org/10.1021/acs.iecr.0c05203.

[115] Phyllis2—MSW (#2920).

[116] M. Azam, S.S. Jahromy, W. Raza, C. Jordan, M. Harasek, F. Winter, Comparison of the combustion characteristics and kinetic study of coal, municipal solid waste, and refuse-derived fuel: Model-fitting methods, Energy Sci. Eng. 7 (2019) 2646–2657, https://doi.org/10.1002/ese3.450.

[117] A.M. Omari, Characterization of Municipal Solid Waste for Energy Recovery. A Case Study of Arusha, Tanzania, vol. 2, 2015.

[118] V.V. Gedam, I. Regupathi, Pyrolysis of municipal solid waste for syngas production by microwave irradiation, Nat. Resour. Res. 21 (2012) 75–82, https://doi.org/10.1007/s11053-011-9161-1.

[119] M.-M. Alhadj-Mallah, Q. Huang, X. Cai, Y. Chi, J. Yan, Vitrification of municipal solid waste incineration fly ash using biomass ash as additives, Environ. Technol. 36 (2015) 654–660, https://doi.org/10.1080/09593330.2014.957245.

[120] D.O. Albina, K. Millrath, N.J. Themelis, Effects of feed composition on boiler corrosion in waste-to-energy plants, in: Proc 12TH Annu North Am Waste to Energy Conf, 2004, pp. 99–109, https://doi.org/10.1115/NAWTEC12-2215.

[121] M. Otero, M.E. Sanchez, X. Gómez, A. Morán, Thermogravimetric analysis of biowastes during combustion, Waste Manag. 30 (2010) 1183–1187, https://doi.org/10.1016/j.wasman.2009.12.010.

[122] M. Hernandez, A. Salimbeni, M. Hitzl, J. Zhang, G. Wang, K. Wang, et al., Evaluation of utilising ingelia hydrochar produced from organic residues for blast furnaces injection—comparison with anthracite and bituminous coal, Eur. Biomass Conf. Exhib. Proc. (2018) 1560–1568, https://doi.org/10.5071/26THEUBCE2018-ICO.8.5.

[123] M. Nasrullah, P. Vainikka, J. Hannula, M. Hurme, P. Oinas, Elemental balance of SRF production process: solid recovered fuel produced from municipal solid waste, Waste Manag. Res. 34 (2016) 38–46, https://doi.org/10.1177/0734242X15615697.

[124] Z. Cao, D. Jung, M.P. Olszewski, P.J. Arauzo, A. Kruse, Hydrothermal carbonization of biogas digestate: effect of digestate origin and process conditions, Waste Manag. 100 (2019) 138–150, https://doi.org/10.1016/j.wasman.2019.09.009.

[125] M. Barbanera, F. Cotana, U. Di Matteo, Co-combustion performance and kinetic study of solid digestate with gasification biochar, Renew. Energy 121 (2018) 597–605, https://doi.org/10.1016/j.renene.2018.01.076.

[126] A. Maharana, An Integrated Design of Hydrothermal Liquefaction and Biogas Plant For The Conversion of Feedstock (Biomass) To Biofuel, 2013, https://doi.org/10.13140/RG.2.2.34782.23368.

[127] Phyllis2—digestate from organic waste (#2907).

[128] K.R. Parmar, A.B. Ross, Integration of hydrothermal carbonisation with anaerobic digestion; opportunities for valorisation of digestate, Energies 12 (2019) 1586, https://doi.org/10.3390/en12091586.

[129] P.K. Ghosh, P. Ramesh, K.K. Bandyopadhyay, A.K. Tripathi, K.M. Hati, A.K. Misra, et al., Comparative effectiveness of cattle manure, poultry manure, phosphocompost and fertilizer-NPK on three cropping systems in vertisols of semi-arid tropics. I. Crop yields and system performance, Bioresour. Technol. 95 (2004) 77–83, https://doi.org/10.1016/j.biortech.2004.02.011.

[130] D.P. Eckhardt, M. Redin, N.A. Santana, C.L. De, J. Dominguez, R.J.S. Jacques, et al., Cattle manure bioconversion effect on the availability of nitrogen, phosphorus, and potassium in soil, Rev. Bras. Ciênc. Solo 42 (2018), https://doi.org/10.1590/18069657rbcs20170327, 170327.

[131] F. Fantozzi, C. Buratti, Biogas production from different substrates in an experimental continuously stirred tank reactor anaerobic digester, Bioresour. Technol. 100 (2009) 5783–5789, https://doi.org/10.1016/j.biortech.2009.06.013.

[132] Z. Yao, Y. Baojun, Y. Qiaoxia, C. Hongliang, Y. Shuiping, Combustion Characteristics of Cattle Manure and Pulverized Coal Co-Firing under Oxy-Fuel Atmosphere in Non-Isothermal and Isothermal Conditions, 2018.

[133] J. Lu, J. Watson, J. Zeng, H. Li, Z. Zhu, M. Wang, et al., Biocrude production and heavy metal migration during hydrothermal liquefaction of swine manure, Process. Saf. Environ. Prot. 115 (2018) 108–115, https://doi.org/10.1016/j.psep.2017.11.001.

[134] A.M. Smith, U. Ekpo, A.B. Ross, The influence of pH on the combustion properties of bio-coal following hydrothermal treatment of swine manure, Energies 13 (2020) 331, https://doi.org/10.3390/en13020331.

[135] Z. Lentz, P. Kolar, J.J. Classen, Valorization of swine manure into hydrochars, Processes 7 (2019) 560, https://doi.org/10.3390/pr7090560.

[136] K.S. Ro, J.A. Libra, S. Bae, N.D. Berge, J.R.V. Flora, R. Pecenka, Combustion behavior of animal-manure-based hydrochar and pyrochar, ACS Sustain. Chem. Eng. 7 (2019) 470–478, https://doi.org/10.1021/acssuschemeng.8b03926.

[137] C. Song, W. Yuan, S. Shan, Q. Ma, H. Zhang, X. Wang, et al., Changes of nutrients and potentially toxic elements during hydrothermal carbonization of pig manure, Chemosphere 243 (2020), https://doi.org/10.1016/j.chemosphere.2019.125331, 125331.

[138] F. Cotana, V. Coccia, A. Petrozzi, G. Cavalaglio, M. Gelosia, M.C. Merico, Energy valorization of poultry manure in a thermal power plant: experimental campaign, Energy Procedia 45 (2014) 315–322, https://doi.org/10.1016/j.egypro.2014.01.034. Elsevier Ltd.

[139] V. Mau, A. Gross, Combustion Behavior of Hydrochar Derived from Poultry Manure, 2017.

[140] M. Tańczuk, R. Junga, A. Kolasa-Więcek, P. Niemiec, Assessment of the energy potential of chicken manure in Poland, Energies 12 (2019) 1244, https://doi.org/10.3390/en12071244.

[141] J. Lee, D. Choi, Y.S. Ok, S.R. Lee, E.E. Kwon, Enhancement of energy recovery from chicken manure by pyrolysis in carbon dioxide, J. Clean. Prod. 164 (2017) 146–152, https://doi.org/10.1016/j.jclepro.2017.06.217.

[142] M. Ihnat, L. Fernandes, Trace elemental characterization of composted poultry manure, Bioresour. Technol. 57 (1996) 143–156, https://doi.org/10.1016/0960-8524(96)00061-2.

[143] O. Sahin, M.B. Taskin, Y.K. Kadioglu, A. Inal, D.J. Pilbeam, A. Gunes, Elemental composition of pepper plants fertilized with pelletized poultry manure, J. Plant Nutr. 37 (2014) 458–468, https://doi.org/10.1080/01904167.2013.864307.

[144] V. Marcantonio, D. Monarca, A. Colantoni, M. Cecchini, Ultrasonic waves for materials evaluation in fatigue, thermal and corrosion damage: a review, Mech. Syst. Signal Process. 120 (2019), https://doi.org/10.1016/j.ymssp.2018.10.012.

[145] J. Wang, Y. Zhang, L. Han, L. Chang, W. Bao, Simultaneous removal of hydrogen sulfide and mercury from simulated syngas by iron-based sorbents, Fuel 103 (2013) 73–79, https://doi.org/10.1016/j.fuel.2011.10.056.

[146] R.P. Gupta, W.S. O'Brien, Desulfurization of hot syngas containing hydrogen chloride vapors using zinc titanate sorbents, Ind. Eng. Chem. Res. 39 (2000) 610–619, https://doi.org/10.1021/ie990533k.

[147] Y. Liu, H. Yu, J. Liu, D. Chen, Catalytic characteristics of innovative Ni/slag catalysts for syngas production and tar removal from biomass pyrolysis, Int. J. Hydrog. Energy 44 (2019) 11848–11860, https://doi.org/10.1016/J.IJHYDENE.2019.03.024.

[148] U. Rhyner, P. Edinger, T.J. Schildhauer, S.M.A. Biollaz, Experimental study on high temperature catalytic conversion of tars and organic sulfur compounds, Int. J. Hydrog. Energy 39 (2014) 4926–4937, https://doi.org/10.1016/J.IJHYDENE.2014.01.082.

[149] A. Ephraim, L.D. Ngo, D. Pham Minh, D. Lebonnois, C. Peregrina, P. Sharrock, et al., Valorization of waste-derived inorganic sorbents for the removal of HCl in syngas, Waste Biomass Valoriz. 0 (2018) 1–12, https://doi.org/10.1007/s12649-018-0355-1.

[150] S. Cheah, Y.O. Parent, W.S. Jablonski, T. Vinzant, J.L. Olstad, Manganese and ceria sorbents for high temperature sulfur removal from biomass-derived syngas—the impact of steam on capacity and sorption mode, Fuel 97 (2012) 612–620, https://doi.org/10.1016/j.fuel.2012.03.007.

[151] J.I. Baek, T.H. Eom, J.B. Lee, S. Jegarl, C.K. Ryu, Y.C. Park, et al., Cleaning of gaseous hydrogen chloride in a syngas by spray-dried potassium-based solid sorbents, Korean J. Chem. Eng. 32 (2015) 845–851, https://doi.org/10.1007/s11814-014-0256-3.

[152] F.N. Cayan, M. Zhi, S.R. Pakalapati, I. Celik, N. Wu, R. Gemmen, Effects of coal syngas impurities on anodes of solid oxide fuel cells, J. Power Sources 185 (2008) 595–602, https://doi.org/10.1016/J.JPOWSOUR.2008.06.058.

[153] N.T. Ngoc Lan Thao, K.-Y. Chiang, H.-P. Wan, W.-C. Hung, C.-F. Liu, Enhanced trace pollutants removal efficiency and hydrogen production in rice straw gasification using hot gas cleaning system, Int. J. Hydrog. Energy 44 (2019) 3363–3372, https://doi.org/10.1016/J.IJHYDENE.2018.07.133.

[154] V. Marcantonio, E. Bocci, J.P. Ouweltjes, L. Del Zotto, D. Monarca, Evaluation of sorbents for high temperature removal of tars, hydrogen sulphide, hydrogen chloride and ammonia from biomass-derived syngas by using Aspen plus, Int. J. Hydrog. Energy (2020), https://doi.org/10.1016/j.ijhydene.2019.12.142.

[155] M. Chomiak, J. Trawczyński, Z. Blok, P. Babiński, Monolithic Zn-Co-Ti based sorbents for hot syngas desulfurization, Fuel Process. Technol. 144 (2016) 64–70, https://doi.org/10.1016/j.fuproc.2015.12.016.

[156] H. Cui, S.Q. Turn, V. Keffer, D. Evans, T. Tran, M. Foley, Contaminant estimates and removal in product gas from biomass steam gasification, Energy Fuel 24 (2010) 1222–1233, https://doi.org/10.1021/ef9010109.

[157] J.A. Rodriguez, A. Maiti, Adsorption and decomposition of H2S on MgO(100), NiMgO(100), arid ZnO(0001) surfaces: a first-principles density functional study, J. Phys. Chem. B 104 (2000) 3630–3638, https://doi.org/10.1021/jp000011e.

[158] S.D. Sharma, M. Dolan, A.Y. Ilyushechkin, K.G. McLennan, T. Nguyen, D. Chase, Recent developments in dry hot syngas cleaning processes, Fuel 89 (2010) 817–826, https://doi.org/10.1016/j.fuel.2009.05.026.

[159] https://www.statista.com/.

[160] https://markets.businessinsider.com/commodities.

[161] R.B. Slimane, J. Abbasian, Copper-based sorbents for coal gas desulfurization at moderate temperatures, Ind. Eng. Chem. Res. 39 (2000) 1338–1344, https://doi.org/10.1021/ie990877a.

[162] Y. Zeng, S. Kaytakoglu, D.P. Harrison, Reduced cerium oxide as an efficient and durable high temperature desulfurization sorbent, Chem. Eng. Sci. 55 (2000) 4893–4900, https://doi.org/10.1016/S0009-2509(00)00117-2.

[163] N.O. Ikenaga, Y. Ohgaito, H. Matsushima, T. Suzuki, Preparation of zinc ferrite in the presence of carbon material and its application to hot-gas cleaning, Fuel 83 (2004) 661–669, https://doi.org/10.1016/j.fuel.2003.08.019. Elsevier.

[164] X. Ren, E. Rokni, Y. Liu, Y.A. Levendis, Reduction of HCl emissions from combustion of biomass by alkali carbonate sorbents or by thermal pretreatment, J. Energy Eng. 144 (2018), https://doi.org/10.1061/(ASCE)EY.1943-7897.0000561.

[165] B. Dou, C. Wang, H. Chen, Y. Song, B. Xie, Y. Xu, et al., HCl neutralization by alkaline sorbents in the gasification of chloride-containing fuel in the filtration combustion mode, ACS Sustain. Chem. Eng. 6 (2018) 13056–13065, https://doi.org/10.1016/j.apenergy.2018.09.060.

[166] J. Partanen, P. Backman, R. Backman, M. Hupa, Absorption of HCl by limestone in hot flue gases. Part II: importance of calcium hydroxychloride, Fuel 84 (2005) 1674–1684, https://doi.org/10.1016/j.fuel.2005.02.012.

[167] C.S. Chyang, Y.L. Han, Z.C. Zhong, Study of HCl absorption by CaO at high temperature, Energy Fuels 23 (2009) 3948–3953, https://doi.org/10.1021/ef900234p.

[168] B. Coda, M. Aho, R. Berger, H. KRG, Behavior of Chlorine and Enrichment of Risky Elements in Bubbling Fluidized Bed Combustion of Biomass and Waste Assisted by Additives, 2001, https://doi.org/10.1021/EF000213.

[169] N. Verdone, P. De Filippis, Reaction kinetics of hydrogen chloride with sodium carbonate, Chem. Eng. Sci. 61 (2006) 7487–7496, https://doi.org/10.1016/j.ces.2006.08.023.

[170] Y. Ohtsuka, N. Tsubouchi, T. Kikuchi, H. Hashimoto, Recent progress in Japan on hot gas cleanup of hydrogen chloride, hydrogen sulfide and ammonia in coal-derived fuel gas, Powder Technol. 190 (2009) 340–347, https://doi.org/10.1016/j.powtec.2008.08.012.

[171] G.N. Krishnan, A. Canizales, R. Gupta, R. Ayala, Development of Disposable Sorbents for Chloride Removal from High-Temperature Coal-Derived Gases. Pittsburgh, PA, and Morgantown, WV, 1996, https://doi.org/10.2172/421952.

[172] Z.M. Shareefdeen, W. Ahmed, A. Aidan, Kinetics and modeling of H2S removal in a novel biofilter, Adv. Chem. Eng. Sci. 1 (2011) 72–76, https://doi.org/10.4236/aces.2011.12012.

[173] X.M. Meng, J.W. De, A.H.M. Verkooijen, Thermodynamic analysis and kinetics model of H2S sorption using different sorbents, Environ. Prog. Sustain. Energy 28 (2009) 360–371, https://doi.org/10.1002/EP.10386.

[174] M. Bal, S. Biswas, S.K. Behera, B.C. Meikap, Modeling and optimization of process variables for HCl gas removal by response surface methodology, J. Environ. Sci. Health A Tox. Hazard. Subst. Environ. Eng. 54 (2019) 359–366, https://doi.org/10.1080/10934529.2018.1551650.

[175] B.L. Dou, J.S. Gao, X.Z. Sha, A study on the reaction kinetics of HCl removal from high-temperature coal gas, Fuel Process. Technol. 72 (2001) 23–33, https://doi.org/10.1016/S0378-3820(01)00176-X.

[176] M. Stemmler, A. Tamburro, M. Müller, Thermodynamic modelling of fate and removal of alkali species and sour gases from biomass gasification for production of biofuels, Biomass Convers. Biorefin. 3 (2013) 187–198, https://doi.org/10.1007/s13399-013-0073-7.

[177] K.Q. Tran, K. Iisa, B.M. Steenari, O. Lindqvist, A kinetic study of gaseous alkali capture by kaolin in the fixed bed reactor equipped with an alkali detector, Fuel 84 (2005) 169–175, https://doi.org/10.1016/j. fuel.2004.08.019.

[178] Y. Wang, H. Tan, X. Wang, R. Cao, B. Wei, The condensation and thermodynamic characteristics of alkali compound vapors on wall during wheat straw combustion, Fuel 187 (2017) 33–42, https://doi.org/10.1016/j. fuel.2016.09.014.

[179] R. Moradi, V. Marcantonio, L. Cioccolanti, E. Bocci, Integrating biomass gasification with a steam-injected micro gas turbine and an organic Rankine cycle unit for combined heat and power production, Energy Convers. Manag. 205 (2020), https://doi.org/10.1016/j.enconman.2019.112464.

[180] F. Kerscher, J. Bolz, I. Stellwag, V. Handy, R. Bier, H. Spliethoff, Alkali removal with mineral sorbents – part II: fixed-bed experiments and model validation, Powder Technol. 389 (2021) 406–415, https://doi.org/ 10.1016/J.POWTEC.2021.05.047.

CHAPTER 10

Modeling, analysis, and simulation of cryogenic distillation processes for syngas purification: Tray and packed types

Sonia Sepahi and Mohammad Reza Rahimpour

Department of Chemical Engineering, Shiraz University, Shiraz, Iran

1. Introduction

Global warming and climate change are significant concerns. Due to the great usage of fossil fuels, greenhouse gases emission to the environment. One of these greenhouse gases that has an important role in global warming and environment consequence is CO_2. Different sources, especially coal-fired power plants, emit roughly 2 billion tons of CO_2 per year [1,2]. By the year 2100, pursuant to a report by the Intergovernmental Panel on climate change, the amount of CO_2 in the atmosphere will reach up to 570 ppm, which will cause an approximately $1.9°C$ increase in mean global temperature and a 3.8 m increase in the average sea level. Thus, CO_2 capture, utilizing, and storage from large discharge sources are appropriate for relieving CO_2 emission [3]. By building on a CO_2 capture and storage unit into power plants, CO_2 emission can be reduced to 80%–90% [4]. Therefore, different CO_2 capture methods and technology have been considered.

There are several strategies for CO_2 capture and separation, such as absorption, adsorption, membrane, chemical looping cycle, hydration, biotechnology, and cryogenic phase change capture of CO_2. The essential target is to select a method with a minimum energy penalty, minimum chemical contamination, and optimal capture efficiency. Selecting a satisfactory technique for CO_2 capture is highly depends on the discharge source and conditions such as flow rate, temperature, composition, and CO_2 partial pressure. Also, the production properties as well as purity of CO_2, flow pressure, and standards for H_2S, SO_x, and NO_x in discharge can have an influence on choosing the method [5–8].

Advances in Synthesis Gas: Methods, Technologies and Applications. https://doi.org/10.1016/B978-0-323-91879-4.00001-1

Cryogenic technology has been noticed due to its advantages, including avoiding problems caused by chemical solvents and physical sorbents. The fundamental challenges in cryogenic technology are cold energy sources, capture costs, and impurities [9–11].

Vapor pressure, boiling point, solubility, adsorption capacity, and diffusivity are properties in which primary operations for the separation and purification of gas can be applied on their base. Separation mechanisms for components that should be separated include phase creation by heat transfer or shaft work, chemical conversion to another compound, adsorption on a solid, absorption in a liquid, and permeation through a membrane [12,13]. An example for chemical direction is dry reforming, where CO_2 reacts with CH_4 and produces syngas, which is a useful product, and it can further be used in the generation of liquid fuels [14].

Separation by creating a phase change can be caused by partial condensation or partial vaporization with different volatility and desublimation point. If the volatility difference of components is not adequately significant as for N_2 and CH_4, for achieving the separation purpose in a single step, multiple stages in a distillation process are required. In nitrogen rejection units (NRU) technologies, the cryogenic distillation method has a separation power more than eight times of N_2 selective membrane and adsorption process [14].

Bio syngas manufactured from biomass gasification has impurities such as sulfur oxides (SOx), hydrogen sulfide (H_2S), ammonia (NH_3), nitrogen (N_2), chlorine (Cl), water (H_2O), alkali compounds, tar, ash, and particulate material. These contaminants may cause serious difficulties in combustion, manufacturing chemicals, and gas-to-liquid (GTL) processes. The impurities can have different compositions due to various carbonaceous feedstocks. Most of these contaminants should be removed or be treated to some degree [15,16].

CO_2 does not limit the direct combustion of syngas. Still, due to no heating value and existence in significant volumetric concentration (about 40%), can considerably impact the overall combustion output energy. CO_2 can make problems in GTL processes and inhibit liquid fuel production. H_2S can convert to SO_2, which is a harmful greenhouse gas and participant in acid rain. Additionally, H_2S can become H_2SO_4, a corrosive acid that can destroy the process equipment. H_2S also negatively affects the catalyst, sorbents, enzymes, and microorganisms activity and performance used for purification and subsequent chemicals production [15]. N_2 is an inert gas and will not freeze or cause corrosion in a cryogenic gas plant but keeping its concentration less than 1% avoids stratification and rollover of the liquid product during shipping [14].

Absorption, adsorption, distillation, membrane separation, and hydrates are technologies for CO_2, H_2S, and N_2 removal from natural gas. Pipeline quality gas should have less than 2% CO_2 and 3% N_2, and the gas that feeds into cryogenic gas plants should have less than 50 ppmv CO_2 to avoid forming dry ice and 1% N_2 [17,18].

Biogas obtained from the anaerobic digestion of biomass mainly consists of methane (CH_4) and carbon dioxide (CO_2). Biogas has impurities such as H_2S, N_2, O_2, and siloxanes. CH_4 content in biogas is between 50% and 70%, and to gain pure biomethane, CO_2 and other contaminants should be removed. Biogas upgrading technologies are water or chemical scrubbing, organic physical scrubbing, membrane separation, pressure swing adsorption, and cryogenic methods [16,19]. According to other literatures, the most commercial method for biogas purification is water scrubbing, then membrane technology and chemical scrubbing come next [20]. Cryogenic techniques provide high-purity products (ranging from 95% to 99%) while avoiding the use of chemical solvents, resulting in no secondary pollution. Despite many advantages, the high investment and operation costs make cryogenic techniques commercially less employed [21–23].

The base of cryogenic separation is the difference in liquefaction temperature for compounds. CH_4 is separated selectively from other components by sequentially decreasing the biogas temperature. Thus, a highly pure biomethane is obtained in this method and is known as liquefied natural gas [24,25]. For removing the impurities from biogas with cryogenic technology, the simplest path is to operate at a constant pressure of 10 bar. By gradually decreasing temperature, each impurity (or some of them) will be removed in different steps. In the first step of this method, the temperature is appointed at $-25^{\circ}C$, in which H_2S, H_2O, and siloxanes will be removed. In the second step, the temperature is set up at $-55^{\circ}C$ to liquefy CO_2 partially and subsequently decrease the temperature to $-85^{\circ}C$ for complete removal of CO_2 by a solidification stage. The liquefied CO_2 has a high purity and can be sold to intensify overall economic efficiency [26–28].

Additionally, another option is to dry the gas first and then apply compression in multiple stages up to 80 bar. This pressure allows using a higher temperature about $-45^{\circ}C$ to $-55^{\circ}C$. In this method, necessarily should use an intermediate cooling between the multistage compression [29]. Overall, this technique requires high energy and has more operational costs than other biogas upgrading methods [30].

2. Syngas purification

In addition to the most popular definition of syngas, a mixture of carbon monoxide, hydrogen, and carbon dioxide, syngas also refers to the mixture of N_2 and H_2 that can be used for ammonia synthesis [31]. Syngas is mainly generated from natural gas by steam reforming (SR) and autothermal reforming (ATR). Synthesis gas can also derive by partial oxidation (PO) of methane, biogas, coal, heavy oils, and petroleum coke. Coal gasification is another route for producing syngas and raw material in countries where natural gas is not available. In this method, with partial oxidation and steam treatment, syngas is produced. A substantial step after syngas production is its purification, which strongly depends on the feedstock and the production procedure. When natural gas is the feedstock, more than half of the total expenditure is required for the purification process and up to 70%–80% of the total spending when coal is the raw material [32–34].

Additionally to the major components, syngas also includes sulfur, particulate ash, tars, halogenated compounds, carbon, and metals, in which they can poison the catalyst and make corrosion in further applications. Ought to this reason, the impurities must be removed to a permissive amount. Due to the different uses of syngas, the purification procedure may vary. The most substantial components that should be removed are sour gases such as sulfur and CO_2. Sulfur and CO_2 may convert to acids and create a poisonous environment. Also in ammonia production purposes, CO_2 can produce carbonates in contact with ammonia and water, and the unwanted product may deposit on the catalyst. In this case, if the sulfur hadn't been removed before entering the process, it targets Fe and poisons the catalyst. In ammonia synthesis, if CO, H_2O, CO_2, and O_2 are present, they will deposit and have negative effects on the catalyst. In the production of synthesis hydrocarbons and fuels, the important major is CO_2 and H_2S removal since they would negatively affect the reaction by catalyst poisoning [35,36].

For removing the sour gases from syngas, many techniques have been provided. Physical and chemical absorption, pressure swing adsorption (PSA), membrane separation, physical and chemical adsorption, adsorption by solid agents, bio fixation, hydrating, chemical looping, and cryogenic separation are the suggested methods [37].

Syngas composition also has NH_3, HCN, COS, and CO. The more the variety of components in the syngas flow, the more the complexity of the purification process. For example, a traditional process for purification of IGCC (integrated gasification combined cycle), which is coupled with a CCS (carbon capture storage) plant, removes H_2S from the syngas by absorption and Selexol (such as monoethanolamine (MEA) and methyldiethanolamine (MDEA)), and CO_2 will be captured by Selexol. This process has many problems: (i) In the desulfurization unit, absorption takes place at low temperature and high pressure, while desorption performs at high temperature and low pressure. Which there is a problem with repeated lifting and fever of temperature; (ii) Absorption of both H_2S and CO_2 needs solvent, in which the solvent requires regeneration; (iii) The hydrolysis removal process for COS/HSN needs catalysts; (iv) For the water-gas shift reaction, based on the conditions, there is a need of low-temperature catalyst, high-temperature catalyst, and a wide temperature sulfur-resistance catalyst; (v) There is a lot of pressure loss and pressure changing while the absorption and desorption processes and also CO_2 need to be pressurized to a required amount by repetitive operations after desorption [38].

Cryogenic separation is a physical phase change method that removes CO_2, H_2S, and other contaminants according to the difference of boiling points of the materials under extremely low temperatures. This process does not require solvent and chemical species, and therefore, there is no solvent regeneration and solvent loss. The captured CO_2 from this process has high pressure and has the demand pressure for transporting and storage; thus, no further compression is needed [11].

2.1 Cryogenic CO₂ capture

Cryogenic separation is one of the new and green technologies in which its best advantage is non using of solvents. However, challenges as well as high cost and cold energy sources make the utilization of this technique encounter difficulties. Cryogenic distillation of CO_2 is a carbon capture approach. In this technique, components separate due to their different boiling temperature. The cryogenic process can obtain more CO_2 purity than other separation methods. Different types of cryogenic distillation have been provided. A short discussion of these methods is given in below [11,27]:

2.1.1 Cryogenic distillation

In 1982, Holmes and Ryan presented a conventional pathway for cryogenic distillation. In a simple and general explanation, as shown in Fig. 1, the feed stream goes into the distillation section after passing two cooling stages. In the distillation column, packing material or trays

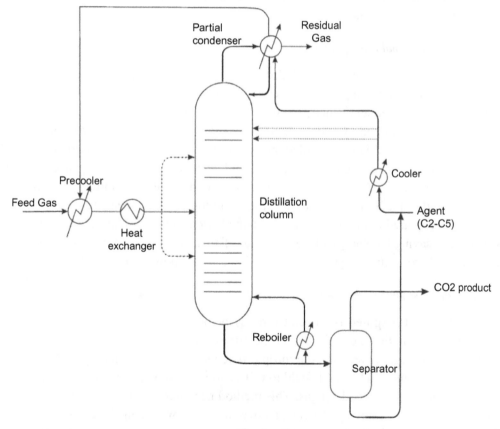

Fig. 1
Cryogenic CO₂ capture technology prepared by Rayan Holmes [39].

have been embedded so that liquid and vapor phases have good contact. The separated syngas will exit the unit by a partial condenser from the top. The stream of rich CO_2 as a bottom product first passes through the reboiler to provide the required heat of vaporization and the recycle. The recovered CO_2 further goes to a separator for more purity [39].

Operation conditions for traditional cryogenic distillation processes are super low (cryogenic) temperatures and high pressures. By operating at high pressures, CO_2 storage costs decrease. Nearby those days, Holmes and Ryan introduced heavier hydrocarbons to the distillation condenser to ward off CO_2 solidification [40].

The energy required for cryogenic distillation can reach up to 50% of plant operating costs. Cyclic distillation, reactive distillation, thermally coupled columns, and heat-integrated distillation columns are energy-efficient solutions [41,42].

Maqsood et al. prospected side-mounted switched cryogenic configuration network that was integrated and intensified. The results showed a considerable reduction in energy consumption and size requirement. Also, by optimizing the intensified cryogenic network, the overall yield increased to about 69% [43,44].

2.1.2 Cryogenic external cooling loop

The main challenge in cryogenic distillation is the cold energy source. Utilizing the thermal swing process, exterior cooling loop, carbon extraction, and cold energy reuse from sources are effective methods for reducing energy consumption. A hybrid cryogenic system with an outer cooling loop has developed by Baxter et al. [45]. The feed entering this cryogenic external cooling loop should be dried and cooled, and then the gas will be compressed and cooled slightly near a point that CO_2 changes phase to solid. Next, the gas will be expanded for further chilling and amounts of CO_2 will be precipitated due to the final temperature. The cold energy of solid CO_2 and residual flue gas is used by the incoming gases and heated again. As shown in Fig. 2, CO_2 is captured in the liquid phase and an N_2-rich stream discharge. The advantage of this method is saving the energy of liquified gas. This technology can be more profitable by integrating with conventional power plants and renewable power sources [46].

2.1.3 Dynamic cryogenic packed bed

Tuinier et al. in 2010, exposed a cryogenic process including a dynamic cooled packed bed. The basis of separation in this method is either the difference in boiling and sublimation points of components. No plugging and pressure drop problems are in this technology. Moving interfaces of cryogenic packed bed, yield to CO_2, and water separation concurrently. The cold energy is supplied by liquified syngas. This method has three main steps in a cycle: cooling down the bed, removing CO_2, and recovering. As shown in Fig. 3, refrigerated N_2 is fed to the cooling cycle in the first step, and then the CO_2 and SO_2 can be deposited on the packing bed surface at different points of the bed. For high H_2S removal from the raw syngas, the temperature must be kept as low as $-150°C$, which increases the operation cost. By the way,

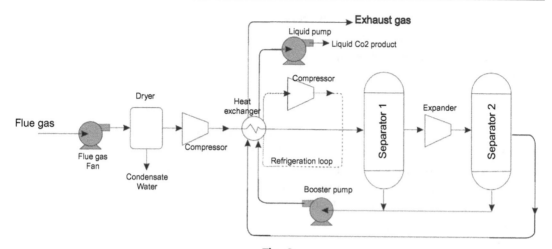

Fig. 2
The flow diagram of the low-temperature method for CO_2 capture [45].

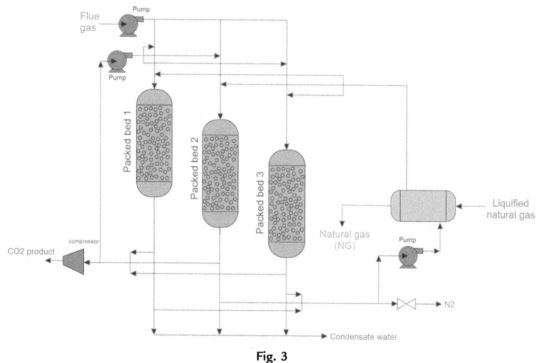

Fig. 3
The dynamic cryogenic packed-bed process [23].

syngas with high purity exits the process without phase change. Installation investment for cryogenic packed bed due to its smaller bed size has a lower cost than vacuum pressure swing adsorption (VPSA) process and also the energy consumption is lower. Despite advantages of cryogenic packed bed proportion to monoethanolamine (MEA) absorption and VPSA, improving the thermal insulation of cryogenic packed beds is a challenge. When liquefied syngas is attainable, the cold energy required in this process becomes competitive, and if it is not able, the energy consumption increases and decreases the advantage of cryogenic energy saving [47–50].

2.1.4 Anti-sublimation (AnSU)

Clodic and Younes in 2003 proposed an anti-sublimation CO_2 capture method. This process can be illustrated in five steps. As shown in Fig. 4, first the flue gas cooled down to $-40\,^\circ C$ and moisture will remove and clean up. In the second step, rich flue gases exchange heat with poor flue gases. Next, there is a refrigeration integrated cascade, and in fourth stage, heat is exchanged until CO_2 freezes and at the final step, CO_2 will be recovered. An advantage of this method is that the heat of fusion from CO_2 defrosting will be recovered owing to phase changing on the heat exchanger surface [51,52].

Heat transfer of the surface of the heat exchanger encountered unfavorable effects due to layer growing of the frosted CO_2. Thus, materials used to prepare the heat exchangers must be selected from those with good thermal conductivity and mechanical stresses. The phase Change point of CO_2 is related to its partial pressure in the flue gas (CO_2 concentration). Pure CO_2 frosts at $-78.5\,^\circ C$ and will be reduced with being in contact with other specious. Thus, decrease of CO_2 concentration in the flue gas increases total energy consumption of AnSU process [53,54].

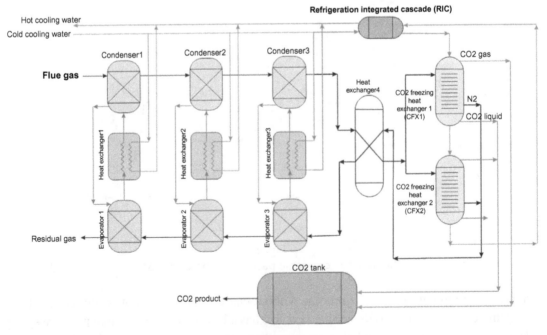

Fig. 4
Anti-sublimation CO_2 capture process (AnSU) process flow diagram [27].

2.1.5 Controlled freezing zone (CFZ)

Sour gases such as acidic impurities such as CO_2 and H_2S have a concentration between 20% and 40% in raw natural gas, and unfortunately, their concentration can sometimes be as high as 70%. An efficient integral solution for removing sour impurities is controlled freezing zone (CFZ). This process consists of three sections. The sections from the top as shown in Fig. 5 are upper distillation section (UD), controlled freezing zone (CFZ), and down distillation (DD), respectively. At first, the feed stream is precooled and then enters CFZ through a spray nozzle. Captured CO_2 can be commercially used for enhanced oil recovery (EOR) and can enhance the economic viability of producing sour natural gas reserves. [14,27,55].

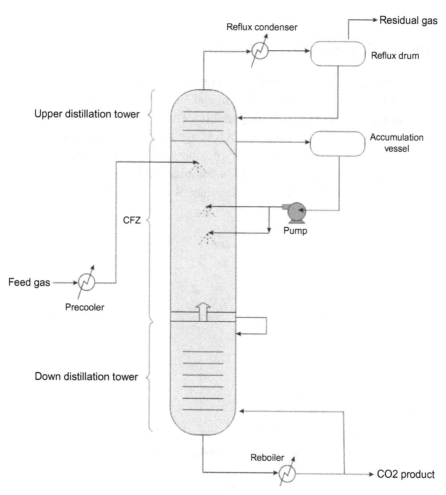

Fig. 5
The schematic diagram of the CFZ process [56].

2.1.6 Stirling cooler system

This system consists of three parts: a pre-freezing tower, main freezing tower and storage tower, respectively, and a free-piston Stirling cooler chilled each part as shown in.

Fig. 6 This system was designed by Song et al. for capturing CO_2 at low temperatures with reasonable waste latent and sensible heat recoveries. H_2O condensate and separate by Stirling cooler-1 in the pre-freezing tower. In the main-freezing tower, to capture CO_2 by anti-sublimation, the temperature is reduced by Stirling cooler-2. Stirling cooler-3 chilled the CO_2 and gathered it on the storage tower in the form of frosted, and the residual gas is discharged to the ambient atmosphere [57–59]. This method is just on a laboratory scale, and no pilot or large-scale application of Stirling cooler-based cryogenic CO_2 capture system is available [27].

2.2 Advantages of cryogenic CO_2 capture technologies

One of the cryogenic CO_2 capture technology advantages is the high pressure and high purity of produced CO_2. This feature benefits utilization and storage, and for these two aims, compression and pipeline transport are required to deliver the captured CO_2. For storage, transportation, and utilization, the purity of CO_2 is a major factor. If the purity of CO_2 is high, it could be simply compressed to the required pressure and then been transported. CO_2 can be captured in different phases such as a liquid, a combination (slurry), or a solid. The CO_2 captured by cryogenic techniques is more pure than other separation methods. Converting

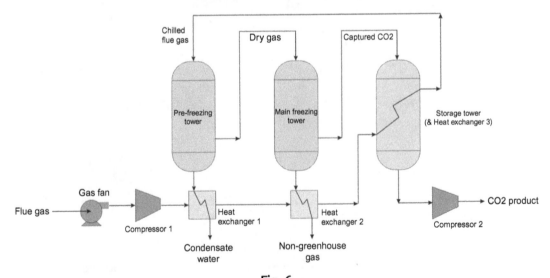

Fig. 6
Process flow diagram of the Stirling cooler-based CO_2 capture process with heat recovery at low temperature [27].

CO_2 to valuable chemicals is more efficient if CO_2 has higher purity. CO_2 can be turned to other chemicals by catalytic or biological reactions, for example, in steam methane reforming and artificial photosynthesis, respectively. Due to the high pressure of captured CO_2 from cryogenic methods, no additional compression treatment is required. Also, the low temperature of the captured CO_2 can be reused in the industry as a cold energy source. Therefore, it could be united with other low-temperature processes.

2.3 Challenges of cryogenic CO_2 capture technologies

The most important challenge that limits the cryogenic methods is the cold energy source required. Refrigeration and compression are the methods to supply the cryogenic conditions. Reusing the cold energy from captured CO_2 and residual gases can reduce refrigeration and compression energy consumption. Therefore, using a suitable heat exchanger to recover the latent and sensible heat of this streams is necessary. The CO_2 capture unit and compression and also the equipment for transporting the compressed materials are the major cost in cryogenic carbon capture plant (80% for carbon capture and compression, 10% for transportation, and 10% for storage). In Table 1, the advantages, limitations, cold energy sources, and CO_2 recovery are listed for each process.

3. Cryogenic distillation modeling

The output syngas from the water-gas shift reaction, a feed for the IGCC plant, in addition to H_2, CO, and CO_2, as shown in Table 2, consists of H_2S, H_2O, N_2, NH_3, CH_4, and Ar. CO_2 percent in this flow is approximately up to 39.85%, which is high enough; thus, a distillation cryogenic temperature process is a suitable method for this syngas purification. This process contains multistage compression, refrigeration, and separation [38].

In cryogenic distillation, the separation of the mixed syngas is carried out based on the difference between the boiling points of the components. By controlling the temperature and pressure in different conditions, the gases reach their boiling point and become liquid and separate from the syngas. As shown in Fig. 7, in a same pressure, the boiling points increase as follow: $H_2 < N_2 < CO < Ar < CH_4 < CO_2 < H_2S < NH_3 < H_2O$. Thus, water is the first component that separates due to its highest boiling point, and then NH_3, H_2S, and CO_2 would be separated, respectively [38].

3.1 Modeling, simulation, and analysis of the cryogenic distillation column with trays

In the first step, the raw syngas is cooled by a heat exchanger and then enters a flash tank in order to remove most of the water from the flow. In the flash tank, the separated water discharge from the bottom and the remaining flows without water left from the top and enters to a desulfurization tower. H_2S and NH_3 with near boiling points left the desulfurization tower from

Table 1 Advantages and limitations of the existing low-temperature CO_2 capture methods.

Process	Cold energy source	CO_2 recovery	Advantages	Limitations	References
Packed bed	Liquid nitrogen gas (LNG)	99%	• Atmosphere • Concurrent H_2O and CO_2 removal • Low risk of high pressure • surface area-to-volume ratio of the column	• It is contingent on LNG availability • Lab-scale	[47,60]
Distillation	Compressor and cooler	85%	• Save money on compression costs • Simple in pumping to a storage location • Potential for energy storage • Potential for water saving • Concurrent removal of other pollutants (Hg, SO_x, NO_2, HCl, etc.)	• Capital cost for pressure difference • High installation cost	[38,45]
AnSU	Liquified natural gas (LNG)	90%	• Atmosphere • Lower energy penalty than MEA absorption • Pilot demonstration	• It is contingent on the location of natural gas station • No H_2O can be withstood • Frost CO_2 has unfavorable effects on heat conduction adverse mechanical stresses	[52]
CryoCell	Chiller	–	• No process heating system needed • No potential for corrosion • No potential for foaming • Save money on compression cost	• More applicable for high CO_2 concentration (higher than 20%) • High compression power demand	[61]
Stirling cooler	Stirling cooler	85%	• Atmosphere • Concurrent H_2O and CO_2 removal • Lower energy penalty than MEA absorption • Potential of energy storage potential	• Exergy loss because of temperature difference • Frost layer scrapping problem • Lab scale	[57,58]

Table 2 Matter composition of syngas fed to IGCC plant after water-gas shift [38].

Component	Vol%
H_2	54.66
CO	1.33
CO_2	39.85
H_2S	0.14
H_2O	0.21
CH_4	0.01
N_2	3.20
NH_3	0.02
Ar	0.58

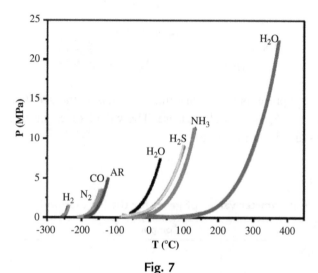

Fig. 7

The relationship between the saturation vapor pressure and temperature of each component in syngas [38].

the bottom and H_2S and NH_3 free gas flowed out from the top. The gas with considerable amounts of CO_2 further passes through compressor and pressurized and then enters to the second column named decarburization for CO_2 removal. The stream rich in CO_2 leaves the bottom of the decarburization tower. This CO_2-rich stream has a high pressure in which can be directly transported. The purified syngas, which has the required purity, exits from the top. This stream can generate electricity by entering the power generation system of gas-steam combined cycle and increases the profit of the cryogenic process [38].

For modeling the cryogenic distillation tower, the equations are the same as a common modeling of a distillation tower, and due to the purity required in each step, the operating properties may be defined. Peng-Robinson equation of state (PR-EOS) can be used for property prediction in the simulation. This equation of state is suitable for mixed hydrocarbon gases and a mixture of weak polar and nonpolar gases such as CO_2, H_2S, H_2, etc. [38].

Peng-Robinson equation of state is as follows [62]:

$$p = \frac{RT}{V-b} - \frac{\alpha a}{V^2 + 2bV - b^2} \tag{1}$$

$$a = 0.45725R^2 T_C^2 / P_c \tag{2}$$

$$b = 0.0778RT_c / P_c \tag{3}$$

$$\alpha = \left[1 + \left(0.37464 + 1.54226\omega - 0.26992\omega^2\right)\left(1 - T_r^{0.5}\right)\right]^2 \tag{4}$$

$$b = \Sigma_i y_i b_i \tag{5}$$

$$\alpha a = \Sigma_i \Sigma_j y_i y_j (\alpha a)_{ij} \tag{6}$$

$$(\alpha a)_{ij} = \left(1 - k_{ij}\right)\sqrt{(\alpha a)_i (\alpha a)_j} \tag{7}$$

k_{ij} is a parameter that represents the binary interaction, ω is the acentric factor, T_c denotes critical temperature, and P_c is critical pressure. The values of k_{ij} parameter for gases in this process for PR-EOS are listed in Table 3.

Table 3 The k_{ij} parameter values of gases for the PR thermodynamic model [38].

Component i	Component j	k_{ij}
H_2S	CO	0.0544
	H_2O	0.0400
	CO_2	0.0974
	N_2	0.1767
NH_3	H_2O	−0.2589
	N_2	0.2193
	Ar	−0.1800
CO	N_2	0.0307
	H_2	0.0919
	CH_4	0.0300
CO_2	N_2	−0.0170
	H_2	−0.1622
	CH_4	0.0919
N_2	H_2	0.1030
	CH_4	0.0311
	Ar	−0.0026
H_2O	CO_2	0.1200
H_2	CH_4	0.0156
CH_4	Ar	0.0230

The major purpose of the cryogenic distillation tower is to remove CO_2 and H_2S from the synthesis gas. CO_2 and H_2S removal rates can be demonstrated in Eq. (8) and Eq. (9), respectively [38].

$$CO_2 \text{ removal rate} = \frac{CO_2 \text{ molar quantity from the bottom of tower}}{CO_2 \text{ molar quantity of the feed into distillation tower}} \qquad (8)$$

$$H_2S \text{ removal rate} = \frac{H_2S \text{ molar quantity from the bottom of tower}}{H_2S \text{ molar quantity of the feed into distillation tower}} \qquad (9)$$

The simulation of this process had been done in Aspen Plus software as shown in Fig. 8 below. For determining the properties of the process such as reflux ratio, number of theoretical plates, load of reboiler, and condenser, we first need to determine the reflux ratio. The reflux ratio is an important parameter because it has effect on other parameters and even on removal rate of impurities and also influences the operating and investment cost of the distillation unit [38].

3.1.1 Influence of column parameters on H_2S removal rate

The H_2S removal rate varies due to the reflux ratio effect. As shown in Fig. 9A that is the result of the simulation, the H_2S removal rate increases when the reflux ratio is less than 2.4 and for reflux ratios greater than 2.4, the H_2S removal rate maintains at about 92.5%. The optimal reflux ratio is 2.4 because if the reflux ratio is larger than 2.4, there is no change in H_2S removal rate and also the reboiler and condenser load will increase; thus the operation and investment cost will increase [38].

As shown in the simulation result in Fig. 9B, by increasing the number of the column stages, the H_2S removal rate increases to a certain value of 44 stages. But after that, the removal rate remains constant, and increasing the number of stages does not affect it. Thus, the optimal number of stages is 44 and in a greater number of stages from this, only operation and investigation cost will increase and has no beneficial effect on removal rate [38].

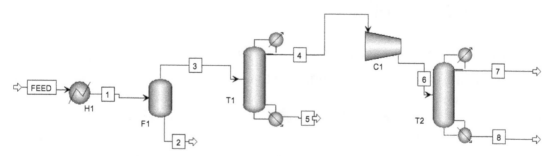

Fig. 8

The cryogenic distillation process. H1:Heat exchanger; F1:Flash tank; T1:Desulfurization tower; C1: Compressor; T2:Decarburization tower [38].

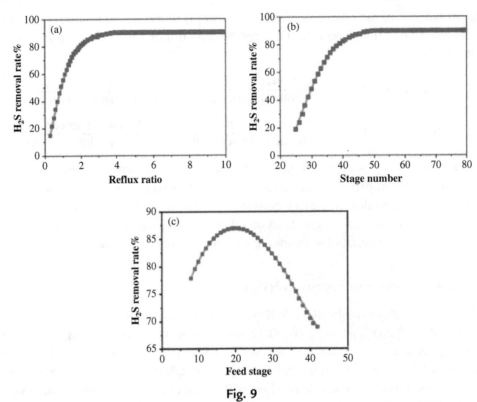

Fig. 9

The effect of desulphurization tower parameters on H_2S removal rate [38].

Also, the feed stage can influence the H_2S separation. As shown in Fig. 9C, the H_2S removal rate first increases with the increase of the feed stage, but then it decreases. When the feed stage raises too high, it will affect the top product purity due to its more content of refractory components. If the feed stage moves down, volatile components increase in the reboiler; thus, the feed stage has a role in determining the quality of products at the top and bottom of the column. H_2S removal reaches its maximum value in feed stage number 20 and the optimal feed stage selected to be 20th [38].

3.1.2 Influence of column parameters on CO_2 removal rate

The simulation results for reflux ratio effect on CO_2 removal rate are shown in Fig. 10A. The CO2 removal rate increases with the increase of reflux ratio, but at a reflux ratio of 2, the CO2 removal rate remains constant at 85%. Thus, as discussed before, the optimal reflux ratio for the CO_2 removal distillation column is 2 [38].

In the distillation unit, the more the number of the stages, the higher the CO2 removal rate, but it increases the operation and investment cost. As shown in Fig. 10B, with increasing in the number of stages, the CO2 removal rate increases until the stage number of 10, and after that it

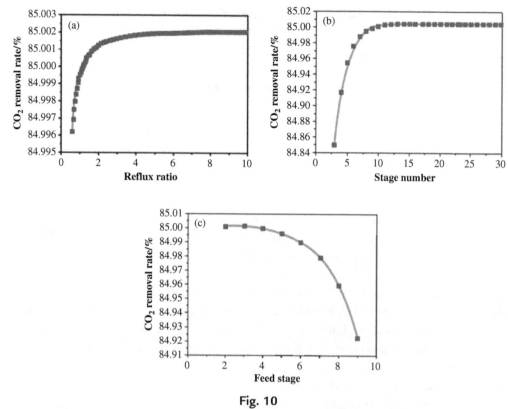

Fig. 10

The effect of decarburization tower parameters on CO_2 removal rate [38].

remains unchanged. The optimal number of stages is 10 due to considering the CO2 removal efficiency and operating and investigation costs [38].

Also, feed stage influences the CO2 removal rate in the decarburization column. As shown in Fig. 10C, the CO_2 removal rate decreases when the feed stage increases. If the feed stage is too down, CO_2 concentration in the reboiler increases and destructively affects the purity of the top and bottom products. Thus, to have an acceptable CO_2 removal, the second stage is selected to be the optimal stage feed [38].

The overall information and setting parameters for the equipments are listed in Table 4.

3.2 Modeling, simulation, and analysis of the packed beds

Cryogenic packed beds (CPBs) work based on the sublimating of the contaminants through the bed. When the gas flow passes through the chilled bed, materials reach their freezing point and are deposited on the surface of the packed bed and removed from the flow. The desired components with a lower freezing point pass the bed and exit from the tower [15].

Table 4 Parameters in cryogenic distillation tower [38].

Model	Model type	Function	Parameter preliminary setting
H1	HEATER	Gas cooling	No pressure change, 0°C of outlet temperature
F1	FLASH	Gas-liquid separation	3.23 MPa of flash pressure, 0°C of flash temperature
T1	RADFRAC	Gas-liquid separation	2.4 (mole ratio) of reflux ratio, 44 stages, kettle reboiler, partial condensation, feed above stage 20th, 1049.74 kg/h of bottom product flow rate, 3.22 MPa of top-stage pressure, 3.23 MPa of 44th-stage pressure
C1	COMPR	Gas supercharging	7 MPa of outlet pressure
T2	RADFRAC	Gas-liquid separation	10 stages, 2.4 (mole ratio) of reflux ratio, feed Above stage 2th, kettle reboiler, partial condensation, 329,476.93 kg/h of bottom product flow rate, 6.99 MPa of top-stage pressure, 7.01 MPa of 10th-stage pressure

Modeling of the CPBs was first presented by Tuinier et al. in 2010. A 1D pseudo-homogeneous model for a mixture of N_2, CO_2, and H_2S was constructed in which removal of H_2S and CO_2 was the aim of the separation. Later in 2011, Tuinier et al. used their model for different N_2, CO_2, and H_2O compositions and investigated the mass deposition and axial temperature profiles. In 2012, Tuinier and Annaland used this model for removing H_2S and CO_2 from biogas, as shown in Fig. 11. DiMaria et al. use the introduced model by Tuinier et al. for simulating syngas purification and compare the model results with the COMSOL software results [9,15,47,48].

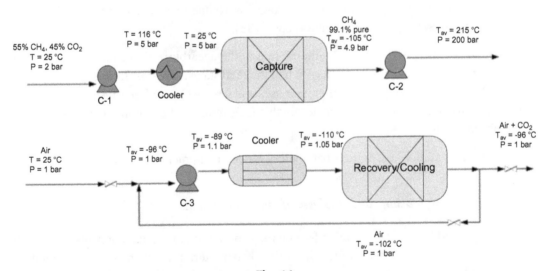

Fig. 11
Simplified process scheme for cryogenic packed-bed method in Tuinier et al. work [48].

The model presented by Tuinier et al. represents a dynamically operated packed bed. This process has three steps including capture cycle, recovery cycle, and cooling cycle, respectively [9].

3.2.1 Equations for 1D pseudo-homogeneous model of the packed bed

Components mass balance for the gas phase:

$$\varepsilon_g \rho_g \frac{\partial x_{i,g}}{\partial t} = -\rho_g v_g \frac{\partial x_{i,g}}{\partial z} + \frac{\partial}{\partial z}\left(\rho_g D_{eff} \frac{\partial x_{i,g}}{\partial z}\right) - \dot{m}_i'' a_s + x_{i,g} \sum_{i=1}^{n_c} \dot{m}_i'' a_s \tag{10}$$

Components mass balance for the solid phase:

$$\frac{\partial m_i}{\partial t} = \dot{m}_i'' a_s \tag{11}$$

Total continuity equation for the gas phase:

$$\frac{\partial\left(\varepsilon_g \rho_g\right)}{\partial t} = -\frac{\partial\left(\rho_g v_g\right)}{\partial z} - \sum_{i=1}^{n_c} \dot{m}_i'' a_s \tag{12}$$

Energy balance for gas and solid phase:

$$\left(\varepsilon_g \rho_g C_{p,g} + \rho_s\left(1 - \varepsilon_g\right)C_{p,s}\right)\frac{\partial T}{\partial t} = -\rho_g v_g C_{p,g} \frac{\partial T}{\partial z} + \frac{\partial T}{\partial z}\left(\lambda_{eff} \frac{\partial T}{\partial z}\right) - \sum_{i=1}^{n_c} \dot{m}_i'' a_s \Delta h_i \tag{13}$$

Heat and mass transfer coefficients:

Effective axial heat dispersion in a transient packed bed [63,64]:

$$\lambda_{eff} = \left(1 - \varepsilon_g\right)\lambda_s + \left(\frac{\rho_g v_g C_{P,g}}{\varepsilon_g}\right)^2 \frac{1}{\alpha_{g,s} a_s} \tag{14}$$

Gas-to-solid heat transfer coefficient [65]:

$$\alpha_{g,s} = \frac{\lambda_g}{d_h} 2.978 \left(1 + 0.095 RePr \frac{d_h}{L_c}\right)^{0.45} \tag{15}$$

$$Re = \frac{\rho_g v_g d_h}{\eta_g \varepsilon_g} \tag{16}$$

Axial mass dispersion [66,67]:

$$D_{ax} = D_{eff,i} + \frac{v_g^2 d_{h,c}^2}{192 D_{eff,i}} \tag{17}$$

$$D_{eff,i} = \frac{1}{\sum_{j=1}^{n_c} \frac{y_{j,g}}{D_{i,j}}}$$ (18)

Mass deposition rate [9]:

$$\dot{m}_i'' = \begin{cases} g(y_{i,s}P - p_i^\sigma), & \text{if } y_{i,s}P > p_i^\sigma \\ g(y_{i,s}P - p_i^\sigma)\dfrac{m_i}{m_i + 0.1}, & \text{if } y_{i,s}P < p_i^\sigma \end{cases}$$ (19)

Gas-solid equilibrium [9]:

$$p^\sigma_{CO_2}(T) = \exp\left(10.257 - \frac{3082.7}{T} + 4.08 \ln(T) - 2.2658 \times 10^{-2}T\right)$$

$$\Delta h^{sub}_{CO_2} = 5.682 \times 10^5 \ J/kg$$ (20)

Pressure drop over packing [48]:

$$\frac{\partial p}{\partial z} = -4\frac{f}{d_{h,c}}\frac{1}{2}\rho_g v_g{}^2$$ (21)

$$f = \frac{14.9}{Re}\sqrt{1 + 0.0445\, Re\, \frac{d_{h,c}}{L}}$$ (22)

The mixture flow of gas-phase properties has been computed according to Reid et al. [68], using the pure component data supplied by Daubert and Danner [69].

3.2.2 Simulation of the packed bed

This simulation had been done in COMSOL software for syngas as a mixture of seven components (i.e., H_2, CO, CO_2, H_2S, H_2O, CH_4, and N_2). The equations used in this simulation are the same model equations from the work of Tuinier et al., but they are used for syngas and the contaminants including it [15].

Cryogenic packed-bed properties are the same as the properties used by Tuinier et al. in 2010 (i.e., bed length (300 mm), bed initial diameter (35 mm), packing particle diameter (4.04 mm), inlet mass flux (0.27 $(\frac{kg}{m^2 s})$), inlet temperature (100°C), bed pressure (1 bar)) [9,15].

Eqs. (23)–(27) are used in this simulation:

Antoine equilibrium vapor pressure:

$$\log_{10}\left(p^\sigma_{i,Antoine}\right) = A - \left(\frac{B}{T+C}\right)$$ (23)

Ideal gas law:

$$\rho_{g,i} = \frac{M_i P}{RT}$$ (24)

Diffusion equation represents by Fuller et al. [70,71]:

$$D_i = \frac{1.00 \times 10^{-7} T^{1.75} \left(\dfrac{1}{M_A} + \dfrac{1}{M_B} \right)^{0.5}}{P \left[(\Sigma \vartheta_A)^{1/3} + (\Sigma \vartheta_B)^{1/3} \right]^2} \tag{25}$$

Component B in the above equation is assumed to be air.

Porosity at each point:

$$\varepsilon_g = \varepsilon_{g,initial} - \sum_c^{i=1} \frac{m_i}{\rho_{i,s}} \tag{26}$$

Mole fraction of substrate i in the solid-phase:

$$y_{i,s} = \frac{y_{i,g} P}{p_i^{\sigma}} \tag{27}$$

And the PDE equation as input for COMSOL is:

$$e_a \frac{\partial^2 u}{\partial t^2} + d_a \frac{\partial u}{\partial t} + \nabla(-c\nabla u - \alpha u + \gamma) + \beta \nabla u + au = f \tag{28}$$

In the above equation, u is a dependent variable and letters a–f are coefficients of the equation. Also, Darcy's law for porous media in which this physics module consists of continuity, pressure, and velocity equations, was used in the simulation. This physics module can consider the changes in bed porosity and gas velocity as a result of impurities sublimation and deposition. In cryogenic packed-bed distillation, materials will sublimate due to their freezing point and then deposit on the surface of the packing and lead to the reduction of the bed porosity. This porosity reduction increases within the bed as the deposited mass increases. The porosity reduction is considered in this simulation in Eq. (26) [15].

In contriving the simulation, five assumptions were made in order to simplify the process: (1) Even though Brinkman's flow equation presented more accurate responses, Darcy's law for porous media was chosen. (2) A simple weighting correlation based on molar and mass fractions was used to determine bulk properties. (3) Heat transfer between the packing was considered negligible. (4) The deposited substrates do not influence the packed-bed thermal properties. (5) Latent heats were assumed to be constant with temperature and pressure changing [15].

The temperature, CO_2 mass deposition, and H_2O mass depositions of each in 50 and 150s of run time were selected to compare the simulation and specified modeling results in order to ensure the simulation accuracy. As reported in the DiMaria et al. [15] work, the temperature curves are almost similar and have an insignificant difference in the posterior lengths, but there are two substantial disagreements. At the beginning of the bed, the replication model did not start from the point that the Tuinier model did, and it can be said that the following discrepancies in the bed are a result of this inconsistency. One can say the differences in component properties or COMSOL solver may cause this incompatibility.

Also, CO_2 and H_2O mass deposition curves vs. the bed length are reported by DiMaria et al. [15]. In cryogenic packed beds, the mass deposition of the components pervades as a front. This front moves and expands through the bed as the contaminants deposit with time. The equilibrium between the solid and gas phase controls the maximum deposition value and is a function of temperature, pressure, and contaminants concentration. The obvious discrepancies between the replication and original model are the position and peak heights of their fronts. This incompatibility can result from differences in component properties or even COMSOL solver or because of considering porosity changes.

The change in porosity is considered in Eq. (26), and the simulation result was reported at a time of 150 s. As provided, the maximum porosity change in the bed is 0.027 units. In other words, porosity only reduced 7.2% compared with the initial value (0.375). This change is small and has a negligible influence on the results, so it is a logical idea to assume the porosity to be constant in the calculations [15].

Syngases from different gasification processes and different feedstocks were used in COMSOL, which have different molar fractions and H_2/CO ratios. These gasification processes are derived from circulating fluidized bed (CFB), bubbling fluidized bed (BFB), entrained flow (EF), and fixed bed (FB) gasifiers. The several feedstocks that were used were pulp, wood, bark, coal, refuse-derived fuel (RDF), and municipal solid waste (MSW). The molar fraction of components, H_2/CO ratio, and heat capacity of the feed gas for each couple of process/feedstock are defined in Table 5 [9,15].

Moreover, an energy analysis was accomplished by using a numerical model in addition to COMSOL simulation. The numerical model calculates the energy and mass model of the system based on COMSOL results. Comparisons were made on the value of applicable parameters such as: (1) breakthrough time (BT), that is the time required in which the concentration of the components in outlet reach 10% of the inlet concentration; (2) absolute ambient higher heating value (AAHHV), that is total higher heating value of the outlet gas at room temperature for one capture cycle; (3) heat input into media (HIM), that is the total heat transferred from gas into packing bed in one capture cycle; (4) heat loading rate (HLR), which is the average rate of heat into the packed media over one capture cycle; (5) H_2/CO ratio, that is the ratio of H_2 molars to CO_2 molars in the gas flow [15].

According to the above definitions, a gas with a higher AAHHV/HIM ratio is a desirable gas. Also, a shorter breakthrough time is preferable if we assume that the required time for recovering the system is approximately identical to the breakthrough time or the time in which the bed is saturated. Due to the further usage of the syngas, the H_2/CO ratio (2.0 and 0.6 are

Table 5 Gas composition for different gasifier type and feedstock [72].

Gasification process	Feedstock	Gasifier type	Feed gas heat capacity (kJ kg⁻¹ K⁻¹)	Molar fraction							
				CH_4 (%)	CO (%)	CO_2 (%)	H_2 (%)	H_2O (%)	H_2S (%)	N_2 (%)	H_2/CO ratio
BCL/FERCO	Wood	CFB[a]	1.6	24.00	47.0	15	15.0	0.0	0.00	0	0.30
Shell	Coal	EF	1.4	0.02	67.0	4	24.0	3.0	1.00	1	0.36
Biogas[b]	–	–	2.0	55.00	0.0	44	0.0	0.0	1.00	0	–
Flue gas[c]	–	–	1.4	0.00	0.0	20	0.0	5.0	0.00	75	–
Lurgi	Bark	CFB	1.6	3.80	20.0	14	20.0	0.0	0.00	43	1.00
Purox	MSW	FB	1.9	11.00	40.0	25	24.0	0.0	0.05	0	0.60
Foster Wheeler	Wood	CFB	1.5	5.50	22.0	11	16.0	0.0	0.00	47	0.70
Aerimpianti	RDF	CFB	1.4	8.10	11.0	13	7.9	12.0	0.00	49	0.70
Tampella	Wood	BFB	1.5	4.80	13.0	13	11.0	18.0	0.00	40	0.80
Sydkraft	Wood	CFB	1.4	6.50	16.0	11	11.0	12.0	0.00	44	0.70
Stein	Wood	BFB	1.9	16.00	26.0	40	5.7	0.0	0.00	13	0.80
MTCI	Pulp	BFB[a]	2.5	14.00	9.2	28	43.0	5.6	0.08	0	4.60
EPI	Wood	BFB	1.4	0.00	24.0	7	18.0	0.0	1.00	50	0.30
GTI	Wood	BFB	1.7	11.00	12.0	22	15.0	0.0	0.10	40	1.60
SEI	Wood	BFB	1.5	8.00	16.0	16	13.0	0.0	0.00	48	0.80
ISU	Wood	BFB	1.4	3.10	24.0	13	4.1	0.0	0.00	56	0.20
TPS	RDF	CFB	1.4	7.40	11.0	13	7.9	12.0	0.00	49	0.70

[a]Indirectly heated gasifier.
[b]Biogas from Tuinier and Annaland [48].
[c]Flue gas from Tuinier et al. [9].

applicable for methanol synthesis and Fischer-Tropsch (FT) process, respectively) is a fair comparison and also if the gas is biomass-derived or not [15,48].

Due to the zero amount of the flue gas heating value, HLR is compared for different flows and a lower HLR is beneficial. These terms of comparison and their values are defined in Table 6.

In packed-bed simulation for different gasification processes in COMSOL software, the syngas manufactured from BCL/FERCO process has a larger AAHHV/HIM ratio and thus could be the best in a cryogenic packed bed. Shell syngas has low concentration of impurities; accordingly, it accomplishes likewise BCL/FERCO syngas. Shell syngas has a longer breakthrough time than BCL/FERCO syngas; therefore, the frequency for bed recovery reduces [15].

Biogas has the third rank in value of AAHHV/HIM ratio, thus both BCL/FERCO and Shell process-derived syngas may have a better accomplishment in CPB than biogas. It was advised that biogas is more energetic in CPB than a modern VPSA [73].

As well as flue gas, the Tampella process gas flow has the least value of AAHHV/HIM ratio. It can be said that this is ascribable to higher amounts of N_2 and H_2O. N_2 as an inert can inset heat into the system but supplies no heating value available for use, and H_2O content is removed and provides latent heat into the system.

4. Conclusion

Based on the difference of boiling points of contaminants in a mixture of impure syngas, cryogenic distillation was used to simulate the CO_2 and H_2S removal. The simulation process for the tray columns and packed beds was done in Aspen Plus and COMSOL software, respectively. In distillation by tray column, reflux ratio, stage number, and feed stage, which are effective parameters on the system, had been optimized in terms of CO_2 and H_2S removal values in decarburization and desulfurization tower. The optimal values for the decarburization tower were reflux ratio of 2.0, 10 number of stages, and the feed stage of second. And the optimal results for the desulfurization tower were reflux ratio of 2.4, 44 number of stages, and the feed stage of 20th. In this method, 85% of CO_2, 98.2% of H_2S were removed from syngas.

Biogas was considered to be a criterion in packed-bed distillation results due to its previous viable results. In different processes for syngas production, biogas had the third rank in AAHHV/HIM ratio value and BCL/FERCO and Shell syngas are energetically more feasible in cryogenic packed bed by reason of 37% and 14% AAHHV/HIM ratio value more than that for biogas, respectively. Also, the breakthrough for BCL/FERCO and Shell syngas is longer than that of biogas. GTL postprocessing, The Purox and Foster Wheeler gases had ideal H_2/CO ratios, but the BCL/FERCO and Shell syngas did not have the ideal H_2/CO ratio. From an AAHHV/HIM ratio point of view, the Purox and Foster Wheeler gases were not examined to be energetically viable.

Table 6 Numerical analysis result for energy and mass terms [15].

Gasification process	Breakthrough time (s)	HIM (kJ)	HLR (W)	AAHHV (kJ$_{out}$)	Ratio of AAHHV/HIM	Difference between AAHHV and HIM (kJ)	Mass balance Mass input (g)	Mass deposited (g)	Mass output (g)
BCL/FERCO	180	22.6	126	767	33.50	745	47	13.0	34.0
Shell	239	26.4	111	746	28.20	719	62	7.4	55.0
Biogas	120	21.9	183	544	24.80	522	31	22.0	9.6
Flue gas	206	24.3	118	0	0.00	−24	54	17.0	37.0
Lurgi	197	22.4	114	307	13.70	284	51	12.0	39.0
Purox	144	20.9	145	423	20.20	402	37	17.0	21.0
Foster Wheeler	221	23.2	105	358	15.50	335	57	11.0	47.0
Aerimpianti	205	30.1	147	257	8.55	227	53	16.0	37.0
Tampella	195	33.7	173	230	6.82	196	51	18.0	33.0
Sydkraft	213	31.3	147	299	9.54	268	55	15.0	40.0
Stein	148	21.7	146	282	13.00	261	38	22.0	17.0
MTCI	105	23.2	221	390	16.80	366	27	19.0	8.1
EPI	231	22.6	98	312	13.80	289	60	8.4	52.0
GTI	165	21.1	128	278	13.20	257	43	16.0	27.0
SEI	197	22.3	113	294	13.20	271	51	14.0	38.0
ISU	236	22.8	97	229	10.00	206	61	12.0	49.0
TPS	205	30.1	147	245	8.14	215	53	16.0	37.0

Abbreviations and symbols

AAHHV	absolute ambient higher heating value
ATR	auto thermal reforming
a_s	specific solid surface area per unit bed volume (m^2/m^3)
$\alpha_{g,\,s}$	heat transfer coefficient solid-gas bulk ($W/m^2/K/s$)
BFB	bubbling fluidized bed
BT	breakthrough time
CCS	carbon capture storage
CFB	circulating fluidized bed
CFZ	controlled freezing zone
CPB	cryogenic packed beds
C_p	heat capacity (J/kg/K)
D	diffusion coefficient (m^2/s)
D_{ax}	axial diffusion coefficient (m^2/s)
D_{eff}	effective diffusion coefficient (m^2/s)
DD	down distillation
Δh_i	enthalpy change related to the phase change of component i (J/kg)
d_h	hydraulic diameter of the monolithic channels (m)
EF	entrained flow
EOR	enhanced oil recovery
ε	bed void fraction/porosity
η	viscosity (kg/m/s)
FB	fixed bed
FT	Fischer-Tropsch
f	fanning friction factor for the monolithic channels
GTL	gas to liquid
HIM	heat input into media
HLR	heat loading rate
IGCC	integrated gasification combined cycle
k_{ij}	binary interaction parameters
L	bed length (m)
LNG	liquid nitrogen gas
λ	conductivity (W/m/K)
λ_{eff}	effective conductivity (W/m/K)
M	molar mass (g/mol)
MDEA	methyl diethanolamine
MEA	monoethanolamine
MSW	municipal solid waste
NRU	nitrogen rejection units
m	mass deposition per unit bed volume (kg/m^3)
n_c	number of components
ω	acentric factor
P	pressure (Pa)
P_c	critical pressure
PO	partial oxidation
PR	Peng-Robinson
PR-EOS	Peng-Robinson equation of state
PSA	pressure swing adsorption
p	partial pressure (Pa)
p^σ	equilibrium partial pressure (Pa)

$p^\sigma_{i,Antoine}$	equilibrium partial pressure of the component i based on Antoine parameters A, B, and C (Pa)
R	ideal gas constant (8.3145 J/mol/K)
RDF	refuse-derived fuel
Re	Reynolds number
ρ	density (kg/m^3)
SR	steam reforming
T	temperature (K or °C)
t	time (s)
T_c	critical temperature
ϑ	atomic diffusion volume
D	upper distillation
VPSA	vacuum pressure swing adsorption
v_g	superficial gas velocity (m/s)
x	mass fraction (kg/kg)
y	mole fraction (mol/mol)
z	axial coordinate (m)

References

[1] IEA Statistics, CO2 Emissions From Fuel Combustion-Highlights, IEA, Paris, 2011. http://www.iea.org/co2highlights/co2highlights.pdf. Cited July.

[2] C. IEA, Emissions from Fuel Combustion Highlights, International Energy Agency, Paris, France, 2012.

[3] L. Li, N. Zhao, W. Wei, Y. Sun, A review of research progress on CO2 capture, storage, and utilization in Chinese Academy of Sciences, Fuel 108 (2013) 112–130.

[4] C. Stewart, M.-A. Hessami, A study of methods of carbon dioxide capture and sequestration—the sustainability of a photosynthetic bioreactor approach, Energy Convers. Manag. 46 (3) (2005) 403–420.

[5] Y. Liu, S. Deng, R. Zhao, J. He, L. Zhao, Energy-saving pathway exploration of CCS integrated with solar energy: a review of innovative concepts, Renew. Sust. Energ. Rev. 77 (2017) 652–669.

[6] R.T. Porter, M. Fairweather, M. Pourkashanian, R.M. Woolley, The range and level of impurities in CO2 streams from different carbon capture sources, Int. J. Greenhouse Gas Control 36 (2015) 161–174.

[7] C. Song, Q. Liu, N. Ji, S. Deng, J. Zhao, Y. Li, Y. Kitamura, Reducing the energy consumption of membrane-cryogenic hybrid CO2 capture by process optimization, Energy 124 (2017) 29–39.

[8] M. Mehdipour, A. Elhambakhsh, P. Keshavarz, M.R. Rahimpour, Y. Hasanzadeh, CO2 separation by rotating liquid sheet contactor: a novel procedure to improve mass transfer characterization, Chem. Eng. Res. Des. 172 (2021) 120–126.

[9] M. Tuinier, M. van Sint Annaland, G.J. Kramer, J. Kuipers, Cryogenic CO2 capture using dynamically operated packed beds, Chem. Eng. Sci. 65 (1) (2010) 114–119.

[10] D. Berstad, R. Anantharaman, P. Nekså, Low-temperature CO2 capture technologies—applications and potential, Int. J. Refrig. 36 (5) (2013) 1403–1416.

[11] A. Keshavarz, M. Ebrahimzadeh Sarvestani, M.R. Rahimpour, Cryogenic CO2 capture, Sustain. Agric. Rev. 38 (2019) 251–277. Springer.

[12] A.L. Kohl, R. Nielsen, Gas Purification, Elsevier, 1997.

[13] J.D. Seader, E.J. Henley, D.K. Roper, Separation Process Principles, Wiley, New York, 1998.

[14] T.E. Rufford, S. Smart, G.C. Watson, B. Graham, J. Boxall, J.D. Da Costa, E. May, The removal of CO2 and N2 from natural gas: a review of conventional and emerging process technologies, J. Pet. Sci. Eng. 94 (2012) 123–154.

[15] P.C. DiMaria, A. Dutta, S. Mahmud, Syngas purification in cryogenic packed beds using a one-dimensional pseudo-homogenous model, Energy Fuel 29 (8) (2015) 5028–5035.

[16] M. Farsi, Biomass conversion to biomethanol, in: Advances in Bioenergy and Microfluidic Applications, Elsevier, 2021, pp. 231–252.

[17] B. Hubbard, New and Emerging Technologies (Petroskills Workshop). 2010 Gas Processors Association Convention, John M. Campbell & Co, Austin, Texas, 2010.

[18] A.J. Kidnay, W. Parrish, Fundamentals of Natural Gas Processing, CRC, New York, 2006.

[19] Q. Sun, H. Li, J. Yan, L. Liu, Z. Yu, X. Yu, Selection of appropriate biogas upgrading technology—a review of biogas cleaning, upgrading and utilisation, Renew. Sust. Energ. Rev. 51 (2015) 521–532.

[20] M. Persson, O. Jönsson, A. Wellinger, Biogas upgrading to vehicle fuel standards and grid injection, IEA Bioenerg. Task (2006) 1–34.

[21] L.A. Pellegrini, G. De Guido, S. Langé, Biogas to liquefied biomethane via cryogenic upgrading technologies, Renew. Energy 124 (2018) 75–83.

[22] C. Song, Q. Liu, N. Ji, S. Deng, J. Zhao, Y. Li, Y. Song, H. Li, Alternative pathways for efficient CO2 capture by hybrid processes—a review, Renew. Sust. Energ. Rev. 82 (2018) 215–231.

[23] M. Tuinier, H. Hamers, M., Van Sint Annaland, techno-economic evaluation of cryogenic CO2 capture—a comparison with absorption and membrane technology, Int. J. Greenhouse Gas Control 5 (6) (2011) 1559–1565.

[24] A.M. Yousef, W.M. El-Maghlany, Y.A. Eldrainy, A. Attia, New approach for biogas purification using cryogenic separation and distillation process for CO2 capture, Energy 156 (2018) 328–351.

[25] Y. Tan, W. Nookuea, H. Li, E. Thorin, J. Yan, Cryogenic technology for biogas upgrading combined with carbon capture—a review of systems and property impacts, Energy Procedia 142 (2017) 3741–3746.

[26] M. Riva, M. Campestrini, J. Toubassy, D. Clodic, P. Stringari, Solid-liquid-vapor equilibrium models for cryogenic biogas upgrading, Ind. Eng. Chem. Res. 53 (44) (2014) 17506–17514.

[27] C. Song, Q. Liu, S. Deng, H. Li, Y. Kitamura, Cryogenic-based CO2 capture technologies: state-of-the-art developments and current challenges, Renew. Sust. Energ. Rev. 101 (2019) 265–278.

[28] M.R. Rahimpour, M. Saidi, M. Baniadam, M. Parhoudeh, Investigation of natural gas sweetening process in corrugated packed bed column using computational fluid dynamics (CFD) model, J. Nat. Gas Sci. Eng. 15 (2013) 127–137.

[29] O.W. Awe, Y. Zhao, A. Nzihou, D.P. Minh, N. Lyczko, A review of biogas utilisation, purification and upgrading technologies, Waste Biomass Valoriz. 8 (2) (2017) 267–283.

[30] S. Langè, L.A. Pellegrini, P. Vergani, M.L. Savio, Energy and economic analysis of a new low-temperature distillation process for the upgrading of high-CO2 content natural gas streams, Ind. Eng. Chem. Res. 54 (40) (2015) 9770–9782.

[31] M. Martin, Industrial Chemical Process Analysis and Design, Elsevier, 2016.

[32] G. Bozzano, F. Manenti, Efficient methanol synthesis: perspectives, technologies and optimization strategies, Prog. Energy Combust. Sci. 56 (2016) 71–105.

[33] S. Akbari-Emadabadi, M. Rahimpour, A. Hafizi, P. Keshavarz, Production of hydrogen-rich syngas using Zr modified Ca-Co bifunctional catalyst-sorbent in chemical looping steam methane reforming, Appl. Energy 206 (2017) 51–62.

[34] D. Karimipourfard, N. Nemati, M. Bayat, F. Samimi, M.R. Rahimpour, Mathematical modeling and optimization of syngas production process: a novel axial flow spherical packed bed tri-reformer, Chem. Prod. Process. Model. 13 (2) (2018).

[35] M. Farniaei, M. Abbasi, H. Rahnama, M.R. Rahimpour, A. Shariati, Syngas production in a novel methane dry reformer by utilizing of tri-reforming process for energy supplying: modeling and simulation, J. Nat. Gas Sci. Eng. 20 (2014) 132–146.

[36] S. Soltanimehr, M.R. Rahimpour, A. Shariati, A. Alipoor, Combination of combustion and catalytic reactors for syngas product: modeling and simulation, Top. Catal. (2022) 1–13.

[37] S. Samipour, M.D. Manshadi, P. Setoodeh, CO2 removal from biogas and syngas, in: Advances in Carbon Capture, Elsevier, 2020, pp. 455–477.

[38] H. Li, R. Zhang, T. Wang, X. Sun, C. Hou, R. Xu, Y. Wu, Z. Tang, Simulation of H2S and CO2 removal from IGCC syngas by cryogenic distillation, carbon capture, Sci. Technol. (2021) 100012.

[39] A.S. Holmes, J.M. Ryan, Cryogenic Distillative Separation of Acid Gases From Methane, Google Patents, 1982.

[40] A. Holmes, B. Price, J. Ryan, R. Styring, Pilot tests prove out cryogenic acid-gas/hydrocarbon separation processes, Oil Gas J. 81 (26) (1983) (United States).

[41] A. Kazemi, M. Hosseini, A. Mehrabani-Zeinabad, V. Faizi, Evaluation of different vapor recompression distillation configurations based on energy requirements and associated costs, Appl. Therm. Eng. 94 (2016) 305–313.

[42] A.K. Jana, A new divided-wall heat integrated distillation column (HIDiC) for batch processing: feasibility and analysis, Appl. Energy 172 (2016) 199–206.

[43] K. Maqsood, A. Mullick, A. Ali, K. Kargupta, S. Ganguly, Cryogenic carbon dioxide separation from natural gas: a review based on conventional and novel emerging technologies, Rev. Chem. Eng. 30 (5) (2014) 453–477.

[44] K. Maqsood, A. Ali, A. Shariff, S. Ganguly, Synthesis of conventional and hybrid cryogenic distillation sequence for purification of natural gas, J. Appl. Sci. 14 (21) (2014) 2722–2729.

[45] L. Baxter, A. Baxter, S. Burt, Cryogenic CO2 capture as a cost-effective CO2 capture process, in: International Pittsburgh Coal Conference, 2009.

[46] M.J. Jensen, Energy Processes Enabled by Cryogenic Carbon Capture, Brigham Young University, 2015.

[47] M. Tuinier, M. van Sint Annaland, J. Kuipers, A novel process for cryogenic CO2 capture using dynamically operated packed beds—an experimental and numerical study, Int. J. Greenhouse Gas Control 5 (4) (2011) 694–701.

[48] M.J. Tuinier, M. van Sint Annaland, Biogas purification using cryogenic packed-bed technology, Ind. Eng. Chem. Res. 51 (15) (2012) 5552–5558.

[49] N. Abatzoglou, S. Boivin, A review of biogas purification processes, Biofuels Bioprod. Biorefin. 3 (1) (2009) 42–71.

[50] M. Parhoudeh, F. Farshchi Tabrizi, M.R. Rahimpour, Auto-thermal chemical looping reforming process in a network of catalytic packed-bed reactors for large-scale syngas production: a comprehensive dynamic modeling and multi-objective optimization, Top. Catal. (2022) 1–28.

[51] D. Clodic, M. Younes, A new method for CO2 capture: frosting CO2 at atmospheric pressure, in: Greenhouse Gas Control Technologies-6th International Conference, Elsevier, 2003, pp. 155–160.

[52] D. Clodic, M. Younes, A. Bill, Test results of CO2 capture by anti-sublimation capture efficiency and energy consumption for boiler plants, in: Greenhouse Gas Control Technologies, vol. 7, Elsevier, 2005, pp. 1775–1780.

[53] D. Clodic, R. El Hitti, M. Younes, A. Bill, F. Casier, CO2 capture by anti-sublimation Thermo-economic process evaluation, 4th annual conference on carbon capture and sequestration, Citeseer (2005) 2–5.

[54] C. Song, Y. Kitamura, S. Li, Energy analysis of the cryogenic CO2 capture process based on Stirling coolers, Energy 65 (2014) 580–589.

[55] M.E. Parker, S. Northrop, J. Foglesong, W. Duncan, CO2 Management at ExxonMobil's LaBarge Field, Wyoming, USA, in: IPTC 2009: International Petroleum Technology Conference, European Association of Geoscientists & Engineers, 2009. pp. cp-151-00052.

[56] E. Thomas, R. Denton, Conceptual studies for CO2/natural gas separation using the controlled freeze zone (CFZ) process, Gas Sep. Purif. 2 (2) (1988) 84–89.

[57] C.-F. Song, Y. Kitamura, S.-H. Li, K. Ogasawara, Design of a cryogenic CO2 capture system based on Stirling coolers, Int. J. Greenhouse Gas Control 7 (2012) 107–114.

[58] C.F. Song, Y. Kitamura, S.H. Li, Evaluation of stirling cooler system for cryogenic CO2 capture, Appl. Energy 98 (2012) 491–501.

[59] C.F. Song, Y. Kitamura, S.H. Li, W.Z. Jiang, Parametric analysis of a novel cryogenic CO2 capture system based on Stirling coolers, Environ. Sci. Technol. 46 (22) (2012) 12735–12741.

[60] R.P. Lively, W.J. Koros, J. Johnson, Enhanced cryogenic CO2 capture using dynamically operated low-cost fiber beds, Chem. Eng. Sci. 71 (2012) 97–103.

[61] A. Hart, N. Gnanendran, Cryogenic CO2 capture in natural gas, Energy Procedia 1 (1) (2009) 697–706.

[62] R. Stryjek, J. Vera, PRSV: an improved Peng—Robinson equation of state for pure compounds and mixtures, Can. J. Chem. Eng. 64 (2) (1986) 323–333.

[63] A.G. Dixon, D.L. Cresswell, Theoretical prediction of effective heat transfer parameters in packed beds, AICHE J. 25 (4) (1979) 663–676.

[64] D. Vortmeyer, R. Schaefer, Equivalence of one-and two-phase models for heat transfer processes in packed beds: one dimensional theory, Chem. Eng. Sci. 29 (2) (1974) 485–491.

[65] R. Hawthorn, Afterburner Catalysts-Effects of Heat and Mass Transfer between Gas and Catalyst Surface, 1974.

[66] M. Edwards, J. Richardson, Gas dispersion in packed beds, Chem. Eng. Sci. 23 (2) (1968) 109–123.

[67] A. Cybulski, J.A. Moulijn, Monoliths in heterogeneous catalysis, Catal. Rev. Sci. Eng. 36 (2) (1994) 179–270.

[68] R.C. Reid, J.M. Prausnitz, B.E. Poling, The Properties of Gases and Liquids, 1987.

[69] T. Daubert, R. Danner, Data Compilation Tables of Properties of Pure Compounds, American Institute of Chemical Engineers, New York, 1985.

[70] C.J. Geankoplis, A.A. Hersel, D.H. Lepek, Transport Processes and Separation Process Principles, Prentice Hall, Boston, MA, USA, 2018.

[71] C.J. Geankoplis, Separation Process Principles, 2003.

[72] J.P. Ciferno, J.J. Marano, Benchmarking Biomass Gasification Technologies for Fuels, Chemicals and Hydrogen Production, US Department of Energy. National Energy Technology Laboratory, 2002.

[73] A. Ali, K. Maqsood, N. Syahera, A.B. Shariff, S. Ganguly, Energy minimization in cryogenic packed beds during purification of natural gas with high CO2 content, Chem. Eng. Technol. 37 (10) (2014) 1675–1685.

CHAPTER 11

Purification of syngas with nanofluid from mathematical modeling viewpoints

Ali Behrad Vakylabad

Department of Materials, Institute of Science and High Technology and Environmental Sciences, Graduate University of Advanced Technology, Kerman, Iran

1. Introduction

Today, hydrogen is considered as an important gas for the production of clean, sulfur-free fuels with zero emissions. In recent years, due to the energy crisis and the completion of fossil fuels, and also new environmental laws for the use of clean fuels, hydrogen production in petrochemical and refinery industries has increased [1–4]. In addition to clean fuel, hydrogen has various applications in the chemical, petrochemical, refining, and fuel cells industries. Due to the increase in hydrogen demand in recent years, research studies have been conducted on improving the methods of hydrogen production, separation, and purification [5,6]. Although conditions in the global economy have pushed oil prices lower for a short period of time, the long-term outlook for high oil demand is facing a tightening in oil prices and growing concerns about its decline. The most important factor that has slowed the movement toward hydrogen to date is the incorrect pricing of energy carriers. If the environmental costs resulting from the production and consumption of other fuels are added to their cost, the price of hydrogen will also be reasonably and economically justified. Therefore, the movement is directed toward hydrogen [7].

1.1 Hydrogen, synthesis gas (syngas), and separation methods

Currently, hydrogen is mainly used as a side feed, intermediate chemical, or a special chemical, and only a small share of hydrogen produced is used as an energy carrier. Further research and development are needed to optimize and diversify the commercial methods of hydrogen production. Also, methods of improvements for separation are necessary to reduce the prices of the hydrogen production and increase the production efficiency [8]. Fossil fuels can be converted directly and indirectly into valuable economic and strategic products. In direct processes, the final product is obtained from fossil fuels without creating stable intermediaries,

Advances in Synthesis Gas: Methods, Technologies and Applications. https://doi.org/10.1016/B978-0-323-91879-4.00011-4

while in indirect methods, it is necessary to convert gas into a mixture with appropriate chemical activities. The obtained mixture should have the ability to be converted into various products under the appropriate processes. Conversion of fossil fuels to synthesis gas, which is a mixture of hydrogen and carbon monoxide gases, is a method that has been considered commercially and industrially in recent years. It is worth noting that indirect methods obtained first by synthesis gas have higher efficiency than direct methods. Therefore, synthesis gas can be used to produce various products with high added value [9]. The abovementioned materials justify the necessity of paying attention to research and development in the field of hydrogen separation optimization from syngas.

1.2 Hydrogen and its applications

The growth of global energy demand during the 21st century along with the necessity of controlling the greenhouse gas emissions has led to attention to the new energy carrier, hydrogen. Hydrogen has a high energy density so that the amount of energy in a kilogram of hydrogen equals 2.6kg of natural gas or 3.1kg of gasoline [10]. The use of hydrogen-rich gas for lighting and heating applications continued from the early 19th century to the mid-20th century. This gas containing 50% hydrogen was known as natural gas. After a while, oil and natural gas exploration succeeded in producing light and heat. In 1911, hydrogen was first used to produce ammonia and fertilizer. Today, ammonia production is one of the most important uses of hydrogen [10]. Twelve trillion cubic feet ($339,802,159,104 m^3$) per year is the global hydrogen production, most of which is used to synthesize ammonia and methanol. Also, an important part of it is used in refineries for processes such as hydrocracking and hydrotreating for the production of gasoline and petrol. Hydrogen gas is also used in metallurgical industries, glass and ceramics, feed and beverages, electronics, pharmaceuticals, plastics, etc. [10]. Hydrogen can be produced from a variety of sources, including fossil fuels such as natural gas, oil, coal, and renewable sources such as biomass and water. Fig. 1 shows the sources of hydrogen production and its amount, as well as the demand of various industries [7,10].

Initially, coal was the main source of the syngas production. Nevertheless, the investment required for producing a unit of the syngas from coal is almost three times of the investment necessary for the unit from the natural gas [11]. But, now hydrogen is mainly produced from fossil sources and by reforming. Natural gas (mainly methane) is the most important and economical feed for reforming. However, in areas where natural gas is not available, heavier hydrocarbons or coal can also be used [12]. About 76%–77% of the hydrogen produced comes from natural gas and oil, while 19%–20% is produced from coal, and only 3%–4% of renewable sources are used for H_2 production [10]. The average purity required by hydrogen in several industrial applications is: desulfurization with hydrogen (hydrodesulfurization) (purity 90%), hydrocracking (purity 70%–80%), and as gas fuel (purity 60%–54%) [4,13].

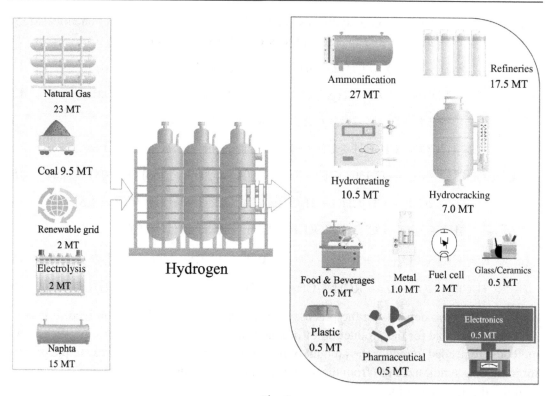

Fig. 1

Sources and amount of hydrogen production and demand of various industries (MT: Million Tons) [7].

In most hydrogen production processes, impurities such as carbon monoxide or carbon dioxide are produced. And, due to the high importance of hydrogen purity in the refining and petrochemical industries mentioned above, the necessity of hydrogen purification processes is more and more evident. Currently, the majority of hydrogen produced is obtained from the syngas [14]. Syngas (hydrogen (H_2) and carbon monoxide (CO) mixture) as an intermediate product plays a pivotal role in the petrochemical industry. In fact, it is a vital gas mixture for a large number of industrial processes including the synthesis of methanol and heavier alcohols, Fisher-Tropsch (FT) process, ammonia and aldehydes, acetylene, dimethyl ether, aromatic and aliphatic hydrocarbons. This mixture is directly used in the production of methanol, ammonia, acetaldehyde, and vinyl acetate. In addition, in gas-to-liquid conversion units, synthesized gas is used as feedstock. The ratio of hydrogen molar to carbon monoxide depends on the type of optimal product and the technology of processing and production of the syngas. The weight ratio of H_2/CO in the synthesized gas is usually between 1 and 3. In order to use the syngas mixture to produce more valuable petrochemical products, this ratio must be reduced [10,14]. Eqs. (1)–(9) determine the weight ratio (H_2/CO) and stoichiometry for the reaction of synthesis gas needed to synthesize different chemicals [15]. The range of the ratio required for most processes is between 0 and 2.

$$2H_2 + 2CO \rightarrow CH_3COOH \qquad \text{(Acetic acid) } (H_2/CO = 1.0) \tag{1}$$

$$CH_3OH + CO \rightarrow CH_3COOH \qquad \text{(Acetic acid) } (H_2/CO = 0.0) \tag{2}$$

$$2H_2 + 2CO \rightarrow CH_2OH \qquad \text{(Methanol) } (H_2/CO = 2.0) \tag{3}$$

$$3H_2 + 2CO \rightarrow CH_2OH - CH_2OH \qquad \text{(Ethylene glycol) } (H_2/CO = 1.5) \tag{4}$$

$$CHCOOCH_3 + CO \rightarrow CH3COOCOCH_3 \qquad \text{(Acetic anhydride) } (H_2/CO = 0.0) \tag{5}$$

$$3H_2 + 2CO \rightarrow CH_2CHO + H_2O \qquad \text{(Acetaldehyde) } (H_2/CO = 1.5) \tag{6}$$

$$4H2 + 2CO \rightarrow CH3CH2OH + H_2O \qquad \text{(Ethanol) } (H_2/CO = 2.0) \tag{7}$$

$$\text{(a) } RCH = CH_2 + CO + H + RCH - CH - CHO^* \qquad \text{(Oxoalcohols) } (H_2/CO = 2.0) \tag{8}$$

$$\text{(b) } RCH_2CH_2CHO + H2 \rightarrow RCH - CH_2CH_2OH^* \qquad \text{(Oxoalcohols) } (H_2/CO = ...) \tag{9}$$

As stated, the ratio depends on the process of gas synthesis and hydrocarbon feed. It is not necessarily suitable for the production of products under optimum conditions. As such, some methods are needed to reduce the amount of hydrogen in the syngas. Fig. 2 is a simple process for producing synthesized gas from the natural gas and separating hydrogen from the syngas, as well as adjusting the ratio.

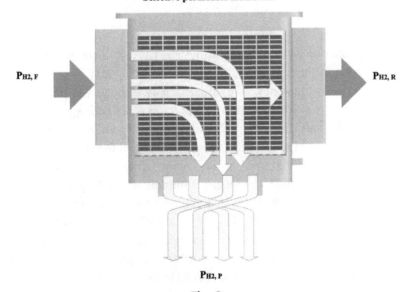

Selective permeable membrane

$P_{H2, F}$

$P_{H2, R}$

$P_{H2, P}$

Fig. 2
Schematic of the membrane separation of the gas feed.

1.3 Hydrogen separation methods

There are various technologies for hydrogen separation, the most important of which include pressure swing adsorption (PSA), cryogenic distillation, and membrane separation [4,16]. These processes are based on different separation rules. Therefore, the characteristics of these processes will be different from each other, which is briefly explained below.

1.3.1 Pressure swing absorption

In this method, purification process is performed using an adsorbent material that selectively acts on a component of a mixture. The adsorption process with pressure fluctuations is based on the principle that the equilibrium value of impurities adsorbed on an adsorbent substrate increases with increasing pressure. Impurities are adsorbed at high pressure and then removed from the solid substrate by reducing the pressure to a lower level (called pressure oscillation and for adsorbent substrate regeneration). Two important determinants of gas adsorption are its volatility and polarity. High-volatile gases such as hydrogen have little interaction with adsorbent and are removed from the substrate with relatively high purity [17]. Adsorption of heavier and polar components is done with a separator. In these substrates, silica, alumina, activated carbon, zeolite, or a combination of these materials is used for suitable functional characteristics [18,19]. When the bed is saturated with impurities, it should be revived or regenerated with reducing pressure and finally cleaning the bed by reverse current to discharge contaminated gases. To supply hydrogen continuously, the PSA system must have several substrate beds, usually 4–16 beds. Each bed in a complete cycle goes through the following steps [17]: (1) adsorbing impurities from gas flow, (2) cocurrent depressurization, (3) countercurrent depressurization, (4) purge at low pressure, and (5) repressurization.

1.3.2 Cryogenic distillation

Cryogenic distillation process is a low-temperature purification process in which separation is based on the difference in boiling temperature (relative volatility coefficient) of the feed components. In the cryogenic distillation process for separation of gases, the gas mixture is first liquidized, which requires high energy consumption [18,20]. In the separation of hydrogen by cryogenic distillation, if the feed contains a large amount of carbon monoxide, a methane washing tower is required to reduce the amount of these gases [4].

1.3.3 Membrane separation

In general, the membrane is a selective surface against the mixed components. The components pass through at different speeds, causing the flow to be separated into two parts: the permeate and the retentate residual. The characteristics of a suitable membrane with commercialization include selectivity, permeability, mechanical strength, and thermal and chemical stability. A suitable membrane must have all the characteristics together in order to be able to meet the industrial conditions for commercialization.

Membrane separation is based on the difference in penetration or diffusion velocity between hydrogen and impurities in passing through the membrane. The gas phase must be dissolved first in the membrane and then spread out in the permeable part of the membrane. For different components, the solubility and penetration speed vary. Also, the driving force for solubility and penetration is the slight pressure difference between the high-pressure side and the low-pressure side of the membrane ($P_{H2,F} > P_{H2,P}$). Fig. 2 shows the overall membrane schematic. In general, the process of separating hydrogen gas from the synthesis gas feed ($P_{H2,F}$) can be said that hydrogen passes through the membrane, and the flow of hydrogen with high purity ($P_{H2,P}$) is produced. Consequently, the ratio of the syngas in the residual flow ($P_{H2,R}$) can be adjusted.

Due to the membrane structure and type of inlet feed gas, there are various mechanisms for transferring gases through the membrane, including the influence of molecular sieve and dissolution-diffusion mechanism, which are mostly used for certain states of the gas-membrane system [21]. It is noteworthy that these transmission mechanisms are different in porous and dense membranes. Additionally, the structural and microstructural conditions of the membrane affect the transmission mechanism. In dense membranes, the mechanism of transfer is considered dissolution-diffusion. Considering that the main membranes used in the separation of gases are polymeric and dense, only the introduction of the dissolution-diffusion model in these membranes is accepted [4]. As mentioned above, membrane contactor systems provide the remarkable advantage of creating the perfect environment for proper contact of the two phases of gas mixture and liquid solvent. This advantage is mainly due to the high surface area per unit volume provided by these systems. This means improved separation efficiency in these systems. Since solvent plays a major role in the CO_2 removal process, further improvement of the solvent means increasing the efficiency of gas feed purification through membrane contactor. Nanomaterials are one of the important levers for further improvement of solvent due to their special characteristics such as promoting heat and mass transfer. Alumina, silica, and carbon nanotubes (CNT) are some examples of nanoparticles used in various fluids to capture CO_2 [22]. Mechanisms have been proposed to improve the efficiency of nanofluids compared with the base fluid, including Grazing phenomenon and creating micro-convection effects [23,24].

1.3.4 Comparison of separation methods

Comparison of advantages and disadvantages of hydrogen separation methods according to the three methods mentioned in Table 1.

The most commonly used process for separating hydrogen from a mixture is the PSA process. The basis of this process is the high adsorption capacity of impurities and hydrogen production with very high purity (up to 99.99%). It should be noted that the PSA process is more expensive

Table 1 Comparison of the advantages and disadvantages of the conventional gas separation methods [25].

Separation methods for hydrogen	Basis	Benefits	Operational constraints
Membrane separation	Difference in capability of gas leakage	Low investment costs; Low energy consumption; Easy operation and control; Low cost of maintenance; Simplicity of capacity increase and increase in number; Membrane Units; Low weight of membrane units	Need to prerefined Feed; Membrane degradation at high temperatures; Inverse relationship of permeability and Selectability
Separation by pressure Swing absorption (PSA)	Surface alternating Difference in capability Adsorption Gases on Adsorbent	High purity of hydrogen to reduce Costs of hydrogen units in streams Downstream; Less use of peripheral systems; Long lifespan due to the selectability (over 20 years)	High energy consumption— High costs
Separation by method Cryogenic distillation	Differences in relative volatility gases in low temperatures	Ability to produce valuable byproducts; Using cold product temperature; Being economical in large scale	Need for prepurification. Feed; Need for insulation; Use of special material. For equipment; High operating cost; Nonapplication for gases with similar volatility

than membrane separation in terms of both initial investment and operating cost. Cryogenic distillation is also a low-temperature separation process, which is a disadvantage of this process. It is also very energetic and does not work to produce hydrogen with very high purity. The disadvantages mentioned for two processes of PSA and cryogenic distillation have led to many research studies to find a more economical and safer process. Thus, the membrane separation has been proposed as a suitable alternative [4,15]. In other studies, these separation processes have been compared, which show the economic advantage of membrane separation technology compared with the other two methods because the investment costs and energy consumption of this process are lower. Elimination of heat exchange equipment and through it, safer form of membrane operation is another advantage of membrane processes than usual processes [26–28]. In addition to comparing the quantitative and qualitative performance of these three methods, their economic comparison has also been done. The membrane process considered in this comparison includes a two-step arrangement with a recycling flow. Although the cost of gas condensation in the membrane process is higher than other processes, due to the low cost of equipment, membrane separation is more economically suitable [29–31]. In addition to lower cost, the selection of membrane methods has a number of other reasons, including the variety of processes (by adjusting the pressures, all the required ratios of hydrogen/carbon monoxide can be produced), simplicity, simple operation, quick setup, and easy installation. But one of the disadvantages of this technology is its product, which is a condensed hydrogen flow. It is removed from the system at low pressure, which needs to be recondensed to apply it [15,32,33].

1.4 A brief overview on the simulation

Before the operational implementation of new ideas for the production and treatment of various gas types (e.g., syngas, natural gas), they can be thoroughly checked using developed simulation software packages to achieve technically accurate operational parameters as well as economically the best process costs [34]. Hydrogen production and purification using the sorption process using ASPEN plus process simulator have been reported [35]. In this process, first, with steam reforming method, biomass feed is converted into gas compounds with the aim of producing hydrogen. Then, the CaO adsorbent is used to remove CO_2 for purification of the gas product. Simulation plays three main roles for process design: (1) explaining the optimal operating conditions for achieving the best reforming reaction according to which suitable reactors can be designed in terms of temperature, pressure, and dimensions; (2) designing absorption and desorption reactors for the product purification; and (3) describing the optimal operating conditions to achieve higher hydrogen yields [35].

Because of its great potential and flexibility, ASPEN Plus can provide a complete reactive environment with all operational parameters for gas sweetening where the absorption of gas impurities is performed with aqueous amine solutions. Coupling mass transfer and reaction along with specific characteristics of the process operation can be accurate enough to be very closely compatible with the executed operations in the pilot plant and industrial scale [36]. This achievement is thanks to the use of a more accurate model of two-film theory in which it considers the acceleration of mass transfer due to a complex system of chemical reactions without the use of simplified concepts of the enhancement factor [36]. Consequently, considering multicomponent mass transfer, the specific properties of electrolyte solutions as well as the absorption column hydrodynamics, this model expresses the operational realities with high precision in the mathematical language. This model provides a great help to understand the process, identify operational parameters to optimize, design, and develop the process [36].

Since energy is a determining parameter in the process economy, achieving operational techniques for gas purification at low temperatures can help to increase gas reserves, so that using these technologies, economic extraction of low-quality gas reserves is provided. To achieve such goals, modeling is the least costly and efficient method. Using energy and exergy analyses, low-temperature processes are modeled for natural gas purification of high CO_2 content [37]. Exergy knowledge helps us to well define these changes for various forms of energies and monitor the optimization of energy consumption. Electrical and mechanical energy can be called gross exergy, while heat energy brings waste on the way to converting to other types of energies. In fact, exergy is a kind of energy modeling based on the first and second law of thermodynamics. The analytical application of exergy helps us to find the amount of energy wasted in the processes that reach equilibrium [38].

Momentum, energy, and mass transfer may be considered in a comprehensive mathematical model. In this model, a contactor module of hollow fiber membrane (HFM) is considered for carbon dioxide absorption from a gas mixture. Latent heat of vaporization and varied absorbent temperatures can result in the amount of water evaporated from the liquid solvent. As a nonwetted mode, the model describes the countercurrent contact of gas with the solvent filled in the membrane pores. In addition, the model covers the axial and radial diffusion in the three main parts of the HFM: (1) inside the HFM; (2) through the membrane skin; (3) and within the shell side of the contactor. The model validation is finally done with polyvinylidene fluoride (PVDF) HFMs. The membranes are fabricated via thermally induced phase separation. They may be used to absorb carbon dioxide from a gas mixture stream. These experimental data are reliable for studying the effect of the inlet liquid absorbent temperature on the membrane performance. The amount of water vapor is a direct function of absorbent inlet temperatures [39]. Since chemical reaction rate between CO_2 and NaOH increases with increasing temperature, the inlet temperature of the aqueous sodium hydroxide solution is directly related to CO_2 elimination from the gas mixture [39]. Likewise, different mathematical models can be developed to describe gas purification (e.g., CO_2 removal from gas mixture). These models should finally be validated with experimental data (laboratory, pilot, industrial) that aim to understand the process and predict the results of variations in the operational parameters of the process on the one hand and the structure of mechanical design of membranes on the other hand on the overall performance of the absorption process. Some of these models describe the chemical absorption of CO_2 from a gas mixture in a membrane [40]. Other models have been developed to simulate the SO_2 absorption in hollow fiber ceramic membrane contactor [41]. Based on resistance-in-series, numerical models may closely describe the effect of wetting on CO_2 sorption in the membrane contactor [42,43]. The structural parameters of the membrane such as pore size, pore size distribution, and effective surface porosity are considered in the models to find the relations between the mass transfer coefficient and the structure of the membrane [44]. Each of these modeling methods is more focused from a special point of view. However, some models have tried to incorporate all the effects affecting the process such as energy balance and momentum transport, as well as partial evaporation of the solvent [39].

In this chapter, the main purpose is to present two various models using ASPEN Plus software from operational viewpoints and a mechanistic model using COMSOL software. In the first model, a generality of operational units necessary for gas production and treatment is proposed. In the second model, the mechanism of absorption of impurities from as-produced syngas is considered in special membranes containing nanofluids. The multiphysical nature of the

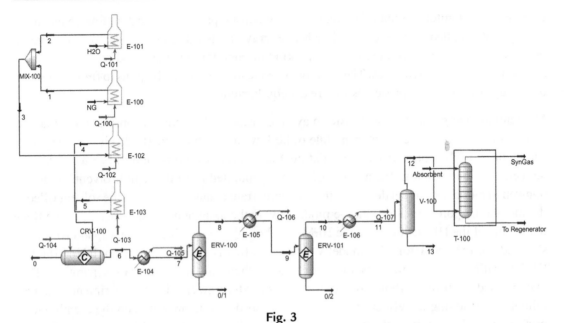

Fig. 3

Designed factory for the production and purification of syngas from the natural gas (NG) [45].

processes is well covered in COMSOL software. As a result, using this software, a comprehensive mathematical model can be developed for the absorption process from the perspectives of momentum, energy, and mass transfer.

2. Design of syngas production and purification: ASPEN model

Fig. 3 shows the simulation for syngas production from steam natural gas (NG).

First of all, the components are: methane (CH_4), carbon dioxide (CO_2), carbon monoxide (CO), water (H_2O), hydrogen (H_2), and nitrogen (N_2). The fluid package is the Peng-Robinson [46].

The two reactions are involved:

(1) first is steam deforming, which is known as conversion reaction (CRV-100) (steam reforming directions).

$$2CH_4 + 3H_2O \rightarrow CO_2 + CO + 7H_2 \tag{10}$$

(2) second reaction will be water gas shift reaction (WGSR) as an equilibrium reaction.

$$CO + H_2O \rightarrow CO_2 + H_2 \tag{11}$$

The temperature to be set for the NG is 38°C and pressure is $40\,\text{kg}\,\text{cm}^{-2}$. The mass flow rate can vary depending on how much production is desired. As such, the fraction of the methane is 0.99; the second component will be nitrogen as 0.05; and the last one is carbon dioxide as 0.01. It is assumed that this combination is the input feed of the NG into the process with the mass flow of $5600\,\text{kgh}^{-1}$. This value can be considered variable based on the design and the production capacity of the process. In the following, heat exchanger is added to produce steam (vapor phase) from liquid phase water. To this end, the temperature is 150°C, and the pressure is $40\,\text{kg}\,\text{cm}^{-2}$ for the heater. Under this designed plan, the second mainstream is water, which is mixed with the NG flow in the phase of vapor at a rate of $1050\,\text{kmol}\,\text{h}^{-1}$. One thing to consider is the ratio of these two streams, which should be carefully selected based on the plant design. Heaters and coolers are used to regulate the temperature of the different streams in the plant. Input, output, energy requirements, pressure difference, and composition in this equipment can be adjusted based on the temperatures required for the downstream processes. In this plant, for example, the temperature required for the conversion reaction (Eq. 10) is 870°C. Having been mixed, the NG and water vapor streams enter the heater (E-103) to achieve this temperature before reforming. Table 2 shows the balance of performance of the participants in Eq. (10) corresponding to the CRV-100 conversion reactor.

Table 3 summarizes the operating conditions of the conversion reactor CRV-100. This reactor is balanced from heat and mass transfer viewpoints.

Table 4 summarizes hydrogen production through reforming (Eq. 10) in the conversion reactor (CRV-100).

Tables 5 and 6 summarize two stages of operation for equilibrium reaction in the first (ERV-100) and second (ERV-101) equilibrium reactors. These two operational stages provide purer hydrogen for later operational stages by eliminating or reducing impurities including CO, CO_2, and H_2O.

Table 6 shows the combination of the feed inflow (Stream 9) and outflow (Stream 10) to the equilibrium reactor ERV-101. In this reactor, with CO and H_2O reaction and H_2 production, the amount of this product is optimally adjusted. The highest possible amount of pure hydrogen in the presence of catalyst is achieved in ERV-101.

Table 2 Mass balancing around the streams of the conversion reactor CRV-100.

Components	Total inflow	Total reaction	Total outflow
Methane	8.641E−002	−8.641E−002	0.0000
CO_2	9.193E−004	4.321E−002	4.413E−002
CO	0.000	4.321E−002	4.321E−002
Hydrogen	0.000	0.3024	0.3024
Nitrogen	4.597E−003	0.0000	4.597E−003
H_2O	0.2917	−0.1296	0.1620

Table 3 Operational condition of the CRV-100 from heat and mass transfer viewpoints [45].

Stream name	5	6	Q-104 (Energy Flow)
Vapor	1.000	1.000	–
Temperature (°C)	870.000	870.000	–
Pressure (kg/cm^2)	34.000	32.000	–
Molar Flow (kgmole/h)	1380.948	2003.131	–
Mass Flow (kg/h)	24,515.856	24,516.089	–
Std Ideal Liq Vol Flow (m^3/h)	36.375	56.448	–
Molar Enthalpy (kJ/kgmole)	−1.660e+005	−8.177e+004	–
Molar Entropy (kJ/kgmole-C)	206.0	169.9	–
Heat Flow (kJ/h)	−2.292e+08	−1.638e+08	6.543e+07

Table 4 The composition of the feed and final product of the reactor CRV-100 (flow basis:molar) [45].

Compositions	Feed stream (5)	Product stream (6)
Flow rate (kgmole/h)	1.240765e+03	8.003e+05
CH_4	0.225	0.000
CO_2	2.397E−003	7.930E−002
CO	0.000	7.765E−002
H_2	0.000	0.544
N_2	1.198E−002	8.261E−003
H_2O	0.760	0.291

Table 5 The first equilibrium reactor ERV-101 to regulate the product stream with H_2 production (flow rate (kgmole h^{-1})).

Compositions	Feed stream (7)	Product stream (8)
CH_4	0.000	0.000
CO_2	7.930E−002	0.128
CO	7.765E−002	2.923E−002
H_2	0.544	0.592
N_2	8.261E−003	8.259E−003
H_2O	0.291	0.243

Table 6 The second equilibrium reactor ERV-101 to regulate the product stream with more H_2 production (flow rate ($kgmole\,h^{-1}$)).

Compositions	Feed stream (9)	Product stream (10)
CH_4	0.000	0.000
CO_2	0.128	0.153
CO	$2.923E-002$	$4.177E-003$
H_2	0.592	0.617
N_2	$8.261E-003$	$8.261E-003$
H_2O	0.243	0.218

The mass flow of the NG is assumed to be $5600\,kg\,h^{-1}$. This value can be considered variable based on the design of the process, as well as the factory production capacity. It is assumed that this combination is the input of the NG into the process. Under this hypothetical plan, the second mainstream is water, which is mixed with the NG flow in the form of vapor at a rate of 1050 kmol per hour ($kmol\,h^{-1}$). One thing to consider is the ratio of these two streams, which must be carefully selected based on the design. Heaters and coolers are used to regulate the temperature of the different streams. Input, output, energy requirements, pressure difference, and composition in this equipment can be adjusted based on the temperatures required for downstream processes. In this plant, for example, the temperature required for the conversion reaction (Eq. 10) is 870°C. As such, having been perfectly mixed, the NG and water vapor streams enter the heater to reach this temperature. After steam reforming in the conversion reactor, for WGSR, it has to react in the higher temperature shift (HTS) and lower temperature shift (LTS).

$$CO + H_2O \rightarrow CO_2 + H_2 \tag{12}$$

This catalytic reaction is started in the HTS at 409.5°C (ERV-100) and continued in the LTS at 239.5°C (ERV-101). The equilibrium reactors (ERV-100 and ERV-101) and the reaction (Eq. 2) are operated typically with the mole fraction. The conversion for the second one is higher compared with the first one.

Now, again putting the cooler to cool down the stream #10 (Fig. 2, the plant flowsheet) before entering to the put separator to separate vapor and liquid. The separator relatively purifies the hydrogen through removing water phase. The product of the separator (V-100) (Table 2) is fed to the absorber (T-100). The main function of the T-100 is to remove carbon dioxide with membrane separation technique. Also, it is possible to remove CO_2 from the stream #12 with cryogenic distillation [47,48]. Table 7 shows the final purified composition of the syngas product.

Table 8 shows the operating conditions of separator V-100 along with three main streams of input feed (11) and output products (12 and 13). The data are summarized after the mass and energy balance.

Table 7 The final regulated syngas stream from the plant (Absorber T-100 (flow basis:molar)) [45].

Compositions	Feed (Stream # 12)	Syngas	To regenerator
CH_4	0.000	0.0000	0.0000
CO_2	0.194	0.0000	0.0004
CO	5.327E−003	0.007	0.0000
H_2	0.787	0.981	0.0000
N_2	1.053E−002	0.011	0.0000
H_2O	2.992E−003	0.002	0.9996

Table 8 Operating conditions of the Separator V-100 of the plant (Fig. 2).

Operational parameter	Feed stream 11	Product stream 12	Retentate Stream 13
Vapor	0.784	1.000	0.000
Temperature (°C)	35.000	35.000	35.000
Pressure (kg/cm^2)	22.000	22.000	22.000
Molar flow (kgmole/h)	2003.1310	1570.976	432.1553
Mass flow (kg/h)	24,516.0445	16,713.7415	7802.3030
Std Ideal Liq Vol Flow (m^3/h)	60.729	52.905	7.824
Molar enthalpy (kJ/kgmole)	−1.225e+005	−7.762e+004	−2.856e+005
Molar entropy (kJ/kgmole-C)	101.2	113.6	56.40
Heat flow (kJ/h)	−2.457e+08	−1.220e+08	−1.234e+08

3. Simulation of capturing gas in nanofluid

3.1 Dissolution-diffusion model

Dissolution-diffusion model is an important accepted and general mechanism for separation of gas through the dense polymer membranes. In this model, gas molecules dissolve on the membrane surface, penetrate into its structure, and then excrete from the other side. Thus, in the dissolution-penetration model, three steps can be considered for passing a component through the membrane [49,50]:

- Adsorption in the upstream face of the membrane.
- Diffusion into the membrane.
- Evaporation or excretion from the downstream part of the membrane.

In this mechanism, the transfer of molecules is due to differences in chemical potential on both sides of the membrane, due mainly to the difference in the concentration.

According to this model, the difference in the diffusion rate of materials causes separation. This difference is due to the difference in solubility of the components within the membrane. As a result, the starting point for mathematical modeling of permeation in the membranes is based on the thermodynamic relationships. Therefore, J_i, which is the transfer flux of component i, is stated with the simple relationship (Eq. 13).

$$J_i = -L_i \times \left(\frac{d\mu_i}{dx}\right) \tag{13}$$

$\frac{d\mu_i}{dx}$ is the chemical potential gradient for the component I, and L_i denotes a proportional coefficient. In order to consider the pressure gradient and concentration, the chemical potential is expressed as Eq. (14).

$$d\mu_i = R \times T \times dln(\gamma_i \times c_i) + v_i \times dp \tag{14}$$

In which c_i is concentration of the species i, γ_i denotes the activity coefficient, v_i is the molar volume of component i, p represents pressure, and T denotes temperature.

In incompressible phases such as solid membranes, the volume does not change with pressure. By taking into account the previous relationship, it is possible to reach a relationship (Eq. 15).

$$\mu^i = \mu_i^0 + R \times T \times ln(\gamma_i \times c_i) + v_i(p - p_i^0) \tag{15}$$

Where the μ_i^0 is the chemical potential of the passing component at the reference pressure p_i^0.

But in gases, the molar volume changes with pressure (Eq. 16), which can be reached by using the ideal gas rules and integration of Eq. (14).

$$\mu_i = \mu_i^0 + R \times T \times ln(\gamma_i \times c_i) + R \times T \times ln\left(\frac{p}{p_i^0}\right) \tag{16}$$

To ensure that the reference chemical potential in the relationships (15) and (16) is the same as the saturated vapor pressure of the component i, p_i^{sat}, Eq. (15) is rewritten as Eq. (17), which is obtained for the incompressible phase in the membrane.

$$\mu_i = \mu_i^0 + R \times T \times ln(\gamma_i \times c_i) + v_i \times (p - p_i^{sat}) \tag{17}$$

And for gases, Eq. (15) is obtained as Eq. (18).

$$\mu_i = \mu_i^0 + R \times T \times ln(\gamma_i \times c_i) + R \times T \times ln\left(\frac{p}{p_i^{sat}}\right) \tag{18}$$

In this model, gaseous components, with pressure on the membrane side (pH), can pass through the membrane as long as the pressure on the permeate side (P_i) is less than the feed pressure. Considering that chemical potentials in the interface of the membrane and the gas stream are the same, using Eqs. (17) and (18), Eq. (19) is achieved.

$$\mu_i^0 + R \times T \times ln(\gamma_{ih} \times c_{ih}) + R \times T \times ln\left(\frac{p}{p_i^{sat}}\right)$$
$$= \mu_i^0 + R \times T \times ln\left(\gamma_{ih(m)} \times c_{ih(m)}\right) + v_i \times (p_h - p_i^{sat}) \tag{19}$$

The index (m) is the interface of the membranes with the gas stream. By rearranging Eq. (19), Eq. (20) is obtained.

$$\gamma_i = \frac{\gamma_{ih}}{\gamma_{ih(m)}} \times \frac{p_h}{p_i^{sat}} \times C_{ih} \times e^{\left(-\frac{v_i \times \left(p_h - p_i^{sat}\right)}{R \times T}\right)} \tag{20}$$

In Eq. (20), the exponential function is close to 1. Therefore, the concentration is expressed with Eq. (21)

$$C_{io(m)} = \frac{\gamma_{ih}}{\gamma_{ih(m)}} \times \frac{p_h}{p_i^{sat}} \times C_{ih} \tag{21}$$

And, considering that the phrase $C_{ih}.P_h$ is the partial pressure in the feed gas P_{ih}, the equation can be easier to write as Eq. (22).

$$C_{ih(m)} = \frac{\gamma_{ih}}{\gamma_{ih(m)}} \times \frac{P_{ih}}{p_i^{sat}} \tag{22}$$

The term $C_{ih(m)} = \frac{\gamma_{ih}}{\gamma_{ih(m)} \times p_i^{sat}}$ is a coefficient of the solubility i.e., K_i^G. Therefore, the concentration of component i in the interface of the membrane with the feed can be described as Eq. (23).

$$C_{ih(m)} = K_i^G \times P_{ih} \tag{23}$$

Similarly, the concentration of the desired component on the permeate side is in the form of Eq. (24).

$$C_{il(m)} = K_i^G \times P_{il} \tag{24}$$

Eq. (25) can also be reached from Eqs. (13) and (14).

$$J_i = -L_i \times \left(\frac{R \times T \times dln(\gamma_i \times c_i)}{dx} + \frac{v_i \times dp}{dx}\right) \tag{25}$$

In this model, it is assumed that the membrane pressure is uniform and the chemical potential within the membrane is expressed as a concentration gradient. As a result, the second part of Eq. (25) can be considered zero. Also, the pressure within the membrane is considered constant. Therefore, it is possible to reach Eq. (26).

$$J_i = -\frac{R \times T \times L_i}{c_i} \times \frac{dc_i}{dx} \tag{26}$$

Eq. (26) is the relationship between Fick law when the term $\frac{R \times T \times L_i}{c_i}$ is substituted with D_i (diffusion coefficient), which results Eq. (27).

$$J_i = -D_i \times \frac{dc_i}{dx} \tag{27}$$

By integration of Eq. (27) to x in membrane lengthwise with thickness l under steady-state conditions, Eq. (28) can be reached.

$$J_i = -D_i \times \frac{(c_{ih} - c_{il})}{l} \tag{28}$$

By combining Eqs. (23) and (24) with the relationship between Fick law, which is established in Eq. (28), Eq. (29) can be reached.

$$J_i = \frac{D_i \times K_i^G (P_{ih} - P_{il})}{l} \tag{29}$$

$D_i \times K_i^G$ is the coefficient of permeability i.e., P_i. Thus, the relationship (Eq. 30) is obtained.

$$J_i = \frac{P_i \times (p_{ih} - p_{il})}{l} \tag{30}$$

The permeability coefficient is a characteristic parameter that is obtained from Eq. (30) with conducting a simple experimental trial on the membrane and also via measuring its thickness (l) [50]. Also, the selectivity coefficient of polymer membrane compared with the two components i and j can be stated as Eq. (31).

$$\alpha_{ij} = \frac{\frac{y_i}{y_j}}{\frac{x_i}{x_j}} \tag{31}$$

The y_i is the mole fraction of species i in the permeate flow stream, and the x_i is the mole fraction in the feed side. Also, the ideal selectivity coefficient is stated as Eq. (32).

$$\alpha_{i/j}^* = \frac{P_i}{P_j} \tag{32}$$

It should be noted that $\alpha_{i/j}^*$ is used to determine membrane selectability [10].

3.1.1 Flow pattern in membrane modeling

After determining and measuring the permeation coefficients of the components, it is necessary to identify probable flow patterns in the membrane modules. The governing equations, which are mainly differential, can model and simulate the membrane modules. According to the entry and exit of the gas streams on both sides of the membrane, the type of membrane (symmetric or asymmetric), and the size of the module, four patterns of perfectly mixing, transverse, aligned, and heterogeneous stream in the membrane module are considered. Then, the modeling could be discussed based on the flow patterns [51,52].

3.1.2 Cross-flow pattern

In the cross-flow pattern, the flow on the feed side is assumed to be a plug flow, and the permeate flow is exiting perpendicular to the membrane surface. In the application of this flow pattern for asymmetric membranes, many empirical and theoretical studies have been

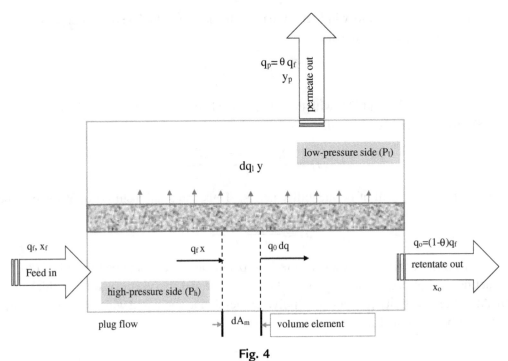

Fig. 4
Parameters of cross-flow pattern from mathematical viewpoint.

presented. The studies prove the applicability of using this model for asymmetric membrane modeling [51,53]. Considering that asymmetric polymer membranes are mainly used in the gas separation, also in order to use hollow fiber modulus, which is feed entry into the shell, cross-flow model is recommended [54]. As such, here the cross-flow pattern is explained in the modeling of the membrane module (Fig. 4).

3.1.3 A model for cross flow

Schematic cross-flow details are shown in Fig. 4. The flow on the high-pressure side of the flow is plug flow. On the low-pressure side, the permeate flow is considered perpendicular to the surface. Some of the important symbols in Fig. 4 include: the speed of the total flow of permeate from the surface (d_q), molar fraction at high pressure side at any point of the membrane (x), molar fraction of the feed (x_f), molar fraction of the retentate flow (x_0), molar fraction at low pressure side (y), molar fraction of the permeate flow current (y_p), feed stream rate (q_f), rate of the retentate-out flow (q_o), rate of permeate flow (q_p), pressure on the high-pressure side (pH), pressure on the low-pressure side (p_l), membrane differentiation area (dA_m), effective membrane thickness (thickness of dense layer of the membrane) (t), membrane permeability relative to the component A (P_A), and membrane permeability relative to the component B (P_B). In this model, it is assumed that there is no pressure drop on both ends of the membrane and the composition of the permeate flow at any point along the membrane is calculated by the relationship between the permeate rate of the feed components at that point. Assumptions of the proposed model are: a two-component mix; membrane permeability independent of the gas flow pressure; negligible pressure loss on both sides across the membrane; no concentration polarization; isotherm operation.

By writing the overall balance on the module, you can reach Eqs. (33) and (34).

$$q_f = q_p + q_0 \tag{33}$$

$$q_f \times y_f = q_p \times y_p + q_0 \times y_0 \tag{34}$$

For each stage of the stage cutting membrane, λ is defined as a fraction of the feed that has been permeate and is obtained from Eq. (35).

$$\theta = \frac{q_p}{q_f} \tag{35}$$

Using Eq. (33) and definition θ, the relationship (Eq. 36) is achieved.

$$y_p = \frac{x_f - x_o(1 - \theta)}{\theta} \tag{36}$$

Local rate of permeation for any point with differentiation area of membrane can be shown with Eq. (37).

$$y \times dq = \frac{P_A}{t} \times [p_h \times x - p_l \times y] \times dA_m \tag{37}$$

And, for the other part, the relationship between the rate of permeate from the membrane surface at any point is in the form of Eq. (38).

$$(1 - y) \times dq = \frac{P_A}{t} \times [p_h \times (1 - x) - p_l \times (1 - y)] \times dA_m \tag{38}$$

By dividing Eq. (37) into Eq. (38), the relationship (Eq. 39) is achieved.

$$\frac{y}{1-y} = \frac{\alpha^* \left[x - \left(\dfrac{P_l}{P_h}\right) \times y \right]}{(1-x) - \left(\dfrac{P_l}{P_h}\right) \times (1-y)} \tag{39}$$

With the transforms and by analytical solution of the above three Eqs. (37)–(39), Eq. (40) is achieved [55,56].

$$\frac{(1-\alpha^*) \times (1-x)}{(1-x_f)} = \left[\frac{U_f - \dfrac{E}{D}}{U - \dfrac{E}{D}} \right]^R \times \left[\frac{U_f - \alpha^* + F}{U - \alpha^* + F} \right]^S \times \left[\frac{U_f - F}{U - F} \right]^T \tag{40}$$

In which, the θ^* is a staged cut calculated from the relationship (Eq. 40), which must be equal to the amount of the membrane stage cut.

Eqs. (41–48) show the parameters of Eq. (40).

$$i = \frac{x}{1-x} \tag{41}$$

By inserting the x_f value in Eq. (41), the value of i_f and by inserting the x_o value in Eq. (41), the i_o value is obtained.

$$U = -D \times i + \left(D^2 \times i^2 + 2E \times i + F^2\right)^{0.5} \tag{42}$$

$$D = 0.5 \times \frac{(1-\alpha^*) \times p_l}{p_h} + \alpha^* \tag{43}$$

$$E = 0.5 \times \alpha^* - D \times F \tag{44}$$

$$F = -0.5 \times \frac{(1-\alpha^*) \times p_l}{p_h} - 1 \tag{45}$$

$$R = \frac{1}{2 \times D - 1} \tag{46}$$

$$S = \frac{\alpha^* \times (D-1) + F}{(2 \times D - 1) - \left(\frac{\alpha^*}{2} - F\right)} \tag{47}$$

$$T = \frac{1}{1 - D - \left(\frac{E}{F}\right)} \tag{48}$$

The required area of the membrane for separation process is obtained by using the transformations of Eqs. (42)–(48) in the form of Eq. (49).

$$A_m = \frac{t \times q_f}{p_h \times P_B} \int_{i_o}^{i_f} \frac{\left(1-\theta^*\right) \times (1-x) \times d \times i}{(f_i - i) \times \left[\frac{1}{1+i} - \left(\frac{p_l}{p_h} \times \frac{1}{1+f_i}\right)\right]} \tag{49}$$

That f_i can be calculated from Eq. (50) [55,56].

$$f_i = (D \times i - F) + \left(D^2 \times i^2 + 2 \times E \times i + F^2\right)^{0.5} \tag{50}$$

Solving this particular integral is not possible analytically due to its complexity. Thus, the numerical methods are the best to solve.

3.1.4 Solution and simulation method

As mentioned earlier, the membrane separation is an important operational unit that can have wide range of applications in industries. Also, due to the industry demand in reducing operating costs and increasing separation efficiency, much research has been done on improving these separation processes. In these processes, after choosing the type of membrane, it is time to configure the operation and also to select the appropriate operating conditions. For these processes, one or more membrane separation steps with different recycling flows can be used. Investigation of the arrangement of the complex as well as the effect of the recycling flow and choosing the appropriate arrangement in this separation process has a significant effect on the selection of optimal mode.

3.1.5 Process operating conditions

The efficiency of the membrane system is not only dependent on the membrane material and its structure, but also depends on the operating conditions of the process. These conditions are the composition of the feed and the phases of the permeate, the inherent selectability of the membrane, the operating pressure, and the flow pattern in the membrane module. To investigate the operational variables on the performance of the process, two experimental and modeling methods can be used. The mathematical models in gas separation processes are expanding day by day. These models provide a suitable and inexpensive way to study and predict the efficiency of the membrane processes and the effect of key parameters on membrane system performance. In the following, numerical simulation of the process is obtained by coding in COMSOL. The results of this model can be compared with available laboratory data. Finally, this model could be generalized for processes with different arrangements. Additionally, the effect of operation variables on these processes may be investigated.

3.1.6 Choosing the membrane material

Choosing the membrane material is the first problem in the design of any membrane process. First, polysulfone or cellulose acetate membranes were used for separation of hydrogen gas. But later, materials such as polyimides have become customary. Many researchers consider this membrane material suitable for the separation of synthesis gas [57]. Among polymer membranes, polyamide has the most attractiveness due to its very good properties, which are: high permeability and selectivity over hydrogen, easy construction, high temperature and chemical resistance, high mechanical strength, and high durability [54].

3.1.7 Mathematical model solving method for membrane

Differentiation model for the cross-flow scheme is used for membrane calculations. The cross-flow differentiation model of input data for solving this model includes pressures of the feed and permeate flow side, input feed, feed composition, and permeability of the membrane components. And, the aim of this model is to obtain the required membrane surface, compressor power, and composition of outputs.

3.1.8 Model solving algorithm

The mathematical model solving algorithm is used to simulate a permeate membrane in Fig. 5.

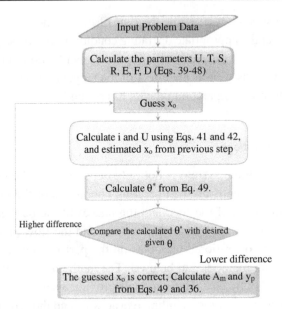

Fig. 5

Algorithm to solve mathematical model to simulate a membrane module.

Guess and check or try and error method is used to solve this model. In this way, a value is assumed for the molar fraction of the retentate flow, x_o. For the convenience of the solution path, the initial value is considered equal to the feed molar fraction, x_f. Each cyclic rotation in the solution algorithm assumes a lesser value for it (x_o) to find the desired value through the algorithm (Fig. 5).

3.1.9 Simulation validation

Validation of the coded program for simulation of the process is compared using the experimental data. The data for separation of hydrogen and carbon monoxide mixtures with polyimide membranes for model validation are specified in Table 4. The experimental results are presented in two parts: results in laboratory scale and semi-industrial scale with the aim of condensing hydrogen in the output stream [58] (Table 9).

According to the analysis of this model, it can be concluded that determining the type of process to reach the optimal point depends on many parameters. The optimal conditions are not straightforward. Therefore, for any type of conditions, all operations must be optimized. For example, in the process of separating hydrogen from synthesized gas and obtaining a product with a concentration of 90% carbon monoxide, a two-step arrangement with the retentate recycling flow of feed with concentrations of 50% hydrogen and 50% carbon monoxide, the

Table 9 Operational conditions considered in experimental trials [58].

	Membrane type	Modulus type	Input feed composition	Temperature (K)	Feed pressure (kPa)	Pressure of the low-pressure Side Membrane (kPa)	Selectability	Permeability (P) (10^{-10} mol s^{-1} m^{-2} Pa^{-1})
separation performances in Scale Laboratory	Membrane Asymmetric Polyimide	(Fibers Hollow) Feed in to Shell	H$_2$: 50.5% CO: 49.5%	373	592	101	52	H: 670 CO: 12.9
separation performances in Scale Pilot Plant	Membrane Asymmetric Polyimide	(Fibers Hollow) Feed in to Shell	H$_2$: 46.5% CO: 53.5%	373	3040	690	62	H: 603 CO: 9.7

system with a pressure ratio equal to 6 of two-sided membrane is more suitable. But, to obtain the product with a concentration of 95% carbon monoxide from this feed, in terms of process cost, the pressure ratio equal to 8 for the two sides of the membrane is more suitable than other pressure drops. Using this feed and comparing the results of the process cost of each various arrangements for hydrogen separation and obtaining the product with a concentration of 90% and 95% carbon monoxide, the use of a single-stage arrangement without recycling flow is more appropriate. Also, the two-staged arrangement with retentate recycling flow can be used to simultaneously obtain permeate stream with hydrogen concentration above 90% and residual flow with carbon monoxide concentration above 90%.

Using feed with concentrations of 75% hydrogen and 25% carbon monoxide for separation of hydrogen and adjusting the ratio of hydrogen and carbon monoxide (1–2), a single-stage arrangement with recycling retentate flow is the best operating conditions. In contrast, for this process, if the simultaneous production of hydrogen flow with high concentration (more than 99%) is also targeted, two-stage arrangements without recycling flow are more suitable [59,60].

3.2 Mathematical modeling of CO_2 absorption in nanofluid

Multiphysics interface in the COMSOL makes the modeling packed beds, monolithic reactors, and other catalytic heterogeneous reactors of the reacting flow in the porous media substantially simplified. In this paradigm, the diffusion, convection, diffusion, and reaction of chemical species for porous media flow are defined without having to set up separate interfaces and couple them. In this modeling, all of the couplings and physics interfaces of heterogeneous catalysis are automatically combined with porous media flow. And, dilute or concentrated chemical species transport through the multiphysics interface. Complementing this multiphysics interface, just similar ones for laminar and turbulent flow, new couplings to other types of flow models can switch or define without having to redefine and set up a new interface for the participating physical phenomena. The type of flow to be modeled as well as the transport of chemical species can be selected from the Settings window without losing any of the defined material properties or reaction kinetics. With this options, it is possible to compare different reactor structures or model flow in both free and porous media in one reactor, even when the two regimes are connected [61–63].

Effective parameters in the nanofluids can be expressed by mathematical equations. At the boundary of gas and nanofluid phases, two phenomena of mass transfer and CO_2 reaction lead to its concentration change, which can be expressed by the Navier-Stokes continuity equation (Fig. 6) [62,63].

$$\frac{\partial C_i}{\partial t} = -(\nabla.C_i V) - (\nabla.J_i) + R_i \tag{51}$$

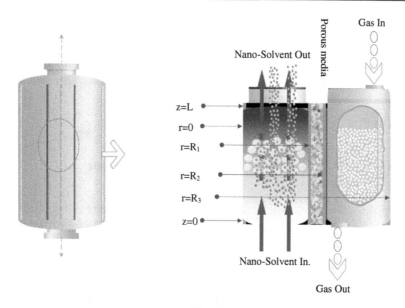

Fig. 6

The schematic of the countercurrent absorption of CO_2 from the syngas flow into the nano-solvent MDEA involving diffusion to the porous environment and absorption in the nano-solvent media along with the modeling parameters [62,63].

Eq. (51) is characterized by concentration parameters (C_i, mol cm^{-3}), time (t, s), velocity (V, ms^{-1}), flux diffusion (J_i, molm^{-2} s^{-1}), and reaction rate of component i (R_i, molm^{-3} s^{-1}). Following Fick's law (J_i), this equation is solved for gas absorption in the nano-solvent with the assumption of steady state.

3.2.1 Balancing

We consider synthetic gas containing a mixture of CO_2, H_2S, H_2, CO, CH_4, etc., which is passing through the shell side countercurrent with the nano-solvent flow. In the opposite direction, the nano-solvent or nanofluid flow from the tube side is flowing. The interface of these two opposite streams is the porous membrane to create a suitable contactor system. Therefore, material exchanges between these three sections can be balanced after reaching a stable state. For this balance, a two-dimensional view of the parameters is assumed as Fig. 6 in which as-synthesized syngas is inserted into the shell side ($z=L$) and on the opposite side, the nano-solvent is flowing through the tube side ($z=0$) in a countercurrent arrangement. Also, to develop the mathematical model, it is assumed that solvent and gas flows have a laminar parabolic velocity profile.

3.2.2 Shell side

At the boundary of gas and nanofluid phases along the lengthwise of the membrane, two phenomena of mass transfer and CO_2 reaction lead to its concentration change, which can be expressed by the continuity equation.

$$\frac{\partial C_i}{\partial t} = -\nabla N_i + R_i \tag{52}$$

Eq. (52) is characterized by the concentration of species i (e.g., CO_2) (C_i), the flux of the species flow (N_i), and the reaction rate (R_i).

The right side of this equation consists of two physical phenomena (mass transfer) and chemical reaction. Mass transfer is characterized by the concept of Fick's law of diffusion, which can determine fluxes of species i (Eq. 53). The rate of the diffusion coefficient of species i and the axial velocity along the length of the module quantify the flux of the species. By combining two Eqs. (52) and (53), three main mechanisms of mass transfer are defined in the form of hybrid Eqs. (52) and (53) for the mass balance of the species (Eq. 54). The diffusion coefficient of species i (D_i) and the axial velocity along the length of the module (V_z) quantify the flux of the species.

$$N_i = -D_i \nabla C_i + C_i V_z \tag{53}$$

The three main parts of mass transfer are defined in the form of combined Eqs. (52) and (53) in order to balance the mass of species (Eq. 54).

$$\frac{\partial C_i}{\partial t} = -\nabla(-D_i \nabla C_i + C_i V_{zi}) + R_i \tag{54}$$

The parameters of Eq. (54) are chemical reaction (R_i), convection (C_i), and concentration-driven diffusion (D_i). If it is assumed that no reaction is done between CO_2 and nano-solvent, CO_2 concentration remains constant (Eq. 55).

$$\left(-D_i \nabla^2 C_i + - \nabla C_i V_{zi}\right) + R_i = 0 \tag{55}$$

In Eq. (55), $C_i V_z$ is the convection. Considering the cylindrical coordinates, Eq. (55) is rewritten as an Eq. (56).

$$D_i \nabla^2 C_i = \nabla . C_i V_z \tag{56}$$

This equation is the mass balance in a stable mode for the transmission of CO_2 in the shell side without reacting which can be obtained in accordance with Happel's free surface model.

According to Fig. 6, the boundary conditions for solving these equations are summarized as follows. Assuming membrane hydrophobicity, only gas molecules exist in its porosity spaces. Therefore, in stable mode, it is only possible to transmit CO_2 inside the membrane with a diffusion mechanism (Eq. 57).

In addition to the overall penetration coefficient, this penetration coefficient depends on the main membrane parameters, i.e., porosity (ε) and tortuosity (τ).

$$D_{CO2-shell}\left[+ \frac{1}{r} \frac{\partial C_{CO2}}{\partial r} + \frac{\partial^2 C_{CO2-shell}}{\partial z^2} \right] = V_{z-shell} \frac{\partial C_{CO2}}{\partial z} \tag{57}$$

The velocity profile ($V_{z-shell}$) can be obtained in accordance with Happel's free surface model (Eq. 58).

$$V_{z-shell} = 2 \times (V)\left[1 - \left(\frac{R_2}{R_3}\right)^2\right] \times \left[\frac{\left(\frac{r}{R_3}\right)^2 - \left(\frac{R_2}{R_3}\right) + 2 \times ln\left(\frac{R_2}{r}\right)}{3 + \left(\frac{R_2}{R_3}\right)^4 - 4 \times \left(\frac{R_2}{R_3}\right)^2 + 4 \times ln\left(\frac{R_2}{R_3}\right)} \right] \tag{58}$$

According to Fig. 6, the boundary conditions for solving these equations are summarized (Eqs. 59–61).

$$z = L, C_{i-shell} = C_{initial} \ (i = CO_2) \tag{59}$$

$$r = R_2, C_{i-shell} = C_{i-membrane} \tag{60}$$

$$r = R_3, \partial C_{i-shell}/\partial r = 0 \ (\text{symmetry}) \tag{61}$$

In the cylindrical coordinates intended for the model, boundary conditions are defined in the positions of the shell side (Fig. 6):

For mass transfer:

(1) At $z=0$, convective flux is considered.
(2) At $z=L$, $C_{CO2}=C_0$
(3) At $r=R_2$, $C_{CO2}=C_{membrane}$
(4) $r=R_3$, Insulated

For momentum transfer:

(1) At $z=0$, Outlet: pressure, no viscous stress, $p=0$
(2) At $z=L$, inlet velocity, $V=V_{0,shell}$
(3) At $r=R_2$, No slip, wall
(4) $r=R_3$, No slip, wall

3.2.3 Membrane

Assuming hydrophobic membrane, only gas molecules exist in its porosity spaces. Thus, in steady state, it is only possible to transmit CO_2 inside the membrane with a diffusion mechanism (Eq. 62).

$$D_{CO2-membrane}\left[\frac{\partial^2 C_{CO2-m}}{\partial r^2} + \frac{1}{r}\frac{\partial C_{CO2-m}}{\partial r} + \frac{\partial^2 C_{CO2-m}}{\partial z^2}\right] = 0 \tag{62}$$

In addition to the overall penetration coefficient (D_i), this diffusion coefficient ($D_{CO2-membrane}$) depends on the main membrane parameters, i.e., porosity (ε) and tortuosity (τ) (Eq. 63).

$$D_{CO2-membrane} = \frac{D_i \times \varepsilon}{\tau} \tag{63}$$

The boundary conditions for solving the differential equation of mass balance in the membrane are Eqs. (64) and (65).

$$r = R_1, C_{CO2-m} = C_{CO2-tube}/m_{CO2} \tag{64}$$

$$r = R_2, C_{CO2-m} = C_{CO2-shell} \tag{65}$$

In these boundary conditions, the solubility of the species (m_{CO2}) in nano-solvent (e.g., monoethanol amine (MEA)-NS) is effective.

Diffusivity of CO_2 in the membrane pores is obtained as Eq. (66) [64]:

$$D_{CO2,m} = \frac{\varepsilon D_{CO2,s}}{\tau} \tag{66}$$

where tortuosity factor (τ) is calculated using the fiber porosity (ε) (Eq. 67) [65]:

$$\tau = \frac{(2-\varepsilon)^2}{\varepsilon} \tag{67}$$

In the cylindrical coordinates intended for the model, boundary conditions are defined in the positions of the membrane (Fig. 7):

For mass transfer:

(1) At $z=0$, Insulated.
(2) At $z=L$, Insulated.
(3) At $r=R_1$, $C_{CO2} = C_{tube}/m$.
(4) $r=R_2$, $C_{CO2}=C_{shell}$.

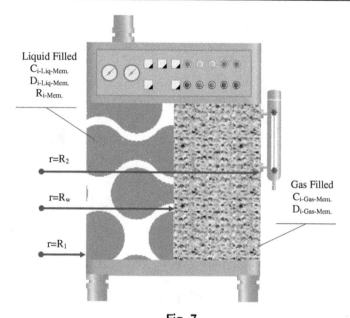

Fig. 7
Schematic of nano-solvent filled membrane and gas-filled membrane [61,62].

3.2.4 Tube side

In terms of mass balancing, the important difference in this section is compared with the previous two parts of transfer with chemical reaction with nano-solvent. Hence, this mass transfer equation will consist of all three cases of diffusion, convection, and reaction (Eq. 68).

$$D_{CO2-t}\left[\frac{\partial^2 C_{CO2-t}}{\partial r^2} + \frac{1}{r}\frac{\partial C_{CO2-t}}{\partial r} + \frac{\partial^2 C_{CO2-t}}{\partial z^2}\right] = V_{z-t}\frac{\partial C_{CO2-t}}{\partial z} \tag{68}$$

In this tube section, according to the Newtonian laminar flow assumption, Eq. (69) indicates the velocity distribution.

$$V_{z-t} = 2(V)\left[1 - \left(\frac{r}{R_1}\right)^2\right] \tag{69}$$

Eq. (69) is solved by using the boundary conditions belonging to the tube side (Eqs. 68–73).

$$z = 0, C_{CO2-t} = 0, C_{MEA-tube} = C_{initial}\ (i = \text{all species}) \tag{70}$$

$$r = 0, \partial C_{CO2-t}/\partial r = 0\ (\text{symmetry}) \tag{71}$$

$$\text{r} = R_1, C_{CO2-t} = C_{CO2-m} \times m_{CO2} \tag{72}$$

$$\partial C_{x-tube}/\partial r = 0\ (x = \text{all other species than CO}_2) \tag{73}$$

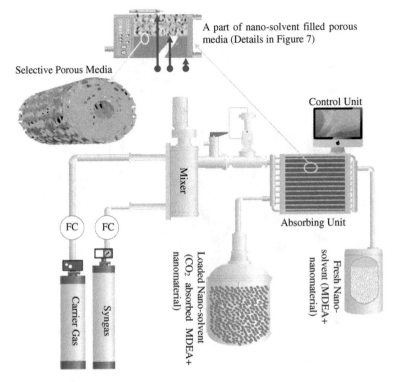

Fig. 8
Schematic of CO_2 gas uptake in nano-solvent containing carbon nanotube nanoparticles (CNT) in methyldiethanolamine (MDEA) [62,63].

In this way, with this modeling, in the general schematic of CO_2 absorption with nano-solvent (Fig. 8), the results can be summarized.

In the cylindrical coordinates intended for the model, boundary conditions are defined in the positions of the tube:

For mass transfer:

(1) At $z=0$, $C_{CO2}=0$, $C_M=C_0$, $q=0$.
(2) At $z=L$, Convective flux.
(3) At $r=0$, Axial symmetry
(4) $r=R_1$, $C_{CO2}=C_{membrane} \times m$, insulated

3.2.5 Parameters for solving differential equations to simulate

The membrane specification may vary based on the defined operation conditions. The assumptions below are considered for the modeling:

(1) The fluid flow is steady state and isothermal.
(2) The profile of gas velocity is completely developed.
(3) The gas has ideal behavior.
(4) The liquid phase flow is laminar.
(5) At all operating conditions, the fiber membrane pores are only filled with the gas mixture.

The three main pillars of differential equations governing the process are as follows:

(1) The architecture of the modular porous environment for CO_2 gas transmission may contain the parameters such as: fiber length (210 mm), number of membranes (11), inner diameter (0.42 mm), outer diameter (1.1 mm), membrane thickness (0.34 mm), membrane porosity (0.4585), module contact area (0.003 m^2), module inner diameter (0.8 cm), and module outer diameter (1.2 cm)

(2) The heart of the reactor for CO_2 absorption with these specifications: membrane inner diameter (d_1) (4.2×10^{-4} m), membrane outer diameter (d_2) (11×10–4 m), module inner diameter (d_3) (0.008 m), membrane thickness (δ) (3.4×10^{-4} m), porosity (ε) (0.4585), tortuosity (η) (5.14), $D_{CO2,shell}$ (1.39×10^{-5} m^2/s), $D_{CO2,membrane}$ ($D_{CO2,shell}(\varepsilon/\eta)$ m^2/s), $D_{CO2,tube}$ (1.45×10^{-9} m^2/s), Henry's law constant (0.891 m), nanoparticle true density (2200 kg/m^3), nanoparticle average size (8 nm), and nanoparticle morphology (tubular).

(3) Operating conditions for the absorption of CO_2 gas in the nano-solvent through porous membranes may depend on some various parameters: temperature (298 K), base fluid (MDEA), nanomaterial (carbon nanotube—CNT), inlet CO_2 concentration (20 vol%), nanoparticle concentration (0.05–0.20 wt%), gas-phase flow rate (10–400 mL/min), and solvent flow rate (10–40 mL/min).

Dispersion of CNT NPs will enhance the mass transfer rate in the system due to the synergistic effects. To consider Brownian and Grazing effects into developed mathematical model and simulation, several possible mechanisms were proposed in the literature for gas absorption in NFs. The diffusion coefficient for the MDEA-based nanofluid can be expressed as Eq. (74) [66,67].

$$D_{nf} = D_{bf} \times \left(1 + m_1 \times Re^{m2} \times Sc^{m3} \times \varphi^{m4}\right) \tag{74}$$

Eq. (75) is the modified diffusion coefficient [66,67].

$$D_{nf} = D_{bf} \times \left(1 + 640 \times Re^{1.7} \times Sc^{\frac{1}{3}} \times \varphi\right) \tag{75}$$

The ϕ is NP volume fraction in the liquid solvent. The dimensionless numbers of Sc (Schmidt) (Eq. 76) and Re (Reynolds) (Eq. 77) are also estimated [68].

$$Re = \sqrt{\frac{18 \times K \times T \times \rho^2}{\pi \times d_p \times \rho_p \times \mu}} \qquad (76)$$

$$Sc = \frac{\mu}{\rho D} \qquad (77)$$

where K is the Boltzmann constants, d_p denotes NP diameter (8nm), and D_{bf} represents carbon dioxide diffusion coefficient (1.45×10^{-9} m²/s).

To introduce the Grazing effect in the model, it is assumed that the MDEA-based solvent has two clear-cut phases: solid and liquid. As such, the mass transfer equations are accordingly derived and used in the model. In the membrane modulus, the continuity equation is derived as Eq. (78) for CO_2 flow in the solid phase [63,69,70].

$$\varphi \times \rho_p \times V_z \times \frac{\partial q}{\partial z} = k_p \times \alpha_p \times (C_{CO2,t} - C_s) \qquad (78)$$

where C_S denotes CO_2 concentration at the solid-liquid interface, and q refers to the amount of CO_2 adsorbed by the CNT, which is determined with Langmuir isotherm (Eq. 79). α_p denotes the specific surface area of the NPs. k_p referes to the mass-transfer coefficient between solid CNT and MDEA solvent (Eq. 80) [70].

$$q = q_m \frac{k_d \times C_s}{1 + k_d \times C_s} \qquad (79)$$

where q_m is the maximum adsorption (29.45mol/kg) on the CNT, and Langmuir constant is k_d (0.00049 m³/mol) [70].

$$Sh = \frac{k_p d_p}{D_{CO2,t}} = 2 \qquad (80)$$

This modeling, along with experimental data, obtains the CO_2 absorption rate in the nano-solvent (Eq. 81) [71] (Fig. 9).

MDEA:
2,2'-(methylazanediyl)bis(ethanol)
Methyl diethanolamine ($CH_3N(C_2H_4OH)_2$)
Chemical Formula: $C_5H_{13}NO_2$
Exact Mass: 119.09
Molecular Weight: 119.16
m/z: 119.09 (100.0%), 120.10 (5.4%)
Elemental Analysis: C, 50.40; H, 11.00; N, 11.75; O, 26.85

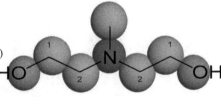

Fig. 9
Specifications of the most famous absorbent solvent methyl diethanolamine (MDEA, CH_3N $(C_2H_4OH)_2$).

$$r_{CO2-MDEA} = -8.741 10^{12} \times \exp^{\frac{-8625}{T}} C_{CO2} \times C_{MDEA} \qquad (81)$$

The effective density of the nanofluid (NF) is marked with Eq. (82) [70].

$$\rho_L^{nf} = \varphi \times \rho_s + (1 - \varphi) \times \rho_L^{bf} \qquad (82)$$

The boundary conditions for the mass and momentum transfer equation are previously determined in the tube, membrane, and shell subdomains of the membrane module.

3.2.6 Numerical analysis

To solve the governing equations, COMSOL Multiphysics software (Proprietary EULA, COMSOL Inc. Sweden) can be used. The modulus of the software is based on the finite element method. The physics (Fig. 6), materials, e.g., MDEA-based nano-solvent and gas phase, and boundary conditions may be easily defined in the software. Then, an adaptive meshing and error control may be applied to minimize the errors of the numerical computation. The suitable solver, e.g., PARDISO can be selected as one of the effective solvers considering the conditions of the membrane systems. By carefully selecting boundary conditions, solver, and proportional meshing, the time of the numerical solution of the problem is significantly reduced [72,73]. These equations and their numerical solution can obtain a robust and consistent phenomenal model to describe the dispersion effects of the CNT in the MDEA solvents for the performance of the HFM module in terms of CO_2 elimination. In these equations, Brownian motion and Grazing effect mechanisms are considered as the predominant factor for increasing the mass transfer of CO_2 to nano-solvent (MDEA-NF) [23,24].

Fig. 10 shows the sites of accumulated CO_2 gas adsorbed on the CNT nanoparticles in the nano-solvent (MDEA-NF). The effects of Brownian motion and Grazing effect mechanisms are well seen in this analysis so that nanoparticles have high efficiency in transferring the gas into the MDEA solvent medium. The bolder (red) dots show the presence of adsorbed CO_2 gas on the CNT nanoparticles.

3.2.7 Model validation

In terms of CO_2 absorption, experimental data of the CO_2 capture in the contactor module for various solvents containing nano-free MDEA fluid and MDEA-based nanofluids confirm the results of the mathematical model. For both membrane contactors, empirical data and simulation results are in good agreement with the presence and absence of the CNT. Elevating the flow rate of MDEA-based nano-solvent increases the removal of CO_2. The flow rate of the nanofluid has a significant effect on the absorption rate. Without nanomaterials, increasing the fluid speed from 10 to 40mL/min results in an increase in carbon dioxide uptake from 38.46% to 46.17%, from 45.71% to 51.12%, and from 53.82% to 57.56% for concentrations of 5%, 10%,

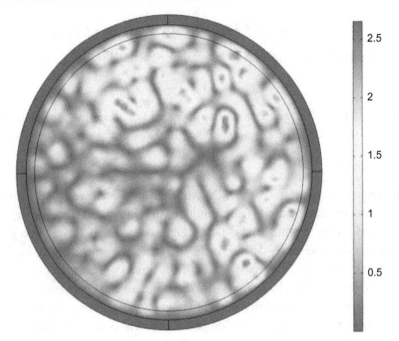

Fig. 10
The adsorption of CO_2 gas in the membrane fiber contains nano-solvent (MDEA-NF). The more bold *red* (dark gray in print versions) dots adsorb CO_2 on the CNT dispersed in the nano-solvent (MDEA-NF).

and 20% MDEA, respectively [74]. By increasing the flow rate of the MDEA-NF (containing 0.5wt% CNT) from 10 to 40mL/min, the CO_2 capture increases up to about 10%. However, the effect of the CNT on the performance of MDEA-NF depends on the operating conditions so that in lower values of the MDEA-based solvent, the effect of the nanomaterials increases.

3.2.8 CO_2 absorption at various gas flow

Simple scalability is one of the salient superiorities of the membrane technology since it can be manufactured depending on the process of capturing in the appropriate dimensions. Increasing the gas flow rate means reducing the mass transfer resistance, resulting in an increase in CO_2 flux and a reduction in the gas residence time in the contactor. In this way, increasing the gas velocity leads to a decrease in CO_2 absorption efficiency. The distribution of diffusion and convection of CO_2 flux with three gas mixture flow rates of 5, 10, and 15mL/min indicates that the convective contribution of the flux through the membrane contactor is significantly considerable compared with the axial diffusion flux. Due to the predominance of the velocity, convective flux is the maximum in the center side of the shell with the maximum gas flow rate. At this elevated gas flow rate, the convective flux may be approximately seven times more than the diffusive one. However, due mainly to the decreased driving force in the z direction, both diffusive and convective fluxes are reduced along the fiber canals [75].

4. Conclusion

Mathematical modeling can help to identify and regulate parameters with the aim of optimizing and developing operating conditions. This chapter deals with this issue to more clearly explore the effect of enhanced mass transfer through nanomaterials on the performance of the absorbing fluid. A synthesis gas (CO and H_2 mixture) as one of the main products of a gasification step from different sources such as NG contains impurities such as CO_2 and H_2S that must be removed in order to prevent, for example, Fischer-Tropsch catalyst poisoning. Therefore, for removal of CO_2, H_2S, etc., and/or adjustment of H_2/CO ratio in the syngas, a multistep purification is required to ensure a deep removal of trace contaminants remaining before feeding into the downstream processing sections. Although absorption by aqueous amine solutions is the most mature and industrially developed technology for capture of CO_2, H_2S, etc., potential exothermic amine degradation in the presence of O_2, CO_2, NO_x, SO_2 leads to solvent loss, equipment corrosion, and generation of volatile degradation compounds. It is here that nanomaterials can help in the promotion of these solvents. In this chapter, it is shown that modeling (COMSOL Multiphysics version 6.0, COMSOL Inc., Stockholm, Sweden) can help to understand the mechanism of the process. By focusing on the finer parts where the chemical processes taking place in different units such as membrane treatment unit, the sorption rate of various gas compounds (e.g., CO_2, CO, H_2S, etc.) can be investigated in the different nano-solvents. More specifically, the proper distribution of nanoparticles in solvent (e.g., carbon nanotube nanoparticles (CNT) in methyldiethanolamine (MDEA)) has been studied. Suitable operating conditions for appropriate dispersion of intrasolvent particles for proper CO_2 capture and determination of the amount of gas absorbed under the operational parameters are indicated. The mathematical model and its numerical solution algorithm are precisely identified. With this paradigm, we can define the suitable configuration and layout of membrane sorption modules and choose the most suitable one. For example, for a single-stage membrane process unit with feed and product conditions, the estimation of process parameters is easily achieved. However, for two-stage processes, optimal conditions depend on different parameters, which will change with each effective parameters. To design a membrane separation process, processes with different arrangements as well as different operational parameters must be compared, and the desired process can be selected. According to the analysis of this model, it can be concluded that determining the type of process to reach the optimal point depends on many parameters. The optimal conditions are not straightforward. Therefore, for any type of conditions, all operations must be optimized. Using feed, for example, with concentrations of 75% hydrogen and 25% carbon monoxide, a single-stage arrangement with recycling retentate flow is the best operating conditions. In contrast, for this process, if the simultaneous production of hydrogen flow with high concentration (greater than 99%) is also targeted, two-stage arrangements without recycling flow are more suitable.

Acknowledgment

This research has been supported by Institute of Science and High Technology and Environmental Sciences, Graduate University of Advanced Technology (Kerman-Iran) under grant number of 7/S/00/2705.

References

[1] Z. Moravvej, M.A. Makarem, M.R. Rahimpour, The fourth generation of biofuel, in: Second and Third Generation of Feedstocks, Elsevier, 2019, pp. 557–597.

[2] M.A. Makarem, et al., Biofuel production from microalgae and process enhancement by metabolic engineering and ultrasound, in: Advances in Bioenergy and Microfluidic Applications, Elsevier, 2021, pp. 209–230.

[3] A. Hassanzadeh, A.B. Vakylabad, Fuel cells based on biomass, in: Advances in Bioenergy and Microfluidic Applications, Elsevier, 2021, pp. 275–301.

[4] S. Adhikari, S. Fernando, Hydrogen membrane separation techniques, Ind. Eng. Chem. Res. 45 (3) (2006) 875–881.

[5] M. Elsherif, Z.A. Manan, M.Z. Kamsah, State-of-the-art of hydrogen management in refinery and industrial process plants, J. Nat. Gas Sci. Eng. 24 (2015) 346–356.

[6] M. Momirlan, T. Veziroglu, Current status of hydrogen energy, Renew. Sustain. Energy Rev. 6 (1–2) (2002) 141–179.

[7] A. Bakenne, W. Nuttall, N. Kazantzis, Sankey-diagram-based insights into the hydrogen economy of today, Int. J. Hydrogen Energy 41 (19) (2016) 7744–7753.

[8] L. Barreto, A. Makihira, K. Riahi, The hydrogen economy in the 21st century: a sustainable development scenario, Int. J. Hydrogen Energy 28 (3) (2003) 267–284.

[9] J.-P. Lange, P. Tijm, Processes for converting methane to liquid fuels: economic screening through energy management, Chem. Eng. Sci. 51 (10) (1996) 2379–2387.

[10] K. Liu, C. Song, V. Subramani, Hydrogen and Syngas Production and Purification Technologies, John Wiley & Sons, 2010.

[11] J.M. Cormier, I. Rusu, Syngas production via methane steam reforming with oxygen: plasma reactors versus chemical reactors, J. Phys. D Appl. Phys. 34 (18) (2001) 2798.

[12] Z. Chen, Y. Yan, S.S. Elnashaie, Modeling and optimization of a novel membrane reformer for higher hydrocarbons, AICHE J. 49 (5) (2003) 1250–1265.

[13] M. Azizi, S.A. Mousavi, CO2/H2 separation using a highly permeable polyurethane membrane: molecular dynamics simulation, J. Mol. Struct. 1100 (2015) 401–414.

[14] J. Gholinezhad, A. Chapoy, B. Tohidi, Separation and capture of carbon dioxide from CO2/H2 syngas mixture using semi-clathrate hydrates, Chem. Eng. Res. Des. 89 (9) (2011) 1747–1751.

[15] S. DiMartino, et al., Hydrogen/carbon monoxide separation with cellulose acetate membranes, Gas Sep. Purif. 2 (3) (1988) 120–125.

[16] L. Shao, et al., Polymeric membranes for the hydrogen economy: contemporary approaches and prospects for the future, J. Membr. Sci. 327 (1–2) (2009) 18–31.

[17] L.S. McLeod, Hydrogen Permeation Through Microfabricated Palladium-Silver Alloy Membranes, Georgia Institute of Technology, 2009.

[18] S. Sircar, Pressure swing adsorption, Ind. Eng. Chem. Res. 41 (6) (2002) 1389–1392.

[19] S. Sircar, T. Golden, Purification of hydrogen by pressure swing adsorption, Sep. Sci. Technol. 35 (5) (2000) 667–687.

[20] A. Hinchliffe, K. Porter, A comparison of membrane separation and distillation, Chem. Eng. Res. Des. 78 (2) (2000) 255–268.

[21] J.G. Wijmans, R.W. Baker, The solution-diffusion model: a unified approach to membrane permeation, in: Materials Science of Membranes for Gas and Vapor Separation, 1, 2006, pp. 159–189.

[22] Y. Cao, et al., Mathematical modeling and numerical simulation of CO2 capture using MDEA-based nanofluids in nanostructure membranes, Process Saf. Environ. Prot. 148 (2021) 1377–1385.

[23] A.B. Vakylabad, Mass transfer mechanisms in nanofluids, in: Nanofluids and Mass Transfer, 2022, pp. 97–113.

[24] A.B. Vakylabad, Mass transfer enhancement in solar stills by nanofluids, in: Nanofluids and Mass Transfer, Elsevier, 2022, pp. 431–447.

[25] O.C. David, Membrane Technologies for Hydrogen and Carbon Monoxide Recovery from Residual Gas Streams, 2012.

[26] F. Lipnizki, R.W. Field, P.-K. Ten, Pervaporation-based hybrid process: a review of process design, applications and economics, J. Membr. Sci. 153 (2) (1999) 183–210.

[27] P.N. Dyer, et al., Ion transport membrane technology for oxygen separation and syngas production, Solid State Ion. 134 (1–2) (2000) 21–33.

[28] N.W. Ockwig, T.M. Nenoff, Membranes for hydrogen separation, Chem. Rev. 107 (10) (2007) 4078–4110.

[29] J.J. Alves, G.P. Towler, Analysis of refinery hydrogen distribution systems, Ind. Eng. Chem. Res. 41 (23) (2002) 5759–5769.

[30] A. Saxena, et al., Membrane-based techniques for the separation and purification of proteins: an overview, Adv. Colloid Interface Sci. 145 (1–2) (2009) 1–22.

[31] V. Hessel, et al., Novel process windows for enabling, accelerating, and uplifting flow chemistry, ChemSusChem 6 (5) (2013) 746–789.

[32] G. Lu, et al., Inorganic membranes for hydrogen production and purification: a critical review and perspective, J. Colloid Interface Sci. 314 (2) (2007) 589–603.

[33] N.N. Li, et al., Advanced Membrane Technology and Applications, John Wiley & Sons, 2011.

[34] M. Sanchez, et al., Aspen plus model of an alkaline electrolysis system for hydrogen production, Int. J. Hydrogen Energy 45 (7) (2020) 3916–3929.

[35] C.M. Kinoshita, S.Q. Turn, Production of hydrogen from bio-oil using CaO as a CO2 sorbent, Int. J. Hydrogen Energy 28 (10) (2003) 1065–1071.

[36] L. Kucka, et al., On the modelling and simulation of sour gas absorption by aqueous amine solutions, Chem. Eng. Sci. 58 (16) (2003) 3571–3578.

[37] M. Baccanelli, et al., Low temperature techniques for natural gas purification and LNG production: an energy and exergy analysis, Appl. Energy 180 (2016) 546–559.

[38] I. Dincer, The role of exergy in energy policy making, Energy Policy 30 (2) (2002) 137–149.

[39] N. Ghasem, M. Al-Marzouqi, N.A. Rahim, Modeling of CO2 absorption in a membrane contactor considering solvent evaporation, Sep. Purif. Technol. 110 (2013) 1–10.

[40] M. Al-Marzouqi, et al., Modeling of chemical absorption of CO2 in membrane contactors, Sep. Purif. Technol. 62 (3) (2008) 499–506.

[41] P. Luis, A. Garea, A. Irabien, Modelling of a hollow fibre ceramic contactor for SO2 absorption, Sep. Purif. Technol. 72 (2) (2010) 174–179.

[42] S. Khaisri, et al., A mathematical model for gas absorption membrane contactors that studies the effect of partially wetted membranes, J. Membr. Sci. 347 (1–2) (2010) 228–239.

[43] S. Atchariyawut, R. Jiraratananon, R. Wang, Mass transfer study and modeling of gas-liquid membrane contacting process by multistage cascade model for CO2 absorption, Sep. Purif. Technol. 63 (1) (2008) 15–22.

[44] K. Li, J. Kong, X. Tan, Design of hollow fibre membrane modules for soluble gas removal, Chem. Eng. Sci. 55 (23) (2000) 5579–5588.

[45] https://www.aspentech.com/en/products/engineering/aspen-plus.

[46] R. Stryjek, J. Vera, PRSV: an improved Peng—Robinson equation of state for pure compounds and mixtures, Can. J. Chem. Eng. 64 (2) (1986) 323–333.

[47] T.E. Rufford, et al., The removal of CO2 and N2 from natural gas: a review of conventional and emerging process technologies, J. Petrol. Sci. Eng. 94 (2012) 123–154.

[48] L.P.R. Pala, et al., Steam gasification of biomass with subsequent syngas adjustment using shift reaction for syngas production: an Aspen Plus model, Renew. Energy 101 (2017) 484–492.

[49] P. Pandey, R. Chauhan, Membranes for gas separation, Prog. Polym. Sci. 26 (6) (2001) 853–893.

[50] J.G. Wijmans, R.W. Baker, The solution-diffusion model: a review, J. Membr. Sci. 107 (1–2) (1995) 1–21.

[51] C.Y. Pan, H. Habgood, Gas separation by permeation part I. Calculation methods and parametric analysis, Can. J. Chem. Eng. 56 (2) (1978) 197–209.

[52] L.K. Wang, Y.-T. Hung, N.K. Shammas, Advanced Physicochemical Treatment Technologies, vol. 5, Springer, 2007.

[53] C.Y. Pan, H. Habgood, Gas separation by permeation part II: effect of permeate pressure drop and choice of permeate pressure, Can. J. Chem. Eng. 56 (2) (1978) 210–217.

[54] R.W. Baker, Future directions of membrane gas separation technology, Ind. Eng. Chem. Res. 41 (6) (2002) 1393–1411.

[55] C.J. Geankoplis, A.A. Hersel, D.H. Lepek, Transport Processes and Separation Process Principles, vol. 4, Prentice Hall, Boston, MA, USA, 2018.

[56] S. Weller, W.A. Steiner, Separation of gases by fractional permeation through membranes, J. Appl. Phys. 21 (4) (1950) 279–283.

[57] R.W. Baker, Membrane Technology and Applications, John Wiley & Sons, 2012.

[58] K. Haraya, et al., Performance of gas separator with high-flux polyimide hollow fiber membrane, Sep. Sci. Technol. 23 (4–5) (1988) 305–319.

[59] G. Piechota, Multi-step biogas quality improving by adsorptive packed column system as application to biomethane upgrading, J. Environ. Chem. Eng. 9 (5) (2021) 105944.

[60] W. Ho, K. Sirkar, Membrane Handbook, Springer Science & Business Media, 2012.

[61] N. Hajilary, M. Rezakazemi, CFD modeling of CO_2 capture by water-based nanofluids using hollow fiber membrane contactor, Int. J. Greenhouse Gas Control 77 (2018) 88–95.

[62] Y. Cao, et al., Intensification of CO_2 absorption using MDEA-based nanofluid in a hollow fibre membrane contactor, Sci. Rep. 11 (1) (2021) 1–12.

[63] A. Golkhar, P. Keshavarz, D. Mowla, Investigation of CO_2 removal by silica and CNT nanofluids in microporous hollow fiber membrane contactors, J. Membr. Sci. 433 (2013) 17–24.

[64] R. Faiz, M. Al-Marzouqi, Mathematical modeling for the simultaneous absorption of CO_2 and H_2S using MEA in hollow fiber membrane contactors, J. Membr. Sci. 342 (1–2) (2009) 269–278.

[65] S. Srisurichan, R. Jiraratananon, A. Fane, Mass transfer mechanisms and transport resistances in direct contact membrane distillation process, J. Membr. Sci. 277 (1–2) (2006) 186–194.

[66] E. Nagy, T. Feczkó, B. Koroknai, Enhancement of oxygen mass transfer rate in the presence of nanosized particles, Chem. Eng. Sci. 62 (24) (2007) 7391–7398.

[67] A. Bahmanyar, et al., Mass transfer from nanofluid drops in a pulsed liquid-liquid extraction column, Chem. Eng. Res. Des. 92 (11) (2014) 2313–2323.

[68] R. Prasher, P. Bhattacharya, P.E. Phelan, Thermal conductivity of nanoscale colloidal solutions (nanofluids), Phys. Rev. Lett. 94 (2) (2005), 025901.

[69] P.Y. Apel, et al., Structure of polycarbonate track-etch membranes: origin of the "paradoxical" pore shape, J. Membr. Sci. 282 (1–2) (2006) 393–400.

[70] L. Sumin, et al., Experimental and theoretical studies of CO_2 absorption enhancement by nano-Al_2O_3 and carbon nanotube particles, Chin. J. Chem. Eng. 21 (9) (2013) 983–990.

[71] N. Haimour, A. Bidarian, O.C. Sandall, Kinetics of the reaction between carbon dioxide and methyldiethanolamine, Chem. Eng. Sci. 42 (6) (1987) 1393–1398.

[72] M. Pishnamazi, et al., Molecular investigation into the effect of carbon nanotubes interaction with CO_2 in molecular separation using microporous polymeric membranes, Sci. Rep. 10 (1) (2020) 1–12.

[73] S.M.R. Razavi, et al., Simulation of CO_2 absorption by solution of ammonium ionic liquid in hollow-fiber contactors, Chem. Eng. Process. Process Intensif. 108 (2016) 27–34.

[74] M. Rezakazemi, et al., CFD simulation of natural gas sweetening in a gas-liquid hollow-fiber membrane contactor, Chem. Eng. J. 168 (3) (2011) 1217–1226.

[75] A. Qatezadeh Deriss, S. Langari, M. Taherian, Computational fluid dynamics modeling of ibuprofen removal using a hollow fiber membrane contactor, Environ. Prog. Sustain. Energy 40 (1) (2021) e13490.

Modelling and simulation of processes using syngas

CHAPTER 12

Modeling, simulation, and optimization of methane production processes

Rafael Nogueira Nakashima[a], Daniel A. Flórez-Orrego[a,b,c], Meire Ellen Gorete Ribeiro Domingos[d], Moisés Teles dos Santos[d], and Silvio de Oliveira Junior[a]

[a]Department of Mechanical Engineering, Polytechnic School of University of São Paulo, São Paulo, Brazil [b]National University of Colombia, School of Processes and Energy, Medellin, Colombia [c]École Polytechnic Fédérale de Lausanne, Lausanne, Switzerland [d]Department of Chemical Engineering, Polytechnic School of University of São Paulo, São Paulo, Brazil

1. Introduction

Methane, the main component of natural gas, is one of the most important input of the energy and industry sectors with a wide range of applications and end uses. The rising consumption of methane on the supply chains of different products and services has led to concerns about the sustainability and energy security of the natural gas infrastructure [1]. In the past decades, different technologies have been proposed and demonstrated to produce methane from different carbon sources, for instance, coal and biomass [2]. These options have the potential to increase the availability of methane around the globe and, in some cases, reduce greenhouse gas emission derived from the natural gas industry. The main processes commercially available today can be classified into two types, methanation and anaerobic digestion, as illustrated in Fig. 1.

The methanation process is based on the Sabatier reaction, shown in Eqs. (1) and (2), which was reported in 1887 by Paul Sabatier and Jean-Baptiste Senderens [3]. According to Eqs. (1) and (2), the methanation process requires hydrogen, which is usually supplied by the gasification of a solid fuel, to produce methane from carbon oxides. The first commercial application of these reactions was proposed to convert coal into synthetic natural gas (SNG) in the 1980s [1]. In recent years, this process has been adapted to convert different types of biomass [4] and to use hydrogen derived from electrolysis, preferably using renewable electricity (e.g., wind and solar). A significant effort also has been devoted to efficiently capturing CO_2 to mitigate emissions from the methanation of fossil fuels (e.g., coal and naphtha).

Advances in Synthesis Gas: Methods, Technologies and Applications. https://doi.org/10.1016/B978-0-323-91879-4.00010-2

Methanation production route

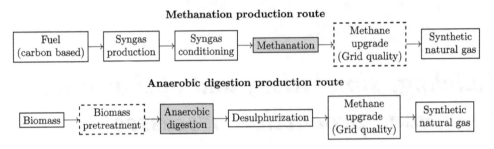

Anaerobic digestion production route

Fig. 1
Main routes of methane production.

$$CO_2 + 4H_2 \leftrightarrow CH_4 + 2H_2O \tag{1}$$

$$CO + 3H_2 \leftrightarrow CH_4 + H_2O \tag{2}$$

On the other hand, anaerobic digestion consists of the degradation of organic substances catalyzed by bacteria in the absence of oxygen, whose chemical reactions can be simplified as in Eqs. (3) and (4) [5]. Although anaerobic digestion has been known for centuries, the commercial application of large reactors for energy supply is a fairly recent achievement. The main strategic advantage of anaerobic digestion relies on the sustainability of combining the treatment of organic wastes with biofuel production [6]. However, as the process relies on living organisms at mild temperatures (35–55°C), biological methane production is susceptible to instabilities and low productivity (i.e., flow rate of methane per volume of reactor and biomass) [7].

$$CH_3COOH \rightarrow CH_4 + CO_2 \tag{3}$$

$$CO_2 + 4H_2 \rightarrow CH_4 + 2H_2O \tag{4}$$

At the moment, the anaerobic digestion route is considered the main source of renewable methane in the world [8]. This dominance can be explained by the simpler design of anaerobic reactors, in comparison with gasification [9], the lower costs of organic wastes, and the preexisting infrastructure of wastewater treatment and solid waste disposal. Thus, in some cases, the production of SNG can receive financial support from the waste treatment service, which employs anaerobic digestion [8]. As for methanation, there are few examples of facilities using this technology to produce SNG [2], but the process is widely employed in ammonia production to remove carbon monoxide from syngas [3]. In recent years, researchers have been proposing hybrid solutions for methane production, which aims to combine methanation and anaerobic digestion to improve efficiency and sustainability with lower costs [10]. These solutions may play an important role in the energy transition to a more sustainable life cycle of natural gas production and usage.

Thus, the following sections present the theoretical background, modeling methods, and optimization opportunities for the main methane production processes, methanation, and anaerobic digestion.

2. Methanation

The methanation reactions, Eqs. (1) and (2), can be rewritten as the reverse methane steam reforming and water gas shift reactions, shown in Eqs. (5) and (6). These reactions can occur in the presence of catalysts metals of the groups 8–10 (e.g., Ru, Ni, Co, and Fe), among which nickel is the industry standard due to its selectivity and costs [11]. In practice, methanation usually employs multiple fixed-bed reactors at temperatures and pressures between 300°C and 700°C and 20 and 30 bar, respectively [1].

$$CO + 3H_2 \leftrightarrow CH_4 + H_2O_{(g)} \qquad \Delta H^0{}_{298K} = -206 \, kJ/mol \qquad (5)$$

$$CO + H_2O_{(g)} \leftrightarrow H_2 + CO_2 \qquad \Delta H^0{}_{298K} = -41 \, kJ/mol \qquad (6)$$

As it can be observed from Eqs. (5) and (6), the reverse steam reforming and water gas shift reactions are exothermic and, therefore, their yield is inversely proportional to the reactor temperature. An analogous observation, also based on Le Chatelier's principle, can be extended to the reactor pressure. Higher pressure levels increase the methane yield, since methanation tends to decrease the volume of the products, according to Eqs. (1) and (2). The conversion yields of hydrogen for a mixture of 3:1 H_2:CO at chemical equilibrium for different temperatures and pressures are illustrated in Fig. 2.

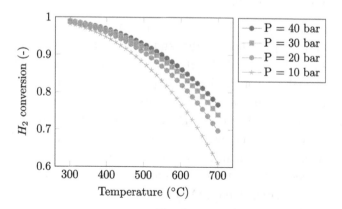

Fig. 2

H_2 conversion for different temperature and pressure based on the chemical equilibrium for a mixture of 3:1 H_2:CO.

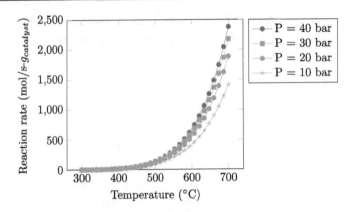

Fig. 3

Influence of temperature in CO_2 conversion for the methanation of a mixture of 4:1 H_2:CO_2 at 4 bar.

It is important to highlight that, although lower reaction temperatures lead to better theoretical conversion yields, the methane production may be limited by the reaction rate of methanation. Fig. 3 exemplifies the CO_2 conversion for a stoichiometric mixture (4:1 CO_2:H_2) at 4 bar according to the kinetic rate equations proposed by Koschany et al. [12]. The rate of CO_2 conversion diminishes as the operating temperature decreases. This can be explained by the exponential relationship between reaction rate and temperature [12]. Therefore, at lower temperatures, the reaction rate is greatly reduced and restricts the chemical equilibrium. Moreover, lower operating temperatures may require more catalyst volume and larger reactors to compensate for the decrease in the methanation reaction rate. A similar trade-off exists for the reacting pressure, since higher pressures require larger compressors and demand more power. Thus, the effective design of methanation would be a compromise between conversion, equipment size, and power consumption.

The methanation reaction requires from 3 to 4 mol of hydrogen per mole of CO or CO_2, respectively. In the same way that temperature and pressure, the initial composition of reactants affects the chemical equilibrium and, therefore, changes the products composition. Fig. 4 exemplifies the effect on the molar fraction of CH_4, CO_2, CO, and H_2 on a dry basis for different ratios of H_2/CO. The maximum concentration of methane is attained for the stoichiometric ratio, while the concentrations of CO_2 and H_2 have opposite trends for variations in the inlet ratio of H_2/CO. At H_2/CO ratios below the stoichiometric value, carbon dioxide tends to be formed as an intermediary product, since there is not sufficient hydrogen for conversion. The H_2/CO ratio is usually controlled by the gasification parameters and a preconditioning process, which may convert a portion of CO into CO_2 (e.g., water gas shift reaction) and separate it from syngas (e.g., physical absorption). In addition, hydrogen can also be produced from another process (e.g., water electrolysis) and mixed to the syngas.

At low temperatures and high pressures, the stoichiometric methanation produces methane that may only require a drying process for injection in the natural gas grid. However, depending on

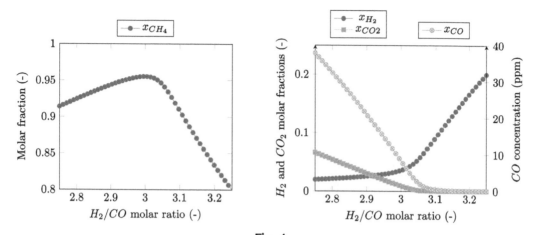

Fig. 4
Chemical equilibrium composition in dry basis for different ratios of H_2/CO at 300°C and 25 bar.

the local legislation and final use of methane, an upgrade process to remove carbon dioxide or hydrogen may still be necessary. In order to minimize the need for additional separation stages, it is usual to operate methanation close to the stoichiometric ratio [13]. It is important to highlight that the presence of ethane or other hydrocarbons in syngas (excluding CH_4) may alter the stoichiometric ratio of methanation [14].

2.1 Catalyst deactivation

As previously mentioned, catalysts are fundamental to enable the conversion of CO and CO_2 into methane through the Sabatier reaction. However, the nickel-based catalysts used in methanation can be very sensitive to sulfur poisoning, thermal degradation, and fouling, which can greatly reduce the conversion efficiency [2]. Thus, in order to avoid the catalyst deactivation, methanation has to operate under certain constraints of syngas composition and temperature.

Syngas produced from gasification may contain traces of ammonia, tars, chlorine, alkalis, and sulfur, which can contaminate nickel-based catalysts. For instance, hydrogen sulfide can react with nickel oxides to form nickel sulfide, Eq. (7), deactivating the catalysts. Thus, the sulfur concentration in syngas should be ideally reduced below 20 ppb to avoid catalyst deactivation [11].

$$H_2S_{(g)} + NiO_{(s)} \leftrightarrow NiS_{(s)} + H_2O_{(g)} \tag{7}$$

Solid carbon can be formed at H_2/CO ratios below the stoichiometric value depending on the operating temperature and pressure, as illustrated in the chemical equilibrium results of Fig. 5. There are two main reactions responsible for the soot formation: the Boudouard reaction,

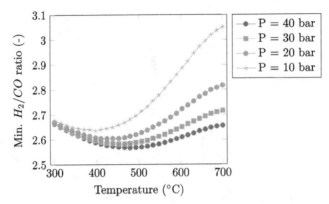

Fig. 5

Minimal H_2/CO ratio required to avoid carbon formation according to chemical equilibrium for different pressures and temperatures.

Eq. (8), and the methane decomposition, Eq. (9). The first reaction is predominant at lower temperatures, while the second is more relevant at high temperatures and lower pressures.

$$2CO_{(g)} \rightarrow C_{(s)} + CO_{2(g)} \qquad \Delta H^0_{298K} = -172\,kJ/mol \qquad (8)$$

$$CH_{4(g)} \rightarrow C_{(s)} + 2H_{2(g)} \qquad \Delta H^0_{298K} = 75\,kJ/mol \qquad (9)$$

Although carbon is an important component in the reaction mechanism [15], this intermediary product may deactivate the catalysts and compromise the reactor efficiency [2]. In order to avoid problems with carbon formation, the methanation can be operated at higher H_2/CO ratios or steam can be mixed with syngas prior to the reactor inlet [16]. In addition, carbon formation can also be inhibited by the presence of carbon dioxide; therefore, the methanation of CO_2 rarely has this type of problems [2].

Lastly, high operating temperatures and H_2O/H_2 ratios may promote the reorganization of catalysts particles into larger structures and, consequently, reduce the surface area and activity of nickel-based catalysts. Since this phenomenon negatively affects the methane production, the methanation is usually restricted to a maximum temperature, which varies from 500°C to 700°C depending on the catalysts materials [3].

2.2 Commercial processes examples

Over the last decades, several concepts for the methanation process were proposed and validated through experimentation, but a limited number of plants have achieved commercial success [17]. In general, the methanation currently employed for SNG production uses multiple fixed-bed reactors with intermediate cooling and recycle stages [1]. Fig. 6 illustrates an example of the methanation process for SNG production [13]. This design aims to reduce the average operating temperature to increase the yield of the methanation reaction. The conversion

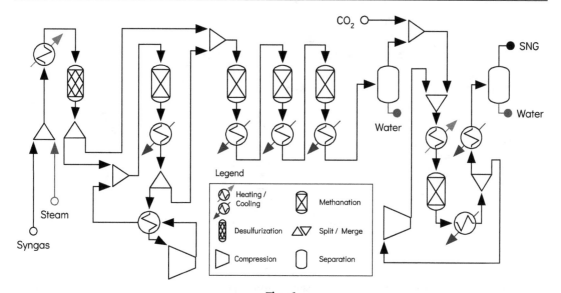

Fig. 6

Methanation process without downstream CO_2 separation. *Adapted from Wix, C. (2013). Process for the Production of Substitute Natural Gas (USPTO Patent No. US 8,530,529 B2).*

of CO and CO_2 into methane is an exothermic reaction and, therefore, it can significantly increase the operating temperature under adiabatic conditions. Fluidized bed and isothermal reactors have also been proposed for methanation, but their application in commercial plants is limited [2].

The methanation process depicted in Fig. 6 is based on the system commercialized by Haldor Topsøe for syngas with a composition at (or slightly above) the stoichiometric ratio (3:1 H_2:CO or 4:1 H_2:CO_2) [13]. Syngas is preheated and passes through a sulfur adsorbent to protect the downstream nickel catalysts from poisoning and deactivation. In addition, steam can be added to syngas to avoid carbon formation at specific operational conditions (e.g., start-up, low pressures, etc.). Next, syngas is sent to the main methanation process, also called wet methanation, which consists of multiple fixed-bed reactors with intercooling. The reactors can be arranged in sequence, parallel, or inside a loop, and the number of beds varies with the product requirements and plant characteristics. After the wet methanation process and water separation, the syngas has high concentration of methane (>80%), which varies depending on the operational temperature and pressure of each step [13]. However, in order to ensure the quality required for grid injection (>94%), an additional methanation reactor can be used for polishing the methane concentration. This process is usually referred to as dry methanation, since it uses carbon dioxide as the main methanation agent, which can be supplied by a downstream acid gas removal plant. After water separation, the final product has a methane concentration between 95% and 98% and can be injected into the natural gas grid [13].

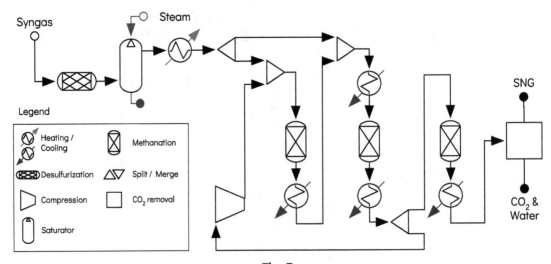

Fig. 7

Methanation process with downstream CO_2 separation. *Adapted from Stroud, H. J. F., Tart, K. R. (1979). Production of Substitute Natural Gas (IPO Patent No. UK1544245).*

It is important to highlight that methanation is a part of a larger process chain, as shown in Fig. 1. Therefore, SNG production plants usually integrate syngas conditioning, steam generation, and methane upgrade with methanation to achieve higher efficiencies [10]. For instance, the design presented in Fig. 6 illustrates some possible integrations with the acid gas removal plant, whereas the intercooling processes can be used for steam generation. There are two major approaches in methanation plants: upstream CO_2 removal, as implied in Fig. 6, and downstream CO_2 removal. Fig. 7 exemplifies a methanation process, which uses downstream CO_2 removal and operates with syngas composition below the stoichiometric ratio [16]. As it can be observed, this alternative process is fairly similar to the one depicted in Fig. 6, but there is an additional reactor in the methanation loop, and the dry methanation is substituted for a CO_2 separation plant.

2.3 Modeling methods

The composition and outlet temperature of the products of fixed-bed reactors used in methanation can be calculated by the chemical equilibrium condition if sufficient catalyst material is employed. The chemical equilibrium condition can be determined by an optimization problem (i.e., minimization of Gibbs free energy or maximization of entropy), which can be simplified by a system of algebraic equations derived from the molar, energy, and entropy balances. For instance, the restricted chemical equilibrium condition for the CO_2 methanation, at a certain temperature and pressure, can be expressed as Eqs. (10)–(12). The fixed-bed reactors are usually considered as adiabatic; therefore, the temperature varies with methane production and can be estimated by including the energy balance, Eq. (13).

$$\ln\left(-\frac{\Delta G^0_{r,T}}{RT_{outlet}}\right) = \frac{x_{CH_4} x_{H_2O}}{x_{CO_2} x_{H_2}^4}\left(\frac{P}{P_0}\right)^{-2} \tag{10}$$

In which,

$$\Delta G^0_{r,T} = g^0_{CH_4,T} + g^0_{H_2O,T} - \left(g^0_{CO_2,T} + 4g^0_{H_2,T}\right) \tag{11}$$

$$x_i = \frac{n_{i,0} + \nu_i \xi}{\sum_i (n_{i,0} + \nu_i \xi)}, \text{ for } i \in \{CO_2, H_2, CH_4 \text{ and } H_2O\} \tag{12}$$

Subjected to:

$$\sum_i [(n_{i,0} + \nu_i \xi) h_{i,To}] = \sum_i (n_{i,0} h_{i,Ti}) \tag{13}$$

In this case, the problem expressed in Eqs. (10)–(12) only considers the CO_2 methanation reaction, but an analogous system of equations can be achieved including the water-gas shift and carbon formation reactions, if necessary. A more general approach to determine chemical equilibrium can be formulated based on the conservation of chemical elements, instead of reaction extents [18]. Anyhow, the chemical equilibrium condition is usually determined by solving a nonlinear system of equations, as exemplified in Eqs. (10)–(12). Different variations of Newton's method have been proposed to solve problems of chemical equilibrium [19]. Moreover, several engineering software have inbuilt routines for chemical equilibrium that can easily determine the outlet composition and temperature (e.g., Aspen, Hysys, DWSim, Cantera). For instance, the chemical equilibrium solutions presented in the figures of this chapter section were calculated using the Cantera open-source library [20].

The chemical equilibrium modeling of methanation can be used in the preliminary design of the process to select working conditions of pressure, temperature, and the arrangement of reactors. However, in order to estimate the size of reactors and the quantity of catalysts, it is necessary to employ a kinetic model for methanation. For general analysis, the methanation reactor can be modeled as a one-dimensional plug-flow reactor, neglecting the radial changes of temperature and composition. There are several global reaction rates proposed in the literature, and the most widely accepted ones are those based on the Langmuir-Hinshelwood kinetics [2]. For instance, Eqs. (14)–(23) describe the kinetic model proposed by Koschany et al. [12] for an isothermal CO_2 methanation, which is based on the Langmuir-Hinshelwood kinetics.

$$\frac{dn_i}{dm_{cat}} = \nu_i r, \qquad \text{for } i \in \{CO_2, H_2, CH_4 \text{ and } H_2O\} \tag{14}$$

For the boundary condition $m_{cat}=0$ (reactor inlet):

$$n_i = n_{i,0} \qquad \text{for } i \in \{CO_2, H_2, CH_4 \text{ and } H_2O\} \tag{15}$$

In which,

$$r = \frac{kP_{H2}^{0.5}P_{CO2}^{0.5}}{BEN^2}\left(1 - \frac{P_{CH4}P_{H2O}^2}{P_{CO2}P_{H2}^4 K_{eq}}\right) \tag{16}$$

$$k = 3.46 \cdot 10^{-4} \exp\left(-\frac{77500}{R}\left(\frac{1}{T} - \frac{1}{555}\right)\right) \tag{17}$$

$$P_i = P\frac{n_i}{\sum_i n_i}, \qquad \text{for } i \in \{CO_2, H_2, CH_4 \text{ and } H_2O\} \tag{18}$$

$$K_{eq} = \ln\left(-\frac{\Delta G^\circ_{r,T}}{RT}\right) \tag{19}$$

$$BEN = 1 + K_{OH}P_{H2O}P_{H2}^{-0.5} + K_{H2}P_{H2}^{0.5} + K_{mix}P_{CO2}^{0.5} \tag{20}$$

$$K_{OH} = 0.5\exp\left(-\frac{22400}{R}\left(\frac{1}{T} - \frac{1}{555}\right)\right) \tag{21}$$

$$K_{H2} = 0.44\exp\left(-\frac{6200}{R}\left(\frac{1}{T} - \frac{1}{555}\right)\right) \tag{22}$$

$$K_{mix} = 0.88\exp\left(-\frac{10000}{R}\left(\frac{1}{T} - \frac{1}{555}\right)\right) \tag{23}$$

Equations (14) and (15) are an initial value problem; in other words, a variation rate described by an ordinary differential equation, namely Eq. (14), with an initial condition, i.e., Eq. (15). Since the kinetic model is one-dimensional and the catalyst is assumed to be equally distributed throughout the reactor volume, the solution of Eqs. (14)–(23) is independent of the reactor geometry (except for its length). This type of problem can be solved using numerical methods adapted to handle instabilities (i.e., stiff problems), such as the Rosenbrock methods [21]. Several programming languages and software have algorithm libraries and inbuilt functions to solve this type of problem. For instance, the kinetic model described in Eqs. (14)–(23) can be solved by using the "DifferentialEquations.jl" package for the Julia programming language [22].

The results for a stoichiometric mixture of CO_2 and H_2 at 4bar, with 25mg of catalyst (58% wt. NiAl(O)$_x$), and volumetric flow rate of 150NL/h are shown in Fig. 8. In addition, the results estimated by the chemical equilibrium condition, Eqs. (10)–(13), are also presented for the sake of comparison. As it can be observed, the chemical equilibrium modeling may overestimate the methane production depending on the operational conditions and the quantity

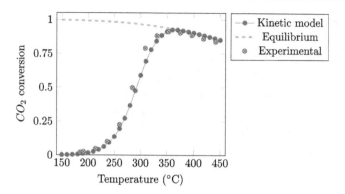

Fig. 8
Kinetic model and chemical equilibrium results for experimental data reported
by Koschany et al. [12].

of catalyst. For instance, in order to achieve the maximum conversion at temperatures below 350°C in this example, it would be required to increase the catalyst mass or reduce the flow rate.

The kinetic model describes a relationship between operational conditions, methane production, and catalyst mass (or volume), which can be used to estimate the reactor cost for an economic analysis. A simple model for the cost of a reactor based on the catalyst mass is described in Eqs. (24)–(27) [23]. The reactor cost is estimated by calculating the cost of the catalyst, assuming a constant specific cost ($\Pi_{catalyst}$), and of the vessel, which is correlated with the reactor volume. The economic model of methanation is important to estimate the trade-offs between the equipment cost and the efficiency necessary to convey optimal solution for methane production.

$$C_{reactor} = C_{catalyst} + C_{vessel} \tag{24}$$

Where,

$$C_{catalyst} = m_{catalyst}\Pi_{catalyst} \tag{25}$$

$$C_{vessel} = F_k C_{ref}\left(\frac{V_{vessel}}{V_{ref}}\right)^{\gamma} \tag{26}$$

$$V_{vessel} = F_V m_{catalyst} \tag{27}$$

The detailed design and analysis of methanation reactors may require the use of complex models based on computer fluid dynamics (CFD). These models can be employed to localize spots with temperatures above technical limits, which may lead to catalyst deactivation and material defects (e.g., high-temperature hydrogen attack) [24]. In general, CFD simulation is applied for reactor designs that include a cooling fluid [24,25], a separation membrane [26], or a fluidized bed [27], since in these problems the mass and energy transfer plays a major role in the conversion efficiency.

Eqs. (28)–(31) describe the governing equations for a reacting gas in a fixed-bed methanation reactor at steady state, adapted from a model proposed by Di Nardo et al. [25]. These equations describe the basic principles of mass (Eq. 28), momentum (Eq. 29), energy conservation (Eq. 30), and species mass balance (Eq. 31), respectively. The source terms associated with turbulent flow in the momentum and energy equations were removed for simplicity, but can be estimated using the k-epsilon model [25]. It is important to notice that Eqs. (28)–(31) were formulated considering the superficial velocity, which is calculated based on the volumetric flow rate, rather than the velocity at each position.

$$\frac{\partial \rho_f}{\partial t} + \nabla \cdot \left(\rho_f \vec{v} \right) = 0 \tag{28}$$

$$\frac{\partial}{\partial t}\left(\rho_f \vec{v} \right) + \nabla \cdot \left(\rho_f \vec{v}\vec{v} \right) = -\nabla P + \nabla \cdot \tau + S_{d,m} \tag{29}$$

$$\frac{\partial}{\partial t}\left[\rho_f E_f + (1-\varepsilon)\rho_s E_s \right] + \nabla \cdot \left[\vec{v}(\rho E + P) \right] = \nabla \cdot \left[k_{eff} \nabla T - \sum_i (h_i J_i) \right] + S_r \tag{30}$$

$$\frac{\partial}{\partial t}\left(\rho_f Y_k \right) + \nabla \cdot \left(\rho_f \vec{v} Y_k \right) = \nabla \cdot J_k + S_k \tag{31}$$

In which,

$$S_{d,m} = -\left(\frac{\mu}{\alpha}\vec{v} + \frac{C_2}{2}\rho_f |\vec{v}|\vec{v}_i \right) \tag{32}$$

$$\alpha = \frac{d_p^2}{150}\frac{\varepsilon^3}{(1-\varepsilon)^2} \tag{33}$$

$$C_2 = \frac{3.5}{d_p}\frac{(1-\varepsilon)}{\varepsilon^3} \tag{34}$$

$$k_{eff} = \varepsilon k_f + (1-\varepsilon)k_s \tag{35}$$

$$S_r = \sum_j (r_j \Delta h_{r,j}) \tag{36}$$

$$S_k = \sum_j r_j v_{j,k} M_k \tag{37}$$

The reactor packed bed is modeled as a porous media by modifying the momentum and energy equations. In the momentum equation, an additional sink term ($S_{d,m}$) based on the Darcy-Forchheimer model, Eqs. (32)–(34), is added to the equation to estimate the flow resistance through a porous media [28]. On the other hand, the energy equation includes the thermal inertia of the porous media and its influence on the effective conductivity (k_{eff}), as shown in Eq. (35) [29].

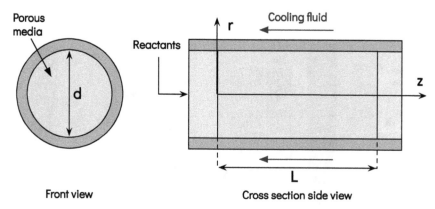

Fig. 9

Geometry and coordinate system for an axis-symmetrical fixed-bed reactor. *Adapted from Chein, R. Y., Chen, W. Y., & Yu, C. T. (2016). Numerical simulation of carbon dioxide methanation reaction for synthetic natural gas production in fixed-bed reactors. J. Nat. Gas Sci. Eng., 29, 243–251. https://doi.org/10.1016/j.jngse.2016.01.019.*

The boundary conditions for the governing Eqs. (28)–(31) depend on the geometry and system of coordinates. Most studies of methanation reactors assume a shell and tube geometry, which can be simplified into a 2D problem assuming axial symmetry in cylindrical coordinates. Fig. 9 illustrates the geometry for an axis-symmetrical 2D model of a shell and tube reactor, with boundary conditions expressed by Eqs. (38)–(41), based on the study of Chein et al. [30].

$$V = V_0\hat{z}, T = T_0 \text{ and } n_i = n_{i,0} \qquad \text{for } z = 0 \text{ (reactor inlet)} \qquad (38)$$

$$P = P_R \text{ and } \frac{\partial V}{\partial z} = \frac{\partial T}{\partial z} = \frac{\partial m_i}{\partial z} = 0 \qquad \text{for } z = L \text{ (reactor outlet)} \qquad (39)$$

$$V = 0, \frac{\partial m_i}{\partial r} = 0 \quad \text{and} \quad k_{eff}\frac{\partial T}{\partial r} = h_c(T - T_c) \qquad \text{for } z = \frac{d}{2} \text{ (reactor outer wall)} \qquad (40)$$

$$\frac{\partial V}{\partial r} = \frac{\partial T}{\partial r} = \frac{\partial m_i}{\partial r} = 0 \qquad \text{for } r = 0 \text{ (reactor centerline)} \qquad (41)$$

Examples of numerical solutions for temperature distributions for different inlet temperatures are illustrated in Fig. 10 [30]. As it can be observed, the temperature is not constant alongside the radius coordinate as it may be assumed in 1D models such as the kinetic model presented in this section. Thus, CFD modeling may be an important tool to identify excessive temperature levels in methanation reactors.

A set of boundary conditions analogous to Eqs. (38)–(41) can be achieved for a 3D model excluding the symmetry assumption. However, the addition of another dimension significantly increases the number of computations and, therefore, should be avoided whenever possible [31]. In general, the governing equations of CFD models require special mathematical and

$T_{in}=300°C$ $T_{in}=350°C$ $T_{in}=400°C$ $T_{in}=450°C$

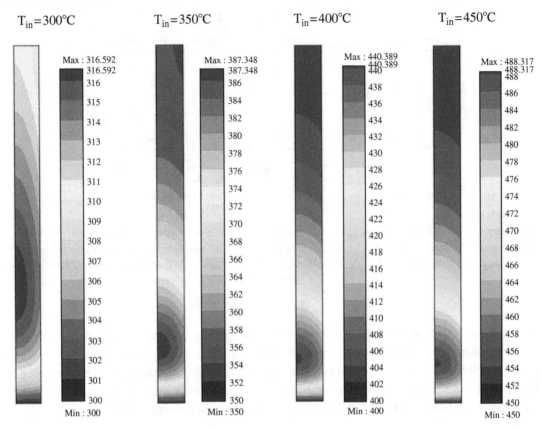

Fig. 10

Temperature distributions in the methanation reactor for various inlet temperatures. *Reprinted from Chein, R. Y., Chen, W. Y., & Yu, C. T. (2016). Numerical simulation of carbon dioxide methanation reaction for synthetic natural gas production in fixed-bed reactors. J. Nat. Gas Sci. Eng., 29, 243–251. https://doi.org/10. 1016/j.jngse.2016.01.019 with permission from Elsevier.*

programming techniques to achieve reliable numerical solutions; therefore, these models are usually solved using dedicated software (e.g., ANSYS Fluent, COMSOL, OpenFOAM) [24,25,30].

2.4 Process optimization

As previously mentioned, the conversion efficiency of methanation can be increased by operating at lower temperatures and higher pressures, but this increases the reactor volume and parasitic energy consumptions (e.g., compression power). The compression of syngas to high pressures (20–30 bar) can also facilitate the removal of CO_2 and sulfur by physical absorption systems (e.g., Rectisol and Selexol) and the injection on the natural gas grid. However, it is preferable to compress syngas after that conversion has occurred (e.g., to SNG) since its volumetric flow rate is significantly lower than that of raw syngas. Thus, the optimization of the operating pressure for methanation should also consider the performance of the acid gas removal plant and the grid injection requirements.

The design of a methanation system should try to optimize the relationship between the cost and the benefits of a higher temperature by separating the conversion process into multiple stages. Fig. 11 illustrates the variation of methane concentration along the multiple stages of a hypothetical methanation process. In general, the stages operate at decreasing levels of temperature to maximize the productivity at the first reactor and adjust the methane concentration in the following steps. Since the maximum temperature is limited by the effect of catalyst sintering and metallurgical limits, it is possible to recycle a portion of the products to increase the initial methane concentration. As it is observed in Fig. 11, this reduces the outlet temperature of the initial reactor, because the methane concentration is limited by the chemical equilibrium. It is important to highlight that the additional complexity and compression equipment cost required for a methanation loop may overcome the benefits of reducing the number of reactors and intercooling steps.

Since the methanation reactions are exothermic, heat is an important by-product of methane production in these systems. Fig. 12 shows the exergy balance for a methanation process using three adiabatic reactors reported by Nogueira Nakashima, et al. [10]. In this example, 12% of the exergy consumption is converted into heat and, if this by-product is not repurposed, the efficiency of methanation is limited to 82%. These results demonstrate the importance of heat management for the efficiency of methanation. In general, the energy recovered in methanation is used to produce steam for syngas conditioning (e.g., water gas shift).

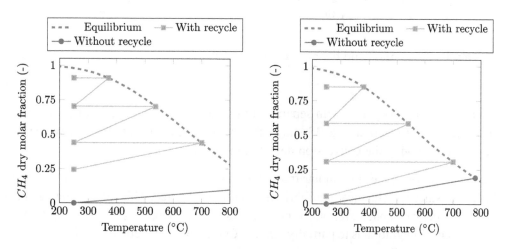

Fig. 11

Diagrams of methane concentration vs. temperature for different compositions at 30 bar. (Left) 3:1 $H_2:CO$. (Right) 4:1 $H_2:CO_2$.

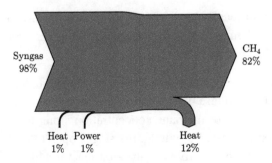

Fig. 12

Exergy balance of syngas methanation. *Adapted from Nakashima, R. N., Flórez-Orrego, D., & De Oliveira Junior, S. (2019). Integrated anaerobic digestion and gasification processes for upgrade of ethanol biorefinery residues. J. Power Technol., 99(2), 104–114.*

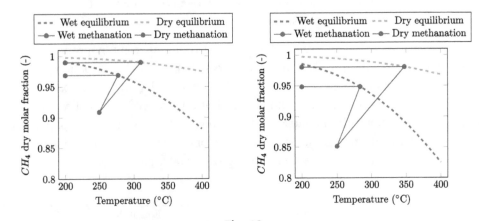

Fig. 13

Differences between wet and dry methanation processes. (A) 3:1 H_2: CO (B) 4:1 H_2: CO_2.

Methanation can be integrated with separation of water to shift the chemical equilibrium in favor of methane production. Fig. 13 shows the methanation concentration for wet and dry methanation, following the conversion depicted in Fig. 11.

As it can be observed, the final molar fraction of methane can achieve higher levels with dry methanation, since the chemical equilibrium condition is shifted. This integration can be used to polish the methane concentration for grid injection, as previously shown in Fig. 6. Another possible process integration is to partially remove CO_2 in the first methanation loop [32], as it is illustrated in Fig. 14. This type of modification aims to minimize the flow rate of the recycled products, which can be significantly high due to the elevated initial methane concentration (as exemplified in Fig. 11).

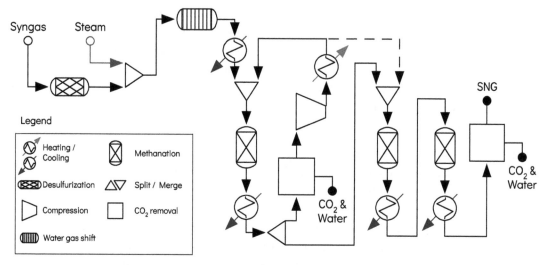

Fig. 14

Methanation process integrated with CO_2 separation. *Adapted from Clomburg Jr., L. A., & Nilekar, A. (2010). Process for Producing a Methane-Rich Gas (UPTO Patent No. US 2010/0162627 A1).*

3. Anaerobic digestion

Biogas is a gaseous mixture of methane, carbon dioxide, and other contaminants (e.g., water, hydrogen sulfide, siloxanes, ammonia, and air) produced by the anaerobic digestion of organic material. This process naturally occurs in the absence of oxygen and consists of the degradation of organic material catalyzed by different bacteria groups. As it is described in Fig. 15, anaerobic digestion can be divided into five major conversion steps: hydrolysis, acidogenesis, acetogenesis, methanogenesis, and sulfate reduction.

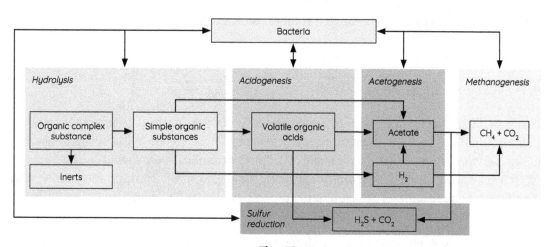

Fig. 15

A simplified diagram of anaerobic digestion.

Solid biomass is a complex organic material that can be converted into simple organic components (e.g., sugars, long-chain fatty acids, and amino acids) by hydrolysis. This process is catalyzed by exoenzymes produced by bacteria and may achieve different degrees of conversion depending on the substrate. For instance, lignocellulose and lignin are examples of recalcitrant organic material in anaerobic digestion reactors due to their strong molecular bonds [5]. Next, the acidogenesis bacteria convert the dissolved organic molecules into short-chained acids (e.g., acetic, propionic, and butyric acids), carbonate, and hydrogen.

Acetate, hydrogen, and carbonate are produced from these intermediary organic substances in a process called acetogenesis. These products are the main substrates for methanogenesis, the process of methane synthesis. The acetogenesis and methanogenesis bacteria groups are strongly connected by the hydrogen concentration. For example, if the methanogenesis process is disturbed, hydrogen concentration increases and inhibits the acetogenesis bacteria, which causes an accumulation of organic acids [33]. On the other hand, sulfur reduction is a process that competes with methanogenesis for substrates to produce hydrogen sulfide (H_2S). Hydrogen sulfide can decrease the biogas production, while the high concentration of sulfur can inhibit other bacteria groups [34]. This alternative conversion step is especially important for industrial wastewater, which can contain high concentrations of sulfur components.

In sum, anaerobic digestion converts biomass into biogas, intermediary organic components (e.g., organic acids, sugars, amino acids, long-chain fatty acids, and inert), and bacteria biomass. Moreover, this process reduces the concentration of organic pollutants (e.g., chemical and biological oxygen demand) and can neutralize the pH of the effluent. However, anaerobic digestion depends on the symbiotic interaction between different bacteria groups, which may be disturbed by the environment, i.e., operating conditions of the diverse components of the biogas plant. Anaerobic reactors are usually designed to enhance the methanogenesis process, since it develops a key role in hydrogen equilibrium, and are quite sensitive to environmental changes [7].

3.1 Environmental requirements

Anaerobic digestion has a number of requirements to develop and sustain its operation under stable conditions. These conditions are directly related with the growth and death of the bacteria groups previously mentioned. As the name of the process implies, there should be an absence of oxygen to avoid the death of anaerobic organisms. Although some bacteria present in anaerobic digestion may be facultative anaerobes, the process cannot be developed with significant concentrations of oxygen.

The bacteria growth and death rate are dependent on the reactor temperature. Since this relationship may vary for each species, the anaerobic digestion has three different temperature ranges with optimal growth conditions: psychrophilic (4–15°C), mesophilic (20–40°C) and thermophilic (45–70°C) [33]. In theory, the higher temperature ranges provide higher bacteria growth rates and, consequently, higher rates of biogas and sludge production. However, the operation at higher temperatures leads to higher energy consumption and less stability, since there are fewer bacteria species that can withstand high temperatures [7]. Thus, the mesophilic temperature range is the usual operation condition for anaerobic digestion reactors.

Bacteria biomass is composed of several chemical elements; thus, it is expected that these elements are required as nutrients for their growth. In general, important nutrients are carbon, nitrogen, phosphor, sulfur, potassium, calcium, and magnesium. The proportion of recommended nutrients varies depending on the source, since the composition of bacteria varies and experiments are subjected to uncertainties. As an example, a mass ratio of 500–1000:15–20:5:3 of C:N:P:S is considered sufficient for anaerobic digestion [35].

The pH influences the bacteria cell homeostasis and the dissociation equilibrium of acids and bases, which directly and indirectly affects the anaerobic digestion. For instance, hydrolysis and acidogenesis bacteria are more active in relatively low pH (5–6), while methanogenesis bacteria develop strictly in pH close to neutral (7) [7]. Thus, in practice, the pH in anaerobic reactors is controlled to be maintained at neutral with a chemical supplement (e.g., CaO, CaOH, NaOH) or by recycling the effluent. In this last scenario, the anaerobic digestion increases the bicarbonate (HCO_3^-) concentration by decomposing organic material [34].

The majority of toxic inhibitions in anaerobic digestion are associated with acids or bases and, thus, indirectly related with the pH. The main examples are undissociated acids (e.g., acetic, butyric, and propionic acids), dissociated hydrogen sulfide (HS^-), and ammonia (NH_3). The concentration equilibrium of these substances is highly dependent on the pH value; thus, a small variation in pH can lead to significant changes in the anaerobic digestion. Other reported toxic materials are heavy metals (e.g., lead, cadmium, copper, zinc) and complex substances found in wastes from specific areas, such as disinfectants, antibiotics, and pesticides.

3.2 Examples of commercial process

There are several designs for anaerobic reactors with commercial application, which can be conveniently summarized in terms of their characteristics [36]. These features are related to some characteristics of feedstock (e.g., solids concentration and feeding frequency), the velocity of anaerobic digestion reactions (e.g., temperature, mixing, and bacteria biomass retention), and the number of stages/phases of the process. An ideal process would be able to continuously convert a feedstock with high concentration of solids at elevated flow rates with high efficiency and stability. In practice, the anaerobic reactor design is strongly dependent on the solids concentration in the feedstock. Table 1 and Fig. 16 show a brief summary and the

Table 1 Main features of selected anaerobic reactors.

Type	Total solids content (TS), % wet mass	Hydraulic retention time (d)	Relative costs
Upflow anaerobic sludge blanket (UASB)	<3%	<5	+++
Continuous stirred tank reactor (CSTR)	3–20%	15–180	++
Anaerobic lagoons	<5%	30–200	+

Sources: Nyns, E.-J., Nikolausz, M., & Liebetrau, J. (2014). Biogas. In Ullmann's Encyclopedia of Industrial Chemistry (pp. 1–14). Wiley-VCH. https://doi.org/10.1002/14356007.a16_453.pub2, Jende, O., Platzer, C., Cabral, C., Hoffmann, H., Rosenfeldt, S., Colturato, L., Burkard, T., Linnenberg, C., Stinner, W., Zörner, F., & Schröder, E. (2015). Relevant Technologies of Anaerobic Digestion for Brazil: Feedstock, Reactors and Uses of Biogas (In Portuguese) and EPA (United States Environmental Protection Agency). (2020). Project Development Handbook: A Handbook for Developing Anaerobic Digestion/Biogas Systems on Farms in the United States (No. 430-B-20–001; p. 132). EPA. https://www.epa.gov/agstar/agstar-project-development-handbook.

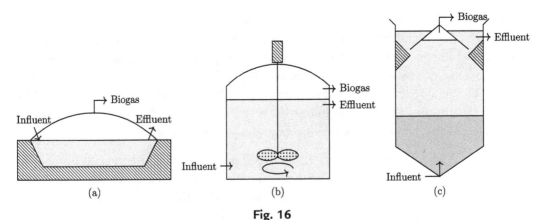

Fig. 16

Schematics of selected anaerobic reactors. (A) Anaerobic lagoons. (B) Continuously stirred tank reactor (CSTR). (C) Upflow anaerobic sludge blanket (UASB).

schematics of the main characteristics of three very common anaerobic reactors: the continuously stirred tank reactor (CSTR), the upflow anaerobic sludge blanket (UASB) and the anaerobic lagoons.

At its core, the anaerobic reactor can be a simple reservoir separated from the atmospheric air, as it is illustrated by the anaerobic lagoon. The conversion efficiency can be improved by adding mechanical mixing, as in the CSTR, and by phase separation, as in the UASB. According to Table 1, these modifications may decrease the retention time of the reactor, but incur additional investment costs. In addition, depending on the solids concentration of the feedstock, some reactor designs are not suitable for anaerobic digestion, such as UASB or anaerobic lagoons without mixing. Thus, the selection of the anaerobic reactor design should consider the characteristics of the feedstock and the technology costs.

3.3 Modeling methods

From the energy perspective, the methane production per unit of organic material, methane yield (MY), is the most important variable in anaerobic digestion, since it is directly proportional to power or biofuel production. Thus, it is important to determine the theoretical limits of biomass conversion based on its general characteristics to design and evaluate biogas production plants.

In wastewater treatment, the total solids are all substances derived from wastewater after water removal (dried wastewater). These solids can be classified depending on whether they can be filtered (suspended or dissolved) or ignited at 500°C (volatile or fixed). Since there is no simple test to verify if a solid has an organic component, it is usual to assume that the solids that can be ignited at 500°C, e.g., the volatile solids (VS), are organic substances [37]. Thus, these classifications are directly linked with the availability of organic material for anaerobic digestion.

Another approach to the classification of wastewater quality is the concentration of oxygen demand, which is the quantity of oxygen necessary to oxidize the wastewater under certain conditions. There are two types of oxygen demand measurements: chemical (COD) and biological (BOD). The first uses potassium dichromate ($K_2Cr_2O_7$), while the second uses microorganisms to oxidize the wastewater. In general, chemical oxygen demand (COD) values are higher than biological oxygen demand (BOD) because potassium dichromate can oxidize complex organic material that microorganisms would take a long time to convert. Thus, COD is an indirect measurement of the total organic concentration, while BOD represents the easily digestible fraction of biomass present in wastewater. Moreover, based on the definition of COD, it is also possible to derive a theoretical oxygen demand (ThOD) for a generic organic substance ($C_cH_hO_oN_nS_s$), as described in Eqs. (42) and (43) [38].

$$C_cH_hO_oN_nS_s + \left(c + \frac{h}{4} - \frac{o}{2} - \frac{3n}{4} + 1.5s\right)O_2 \rightarrow cCO_2 + \left(\frac{h}{2} - \frac{3n}{2} - s\right)H_2O + nNH_3$$
$$+ sH_2SO_4$$

$$(42)$$

$$ThOD = \left(c + \frac{h}{4} - \frac{o}{2} - \frac{3n}{4} + 1.5s\right)MW_{O2} \qquad (43)$$

The COD is the theoretical basis for most anaerobic digestion models, since it is commonly agreed that COD is conservative for ideal anaerobic processes. This assumption allows us to determine the theoretical limit of conversion of biomass to methane, since methane (CH_4) ideally would correspond to 64 g of COD/mol of CH_4. Thus, by converting the units with the ideal gas law, the maximum methane yield is approximately 0.35 Nm³/kg of COD. It is also possible to extend the analysis to common organic substances in order to evaluate the theoretical yield of different groups. Fig. 17 shows the specific methane yield (Nm³ of CH_4/kg

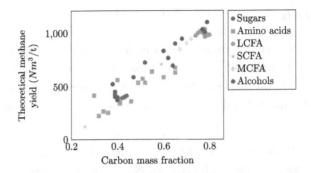

Fig. 17
Theoretical results of specific methane yield as a function of carbon mass fraction.

of biomass) as a function of the carbon fraction (kg of C/kg of biomass) of the major organic substances present in anaerobic digestion. The overall trend shows that organic matter with high concentration of carbon has a higher specific methane yield.

It is important to highlight that, especially for anaerobic digestion, biomass is often available at high moisture content or significantly diluted. Thus, the maximum specific methane production can be significantly lower in total mass basis depending on the water content, as it can be observed in Fig. 18. Therein, the average values of theoretical methane yields reported in Fig. 17 for sugars, amino acids, and long-chain fatty acids are assumed. These results are in accordance with values reported by Jende et al. [39] for methane yields of carbohydrates, proteins, and fats. In addition, as it can be deduced from Fig. 15, COD cannot be totally converted into methane, since at least a portion of the organic material has to be converted into bacteria biomass and intermediary compounds (e.g., acids, sugar, amino acids). Moreover, a significant fraction of organic matter in anaerobic reactors has a slow conversion or is difficult to hydrolyze (e.g., cellulose, hemicellulose, and lignin). Thus, the actual biogas production of a certain feedstock is only a fraction of its theoretical methane yield.

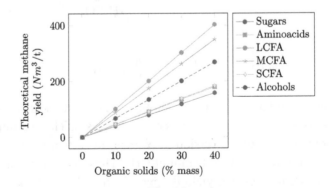

Fig. 18
Effect of the biomass moisture on the maximum specific methane production.

3.3.1 Feedstock supply and production scale

Feedstock can be classified by its origin (e.g., industrial, agriculture, etc.), by its properties (e.g., percentage of solids, concentration of organic compounds), or by its productivity (e.g., methane or biogas yield). In theory, any organic substance or mixture could be used to produce biogas via anaerobic digestion under the suitable conditions. In practice, the organic wastes from agriculture, livestock, industries, and cities are the main sources of feedstock for anaerobic digestion. These options are usually available free of charge or with negative costs (e.g., waste disposal in landfills), and their conversion may incur into other benefits or revenues (e.g., energy, fertilizer, or irrigation).

Differently from fossil fuel resources, biomass is a seasonal product and wastes have a variable production and quality. Uncontrolled changes in quality and quantity of organic material are not ideal for operation of anaerobic reactors. These uncertainties could lead to reductions on methane production, frequent start-ups, and partial load inefficiencies [40]. In order to overcome this issue and to achieve a constant supply of feedstock, it is common to use materials from different sources (co-digestion) and, if possible, with high specific methane production (e.g., energy crops and industrial wastes). Apart from the reduction of uncertainties, this practice can also enhance nutrient proportions (C:N:P:S) and increase the reactor specific production (Nm^3 of CH_4/m^3 of reactor), thus providing higher economic return.

Table 2 shows the biogas and methane yield for different feedstock used in anaerobic digestion. It is important to notice that the reported values of methane yield for a feedstock type may vary due to the concentration of water or organic solids (i.e., volatile solids). For instance, the methane yields presented in Table 2 are relatively lower than most of the theoretical values presented in Fig. 17. The main reason for these differences is the high moisture content of this feedstock, as previously exemplified in Fig. 18. Other losses can be traced back to the organic fraction in total solids and the production of intermediary substances in anaerobic digestion (e.g., bacteria, volatile acids). Moreover, the experimental conditions for each reference source may not be equal, thus average values of methane yield should be always validated. Since most feedstock has a low specific production of methane, a significant amount of wastes is necessary to achieve high biogas production rates. Thus, in general, biogas plants rely on different feedstock sources (e.g., co-digestion and centralized plants) to achieve high and stable energy production.

3.3.2 Influence of operational variables in methane production

As it may be expected, the production rate of methane is directly related with the amount of biomass available for anaerobic digestion. Thus, the concentration of organic substances and their residence time are important variables in the operation of anaerobic reactors. The first parameter is usually given by the concentration of volatile solids (g of volatile solids/m^3) or the COD (kg of O_2/m^3), while the second one is represented by the hydraulic retention time (HRT).

Table 2 Average methane yield of some common feedstock.

Feedstock	Organic solids (% wet mass)	Average CH$_4$ yield (Nm^3/kg$_{OS}$)	(Nm^3/t$_{wet}$)	Sources
Cattle slurry	6.3–8.0	0.15–0.21	9.6–16.8	A, B, C, and D
Poultry manure	16.0–30.0	0.28–0.31	48.0–91.6	A, B, and E
Pig slurry	2.9–6.0	0.25–0.35	9.9–15.0	A, B, and F
Maize silage	31.4–32.2	0.25–0.34	80.8–106.6	B and G
Municipal solid waste	19.3–36.1	0.19–0.33	63.2–68.6	G and H

Source: [A] Biosantech, T. A. S., Rutz, D., Janssen, R., & Drosg, B. (2013). Biomass resources for biogas production. In The Biogas Handbook (pp. 19–51). Elsevier. https://doi.org/10.1533/9780857097415.1.19. [B] Fachagentur Nachwachsende Rohstoffe E. V (FNR). (2010). Practical Guide of Biogas: Generation and Use (in Portuguese) (fifth ed.). FNR. [C] Dong, L., Cao, G., Guo, X., Liu, T., Wu, J., & Ren, N. (2019). Efficient biogas production from cattle manure in a plug flow reactor: a large scale long term study. Bioresour. Technol., 278, 450–455. https://doi.org/10.1016/j.biortech.2019.01.100. [D] Moset, V., Poulsen, M., Wahid, R., Højberg, O., & Møller, H. B. (2015). Mesophilic versus thermophilic anaerobic digestion of cattle manure: methane productivity and microbial ecology. J. Microbial. Biotechnol., 8(5), 787–800. https://doi.org/10.1111/1751-7915.12271. [E] Jurgutis, L., Slepetiene, A., Volungevicius, J., & Amaleviciute-Volunge, K. (2020). Biogas production from chicken manure at different organic loading rates in a mesophilic full scale anaerobic digestion plant. Biomass Bioenergy, 141, 105693. https://doi.org/10.1016/j.biombioe.2020.105693. [F] Bonmatí, A., Flotats, X., Mateu, L., & Campos, E. (2001). Study of thermal hydrolysis as a pretreatment to mesophilic anaerobic digestion of pig slurry. Water Sci. Technol., 44(4), 109–116. https://doi.org/10.2166/wst.2001.0193. [G] Jende, O., Platzer, C., Cabral, C., Hoffmann, H., Rosenfeldt, S., Colturato, L., Burkard, T., Linnenberg, C., Stinner, W., Zörner, F., & Schröder, E. (2015). Relevant Technologies of Anaerobic Digestion for Brazil: Feedstock, Reactors and Uses of Biogas (In Portuguese). [H] Zhu, B., Zhang, R., Gikas, P., Rapport, J., Jenkins, B., & Li, X. (2010). Biogas production from municipal solid wastes using an integrated rotary drum and anaerobic-phased solids digester system. Bioresour. Technol., 101 (16), 6374–6380. https://doi.org/10.1016/j.biortech.2010.03.075.

The HRT, Eq. (44), is the ratio of the reactor volume per volumetric flow rate of wastes entering the reactor.

$$HRT = \frac{V_{reactor}}{\dot{Q}_{in}} \tag{44}$$

Although higher production rates of methane can be achieved with higher influent flow rate, the specific methane yield tends to decrease at low HRT. Fig. 19 illustrates the influence of HRT and COD in the methane production per organic load (methane yield, Nm^3/kgCOD) and per reactor volume (volumetric yield, Nm^3/(dm$^3_{reactor}$)). As it can be observed, high HRT values can produce more methane from a fixed feedstock source, but this fact limits the volumetric flow rate of influent and reduces the volumetric yield. On the other hand, higher COD concentrations favor both methane and volumetric yields.

Bacteria have a limited rate of conversion of biomass that is proportional to their concentration, which has a maximum value depending on the reactor design [36]. Therefore, for high flow rates, the existing bacteria may not be sufficient to convert all degradable organic substances. This situation may also occur for sudden increases in the feedstock concentration, which may lead to a disturbance in the equilibrium of acidogenesis and methanogenesis. In drastic situations, the biogas production can be diminished and a reconditioning time may be necessary.

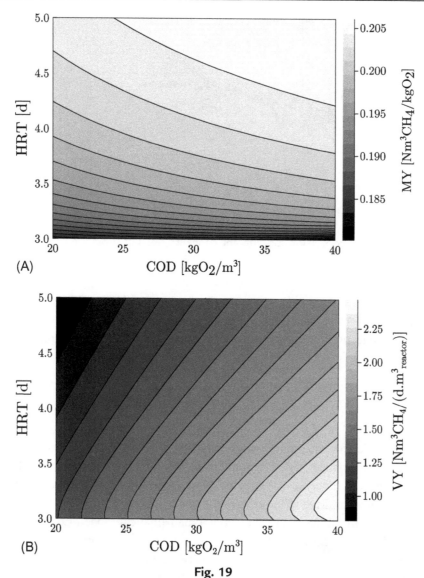

Fig. 19

Influence of HRT and COD on the methane yield and volumetric methane yield for vinasse at 35°C, 1.01 bar, pH7.0, and 0.05 $gSO_4^{2-}/gCOD$. (A) Methane yield [$Nm^3_{CH4}/kgCOD$]. (B) Volumetric yield [$Nm^3_{CH4}/(dm^3_{reactor})$]. *Adapted from Nogueira Nakashima, R. (2018). Exergy Assessment of Biogas Generation and Usage in the Sugarcane Industry (In Portuguese) Master thesis, University of São Paulo. https://doi.org/10.11606/D.3.2018.tde-27082018-153742.*

Since the reaction rates are proportional to the concentration of bacteria, some anaerobic reactors with continuous operation are designed for maximum retention of bacteria biomass (e.g., UASB). The effect of bacteria biomass concentration (Z^x) is illustrated in Fig. 20 using the Monod equation Eq. (45) according to data reported by Rosen and Jeppsson [41]. It is important

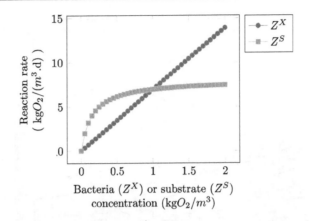

Fig. 20

Reaction rate of acetate conversion for different concentrations of substrate (Z^S) and bacteria (Z^X).

to notice that an increase in solids retention may lead to a reduction in the reactor useful volume and, consequently, reduce the hydraulic retention time (or increase the organic load). Thus, the concentration of bacteria biomass cannot provide an unlimited increase in the reaction rate.

$$r = k \frac{Z^S}{K_S + Z^S} Z^X \tag{45}$$

Another design choice for anaerobic reactors related with the kinetics of anaerobic digestion is the temperature. For instance, high temperatures have higher rates of conversion for chemical, biological, and physical processes (as demonstrated by the Arrhenius equation) [33]. However, the bacteria are quite sensitive to temperature changes and have narrow intervals with high growth rate. Fig. 21 illustrates the results from a mathematical model, Eq. (46), validated by Zwietering, et al. [42] for *Lactobacillus plantarum*. As it has been previously mentioned, this pattern can be explained by the competitiveness between bacteria growth and decay. Growth is the dominant behavior for relatively lower temperatures, while bacteria decay is dominant at high temperatures.

$$k = 0.041(T - T_{min})^2 [1 - \exp(0.161(T - T_{max}))] \tag{46}$$

All in all, the reactor thermal insulation and heat management directly impacts the production rate of biogas. Higher temperatures are able to increase the productivity of biogas production and reduce the specific costs of the reactor, but it also requires more energy for its operation. Thus, in most of the cases the mesophilic temperature range (20–40°C) is employed in commercial applications.

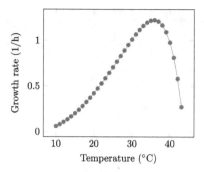

Fig. 21
Temperature influence in the bacteria growth.

3.3.3 Modeling anaerobic reactors

Anaerobic digestion is a relatively difficult process to model due to the variability of organic feedstock and the several reactions involved in the process. The simplest approach is to use an average methane yield, according to the feedstock type as exemplified in the Table 2, in order to estimate the methane production, Eq. (47). This type of model can be used to roughly estimate the required amount of organic feedstock required for a specific methane production. However, since the average methane yields reported in literature may significantly vary according to the source, the methane production may be imprecise.

$$\dot{Q}_{CH4} = MY_{feedstock}\left(\dot{m}_{feedstock}Z_{organic}\right) \tag{47}$$

A generic approach to model anaerobic digestion can be obtained based on the COD and mass balance. For instance, Eq. (48) estimates the methane production based on the ThOD and the flow rate of a generic organic molecule $C_aH_bO_cN_dS_s$, according to Eq. (43). This model assumes that the ThOD present in the organic substances is conservative and can be totally converted into methane. In practical applications, only a fraction of the total COD present in wastewater is converted into methane, therefore values obtained from Eq. (48) can only determine an upper limit for methane production. Fig. 22 illustrates an example of COD balance for anaerobic digestion. The methane production calculated from Eq. (48) can be partially corrected by considering the efficiency of COD removal, Eqs. (49)–(50). However, it is important to highlight that the difference between the influent and effluent COD concentrations may not reflect the actual COD removed from the wastewater, because organic substances can accumulate (and/or decay) in the anaerobic reactor (e.g., solids retention).

$$\dot{Q}_{CH4,\,max} = MY_{max}(\dot{m}_{in}ThOD_{in}) \tag{48}$$

$$\dot{Q}_{CH4} = \dot{Q}_{CH4,\,max}\eta_{COD} \tag{49}$$

$$\eta_{COD} = 1 - \frac{COD_{out}}{COD_{in}} \tag{50}$$

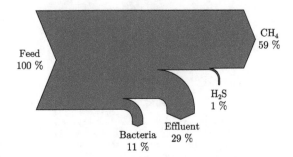

Fig. 22

COD balance for anaerobic digestion of vinasse (COD=40 gCOD/m^3 and 0.05 gSO$_4$$^{2-}$/gCOD) at 35°C, 1.01 bar, pH 7.0, and HRT=5 d. *Adapted from Nogueira Nakashima, R. (2018). Exergy Assessment of Biogas Generation and Usage in the Sugarcane Industry (In Portuguese) Master thesis, University of São Paulo. https://doi.org/10.11606/D.3.2018.tde-27082018-153742.*

Eqs. (38)–(50) are very useful because their variables are widely reported in the literature. However, the design and control of anaerobic digestion reactors often require more information about the effect of the reactor dimensions and operational variables in the mass balance. In these cases, it is possible to use a generic kinetic model to describe anaerobic digestion. The most widely accepted model is the Anaerobic Digestion Model N° 1 (ADM1), which consists of 19 reactions (hydrolysis, acidogenesis, acetogenesis, and methanogenesis) and 32 main variables [33]. Eqs. (51)–(55) exemplify the use of the ADM1 to model a perfectly mixed reactor, as illustrated in Fig. 23, assuming liquid-gas phase equilibrium. Differently from conventional reactors, the balances and stoichiometric coefficients of ADM1 are written in COD basis for organic substances, and the reaction rate is described by a Monod equation. In addition, inhibitions from pH, NH$_3$, and organic acids influence the reaction rate. Fig. 24 illustrates the ADM1 results of methane production for the anaerobic digestion of sugarcane vinasse, an effluent from ethanol distillation.

$$\frac{dZ_i}{dt} = \left(Z_{in,i} - Z_{liq,i}\right)\frac{\dot{Q}_{liq}}{V_{liq}} - Z_{gas,i}\frac{\dot{Q}_{gas}}{V_{gas}} + \sum_j \nu_{i,j} r_j \tag{51}$$

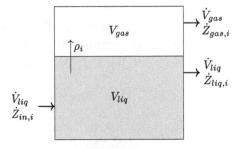

Fig. 23

Anaerobic digestion model for perfectly mixed reactor using ADM1 kinetic model.

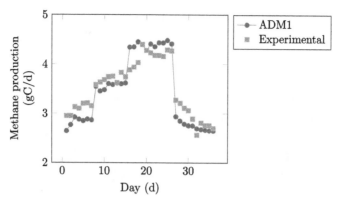

Fig. 24
Comparison between model results from Nogueira Nakashima [43] and experimental results of Barrera et al. [34]. *Adapted from Nogueira Nakashima, R. (2018). Exergy Assessment of Biogas Generation and Usage in the Sugarcane Industry (In Portuguese) Master thesis, University of São Paulo. https://doi.org/10.11606/ D.3.2018.tde-27082018-153742.*

Where,

$$\dot{Q}_{gas} = \frac{RTV_{liq}}{P - P_{H2O}} \left(\frac{\varphi_{H2}}{16} + \frac{\varphi_{CH4}}{64} + \varphi_{CO2} \right) \tag{52}$$

$$\varphi_i = k_L a \left(Z_i - K_{H,i} Z_{gas,i} RT \right) \qquad \text{for } i \in \{H_2, CH_4, CO_2\} \tag{53}$$

$$r_j = k_j X_j \text{ for } j \in \{\text{hydrolysis, bacteria decay and disintegration reactions}\} \tag{54}$$

$$r_j = k_{m,j} \left(\frac{Z_j^S}{K_{S,j} + Z_j^S} \right) Z_j^X I_j \qquad \text{for } j \in \{\text{other reactions}\} \tag{55}$$

The ADM1 can model the dynamic response of anaerobic digestion to feedstock variations, such as changes in flow rate, concentration, pH, or temperature. Moreover, the composition of biogas and effluent can be estimated in detail (e.g., composition and concentration). However, this model approach requires a detailed calibration process, since there are a large number of model parameters. Thus, calibrated parameters and validation data are not as available as the previous modeling approaches.

3.4 Optimization

The most simple and effective optimization technique for methane production using anaerobic digestion is to control the quality of the feedstock. The feedstock composition determines the maximum methane production, and its moisture content is inversely proportional to the reactor productivity. Increasing the concentration of organic substances or feedstock pretreatment can greatly improve the methane production of an anaerobic reactor, as previously shown in Fig. 19.

In addition, maintaining a specific ratio of macronutrients (e.g., C, N, S, P) and avoiding inhibitory substances (e.g., NH_4^+, SO_4^-, organic acids) or operational parameters (e.g., too low or too high temperatures) are essential to maximize bacteria growth and, consequently, the reaction rate.

If the quality of the feedstock cannot be improved, the anaerobic reactor size can be optimized to reduce methane production costs. High hydraulic retention times can improve the efficiency of methane conversion, but at a certain point this benefit is outweighed by the reactor costs and other equipment costs. Fig. 25 illustrates the impact of the hydraulic retention time in the electricity costs and exergy efficiency of a biogas plant, derived from the results of an ADM1 model [44]. As it can be observed, the minimal specific costs are close to the inflection point in the methane yield, since after this point the changes in methane yield are relatively small. In practice, anaerobic reactors are usually oversized to favor stability of methane conversion, since feedstock quality or quantity can vary over time.

Another improvement opportunity is to recover or repurpose the effluent derived from anaerobic digestion, called digestate. Anaerobic digestion cannot convert complex organic molecules, which may represent a significant portion of the wastewater exergy. Fig. 26 shows the results observed by Nogueira Nakashima et al. [10] for the exergy balance of anaerobic digestion. In order to avoid losses, the anaerobic digestion can be operated in stages and the solid portion of the digestate can be separated and recycled.

Lastly, anaerobic digestion can also be integrated with other processes to improve heat management. In general, anaerobic digestion requires energy to operate at mild temperatures, which has to be supplied from another process. Heat is usually supplied by the power

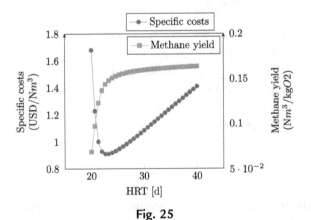

Fig. 25

Effect of hydraulic retention time and concentration in methane specific cost. *Adapted from Nogueira Nakashima, R., & de Oliveira Junior, S. (2021). Trade-offs between productivity, efficiency and costs of biogas plants for agriculture wastes. Proceedings of the 34rd International Conference on Efficiency, Cost, Optimization, Simulation and Environmental Impact of Energy Systems—ECOS 2021, Taormina, Italy.*

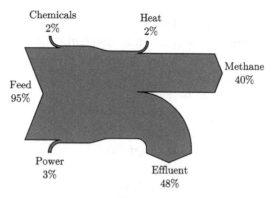

Fig. 26

Exergy balance for anaerobic digestion. *Adapted from Nakashima, R. N., Flórez-Orrego, D., & De Oliveira Junior, S. (2019). Integrated anaerobic digestion and gasification processes for upgrade of ethanol biorefinery residues. J. Power Technol., 99(2), 104–114.*

conversion process through cogeneration, but other integration techniques with separation stages, hydrogen production, or heat pumps are also possible [44]. Another type of process integration is the inhibition of sulfur reduction by injecting small doses of oxygen into anaerobic digestion. This fact effectively reduces the production of H_2S and acts as a gross desulfurization process inside the anaerobic reactor [7].

4. Conclusion

In this chapter, a brief overview of the theoretical background, commercial process, modeling methods, and optimization opportunities for methane production were presented and discussed. Today, there are two main routes for methane production, methanation and anaerobic digestion, each one having its own technical particularities. The numerical modeling and simulation of these processes are important tools to understand the effect of engineering design in efficiency and costs. In addition, these tools can be used to identify and determine optimization opportunities for novel and future methane production processes.

Abbreviations and symbols

C	costs (USD)
CFD	computer fluid dynamics
COD	chemical oxygen demand concentration (kgCOD/m^3)
d_p	pore diameter (m)
E	total energy (J)
F_v	volume factor

F_k	pressure and material factor
$G^0_{r,T}$	Gibbs free energy of reaction at standard conditions (J/mol)
$g^0_{i,T}$	specific Gibbs free energy of species "i" at temperature T and standard pressure (J/mol)
HRT	hydraulic retention time (d)
h	specific enthalpy (J/mol or J/kg)
Δh_r	specific enthalpy of reaction "r" (J/kg)
J	diffusion flux of species (kg/m^2 s)
K	equilibrium constant
K_s	half-velocity constant (kgCOD/m^3)
k	reaction rate (mol/bar s g$_{cat}$ or 1/h)
k_f, k_s, and k_{eff}	fluid, solid, and effective thermal conductivity (W/mK)
$k_L a$	gas-liquid mass transfer constant (1/d)
M_k	molecular weight (kg/mol)
MY	methane yield (Nm^3/kgCOD or Nm^3/kgVS)
m	mass (g)
n_i	molar flow rate (mol/s)
P	pressure (bar)
P_0	pressure at standard conditions (bar)
$\dot{Q}$	volumetric flow rate (m^3/d)
R	universal gas constant (J/mol-K)
r	reaction rate (mol/m^3 s)
S	soluble material concentration (kgCOD/m^3)
$S_{d,m}$	momentum sink term (kg/m^2 s^2)
S_k	source term of specie "k" (kg/m^3 s)
S_r	energy source term (J/m^3 s)
SNG	synthetic natural gas
T	temperature (K)
ThOD	theoretical oxygen demand concentration (kgCOD/m^3)
V	volume (m^3)
VS	volatile solids concentration (kgVS/m^3)
$\vec{v}$	velocity vector (m/s)
X	solid material concentration (kgCOD/m^3)
x	molar fraction
Y	mass fraction
Z	concentration (kgCOD/m^3 or kmol/m^3)
Z^S and Z^X	concentration of substrate and bacteria (kgO$_2$/m^3)
α	permeability (m^2)
ε	porosity

η	efficiency
μ	molecular viscosity (kg/ms)
ν	stoichiometric coefficient
ξ	reaction extent
Π_{cat}	specific catalyst cost (USD/g_{cat})
ρ	fluid and solid density (kg/m^3)
τ	tension matrix (N/m)
ϕ	gas-liquid mass transfer rate (kgCOD/m^{3}d or kmol/m^{3}d)

References

[1] J. Kopyscinski, T.J. Schildhauer, S.M.A. Biollaz, Production of synthetic natural gas (SNG) from coal and dry biomass—a technology review from 1950 to 2009, Fuel 89 (8) (2010) 1763–1783, https://doi.org/10.1016/j. fuel.2010.01.027.

[2] S. Rönsch, J. Schneider, S. Matthischke, M. Schlüter, M. Götz, J. Lefebvre, P. Prabhakaran, S. Bajohr, Review on methanation—from fundamentals to current projects, Fuel 166 (2016) 276–296, https://doi.org/10.1016/j. fuel.2015.10.111.

[3] W. Boll, G. Hochgesand, C. Higman, E. Supp, P. Kalteier, W.-D. Müller, M. Kriebel, H. Schlichting, H. Tanz, Gas production, 3. Gas treating, Wiley-VCH Verlag GmbH & Co. KGaA (Ed.), in: Ullmann's Encyclopedia of Industrial Chemistry, Wiley-VCH Verlag GmbH & Co. KGaA, 2011, p. o12_o02, https://doi.org/10.1002/14356007.o12_o02.

[4] D. Flórez-Orrego, F. Maréchal, S. de Oliveira Junior, Comparative exergy and economic assessment of fossil and biomass-based routes for ammonia production, Energ. Conver. Manage. 194 (2019) 22–36, https://doi.org/10.1016/j.enconman.2019.04.072.

[5] C.A.L. de Chernicharo, Anaerobic Reactors, vol. 4, IWA publishing, 2007.

[6] R. Nogueira Nakashima, D. Flórez-Orrego, H.I. Velásquez, S. de Oliveira Junior, Sugarcane bagasse and vinasse conversion to electricity and biofuels: an exergoeconomic and environmental assessment, Int. J. Exergy 33 (1) (2020) 44, https://doi.org/10.1504/IJEX.2020.109623.

[7] Fachagentur Nachwachsende Rohstoffe E. V (FNR), Practical Guide of Biogas: Generation and Use (in Portuguese), fifth ed., FNR, 2010.

[8] IEA, Outlook for Biogas and Biomethane: Prospects for Organic Growth, IEA, 2020. https://www.iea.org/reports/outlook-for-biogas-and-biomethane-prospects-for-organic-growth.

[9] R. Telini, D. Florez-Orrego, S. de Oliveira Junior, Techno-economic and environmental assessment of ammonia production from residual bagasse gasification: a decarbonization pathway for nitrogen fertilizers, Front. Energy Res. 10 (881263) (2022) 1–15, https://doi.org/10.3389/fenrg.2022.881263.

[10] R.N. Nakashima, D. Flórez-Orrego, S. de Oliveira Junior, Integrated anaerobic digestion and gasification processes for upgrade of ethanol biorefinery residues, J. Power Technol. 99 (2) (2019) 104–114.

[11] DOE, Practical Experience Gained During the First Twenty Years of Operation of the Great Plains Gasification Plant and Implications for Future Projects, U.S. Department of Energy, 2006, p. 76.

[12] F. Koschany, D. Schlereth, O. Hinrichsen, On the kinetics of the methanation of carbon dioxide on coprecipitated NiAl(O), Appl. Catal. Environ. 181 (2016) 504–516, https://doi.org/10.1016/j.apcatb.2015.07.026.

[13] C. Wix, Process for the Production of Substitute Natural Gas, 2013. USPTO Patent No. US 8,530,529 B2.

[14] M. Gassner, F. Maréchal, Thermo-economic process model for thermochemical production of synthetic natural gas (SNG) from lignocellulosic biomass, Biomass Bioenergy 33 (11) (2009) 1587–1604, https://doi.org/10.1016/j.biombioe.2009.08.004.

[15] D. Schmider, L. Maier, O. Deutschmann, Reaction kinetics of CO and CO2 methanation over nickel, Ind. Eng. Chem. Res. 60 (16) (2021) 5792–5805, https://doi.org/10.1021/acs.iecr.1c00389.

[16] H.J.F. Stroud, K.R. Tart, Production of Substitute Natural Gas (IPO Patent No. UK1544245), 1979.

[17] A. Bolt, I. Dincer, M. Agelin-Chaab, A critical review of synthetic natural gas production techniques and technologies, J. Nat. Gas Sci. Eng. 84 (2020), https://doi.org/10.1016/j.jngse.2020.103670, 103670.

[18] S. Gordon, B.J. McBride, Computer Program for Calculation of Complex Chemical Equilibrium Compositions and Applications: I. Analysis, Reference Publication No. 1311, NASA, 1994, p. 64.

[19] F.J. Zeleznik, S. Gordon, An Analytical Investigation of Three General Methods of Calculating Chemical-Equilibrium Compositions, Technical Note D-473, NASA, 1960, p. 37.

[20] D.G. Goodwin, R.L. Speth, H.K. Moffat, B.W. Weber, Cantera: An Object-oriented Software Toolkit for Chemical Kinetics, Thermodynamics, and Transport Processes, 2021, https://doi.org/10.5281/zenodo.4527812.

[21] W.H. Press, S.A. Teukolsky, W.T. Vetterling, B.P. Flannery, Numerical Recipes: The Art of Scientific Computing, third ed., Cambridge University Press, 2007.

[22] C. Rackauckas, Q. Nie, Differentialequations. jl–a performant and feature-rich ecosystem for solving differential equations in julia, J. Open Res. Softw. 5 (1) (2017), https://doi.org/10.5334/jors.151.

[23] F. Marechal, F. Palazzi, J. Godat, D. Favrat, Thermo-economic modelling and optimisation of fuel cell systems, Fuel Cells 5 (1) (2005) 5–24, https://doi.org/10.1002/fuce.200400055.

[24] W. Zhang, H. Machida, H. Takano, K. Izumiya, K. Noringa, Computational fluid dynamics simulation of CO2 methanation in a shell-and-tube reactor with multi-region conjugate heat transfer, Chem. Eng. Sci. 211 (2020), https://doi.org/10.1016/j.ces.2019.115276, 115276.

[25] A. Di Nardo, G. Calchetti, C. Bassano, P. Deiana, CO2 methanation in a shell and tube reactor CFD simulations: high temperatures mitigation analysis, Chem. Eng. Sci. 246 (2021), https://doi.org/10.1016/j.ces.2021.116871, 116871.

[26] D. Schlereth, O. Hinrichsen, A fixed-bed reactor modeling study on the methanation of CO2, Chem. Eng. Res. Des. 92 (4) (2014) 702–712, https://doi.org/10.1016/j.cherd.2013.11.014.

[27] J. Li, B. Yang, Multi-scale CFD simulations of bubbling fluidized bed methanation process, Chem. Eng. J. 377 (2019), https://doi.org/10.1016/j.cej.2018.08.204, 119818.

[28] K. Vafai, S.J. Kim, On the limitations of the Brinkman-Forchheimer-extended Darcy equation, Int. J. Heat Fluid Flow 16 (1) (1995) 11–15, https://doi.org/10.1016/0142-727X(94)00002-T.

[29] ANSYS Inc, Ansys Fluent 12.0/12.1 Documentation, 2009. https://www.afs.enea.it/project/neptunius/docs/fluent/index.htm.

[30] R.Y. Chein, W.Y. Chen, C.T. Yu, Numerical simulation of carbon dioxide methanation reaction for synthetic natural gas production in fixed-bed reactors, J. Nat. Gas Sci. Eng. 29 (2016) 243–251, https://doi.org/10.1016/j.jngse.2016.01.019.

[31] Y. Liu, O. Hinrichsen, CFD simulation of hydrodynamics and methanation reactions in a fluidized-bed reactor for the production of synthetic natural gas, Ind. Eng. Chem. Res. 53 (22) (2014) 9348–9356, https://doi.org/10.1021/ie500774s.

[32] L.A. Clomburg Jr., A. Nilekar, Process for Producing a Methane-Rich Gas, 2010 (UPTO Patent No. US 2010/0162627 A1).

[33] D.J. Batstone, International Water Association (Eds.), Anaerobic Digestion Model No. 1: (ADM1), IWA Publication, 2002.

[34] E.L. Barrera, H. Spanjers, K. Solon, Y. Amerlinck, I. Nopens, J. Dewulf, Modeling the anaerobic digestion of cane-molasses vinasse: extension of the anaerobic digestion model no. 1 (ADM1) with sulfate reduction for a very high strength and sulfate rich wastewater, Water Res. 71 (2015) 42–54, https://doi.org/10.1016/j.watres.2014.12.026.

[35] D. Deublein, A. Steinhauser, Biogas from Waste and Renewable Resources, Wiley-VCH Verlag GmbH & Co. KGaA, 2008, https://doi.org/10.1002/9783527621705.

[36] E.-J. Nyns, M. Nikolausz, J. Liebetrau, Biogas, in: Ullmann's Encyclopedia of Industrial Chemistry, Wiley-VCH, 2014, pp. 1–14, https://doi.org/10.1002/14356007.a16_453.pub2.

[37] G. Tchobanoglous, H.D. Stensel, R. Tsuchihashi, F.L. Burton, M. Abu-Orf, G. Bowden, Eddy (Eds.), Wastewater Engineering: Treatment and Resource Recovery, Fifth ed., McGraw-Hill Education, 2014.

[38] J.R. Baker, M.W. Milke, J.R. Mihelcic, Relationship between chemical and theoretical oxygen demand for specific classes of organic chemicals, Water Res. 33 (2) (1999) 327–334, https://doi.org/10.1016/S0043-1354(98)00231-0.

[39] O. Jende, C. Platzer, C. Cabral, H. Hoffmann, S. Rosenfeldt, L. Colturato, T. Burkard, C. Linnenberg, W. Stinner, F. Zörner, E. Schröder, Relevant Technologies of Anaerobic Digestion for Brazil: Feedstock, Reactors and Uses of Biogas, 2015 (In Portuguese).

[40] R. Nogueira Nakashima, S. de Oliveira Junior, Exergy analysis of biogas production in ethanol distilleries, in: Proceedings of the 24th ABCM International Congress of Mechanical Engineering. 24th ABCM International Congress of Mechanical Engineering, 2017, https://doi.org/10.26678/ABCM.COBEM2017.COB17-0185.

[41] C. Rosen, U. Jeppsson, Aspects on ADM1 Implementation within the BSM2 Framework, Lund University, 2006.

[42] M.H. Zwietering, J.T. de Koos, B.E. Hasenack, J.C. de Witt, K.V.'t. Riet, Modeling of bacterial growth as a function of temperature, Appl. Environ. Microbiol. 57 (4) (1991) 1094–1101.

[43] R. Nogueira Nakashima, Exergy Assessment of Biogas Generation and Usage in the Sugarcane Industry, (In Portuguese) Master thesis, University of São Paulo, 2018.

[44] R. Nogueira Nakashima, S. de Oliveira Junior, Trade-offs between productivity, efficiency and costs of biogas plants for agriculture wastes, in: Proceedings of the 34rd International Conference on Efficiency, Cost, Optimization, Simulation and Environmental Impact of Energy Systems—ECOS 2021, Taormina, Italy, 2021.

CHAPTER 13

Ammonia production from syngas: Plant design and simulation

Mohammad Farsi

Department of Chemical Engineering, School of Chemical and Petroleum Engineering, Shiraz University, Shiraz, Iran

1. Introduction

Ammonia as a colorless gas with a distinct odor is a compound of nitrogen and hydrogen with the chemical formula NH_3. It is one among the world's most widely used synthetic compounds, and it is used to make urea, nitric acid, ammonium nitrate, ammonium sulfate, acrylonitrile, methylamine, and hydrogen cyanide, among other things [1]. Since ammonia is a hydrogen-rich compound, it is a zero-carbon emission fuel for fuel cells [2]. The fertilizer, cleaning products, water, and wastewater treatment sectors, as well as the rubber, pulp and paper, food, leather, and pharmaceutical industries, all utilize ammonia [3]. At the moment, approximately 88% of produced ammonia in the world is converted to fertilizers such as urea, ammonium phosphate, ammonium nitrate, and other nitrates [4]. The Haber–Bosch is the main technology to produce ammonia from hydrogen and nitrogen at pressures up to 100 bar and temperature range of 400–500°C on heterogeneous catalysts [5]. This process was developed in the 1900s by F. Haber and modified by C. Bosch in the BASF company. Although alternative technologies such as electrochemical and photocatalytic processes are appealing, the Haber–Bosch process is still the fundamental technology for ammonia production, according to the defined roadmap for ammonia economy [6,7]. In electrochemical processing, water and nitrogen react directly to form ammonia in an electrolytic cell. [8]. The electrochemical pathway eliminates the independent hydrogen generation step and has a high coulombic efficiency.

The reforming of fossil fuels, coal gasification, thermochemical and biochemical processing of biomass, and water electrolysis are commonly used methods for the production of hydrogen in the Haber–Bosch ammonia plants [9]. Hydrogen production through methane reforming and coke gasification are mature technologies and widely used in the methanol, dimethyl ether, gas to liquid, and ammonia plants. The ammonia production by hydrogen provided from the electrolysis of water offers significant advantages including availability, simplicity and flexibility in operation, and reliability [10]. However, water electrolysis needs more

Advances in Synthesis Gas: Methods, Technologies and Applications. https://doi.org/10.1016/B978-0-323-91879-4.00012-6

enhancement in energy efficiency, safety, durability, portability, and capital costs [11]. Currently, biomass has been introduced as a renewable feedstock to produce ammonia [12]. Although the biomass could be directly converted to bio-hydrogen through gasification, pyrolysis, plasma processing, and anaerobic digestion techniques, the biological conversion of biomass to bio-methane followed by methane to syngas is an efficient indirect method to provide hydrogen in the ammonia plant [13].

The economic returns, energy efficiency, and environmental impacts of ammonia production depend on the applied methods to provide hydrogen and nitrogen [14]. The equivalent carbon dioxide emission of ammonia synthesis through methane reforming, coal gasification, biomass processing, and water electrolysis routes are 1.64, 2.36, 0.55, and 0.46 $kg_{CO_2} kg_{NH_3}^{-1}$, respectively [15,16]. In general, ammonia production with zero-carbon emission is available, when the solar or wind energy sources are applied to produce hydrogen in the water electrolysis route [17]. Fig. 1 shows the energy efficiency of ammonia plants over the last decades. Energy efficiency is defined as the ratio of lower heating value of ammonia to the required energy to produce ammonia in the plant. It appears that the Haber–Bosch technology has been optimized and the energy efficiency of plant has been increased from 30% to 65% by substation of coal gasification units with methane reforming, over the last century [18]. At the moment, up to 96% of hydrogen demand for ammonia synthesis is provided by fossil fuel reforming, particularly methane [19]. The Haldor Topsoe, Kellogg Brown and Roots, Lurgi, Linde, ThyssenKrupp, and Casale are the main licensors of ammonia plants, and they used fossil fuel reforming technique to produce hydrogen.

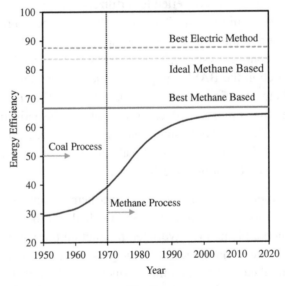

Fig. 1
Energy efficiency of the ammonia synthesis plants over the last decades.

Syngas generation, hydrogen purification, ammonia synthesis, and separation are all part of the ammonia plant's complicated operations. Although all ammonia plants are designed based on the Haber–Bosch technology, different syngas preparation, and purification methods, and separation configurations are used in the plants. The modern ammonia plants could be classified into first- and second-generation based on the applied hydrogen purification method. The hydrogen purity in the generated syngas is managed in the high and low-temperature shift reactors in the first-generation technology, which uses steam and autothermal reforming of light hydrocarbons to create syngas. The syngas is next filtered, and the concentrations of carbon monoxide and carbon dioxide in the absorber column and methanation reactor are reduced to an acceptable level. In second-generation technology, the carbon dioxide removal and methanation stages are replaced by pressure swing adsorption (PSA) units, and a gas plant is used to supply pure nitrogen. In both technologies, the produced hydrogen and nitrogen are converted to ammonia in a catalytic reactor.

2. Ammonia production processes

2.1 First-generation technology

The first-generation technology is applied by KBR, Haldor Topsoe, ThyssenKrupp, and Casale companies. Preheated desulfurized natural gas is combined with steam and fed into the primary reformer. Methane and steam react in the reformer tubes over a nickel catalyst to create hydrogen-rich syngas. The reforming processes are endothermic and equilibrium-constrained, with an external heat source supplying the needed heat. The steam reforming reactions are as follows [20]:

$$CH_4 + H_2O \leftrightarrow CO + 3H_2 \tag{1}$$

$$CH_4 + 2H_2O \leftrightarrow CO_2 + 4H_2 \tag{2}$$

$$CO + H_2O \leftrightarrow CO_2 + H_2 \tag{3}$$

The steam reformer is a fired reactor consisting of a catalyst-filled coil tube hanged in a firebox [21]. The main reformer's exit stream is combined with preheated air and sent to the secondary reformer. The secondary reformer consists of three zones including the burner, combustion zone, and catalytic zone [22]. In the combustion section, oxidation of carbon monoxide, hydrogen, and methane supplied the required heat to progress steam reforming reactions in the catalytic part. The main oxidation reactions are as follows [23]:

$$CH_4 + 1.5O_2 \leftrightarrow CO + 2\,H_2O \tag{4}$$

$$CO + 0.5O_2 \leftrightarrow CO_2 \tag{5}$$

$$H_2 + 0.5O_2 \leftrightarrow H_2O \tag{6}$$

The injected air into the secondary reformer provides the nitrogen in Haber–Bosch reaction. Syngas is created when the combustion section's exhaust stream passes through the catalytic component. The exit stream from the secondary reformer feeds to the shift reactors after cooling to an acceptable temperature in a waste heat boiler and condensing of steam in a separator. Carbon monoxide and steam react catalytically in the water–gas shift process, producing hydrogen and carbon dioxide. Presently, most ammonia plants use a combination of high and low-temperature shift reactors. The iron oxide and copper oxide–zinc oxide catalysts are used in high and low shift converters, respectively. Because CO and CO_2 deactivate the iron catalyst in the ammonia converter, an absorber column and a methanator are used to lower CO and CO_2 concentrations to a safe level. As a result, the low-temperature shift's output product is cooled and fed into the absorption column [24]. The CO_2 is absorbed by the commercial solvents in the absorber, and the outlet stream from the absorber is heated up and fed to the methanation reactor. In the methanator, the residual CO_2 and CO are converted to methane through reverse reforming reactions at temperatures of 200–500°C. The catalytic hydrogenation of carbon dioxide and carbon monoxide to methane is thermodynamically favorable at low temperatures and high pressure. The methanator's output stream is cooled, compressed, and dried before being sent to the ammonia converter. At temperatures ranging from 450°C to 600°C and pressures up to 100 bar, hydrogen and nitrogen react over a heterogeneous catalyst such as fused iron in the ammonia reactor, and ammonia is produced as follows [25]:

$$N_2 + 3H_2 \leftrightarrow 2\,NH_3 \qquad (7)$$

The autothermal reactor or multibed structure fitted with interstage coolers is employed in ammonia synthesis facilities because the ammonia synthesis process is very exothermic and equilibrium-constrained. Fig. 2 shows the schematic diagram of a first-generation ammonia plant. The KBR's ammonia plants use a purified section after CO_2 absorber to remove impurities such as methane and argon to produce inert-free hydrogen. In general, KBR's ammonia process offers greater energy efficiency compared with Halder Topsoe. Table 1 shows the applied catalysts and operation conditions in modern ammonia plants.

2.2 Second-generation technology

The second-generation technology for ammonia synthesis is applied by the Linde company. In the second-generation units, the only primary reformer is applied to produce hydrogen, and the outlet product from shift converters feeds to the PSA unit to separate impurities. In addition, a nitrogen plant provides the required nitrogen in the ammonia plant. The outlet hydrogen from the PSA unit and nitrogen from the gas plant are mixed, compressed, and feed to the ammonia converter. The main advantages of second-generation technology are ultra-pure hydrogen and nitrogen production for ammonia synthesis, low compression cost of recycle

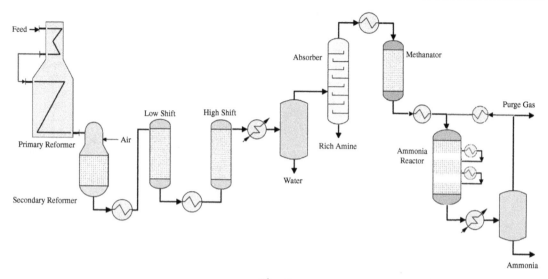

Fig. 2
The schematic diagram of a first-generation ammonia plant.

Table 1 Applied catalysts and operation conditions in ammonia plants.

Reactor	Catalyst	Pressure (bar)	Temperature (°C)	Catalyst lifetime
Desulfurization	ZnO	20–40	300–400	2–4
Primary reforming	Ni Al$_2$O$_3$	20–40	600–900	3–4
Secondary reforming	Ni/Al$_2$O$_3$	20–40	850–1300	3–4
High shift	Fe$_3$O$_4$/Cr$_2$O$_2$	20–40	300–450	3–5
Low shift	Cu/ZnO/Al$_2$O$_3$	20–40	200–300	3–5
Methanation	Ni/Al$_2$O$_3$	20–40	200–500	2–5
Ammonia synthesis	Fe and Ru	150–300	350–550	6–10

stream, and the minimum hydrogen loss through purge gas and methanation steps due to applied PSA unit. Fig. 3 shows the schematic diagram of a second-generation ammonia plant.

2.3 Coal-based ammonia

The coal-based ammonia synthesis process consists of a coal gasification unit to produce syngas, air separation, acid gas removal, and ammonia synthesis units [26]. In the coal-based process, the produced oxygen in the gas plant feeds to the gasifier, and coal is converted into hydrogen, carbon monoxide, carbon dioxide, and methane. The particulate matter in the outlet stream from the gasifier is removed in a scrubber. The outlet stream from the scrubber is mixed with steam and fed to a sour shift reactor to convert carbon monoxide and steam to hydrogen. Since the produced syngas includes sulfurous compounds, the specially designed

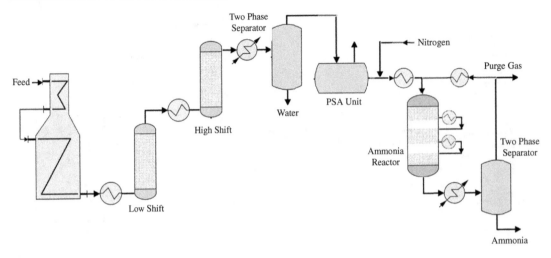

Fig. 3
The schematic diagram of a second-generation ammonia plant.

cobalt–molybdenum catalyst is applied in the sour shift reactor. The outlet stream from sour shift reactor feeds to acid gas removal unit to remove the carbon dioxide and sulfurous compound by effective solvents. The residual carbon monoxide, carbon dioxide, and methane are removed from syngas in a PSA system. Finally, hydrogen and produced nitrogen in the gas plant are mixed and fed to the ammonia synthesis converter. In the ammonia reactor, hydrogen and nitrogen react over the heterogeneous catalyst, and ammonia is produced.

3. Ammonia synthesis catalyst

The activation energy of homogeneous ammonia synthesis reaction is in the range of 230–420 $kJ\,mol^{-1}$ [27]. Since the Haber reaction is equilibrium limited, applying high temperatures to overcome the activation energy barrier decreases the equilibrium yield considerably. Applying the heterogeneous catalysts reduces the activation energy of ammonia synthesis reaction down to $100\,kJ\,mol^{-1}$, and the reaction is progressed at temperatures of 250–400°C [28]. In this regard, the catalyst is the heart of the ammonia plant, and it has a significant effect on the operating temperature, pressure, recycle ratio, and ammonia yield. The catalysts used in the ammonia synthesis process could be categorized into fused-iron and supported metallic catalysts groups [29]. Historically, iron has been known as an effective catalyst for ammonia synthesis since 1905. To enhance the activity of the iron catalyst, it is promoted with various nonreducible oxides such as K_2O, MgO, BaO, SiO_2, and Al_2O_3 [30]. The commercial fused iron catalyst as the first generation is prepared by melting iron oxides with structural promoters such as Al_2O_3, CaO, and an activating promoter such as K_2O. The sulfurous, phosphorus, arsenic, and chlorine compounds are permanent poisons of iron catalyst, while water, carbon monoxide, carbon dioxide, and oxygen are temporary poisons [31,32]. The second- and third-generation catalysts are based on ruthenium and cobalt molybdenum nitride [33]. The ruthenium supported on a proprietary graphitic structure as a noniron catalyst was commercialized in Kellogg Advanced Ammonia Process in 1992 [34]. The higher activity of ruthenium catalyst allows ammonia synthesis at lower operating pressure, which reduces design complexity and capital costs. Although ruthenium-supported graphitic catalysts are more active and efficient compared with fused iron, the majority of ammonia plants nowadays utilize iron catalysts. The main drawbacks of ruthenium catalysts are the methanation of support in the presence of hydrogen and the risk of explosion [35]. In general, alkaline earth metal oxides and hydroxides could be used to improve the stability and reactivity of ruthenium based catalysts [36]. Currently, the electride, hydride, amide, perovskite oxide hydride/oxynitride hydride, nitride, and oxide promoted metals such as Fe, Co, and Ni are proposed to apply in the ammonia synthesis converter [29].

4. Fossil-fuel-based ammonia synthesis

4.1 Process design

The ammonia synthesis loop includes the catalytic converter, ammonia separation section, and compression stage. Based on the situation of the ammonia separation stage and makeup gas feeding, three different configurations are proposed for the ammonia synthesis loop. Fig. 4 shows the ammonia synthesis loops based on the ammonia separation position. In the optimal configuration, the produced ammonia is separated from the outlet stream from the reactor by cooling and condensation. Then, the unreacted compounds are compressed and mixed with fresh makeup gas and recycled to the catalytic converter. Since, once-through pass conversion of nitrogen is only 15%–30% in the commercial ammonia plants, the main part of inlet feed to the reactor is compressed and recycled. The main advantages of optimal configuration are high

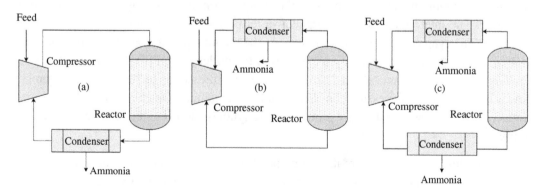

Fig. 4
The ammonia synthesis loops based on the ammonia removal position [37].

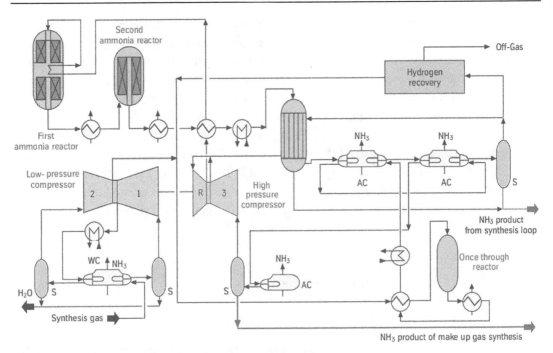

Fig. 5
The dual-pressure ammonia synthesis loop commercialized by Uhde.

energy efficiency, high nitrogen conversion due to low ammonia content at the entrance of the converter, and high ammonia concentration in the condensation stage.

To reduce the compression demand and increase the ammonia yield in the large-capacity plants, Uhde and Synetix proposed the Dual Pressure Process in 2001 as a highly efficient ammonia synthesis process. The key innovation in the dual-pressure process is applying a medium-pressure ammonia synthesis reactor in the conventional synthesis loop. The once-through ammonia synthesis involves the compression of makeup gas in a compressor up to 100 bar. The intercooled once-through converter produces approximately one-third of ammonia. The syngas-ammonia mixture leaving the medium pressure converter is cooled, and 85% of the ammonia produced is separated from the outlet stream from the converter. The remaining syngas is compressed in the high-pressure compressor up to 200 bar. The compressed syngas feeds to the high-pressure converters, and the remaining two-thirds of total ammonia is produced. Fig. 5 shows the dual-pressure ammonia synthesis loop commercialized by the Uhde company.

The temperature and pressure are the main operating variables that affect the ammonia yield and hydrogen conversion. Fig. 6 shows the equilibrium ammonia yield versus operating temperature and pressure in the Haber reaction. Since the ammonia synthesis reaction is accompanied by a volume reduction, the high-pressure condition increases the equilibrium

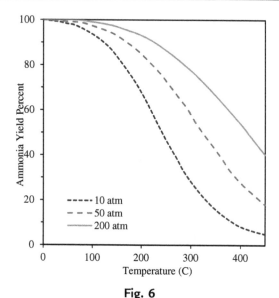

Fig. 6

The equilibrium ammonia yield versus temperature and pressure.

conversion of the Haber reaction and reduces the difficulty of ammonia condensation from unreacted hydrogen and nitrogen [38]. Hoverer, operating at high pressure increases the plant capital cost. Thus, there is an optimal operating pressure to achieve maximum ammonia yield at the minimum cost. Temperature as the most important input in the ammonia plant could change both reaction rate and equilibrium conversion. Although decreasing temperature increases the equilibrium conversion, it decreases the rate of ammonia synthesis reaction. On the other hand, increasing temperature could progress the deactivation of catalyst by thermal sintering. Thus, there is an optimal operating temperature to achieve maximum ammonia yield in the synthesis reactor. To apply the optimal temperature on the reactor, the ammonia reactor requires a preheater to heat the feed stream and a cooling system to cool down the reactor.

The pressure drop is one of the main challenges in ammonia synthesis reactors. When a significant change in total mole flow occurs in a reaction, the radial flow reactors are the best candidate due to low-pressure drop. Although the axial flow, radial flow, and mixed flow reactors are commonly applied in ammonia plants, most large-scale plants use radial-flow reactors. As a disadvantage, the size of radial flow reactors is bigger compared with the axial flow reactors at the same catalyst loading, and therefore, they are more costly.

4.2 Reactor selection

Based on the implemented cooling methods, there are three types of ammonia synthesis reactor including internal direct cooling, quench cooling, and indirect cooling reactors [39]. Fig. 7 shows a simplified diagram of internal direct cooling, quench cooling, and indirect cooling

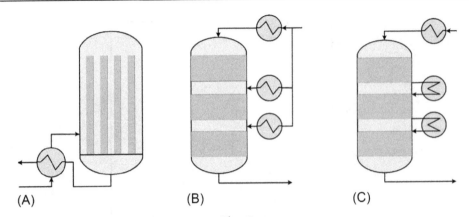

Fig. 7

The simplified diagram of (A) internal direct cooling, (B) adiabatic quench cooling, and (C) indirect cooling convertors.

configurations in the ammonia plant. The internal direct cooling reactor is a shell and tube exchanger in which one side is packed by the catalyst and a cooling medium flows on the other side to remove the heat of reaction [39]. In most cases, the shell side is packed with catalyst and cold feed as the cooling medium flows in the tube side. The Tennessee Valley Authority converter is the most important reactor based on internal cooling. In the cooling section, the cold feed stream flows upward inside the tubes, and a part of the heat generated in the catalyst bed is transferred to the feed stream. At the top of the reactor, the flow direction is reversed and the feed stream enters the catalytic bed. The reactants react over the catalyst and ammonia is produced in the catalytic section. Although the main advantages of internal direct cooling configuration are compact in design and low investment cost, the poor controllability makes the process control difficult. In the adiabatic quench cooling configuration, the reactor is divided into some series catalytic beds and a cold feed stream is injected between beds. The inlet cold feed is mixed with the outlet stream from the catalytic bed and the mixture feeds to the next bed. The Kellogg ammonia process uses the adiabatic quench cooling with three or four beds in series. In the adiabatic indirect cooling configuration, the reactor is divided into some series catalytic beds and interstage coolers are placed between beds. In this configuration, the optimal temperature to shift the equilibrium conversion of nitrogen is provided by cooling of outlet stream from each bed in a heat exchanger.

5. Ammonia plant simulation

5.1 Equilibrium-based models

Generally, two thermodynamic frameworks including stoichiometric and nonstoichiometric approaches are established to calculate the equilibrium concentration of species in a chemical system. In general, the stoichiometric approach is a practical method to model the ammonia process based on minimizing the Gibbs free energy, subject to the predefined reactions [40]. In this regard, the equilibrium constant of reactions is explained based on the Gibbs free energy. The heat balance equation is coupled with the equilibrium model, and the equilibrium concentration of components is determined. Under the ideal condition, the equilibrium constant of reactions is explained based on the equilibrium concentration of compounds as:

$$K_e = \prod_{i=1}^{N} P_i^{\vartheta_i} \tag{8}$$

The equilibrium constant is calculated based on Gibbs free energy as:

$$K_e = e^{-\frac{\Delta G^o}{RT}} \tag{9}$$

When the system operates under the adiabatic condition, the heat balance could be explained as:

$$\sum_{i=1}^{n} n_i H_i \bigg|_{Final} - \sum_{i=1}^{n} n_i H_i \bigg|_{Initial} = 0 \tag{10}$$

In the stoichiometric approach, the equilibrium model and heat balance equation are solved simultaneously and the equilibrium temperature and concentrations are determined. In the nonstoichiometric approach, the Gibbs free energy is minimized, subject to the mass and heat balance equations as:

$$G_t = \sum_{i=1}^{N} n_i G_i^o + \sum_{i=1}^{N} n_i RT \ln\left(\frac{y_i P \emptyset_i}{P^o}\right) \tag{11}$$

According to the mass conservation law, the number of atoms at the initial and final states remains constant.

$$\sum_{i}^{N} n_i a_{i,j} - N_j = 0 \tag{12}$$

where $a_{i,j}$ is the number of atoms of element j in the species of i, and N_j is the total number of atoms of the element.

5.2 Kinetic-based model

Since the kinetic approach incorporates the heat and mass balance equations, reaction kinetics, and transfer resistances in the model could predict the process condition and intermediate products at a specified time. The models applied to describe the catalytic reactors could be classified into two main classes including homogeneous and heterogeneous models [41]. In the heterogeneous model, the heat and mass transfer resistances are considered in the models, while in the homogeneous models, the system is considered as a single phase. In general, external and internal heat and mass diffusions are the major limiting mechanisms in commercial reformers. To develop a simple and accurate model for applied reactors in the ammonia plant, the importance of axial dispersion and radial gradients of concentration and temperature could be investigated. The concentration gradient in the catalyst and gas phases and temperature gradient in the gas are the main resistances in the industrial reactors applied in the ammonia synthesis process. Since the flow regime is turbulence, the plug flow assumption has good accuracy. The mass balance equations in the gas and solid phases of reactors are as follows:

$$-u_g \frac{1}{x^s} \frac{d(x^s C_i^g)}{dx} - a_v k_{gi}(C_i^g - C_i^s) = 0 \tag{13}$$

$$a_v k_{gi}(C_i^g - C_i^s) + \rho_B \sum_{i=1}^{N} \eta_j\, r_{ij} = 0 \tag{14}$$

The energy balance equation in the gas and solid phases of reactors is as follows:

$$-u_g C_t C_p \frac{1}{x^s} \frac{d(x^s T_g)}{dx} - a_v h_f(T_g - T_s) \pm \frac{P}{A_c} U(T_w - T_g) = 0 \tag{15}$$

$$a_v h(T_g - T_s) + \rho_B \sum_{i=1}^{N} \eta_j\, r_{ij}(-\Delta H_j) = 0 \tag{16}$$

The Ergun equation is used to compute the pressure drop along the catalytic reactors [21]:

$$\frac{dp}{dz} = \frac{150 \mu_g (1 - \varepsilon_b)}{d_p^2 \varepsilon_b^3} u_g + \frac{1.75(1 - \varepsilon_b)\rho_g}{d_p \varepsilon_b^3} u_g^2 \tag{17}$$

The superscript, s, in x^s is set 0 for the axial flow and 1 for the radial flow reactors. The effect of internal mass transfer resistance inside the catalyst is applied in the model by the effectiveness factor. The effectiveness factor is defined as the ratio of reaction rate considering mass transfer resistance in the particle to the reaction rate applying the surface concentration in the particle [42]. The applied absorption column could be simulated using kinetic and equilibrium approaches. In the equilibrium-based models, the outlet streams from a tray are assumed to be under the equilibrium condition [23]. The equilibrium-based model for a tray in the contactor is as follows:

$$L_{n-1} x_{i,n-1} + V_{n+1} y_{i,n+1} - V_n y_{i,n} - L_n x_{i,n} + r_i h_l = 0 \tag{18}$$

$$L_{n-1} H_{n-1} + V_{n+1} H_{n+1} - L_n H_n - V_n H_n + \Delta H_j = 0 \tag{19}$$

$$y_{i,n} = K_{i,n} x_{i,n} \tag{20}$$

The solubility coefficient could be calculated as:

$$K_{i,n} = \frac{P_i^* \, \delta_i^l}{P \, \varphi_i^v}$$ (21)

The fugacity of each component is determined by the equation of states. The equilibrium model is more popular compared with the kinetic model to simulate contractors, and it could predict the performance of absorption column with acceptable accuracy. The precision of equilibrium model could be modified by tray efficiency.

5.3 Process flowsheeting

Process simulation is the representation of chemical processes and unit operations in software. Currently, various traditional simulators, such as Aspen Hysys, Aspen Plus, Design II, gPROMS, ProSim, and Petro-Sim, are developed to simulate and optimize the chemical processes under steady and dynamic conditions. The simulators describe a process in a flow diagrams form where the units are connected by heat and mass streams. The software solves the mass and energy balance equations to find the operating points. The main steps to simulate the ammonia process are a selection of component lists, selection of fluid package, defining the reactions sets, and connecting streams and equipment. In this section, a first-generation ammonia synthesis plant with an operating capacity of 1300 tons per day is simulated in Aspen Hysys. Based on the recommendation of Aspen Plus, the Redlich–Kwong modification RKS-BM model provides an accurate description of the thermodynamic properties of ammonia plants. However, the Peng Robinson equation of state could predict the specification of hydrocarbon molecules in the reforming section, and the NRTL could be used in the ammonia synthesis section, due to the polarity of ammonia [43]. Fig. 8 shows the hydrogen mole fraction and temperature profile along with the primary reformer in the simulated ammonia plant. The steam reformer is a fired reactor consisting of a catalyst-filled coil tube hanged in a firebox to supply heat of reactions. Since the reforming reactions are endothermic, the heat of reaction is dominant over the heat transfer from the firebox toward the coil at the initial part of the reformer, and the temperature decreases gradually. In the second part, reducing the reactant concentration decreases the rate of reforming reactions, and temperature increases continuously. The effectiveness factor of catalyst in the primary and secondary reformers is in the range of 0.01–0.03, which indicates that the mass transfer resistance inside the catalyst particles controls the syngas production rate [44]. Fig. 9 shows the profile of temperature and hydrogen mole fraction along the catalytic part of the secondary reformer, respectively. The secondary reformer is separated into two parts, one for combustion and the other for catalysis. The air is combined with the primary reformer's output stream in the combustion section, and a part of hydrogen, carbon monoxide, and methane is oxidized. The heat generated raises the temperature of the incoming stream to the catalytic component and provides the necessary heat for the steam reforming processes to proceed. Because steam reforming processes are endothermic, the temperature lowers and reaches

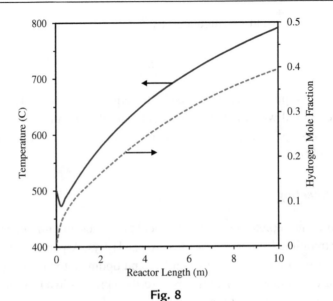

Fig. 8

Hydrogen model fraction and temperature along with the primary reformer.

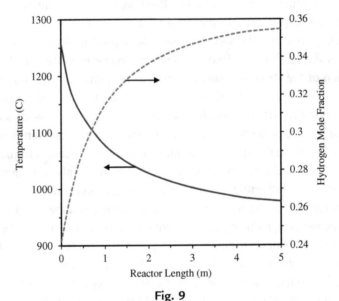

Fig. 9

Hydrogen model fraction and temperature along with the secondary.

equilibrium along the secondary reformer. It appears that temperature is a key parameter in the reforming process, and operating at high temperatures could enhance hydrogen production.

Fig. 10 shows the ammonia mole fraction versus operating temperature in the simulated external cooling ammonia reactor at pressure 290 bar. The considered converter includes four

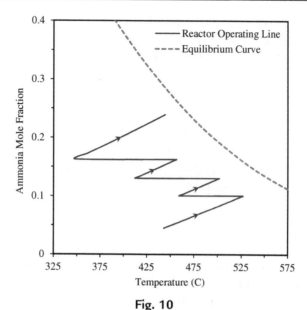

Fig. 10

The ammonia mole fraction versus operating temperature.

adiabatic beds in series equipped by the interstage coolers to cool down the reactor. Since the ammonia reactor is adiabatic and reaction is equilibrium limited, the temperature and ammonia yield increase along the beds, and the process approaches the equilibrium condition. Although cooling the feed temperature could enhance the conversion and ammonia yield in the beds, it decreases the rate of reaction, and more catalysts should be loaded into the system. In general, there is an optimum feed temperature to achieve the maximum ammonia yield and minimum catalyst loading. The presented research in the literature shows that optimization of the operating condition of an ammonia plant could enhance the production rate up to 15% [1].

6. Process optimization

The process optimization is performed in two different modes of operation including steady-state and dynamic optimization. Due to the deactivation of applied catalysts in the ammonia plant and decreasing the ammonia production rate and quality during the process run time, the real-time optimization of the process is a practical solution to obtain the optimal trajectory of inputs and maintain the production capacity at the desired level [45]. The operating temperature and pressure of primary and secondary reformers, shift converters, methanator, and ammonia synthesis reactor could be used as decision variables in the ammonia plant. In addition, the steam-to-methane ratio in the primary reformer and oxygen-to-hydrocarbon ratio in the secondary reformer have a considerable effect on the hydrogen production rate. When an optimization problem is formulated to optimize the operation condition of an ammonia plant, to avoid deactivation of reforming catalysts in the reformers, and reduce the operational cost, the steam-to-methane ratio is set in the range of 2.5–4.5 [46]. In addition, the oxygen-to-methane ratio in commercial autothermal reformers is set in the range of 0.25–0.35. Due to safety and material limitations, the upper bound of skin temperature is limited to 1300K in reformers [47]. Applying high-resistance metals could improve the thermal resistance of the coil. In addition, to avoid the deactivation of iron catalyst in the ammonia converter, the upper bound of total carbon monoxide and carbon dioxide in the inlet stream to the reactor is set to be 10 ppm. In general, applying more stable and active catalysts in syngas production and ammonia synthesis units could enhance the process productivity and lifetime. Finally, the programmed optimization problem could be handled by heuristic optimization methods such as genetic algorithm and differential evolution [48].

7. Conclusion and future outlook

According to the determined roadmap for ammonia economy, the Haber–Bosch process is the main technology for ammonia production, and reforming of hydrocarbons, coal gasification, biomass processing, and water electrolysis could provide the hydrogen in ammonia synthesis plants. Providing up to 96% of hydrogen demand through reforming of hydrocarbons has made the ammonia synthesis process one of the main sectors for global greenhouse gas emission. Since zero-carbon emission is available when the solar or wind energy sources are applied to produce hydrogen in the water electrolysis, it is an attractive route for hydrogen providing in ammonia plants. On the other hand, electrochemical processing has been proposed to produce ammonia through direct conversion of water and nitrogen with the elimination of independent hydrogen generation steps. In general, catalyst synthesis is an active research area, and applying more stable and active catalysts in syngas production and ammonia synthesis units could enhance the process productivity and lifetime. Due to the deactivation of applied catalysts in the ammonia plant and decreasing the ammonia production rate and quality during the process run time, the real-time optimization of the process is a practical solution to obtain the optimal trajectory of inputs and maintain the production capacity at the desired level.

Abbreviations and symbols

A_c cross-sectional area (m^2)
a_v specific surface area (m^2 m^{-3})
C concentration (mole m^{-3})
C_p special heat capacity (J mol^{-1} K)
D_i inner diameter of reformer tube (m)
ΔH_j heat of reaction j (J mol^{-1})
η effectiveness factor
F_g gas flow rate (mole s^{-1})
g gas

H_n enthalpy of stream exit nth tray (J mol^{-1})

h_g heat transfer coefficient (W m^{-2} k^{-1})

K_e phase equilibrium constant

k_{gi} mass transfer coefficient for species i (m s^{-1})

L reactor length (m)

L_n liquid flow rate (mole s^{-1})

μ viscosity (Pa s)

r_i reaction rate (mol kg$_{cat}^{-1}$ s^{-1})

ρ density (kg m^{-3})

s   catalyst surface

T temperature (K)

U overall heat transfer coefficient (W m^{-2} k^{-1})

u_g superficial velocity (m s^{-1})

V_n vapor flow rate (mole s^{-1})

x liquid mole fraction

y vapor mole fraction

z axial direction (m)

References

[1] M. Farsi, N. Chabi, M. Rahimpour, Modeling and optimization of ammonia process: effect of hydrogen unit performance on the ammonia yield, Int. J. Hydrogen Energy 46 (2021) 39011–39022.

[2] R. Lan, S. Tao, Ammonia as a suitable fuel for fuel cells, Front. Energy Res. 2 (2014) 35.

[3] C. Zhou, F. Chen, D. Yang, X. Jia, L. Zhang, J. Cheng, Ammonia: principles and industrial practice ammonia: principles and industrial practice, 1999, Chem. Lett. 38 (7) (2009) 708–709.

[4] V. Pattabathula, J. Richardson, Introduction to ammonia production, CEP Mag. 2 (2016) 69–75.

[5] T. Kandemir, M.E. Schuster, A. Senyshyn, M. Behrens, R. Schlögl, The Haber–Bosch process revisited: on the real structure and stability of "ammonia iron" under working conditions, Angew. Chem. Int. Ed. 52 (48) (2013) 12723–12726.

[6] D.R. MacFarlane, P.V. Cherepanov, J. Choi, B.H. Suryanto, R.Y. Hodgetts, J.M. Bakker, F.M.F. Vallana, A.N. Simonov, A roadmap to the ammonia economy, Joule 4 (6) (2020) 1186–1205.

[7] R.M. Nayak-Luke, R. Bañares-Alcántara, Techno-economic viability of islanded green ammonia as a carbon-free energy vector and as a substitute for conventional production, Energ. Environ. Sci. 13 (9) (2020) 2957–2966.

[8] X. Liu, S. Licht, Iron catalyst facilitated synthesis of Ammonia from N2 and water at high efficiency by electrolysis, in: ECS Meeting Abstracts, IOP Publishing, 2019.

[9] N. Shah, T. Lipman, Ammonia as an Alternative Energy Storage Medium for Hydrogen Fuel Cells: Scientific and Technical Review for Near-Term Stationary Power Demostration Projects, Final Report, UC Berkeley Transportation and Sustainability Research Center, 2007.

[10] D.-A. Chisalita, L. Petrescu, C.-C. Cormos, Environmental evaluation of european ammonia production considering various hydrogen supply chains, Renew. Sustain. Energy Rev. 130 (2020), 109964.

[11] D.M. Santos, C.A. Sequeira, J.L. Figueiredo, Hydrogen production by alkaline water electrolysis, Quím. Nova 36 (2013) 1176–1193.

[12] J. Zakzeski, AD ebczak, PCA Bruijnincx and BM Weckhuysen, Appl. Catal. A 394 (2011) 79–85.

[13] M. Farsi, Biomass conversion to biomethanol, in: M.R. Rahimpour, R. Kamali, M.A. Makarem, M.K. Dehghan Manshadi (Eds.), Advances in Bioenergy and Microfluidic Applications, Elsevier, 2021, pp. 231–252.

[14] C. Philibert, Renewable Energy for Industry, International Energy Agency, Paris, 2017.

[15] V. Singh, I. Dincer, M.A. Rosen, Life cycle assessment of ammonia production methods, in: I. Dincer, C.O. Colpan, O. Kizilkan (Eds.), Exergetic, Energetic and Environmental Dimensions, Elsevier, 2018, pp. 935–959.

[16] X. Liu, A. Elgowainy, M. Wang, Life cycle energy use and greenhouse gas emissions of ammonia production from renewable resources and industrial by-products, Green Chem. 22 (17) (2020) 5751–5761.

[17] N. Esteves, A. Sigal, E. Leiva, C. Rodríguez, F. Cavalcante, L. De Lima, Wind and solar hydrogen for the potential production of ammonia in the state of Ceará–Brazil, Int. J. Hydrogen Energy 40 (32) (2015) 9917–9923.

[18] C. Smith, A.K. Hill, L. Torrente-Murciano, Current and future role of Haber–Bosch ammonia in a carbon-free energy landscape, Energ. Environ. Sci. 13 (2) (2020) 331–344.

[19] B. Parkinson, M. Tabatabaei, D.C. Upham, B. Ballinger, C. Greig, S. Smart, E. McFarland, Hydrogen production using methane: techno-economics of decarbonizing fuels and chemicals, Int. J. Hydrogen Energy 43 (5) (2018) 2540–2555.

[20] M. Farsi, H. Shahhosseini, A modified membrane SMR reactor to produce large-scale syngas: modeling and multi objective optimization, Chem. Eng. Process. Process Intensif. 97 (2015) 169–179.

[21] M. Korobitsyn, F.P. Van Berkel, G. Christie, SOFC as a gas separator, Final report, 2000.

[22] S. Ahmed, M. Krumpelt, Hydrogen from hydrocarbon fuels for fuel cells, Int. J. Hydrogen Energy 26 (4) (2001) 291–301.

[23] Y. Ma, Y. Ma, G. Long, J. Li, X. Hu, Z. Ye, Z. Wang, C. Buckley, D. Dong, Synergistic promotion effect of MgO and CeO2 on nanofibrous Ni/Al2O3 catalysts for methane partial oxidation, Fuel 258 (2019), 116103.

[24] M.R. Rahimpour, M. Farsi, M.A. Makarem, Advances in Carbon Capture: Methods, Technologies and Applications, Woodhead Publishing, 2020.

[25] I. Rossetti, Reactor design, modelling and process intensification for ammonia synthesis, in: Inamuddin, R. Boddula, A.M. Asiri (Eds.), Sustainable Ammonia Production, Springer, 2020, pp. 17–48.

[26] D.A. Bell, B.F. Towler, M. Fan, Coal Gasification and Its Applications, William Andrew, 2010.

[27] J.M. Modak, Haber process for ammonia synthesis, Resonance 16 (12) (2011) 1159–1167.

[28] V.S. Marakatti, E.M. Gaigneaux, Recent advances in heterogeneous catalysis for ammonia synthesis, Synthesis 17 (2020) 18.

[29] J. Humphreys, R. Lan, S. Tao, Development and recent progress on ammonia synthesis catalysts for Haber–Bosch process, Adv. Energy Sustain. Res. 2 (1) (2021) 2000043.

[30] M. Penkuhn, G. Tsatsaronis, Comparison of different ammonia synthesis loop configurations with the aid of advanced exergy analysis, Energy 137 (2017) 854–864.

[31] A. Attari Moghaddam, U. Krewer, Poisoning of ammonia synthesis catalyst considering off-design feed compositions, Catalysts 10 (11) (2020) 1225.

[32] K. Waugh, D. Butler, B. Hayden, The mechanism of the poisoning of ammonia synthesis catalysts by oxygenates O2, CO and H2O: an in situ method for active surface determination, Catal. Lett. 24 (1) (1994) 197–210.

[33] P. Adamski, M. Nadziejko, A. Komorowska, A. Sarnecki, A. Albrecht, D. Moszyński, Chromium-modified cobalt molybdenum nitrides as catalysts for ammonia synthesis, Open Chem. 17 (1) (2019) 127–131.

[34] D.E. Brown, T. Edmonds, R.W. Joyner, J.J. McCarroll, S.R. Tennison, The genesis and development of the commercial BP doubly promoted catalyst for ammonia synthesis, Catal. Lett. 144 (4) (2014) 545–552.

[35] N. Junfang, X. Jiahan, L. Haichao, Activated carbon-supported ruthenium as an efficient catalyst for selective aerobic oxidation of 5-hydroxymethylfurfural to 2,5-diformylfuran, Chin. J. Catal. 34 (5) (2013) 871–875.

[36] H. Bielawa, O. Hinrichsen, A. Birkner, M. Muhler, The ammonia-synthesis catalyst of the next generation: barium-promoted oxide-supported ruthenium, Angew. Chem. Int. Ed. 40 (6) (2001) 1061–1063.

[37] G.R. Maxwell, Synthetic nitrogen products, in: J.A. Kent (Ed.), Handbook of Industrial Chemistry and Biotechnology, Springer, 2017, pp. 1125–1210.

[38] M.J. Palys, A. McCormick, E. Cussler, P. Daoutidis, Modeling and optimal design of absorbent enhanced ammonia synthesis, Processes 6 (7) (2018) 91.

[39] M.H. Khademi, R.S. Sabbaghi, Comparison between three types of ammonia synthesis reactor configurations in terms of cooling methods, Chem. Eng. Res. Des. 128 (2017) 306–317.

[40] H. Atashi, J. Gholizadeh, F.F. Tabrizi, J. Tayebi, S.A.H.S. Mousavi, Thermodynamic analysis of carbon dioxide reforming of methane to syngas with statistical methods, Int. J. Hydrogen Energy 42 (8) (2017) 5464–5471.

[41] G. Froment, Analysis and Design of Fixed Bed Catalytic Reactors, ACS Publications, 1972.

[42] M.F. Lari, M. Farsi, M. Rahimpour, Modification of a tri-reforming reactor based on the feeding policy to couple with methanol and GTL units, Chem. Eng. Res. Des. 144 (2019) 107–114.

[43] D. Frattini, G. Cinti, G. Bidini, U. Desideri, R. Cioffi, E. Jannelli, A system approach in energy evaluation of different renewable energies sources integration in ammonia production plants, Renew. Energy 99 (2016) 472–482.

[44] M. Sarkarzadeh, M. Farsi, M. Rahimpour, Modeling and optimization of an industrial hydrogen unit in a crude oil refinery, Int. J. Hydrogen Energy 44 (21) (2019) 10415–10426.

[45] M. Taji, M. Farsi, P. Keshavarz, Real time optimization of steam reforming of methane in an industrial hydrogen plant, Int. J. Hydrogen Energy 43 (29) (2018) 13110–13121.

[46] M. Pen, J. Gomez, J.G. Fierro, New catalytic routes for syngas and hydrogen production, Appl. Catal. A. Gen. 144 (1–2) (1996) 7–57.

[47] J. Rajesh, S.K. Gupta, G.P. Rangaiah, A.K. Ray, Multiobjective optimization of steam reformer performance using genetic algorithm, Ind. Eng. Chem. Res. 39 (3) (2000) 706–717.

[48] S.S. Rao, Engineering Optimization: Theory and Practice, John Wiley & Sons, 2019.

CHAPTER 14

Alcohols synthesis using syngas: Plant design and simulation

Mohammad Hasan Khademi, Afshar Alipour-Dehkordi, and Fereshteh Nalchifard

Department of Chemical Engineering, College of Engineering, University of Isfahan, Isfahan, Iran

1. Introduction

Nowadays, fossil fuel consumption has increased due to increasing industrial demand and population growth. Fossil fuel combustion releases large amounts of CO_2 as a greenhouse gas into the atmosphere, leading to several global environmental issues such as climate change and global warming. Thus, it is necessary to use an alternative clean energy to prevent these environmental issues. Methanol and higher alcohols (HAs) are promising chemical products that can be used as clean energy and hydrogen carriers [1,2].

Methanol is used to produce some valuable intermediate products such as formaldehyde, acetic acid (ethanoic acid), methyl tertiary butyl ether (MTBE), and silicone and acrylic polymers. Formaldehyde produces spandex fibers, urethane foam, and adhesives in laminated wood. Solvent esters, vinyl acetate monomer, and purified terephthalic acid (PTA) are produced from acetic acid. MTBE is used to enhance the gasoline octane number and acts as a source of oxygen. Furthermore, methanol could be utilized as a fuel in fuel cells and Indy race cars; and a blend of methanol and gasoline is considered as a flexible vehicle fuel [3].

A mixture of carbon monoxide and hydrogen, which is known as syngas, as well as CO_2, is used as feedstock for the methanol synthesis in large industrial scales. Eqs. (1)–(3) present the chemical reactions engaged in this process [4].

$$CO + 2H_2 \leftrightarrow CH_3OH \tag{1}$$

$$CO_2 + H_2 \leftrightarrow CO + H_2O \tag{2}$$

$$CO_2 + 3H_2 \leftrightarrow CH_3OH + H_2O \tag{3}$$

Advances in Synthesis Gas: Methods, Technologies and Applications. https://doi.org/10.1016/B978-0-323-91879-4.00006-0

The methanol synthesis process from natural gas includes three basic steps, which are briefly mentioned as follows [5]:

(i)   Syngas production.
(ii) Methanol production from syngas.
(iii) Distillation of reactor products to achieve the desired product specification.

HAs are valuable products that are used as feedstock in several chemical industries such as pharmaceutical. They could be utilized as promising fuel additives or H_2 carriers for fuel cells, assisting to clean energy production. C_2–C_5 alcohols (linear or branched) are the most practical HAs, which have different uses such as improving the octane number of transportation fuels, either alone or in gasoline mixtures, which improves the performance of engine. Short-chain alcohols also may be used as solvents in a wide range of products, for instance, coatings, paints, printing inks, cleaning agents, lacquer thinners, and adhesives. C_6–C_{22} alcohols (long-chain) mainly play the role of intermediates to produce chemical products. For example, these are extensively utilized in the detergents as well as surfactants synthesis. Despite the wide application of HAs, there is no motivation to commercialize the Fischer–Tropsch (FT) process for the synthesis of HAs. Therefore, HAs are not produced in an industrial scale through FT process, and most researchers have focused on experimental studies in a laboratory or pilot scale [6].

In this chapter, given the importance of producing a variety of alcohols, an overview of commercial and novel configurations for methanol, ethanol, and HAs synthesis processes from syngas is given based on the process simulation and modeling approach. The description of these processes is also discussed along with the analysis of different types of catalytic reactors and the interpretation of the corresponding kinetic models.

2. Methanol and higher alcohols production

2.1 Commercial methanol plants

In 1830, the first methanol synthesis plant was presented, which operated on the basis of wood distillation. Because of high operating cost of wood distillation plants and the low methanol production rate, this method was replaced with the high-pressure synthesis route from syngas. The Baden Aniline and Soda Factory (BASF) Company established the commercialized high-pressure methanol synthesis plant from syngas using ZnO–Cr_2O_3 catalyst. After that, the Imperial Chemical Industries (ICI) Company designed the low-pressure syngas to methanol plant under the conditions of 200–300 °C and 50–100 bar. The required syngas for these processes was provided by steam reforming of methane [7]. A more detailed summary is given as follows:

2.1.1 High-pressure method (BASF process)

In 1923, the production of methanol was firstly launched through the BASF process in Leuna (Germany). Due to the high-pressure operating condition, 250–350 bar, this process was called high-pressure process, in which the temperature was in the range of 320–450°C. The BASF process was the main methanol production technology for over 45 years [8]. The catalyst employed for this process was ZnO–Cr$_2$O$_3$, which could hydrogenate CO to methanol. Additionally, this catalyst showed the selectivity of methane and other light hydrocarbons in the range of 2–5 wt% [9] and the stability to the sulfur, as well as chlorine compounds, which exist in the synthesis gas [10]. The BASF process had some disadvantages such as relatively low catalyst activity and high temperatures requirement, which led to lower equilibrium methanol level [11]. The last high-pressure methanol plants were closed in the mid-1980s [10].

2.1.2 Low-pressure method

Methanol synthesis from syngas could be a relatively expensive operation because of high-pressure conditions and high recycle ratio. In this regard, the economics of this process would be sensitive to changes in the plant design [12]. In the 1960s, the low-pressure methanol production plants were introduced instead of high-pressure plants. This process operates at pressures of 50–100 bar [3]. A new generation of copper-containing catalysts was unveiled for the low-pressure process with better selectivity and higher activity than the high-pressure catalyst [10]. ICI and Lurgi processes are the two main methanol production technologies that dominate the market [3]. By the year 2000, the ICI and Lurgi processes dedicated 60% and 30%, respectively, of the worldwide methanol production to themselves [10].

2.1.2.1 ICI process

ICI designed a method for methanol production in which sulfur-free syngas including CO and H$_2$ was reacted over highly selective Cu/ZnO/Al$_2$O$_3$ catalyst. This so-called "low-pressure" is characterized by mild reaction conditions, 35–55 bar and 200–300°C. This process could produce methanol more economically than the high-pressure method [10,11]. The highly active catalyst causes lower pressure and temperature condition, which results in the diminishing of the required duty for heat exchange and compression in the recycle loop. Besides, lower temperature decreases the coproduction rate of light hydrocarbons. Also, reducing the consumption of syngas and saving 5%–10% of cooling duty via elimination of the heat generated by the side reactions are another achievement of this process [9]. Furthermore, it prevents the premature aging caused by copper sintering [10].

Methanol production rate of this type of plants varies in the range of 50–2500 metric tons/day, and the typical one has a 1000 tons/day capacity. According to exothermicity of

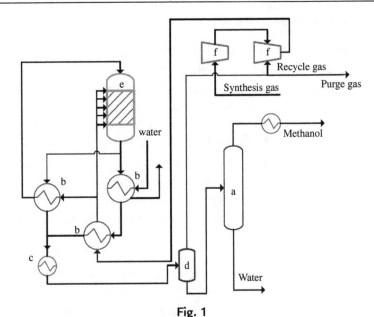

Fig. 1
The ICI low-pressure methanol production method: (A) methanol column, (B) heat exchanger, (C) cooler, (D) separator, (E) reactor, and (F) compressor.

methanol synthesis, ICI process plants are composed of a series of adiabatic beds with interbed cooling pattern via injection of cold syngas. The temperature of the beds changes in the range of 220–300°C depending on the position in the bed and activity of the catalyst. To compensate the reduction in catalysts activity due to the catalyst age, the temperature in the catalyst bed is gradually increased with time. Typically, industrial plants have 2 years of catalyst lifetime, but with proper operation conditions, 8 years of catalyst lifetime is acquirable [11]. Fig. 1 shows a typical ICI process with single-bed adiabatic reactor. The reaction is quenched via feeding cold gas at various points. So, the temperature profile along the reactor axis becomes the shape of a saw tooth [10]. In this type of reactor, different parts of reactor need to operate under different temperatures. This proper temperature distribution is obtained by controlling the feed temperature at each point. Small changes in feed temperature could make changes in the system performance such as catalyst lifetime reduction due to thermal fluctuations [12].

2.1.2.2 Lurgi process

The flow sheet of the methanol production loop designed by Lurgi Company is presented in Fig. 2. Lurgi-type reactor consists of vertical tubes, which are surrounded by boiling water. The methanol synthesis reaction occurs on the catalyst surface in the tube side, and the heat released by this reaction is transferred through the wall to the shell side (boiling water), and subsequently, steam is produced. Small temperature gradient along the reactor axis is obtained

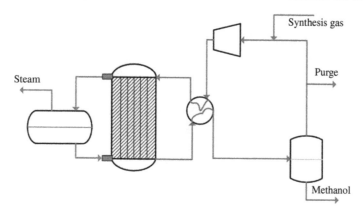

Fig. 2
Flow sheet of Lurgi's methanol synthesis loop [13].

by efficient heat transfer. This type of reactor typically operates at 523 K and 80 bar, and the temperature along the reactor is managed by changing the boiling water pressure. Only small amounts of by-products are produced due to the quasi-isothermal reaction conditions and high selectivity of catalyst [3]. The maximum temperature difference between the boiling water and the center of tube is 10–12°C, which indicates that the energy recovery is nearly optimum from the thermodynamic viewpoint [12]. Unreacted syngas is separated from crude methanol in the next steps, then compressed and recycled to the reactor [3].

Lurgi reactor design has some advantages such as high heat transfer area to catalyst bed volume ratio (ca. 80 m^2/m^3), which causes lower temperature drop across the tube wall. This leads to a better control of the catalyst bed temperature and products selectivity. Also, suitable thermal conditions are provided for the catalyst, which prevents fast deactivation. This system is not very sensitive to changes in the temperature of feed and changes can be controlled by shell-side pressure [12].

2.2 Ethanol production methods

The use of ethanol firstly dates back to the Neolithic period. In this era, ethanol was produced by the fermentation of fruits. In 1908, ethanol was introduced as a vehicle fuel, and firstly Ford's Model T used ethanol as engine fuel. After that, during World War I, demand for ethanol declined due to the production of gasoline, which was cheaper than ethanol, and gasoline became an interesting option. In this time, ethanol was mandatorily blended at higher ratios than 20 v/v% in vehicle fuels. Even though, the declining amount of sugarcane as a source of ethanol production and alternative lower price petroleum were challenges for ethanol industry. Before 2000s, there was no significant progress in ethanol production. After that, by controlling the use of MTBE, a main fuel additive for gasoline, ethanol was again introduced as an attractive additive for fuels. Since then, many efforts have been done to increase the efficiency

Table 1 Different methods for ethanol production [14].

Process	Intermediates	Advantages	Problems	Status
Sugars fermentation	Sugar	Efficiency > 90%	Elevating feedstock cost and conflicting with food supply	Commercialized
Lignocellulose fermentation		Abundant Feedstock	Low efficiency caused high cost	Precommercial
Ethylene hydration	Ethylene		Low efficiency and high cost	Vanished
Direct thermal conversion of syngas	Syngas	Abundant Feedstock	Wide products distribution	Precommercial
Fermentation of syngas	Syngas	Feedstock compatibility	Low efficiency	Pilot
Methanol Carbonylation	Acetic acid	High Efficiency	Homogenous catalysts	Commercialized
Dimethyl ether Carbonylation	Methyl acetate	High Efficiency		Pilot/ Commercializing
Syngas coupled with methanol	DMO	Multiproduct	Low atom economy	

of ethanol synthesis methods and diminish the investment costs. In this regard, different routes of ethanol production, such as fermentation of crops, ethylene hydration, and syngas to ethanol, have been employed. Ethanol can be produced by direct and indirect processes. Direct conversion in the container is performed using metal catalysts or by the fermentation process, whereas in the indirect conversing method, acetic acid/ester or dimethyl oxalate is used as intermediate for the ethanol synthesis. Table 1 shows the different methods of ethanol production and their challenges [14].

Ethanol is currently produced via two main processes, including the hydration of petroleum-based ethylene and the fermentation of sugarcane- or corn-derived sugars. Pure ethanol is produced in industrial scale by the first method, hydration of ethylene over a solid acid catalyst. However, the most ethanol is produced by the fermentation method, which is the commercial route for ethanol synthesis [15].

2.2.1 Production route of ethanol and higher alcohols from syngas

Fig. 3 shows the different pathways in which ethanol and HAs can be produced from syngas either directly or through methanol as an intermediate. The reaction network includes a set of complex reactions with different pathways in which various products can be produce depending on the catalyst, kinetic, and thermodynamic constraints. To show the complexity of the production of ethanol and HAs from syngas, some of the main reactions that can occur are introduced by Eqs. (4)–(12) [15].

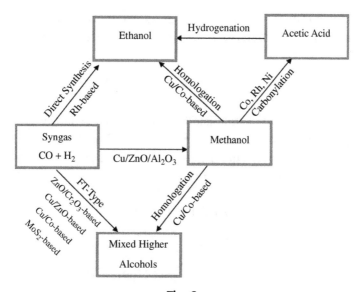

Fig. 3
Pathways for the ethanol and mixed higher alcohols production from synthesis gas [15].

$$CO + 2H_2 \rightarrow CH_3OH \qquad \text{Methanol synthesis} \qquad (4)$$

$$CO + H_2O \rightarrow CO_2 + H_2 \qquad \text{Water} - \text{gas shift} \qquad (5)$$

$$CH_3OH + CO + H_2 \rightarrow CH_3CHO + H_2O \qquad \text{CO Beta addition} \qquad (6)$$

$$CH_3OH + CO + 2H_2 \rightarrow CH_3CH_2OH + H_2O \qquad \text{Methanol homologation} \qquad (7)$$

$$C_nH_{2n-1}OH + CO + 2H_2 \rightarrow CH_3(CH_2)_nOH + H_2O \qquad \text{Higher alcohols homologation} \qquad (8)$$

$$2CH_3OH \rightarrow CH_3CH_2OH + H_2O \qquad \text{Condensation/coupling} \qquad (9)$$

$$2CH_3OH \rightarrow (CH_3)_2O + H_2O \qquad \text{Dehydration/DME formation} \qquad (10)$$

$$(CH_3)_2CO + H_2 \rightarrow (CH_3)_2CHOH \qquad \text{Branched iso-alcohols formation} \qquad (11)$$

$$2CH_3CHO \rightarrow CH_3COOCH_2CH_3 \qquad \text{Methyl ester formation} \qquad (12)$$

There are at least three different routes for catalytic ethanol and HAs production from syngas, which are mentioned below:

(i) Direct catalytic ethanol production from syngas where H_2 and CO react on the surface of the catalyst to produce ethanol. This reaction is shown by Eq. (13).

$$2CO + 4H_2 \rightarrow C_2H_5OH + H_2O \qquad (13)$$

(ii)   The second route is methanol homologation, shown by Eq. (14). This reaction describes reductive carbonylation of methanol over a redox catalyst surface to make a C—C bond.

$$CH_3OH + CO + 2H_2 \rightarrow C_2H_5OH + H_2O \tag{14}$$

(iii)   The third pathway is called the multistep ENSOL process, which including three steps. In the first step, methanol is produced from the syngas (Eq. (15)). After that, carbonylation of methanol to acetic acid occurs (Eq. (16)), and then, in the final step, acetic acid hydrogenation to ethanol takes place (Eq. (17)) [15].

$$CO + 2H_2 \rightarrow CH_3OH \tag{15}$$

$$CH_3OH + CO \rightarrow CH_3COOH \tag{16}$$

$$CH_3COOH + 2H_2 \rightarrow C_2H_5OH + H_2O \tag{17}$$

It should be noted that none of the above methods have been commercialized for the catalytic ethanol synthesis from syngas. Direct catalytic conversion of syngas is the most widely method to produce ethanol. Depending on the catalyst type, by-products consisting methane, olefins and C_2-C_5 alkanes, acetic acid, esters, aldehydes, and ketones are formed during both routes of (i) and (ii). Methanation (Eq. (18)) is one of the possible reactions that can affect the quality of the product. Therefore, to increase the selectivity and yield of ethanol, the catalyst and operating conditions should be modified to diminish the methanation reaction. Another side reaction that may occur is the water–gas shift (WGS) reaction (Eq. (19)), which is suitable for synthesis gas with a low H_2/CO ratio, because this reaction produces extra hydrogen [15].

$$CO + 3H_2 \rightarrow CH_4 + H_2O \tag{18}$$

$$CO + H_2O \rightarrow CO_2 + H_2 \tag{19}$$

Another method that converts synthesis gas to ethanol and other alcohols is syngas fermentation. In this method, syngas is converted to alcohols and organic acids by acetogenic microorganisms through the Wood–Ljungdahl metabolic pathway. Various alcohols, such as ethanol, butanol, propanol, and hexanol, and their corresponding organic acids can be produced by various acetogenic microorganisms. Nevertheless, ethanol and acetic acid are mostly produced, which are presented by Eqs. (20)–(23) [16].

$$4CO + 2H_2O \rightarrow CH_3COOH + 2CO_2 \tag{20}$$

$$2CO_2 + 4H_2 \rightarrow CH_3COOH + 2H_2O \tag{21}$$

$$6CO + 3H_2O \rightarrow C_2H_5OH + 4CO_2 \tag{22}$$

$$2CO_2 + 6H_2 \rightarrow C_2H_5OH + 3H_2O \tag{23}$$

Syngas fermentation has some significant advantages such as high product yield owing to the usage of all biomass fractions, including lignin and also, operating at mild temperatures and pressures. Even so, two of the main issues facing fermentation of syngas are low gas–liquid mass transfer rates and medium cost [16].

Contrary to favorable conditions for methanol synthesis, high pressures and low temperatures, higher alcohols are produced at high temperatures and pressures. For example, direct conversion of syngas to ethanol occurs at 573 K and 200 bar. To achieve maximum HAs synthesis rate, the H_2/CO ratio in the syngas should be approximately in the range of 1–2. Lower H_2/CO ratio could lead to deactivation of catalysts or active sites modification by carbide formation or deposition [15].

Rh, Mo, and Cu-based catalysts are the most proper catalysts that can be used to synthesize ethanol from syngas [14]. Furthermore, many studies have investigated the performance of HAs production on the basis of various catalysts. Different types of catalysts have been suggested for HAs synthesis such as noble metal catalysts, modified methanol catalyst, modified FT synthesis catalysts (nonnoble metals catalysts), and Mo-based catalysts. Cobalt-based catalysts can enhance the selectivity of alcohols, if proper additives were added. Besides, the addition of Mn could lead to an increase in the CO conversion and improvement in the HAs formation. The Co/Mn ratio has a considerable impact on the selectivity of alcohols and CO conversion [2].

3. Reactor modeling

Catalytic chemical reactors are one of the most important parts of chemical processes. The importance of this part is such that if this part can be modeled, simulated, and then optimized, the effect of this optimization on all equipment and downstream parts of the process can be clearly seen. Rigorous mathematical models are useful tools to study the reactor behavior in detail. Such an investigation could lead to significant savings in time and money at expensive pilot plant development stage. The use of the mathematical models together with the optimization tools could also attain appreciable benefits on the road toward the commercialization of new ideas.

3.1 Kinetic models

Kinetic modeling is one the main aspects of reactor design and optimization. Using the appropriate kinetic model to describe the reaction mechanism helps to have more accurate simulation and precise validation of the reactor with real data.

3.1.1 Kinetic models for methanol synthesis

Up to now, several investigators have studied the kinetic modeling of methanol synthesis from syngas. After decades, the most accurate kinetic pathways are still disputed. Even so, the hydrogenation of CO and CO_2 is the most common methanol synthesis pathway. In this section, the most usual kinetic models of syngas conversion to methanol over the Cu/ZnO catalyst are discussed. The methanol carbon source, the rate expression, and operating conditions used to derive the kinetic models are summarized in Table 2.

Natta [30] conducted the first study on the kinetic model of high-pressure methanol production over the ZnO/Cr_2O_3 catalyst in the temperature range of 300–360°C. Leonov et al. [17] and Agny and Takoudis [20] proposed kinetic rate equations for the low-pressure methanol synthesis. They did not consider CO_2 dependence in the rate expression and assumed that CO was the main source of methanol synthesis. Rozovskii and Lin [27] investigated the methanol synthesis over Cu-based catalyst on the basis of CO_2 hydrogenation as the main reaction. Klier et al. [18] and McNeil et al. [23] used CO, CO_2, and H_2 as the feedstock to obtain a kinetic model for methanol production over copper–zinc oxide catalysts based on hydrogenation of CO_2 and CO. They found that catalyst deactivation occurs at low CO_2 concentrations, and methanol production rate slows down at high CO_2 concentrations.

Villa et al. [19] considered the CO hydrogenation and WGS reaction as the only two pathways to produce methanol. Skrzypek et al. [24], Bussche and Froment [25], and Kubota et al. [26] investigated the rate expression for methanol production from the feed containing H_2, CO, and CO_2 on the CuO-ZnO-Al_2O_3 catalyst. The kinetic model was developed based on the Langmuir–Hinshelwood theory. They reported that the main reactions for methanol formation include WGS and CO_2 hydrogenation, and CO participates only in the WGS reaction.

Grabow and Mavrikakis [31] presented an extended microkinetic model for methanol synthesis and WGS reaction over a commercial $Cu/ZnO/Al_2O_3$ catalyst. They found that hydrogenation of CO and CO_2 is possible for methanol synthesis. In addition, under typical industrial operating condition, they mentioned that CO_2 was the source of 2/3 methanol produced.

Takagawa and Ohsugi [21], Graaf et al. [22], Lim et al. [28], and Park et al. [29] developed various kinetic models for the methanol production over commercial catalysts. They considered the hydrogenation of CO and CO_2 as well as WGS reaction under specified operating conditions to drive the rate equation. The kinetic model reported by Graaf et al. [22] is the most widely used kinetic model for simulation of industrial reactors, which has been confirmed by real data in many research studies. Lim et al. [28] mentioned that the rate of CO_2 hydrogenation is lower than the rate of CO hydrogenation. Park et al. [29] showed that although changes in pressure, temperature, and space velocity affect the CO conversion because of CO and CO_2 hydrogenation, these factors have no significant effect on CO_2 conversion due to the

Table 2 Rate expression of methanol synthesis, corresponding operating conditions, and methanol carbon source.

Rate expression	Operating conditions		Carbon source	Reference
	Temperature (K)	Pressure (MPa)		
$$r_{CO \to CH_3OH} = k\left(\frac{p_{CO}^{0.5} p_{H_2}}{p_{CH_3OH}^{0.66}} - \frac{p_{CH_3OH}^{0.34}}{p_{CO}^{0.5} p_{H_2} K_{eq}}\right)$$	493–533	4.0–5.5	CO	Leonov et al. [17]
$$r_{CO \to CH_3OH} = k_{r_{CO \to CH_3OH}}\left(1 + \frac{1}{K_{redox}^{eq}} \frac{P_{CO}}{P_{CO_2}}\right)^{-3} \frac{k_{CO} K_{H_2}^2 \left(P_{CO} P_{H_2}^2 - P_{CH_3OH}/K_{CO \to CH_3OH}^{eq}\right)}{\left(1 + K_{CO} P_{CO} + K_{CO_2} P_{CO_2} + K_{H_2} P_{H_2}\right)}$$	498–523	7.5	CO + CO$_2$	Klier et al. [18]
$$\left.\begin{aligned} &r_{CO_2 \to CH_3OH} = k_{r_{CO_2 \to CH_3OH}}\left(P_{CO_2} - \frac{1}{K_{r_{CO_2 \to CH_3OH}}^{eq}} \frac{P_{CH_3OH} P_{H_2O}}{P_{H_2}^3}\right) \\ &r_{CO \to CH_3OH} = \frac{f_{CO} f_{H_2}^2 - f_{CH_3OH}/K_{eq,1}}{(A + Bf_{CO} + Cf_{H_2} + Gf_{CO_2})^3} \\ &r_{RWGS} = \frac{f_{CO_2} f_{H_2} - f_{H_2O} f_{CO}/K_{eq,2}}{M^2} \end{aligned}\right\}$$	488–518	3.0–9.5	CO	Villa et al. [19]
$$r_{CO \to CH_3OH} = k\left(P_{CO} P_{H_2}^{0.5}\right)^{-1.3}\left(P_{CO} P_{H_2}^2 - \frac{P_{CH_3OH}}{K_{eq}}\right)$$	523–563	0.3–1.5	CO	Agny and Takoudis [20]
$$\left.\begin{aligned} &r_{CO \to CH_3OH} = k_1\left(f_{CO} f_{H_2}^{2.5}\right)^{0.35}\left[1 - \left(\frac{f_{CH_3OH}}{K_{eq,1} f_{CO} f_{H_2}^2}\right)^{0.8}\right]/\left(1 + K_{CO_2} f_{CO_2} + K_{H_2O} f_{H_2O}\right) \\ &r_{WGS} = k_2 f_{H_2}\left(1 - \frac{f_{CO} f_{H_2O}}{K_{eq,2} f_{CO_2} f_{H_2}}\right) \\ &r_{CO_2 \to CH_3OH} = k_3 f_{CO_2}\left[1 - \frac{f_{CH_3OH} f_{H_2O}}{K_{eq,3} f_{CO_2} f_{H_2}^3}\right]/\left(1 + K_{H_2O} f_{H_2O}\right) \end{aligned}\right\}$$	488–574	4.0–9.5	CO	Takagawa and Ohsugi [21]
	483–518	1.5–5.0	CO + CO$_2$	Graaf et al. [22]

Continued

Table 2 Rate expression of methanol synthesis, corresponding operating conditions, and methanol carbon source—Cont'd

Rate expression	Operating conditions		Carbon source	Reference
	Temperature (K)	Pressure (MPa)		
$r_{CO\to CH_3OH} = \dfrac{k_1 K_{CO}\left(f_{CO}f_{H_2}^{3/2} - f_{CH_3OH}/f_{H_2}^{1/2}K_1^{eq}\right)}{\left(1 + K_{CO}f_{CO} + K_{CO_2}f_{CO_2}\right)\left[f_{H_2}^{1/2} + \left(K_{H_2O}/k_{H_2}^{1/2}\right)f_{H_2O}\right]}$ $r_{RWGS} = \dfrac{k_2 K_{CO_2}\left(f_{CO_2}f_{H_2} - f_{CO}f_{H_2O}/K_2^{eq}\right)}{\left(1 + K_{CO}f_{CO} + K_{CO_2}f_{CO_2}\right)\left[f_{H_2}^{1/2} + \left(K_{H_2O}/k_{H_2}^{1/2}\right)f_{H_2O}\right]}$ $r_{CO_2\to CH_3OH} = \dfrac{k_3 K_{CO_2}\left(f_{CO_2}f_{H_2}^{3/2} - f_{CH_3OH}f_{H_2O}/f_{H_2}^{3/2}K_3^{eq}\right)}{\left(1 + K_{CO}f_{CO} + K_{CO_2}f_{CO_2}\right)\left[f_{H_2}^{1/2} + \left(K_{H_2O}/k_{H_2}^{1/2}\right)f_{H_2O}\right]}$	513	2.89–4.38	CO + CO$_2$	McNeil et al. [23]
$r_{CO/CO_2\to CH_3OH} = \dfrac{\left(P_{CO}P_{H_2}^2 - P_{CH_3OH}/K_{eq}\right)}{aP_{CO}P_{H_2}^{1.5}d' + bP_{H_2} + cP_{CO_2}} + \dfrac{\left(P_{CO_2}P_{H_2} - P_{CH_3OH}P_{H_2O}/K_{eq}''P_{H_2}^2\right)}{P_{CO_2}P_{H_2}^{0.5}d' + P_{CO_2}^2 b' + c'P_{H_2O}^3}$ $r_{CO_2\to CH_3OH} = k_1 K_{CO_2}K_{H_2}^2\,\dfrac{P_{H_2}^2 P_{CO_2} - \frac{P_{CH_3OH}P_{H_2O}}{P_{H_2}K_{eq}^{CO_2-CH_3OH}}}{\left(1 + P_{H_2}K_{H_2} + P_{CO}K_{CO} + P_{CO_2}K_{CO_2} + K_{CH_3OH}P_{CH_3OH} + K_{H_2O}P_{H_2O} + K_{CO}P_{CO}\right)^3}$ $r_{RWGS} = k_2 K_{CO_2}K_{H_2}\,\dfrac{P_{H_2}P_{CO_2} - \frac{P_{CO}P_{H_2O}}{P_{H_2}K_{eq}^{RWGS}}}{\left(1 + P_{H_2}K_{H_2} + P_{CO}K_{CO} + P_{CO_2}K_{CO_2} + K_{CH_3OH}P_{CH_3OH} + K_{H_2O}P_{H_2O} + K_{CO}P_{CO}\right)^2}$	460–550	3.0–9.0	CO$_2$	Skrzypek et al. [24]
$r_{CO_2\to CH_3OH} = \dfrac{k'_{5a}K'_2K_3K_4K_{H_2}P_{CO_2}P_{H_2}\left(1 - \frac{P_{CH_3OH}P_{H_2O}}{K^*P_{CO_2}P_{H_2}^3}\right)}{\left(1 + (K_{H_2O}/K_8K_9K_{H_2})(P_{H_2O}/P_{H_2})\sqrt{P_{H_2}K_{H_2}} + K_{H_2O}P_{H_2O}\right)^3}$ $r_{RWGS} = \dfrac{K'_1 P_{CO_2}\left[1 - K_3^*(P_{H_2O}P_{CO}/P_{CO_2}P_{H_2})\right]}{1 + (K_{H_2O}/K_8K_9K_{H_2})(P_{H_2O}/P_{H_2})\sqrt{P_{H_2}K_{H_2}} + K_{H_2O}P_{H_2O}}$	453–553	1.5–5.1	CO$_2$	Bussche and Froment [25]
$r_{CO_2\to CH_3OH} = \dfrac{k_M\left(P_{CO_2}P_{H_2} - \frac{P_{CH_3OH}P_{H_2O}}{K_M P_{H_2}^2}\right)}{\left(1 + P_{CO_2}K_{CO_2} + K_{H_2O}P_{H_2O}\right)^2}$ $r_{RWGS} = \dfrac{k_R\left(P_{CO_2} - \frac{P_{CO}P_{H_2O}}{P_{H_2}K_R}\right)}{1 + P_{CO_2}K_{CO_2} + K_{H_2O}P_{H_2O}}$	513	4.9	CO$_2$	Kubota et al. [26]

Reference	Reactants	Pressure	Temperature	Rate equations
Rozovskii and Lin [27]	CO + CO$_2$	5.2	513	$$r_{CO_2 \to CH_3OH} = \frac{P_{H_2} k_3 \left(1 - \frac{P_{CH_3OH}P_{H_2O}}{K_{p(CH_3OH)}P_{H_2}^3 P_{CO_2}}\right)}{1 + P_{H_2O}K_{-2} + K_{-2}P_{H_2O}/P_{CO_2}K_1}$$
Lim et al. [28]	CO + CO$_2$	5	523–553	$$r_{CO \to CH_3OH} = \frac{K_{H_2}^2 K_{CO} K_{r_{CO \to CH_3OH}} K_{CH,CO} \left[P_{CO}P_{H_2}^2 - \frac{P_{CH_3OH}}{K_{pr_{CO \to CH_3OH}}}\right]}{(1 + K_{CO}P_{CO})(1 + \sqrt{P_{H_2}}K_{H_2} + K_{H_2O}P_{H_2O})}$$ $$r_{WGS} = \frac{K_{H_2}^{0.5} K_{CO_2} K_{r_{WGS}} \left[P_{CO_2}P_{H_2} - \frac{P_{H_2O}P_{CO}}{K_{pr_{WGS}}}\right]}{\left(1 + K_{H_2}^{0.5}P_{H_2}^{0.5} + K_{H_2O}P_{H_2O}\right)(1 + P_{CO_2}K_{CO_2})P_{H_2}^{0.5}}$$ $$r_{CO_2 \to CH_3OH} = \frac{K_{H_2}K_{CO_2}K_{r_{CO_2 \to CH_3OH}}K_{CH,CO_2} \left[P_{CO_2}P_{H_2}^3 - \frac{P_{H_2O}P_{CH_3OH}}{K_{pr_{CO_2 \to CH_3OH}}}\right]}{(1 + K_{CO_2}P_{CO_2})(1 + \sqrt{P_{H_2}}K_{H_2} + K_{H_2O}P_{H_2O})P_{H_2}^2}$$
Park et al. [29]	CO + CO$_2$	5.0–9.0	503–613	$$r_{CO \to CH_3OH} = \frac{K'_{CO \to CH_3OH}K_{CO}\left[f_{CO}f_{H_2}^{1.5} - \frac{f_{CH_3OH}}{f_{H_2}^{0.5}K_{pr_{CO \to CH_3OH}}}\right]}{(1 + K_{CO}f_{CO})(1 + \sqrt{f_{H_2}}K_{H_2} + K_{H_2O}f_{H_2O})}$$ $$r_{WGS} = \frac{K_{CO_2}K'_{r_{WGS}}\left[f_{CO_2}f_{H_2} - \frac{f_{H_2O}f_{CO}}{K_{pr_{WGS}}}\right]}{\left(1 + K_{H_2}^{0.5}f_{H_2}^{0.5} + K_{H_2O}f_{H_2O}\right)(1 + f_{CO_2}K_{CO_2})}$$ $$r_{CO_2 \to CH_3OH} = \frac{K_{CO_2}K'_{r_{CO_2 \to CH_3OH}}\left[f_{CO_2}f_{H_2}^{1.5} - \frac{f_{H_2O}f_{CH_3OH}}{f_{H_2}^{1.5}K_{pr_{CO_2 \to CH_3OH}}}\right]}{(1 + K_{CO_2}f_{CO_2})(1 + \sqrt{f_{H_2}}K_{H_2} + K_{H_2O}f_{H_2O})}$$

WGS reaction. Besides, the content of CO and H_2 has a considerable influence on the conversion of CO and CO_2.

Recently, Leonzio [32] developed a mathematical model to investigate the methanol synthesis reactor performance using different kinetic models. They considered different rate expressions for methanol production from syngas, including Graaf et al. [22], Bussche and Froment [25], pseudo first-order and pseudo zero-order. Comparing the results of mathematical modeling with real data, they concluded that the kinetic model proposed by Bussche and Froment [25] predicts the performance of the methanol industrial reactor better than other models.

3.1.2 Kinetic models for higher alcohols synthesis

Kinetic rate expressions for HAs synthesis have been investigated by some researchers. Park et al. [33] developed a mechanistic kinetic model including initiation, chain growth, and termination steps to produce mixed alcohols from syngas. They performed the experiments at 523–623 K and 1.52–9.12 MPa over K_2CO_3-promoted MoS_2 catalyst. Beretta et al. [34] proposed a lumped kinetic model for HAs production over K-promoted Zn/Cr/O catalyst, which consists of methanol synthesis, WGS reaction, and formation of ethanol, propanol, isobutyl alcohol, C_{4+} alcohols, dimethyl ether, and methane. They also found that the HAs synthesis could be carried out in an adiabatic pilot reactor without uncontrolled risk of heat build-up by controlling the amount of methanol in the feed. Gunturu et al. [35] introduced kinetic models for syngas conversion to HAs and by-products over Co—Mo sulfide catalyst. The main products were linear alcohols up to n-butanol, but higher carbon alcohols and ethers were also found. They presented exponential model and Langmuir–Hinshelwood type rate expressions to describe methanol, ethanol, propanol, and hydrocarbons production rate. Kulawska and Skrzypek [36] reported a simple power law kinetic model to describe production of a mixture of methanol and higher aliphatic alcohols containing ethanol, n-propyl alcohol, linear C_4–C_7 aliphatic alcohols, and branched C_3–C_5 aliphatic alcohols from synthesis gas. Christensen et al. [37] derived a kinetic model for production of mixed alcohols from syngas and investigated the influence of operating conditions on the process performance with emphasis on the presence of H_2S in the feed gas. Increasing the H_2S concentration above 103 ppmv results in a fast stabilization of product distribution and makes the HAs as the dominant products, while with H_2S concentration less than 57 ppmv, methanol was a dominant product. Additionally, H_2S increased chain growth for both hydrocarbons and alcohols, but diminished alcohol selectivity by increasing the hydrocarbon formation. They reported a power-law rate expression for kinetic model of methanol, ethanol, 1-propanol, 1-butanol, methane, ethane, and propane as well as their kinetic parameters. Additionally, Surisetty et al. [38], Zaman and Smith [39], Guo et al. [40], Su et al. [41], Portillo et al. [42], Boahene and Dalai [43], and Göbel et al. [44] developed kinetic models for HAs synthesis form syngas and mixture of methanol/syngas under operating conditions and catalysts summarized in Table 3.

Table 3 Rate expressions of higher alcohols synthesis and corresponding operating conditions and catalyst.

Rate expression	Operating conditions		Catalyst	Reference
	Temperature (K)	Pressure (MPa)		
$r_{MeOHsynthesis} = K_{MeOH}\left(P_{CO}P_{H_2}^2 - \dfrac{P_{MeOH}}{K_{MeOH}}\right)K_{\gamma,MeOH}$ $\left\{\begin{array}{l} K_{MeOH} = exp\left(\dfrac{-\Delta G_{MeOH}^\circ}{RT}\right) \\ \Delta G_{MeOH}^\circ = -24306 + 58.57T \\ K_{\gamma,MeOH} = 1 - P(6.713 \times 10^{-5})\,exp\left(1.7308 \times \dfrac{10^3}{T}\right) \end{array}\right.$ $r_{WCS\,reaction} = K_{SHIFT}\left(P_{CO}P_{H_2O} - \dfrac{P_{H_2}P_{CO_2}}{K_{SHIFT}}\right)$ $\left\{\begin{array}{l} K_{SHIFT} = exp\left(\dfrac{-\Delta G_{SHIFT}^\circ}{RT}\right) \\ \Delta G_{SHIFT}^\circ = -8514 + 7.71T \end{array}\right.$ $r_{Ethanol\,formation} = k_{C1-C2}\dfrac{P_{MeOH}}{P_{H_2}}$ $r_{Propanol\,formation} = k_{C2-C3}\dfrac{P_{EtOH}/P_{H_2}}{1 + K_{H_2O}P_{H_2O}}$ $r_{Isobutyl\,alcohol\,formation} = k_{C3-iC4}\dfrac{P_{PrOH}/P_{H_2}}{\left(1 + K_{H_2O}P_{H_2O} + K_{HA}P_{HA}\right)^2}$ $r_{HA_{C4+}\,formation} = k_{C3-HA}\,sup\,P_{PrOH}$ $r_{DME\,formation} = k_{DME}P_{MeOH}$ $r_{Methane\,synthesis} = k_{CH_4}P_{H_2}$ $r_{HYD_{C2+}\,formation} = k_{HYD}P_{H_2}$	633–693	10–18	K-promoted Zn/Cr/O	Beretta et al. [34]
$r_{C_2H_5OH}^{gross} = A_1\,exp\left(\dfrac{-E_1}{R}\left(\dfrac{1}{T} - \dfrac{1}{T_{CP}}\right)\right)\left(\dfrac{P_{CO}}{P_{CO}^{op}}\right)^{a_1}\left(\dfrac{P_{H_2}}{P_{H_2}^{op}}\right)^{b_1}\left(\dfrac{P_1}{P_1^{op}}\right)^{c_1}\left(\dfrac{P_{CH_3OH}}{P_{CH_3OH}^{op}}\right)^{d_1}\left(\dfrac{t}{t^{op}}\right)^{-\lambda_1}$ $r_{C_3H_7OH}^{gross} = A_2\,exp\left(\dfrac{-E_2}{R}\left(\dfrac{1}{T} - \dfrac{1}{T_{CP}}\right)\right)\left(\dfrac{P_{CO}}{P_{CO}^{op}}\right)^{a_2}\left(\dfrac{P_{H_2}}{P_{H_2}^{op}}\right)^{b_2}\left(\dfrac{P_1}{P_1^{op}}\right)^{c_2}\left(\dfrac{P_{C_2H_5OH}}{P_{C_2H_5OH}^{op}}\right)^{d_2}\left(\dfrac{t}{t^{op}}\right)^{-\lambda_2}$ $r_{HC}^{gross} = A_3\,exp\left(\dfrac{-E_3}{R}\left(\dfrac{1}{T} - \dfrac{1}{T_{CP}}\right)\right)\left(\dfrac{P_{CO}}{P_{CO}^{op}}\right)^{a_3}\left(\dfrac{P_{H_2}}{P_{H_2}^{op}}\right)^{b_3}\left(\dfrac{P_1}{P_1^{op}}\right)^{c_3}\left(\dfrac{P_{CH_3OH}}{P_{CH_3OH}^{op}}\right)^{d_3}\left(\dfrac{t}{t^{op}}\right)^{-\lambda_3}$	573–623	2.76–6.89	Co–Mo sulfide	Gunturu et al. [35]
$r = k_0\,exp\left(-\dfrac{E}{RT}\right)P_{H_2}^n$	593–623	4–10	Cs-doped Cu—Zn composite	Kulawska and Skrzypek [36]

Continued

Table 3 Rate expressions of higher alcohols synthesis and corresponding operating conditions and catalyst—Cont'd

Rate expression	Operating conditions		Catalyst	Reference
	Temperature (K)	Pressure (MPa)		
$$R_i = A_i \exp\left(\frac{-E_{ai}}{RT}\right)\left(P_{H_2}^{pai}\, P_{CO}^{pbi}\, P_{H_2S}^{pci}\right)$$	548–623	10	K_2CO_3/Co/ MoS_2/C	Christensen et al. [37]
$$r_{CH_3OH} = k_{CH_3OH}\left(P_{CO}^a P_{H_2}^b - P_{CH_3OH}^c / K_{CH_3OH}\right)$$ $$\left\{ K_{MeOH} = \exp\left(\frac{-\Delta G_{MeOH}^\circ}{RT}\right) K_{\gamma,MeOH} \right.$$ $$\Delta G_{MeOH}^\circ = -24306 + 58.57T$$ $$K_{\gamma,MeOH} = 1 - P(6.713 \times 10^{-5})\, \exp\left(1.7308 \times \frac{10^3}{T}\right)$$	548–623	5.52–9.65	Co-Rh-Mo	Surisetty et al. [38]
$$r_{C_2H_5OH} = k_{C_2H_5OH}\, P_{CH_3OH}^d$$ $$r_{HA} = k_{HA}\, P_{CH_3OH}^e P_{C_2H_5OH}^f$$ $$r_{HC} = k_{Hc}\, P_{CO}^g P_{H_2}^h$$ $$r_{CO_2} = k_{CO_2}\left(P_{CO}^i P_{H_2}^j - P_{CO_2}^k P_{H_2}^l / K_{WS}\right)$$ $$r_i = A_i\, \exp\left(\frac{-E_i}{RT}\right) P_{CO}^m P_{H_2}^n$$	548–613	–	Rh-K-MoP/ SiO_2	Zaman and Smith [39]
$$r_{CH_3OH} = k_{CH_3OH}\left(P_{CO}P_{H_2}^2 - P_{CH_3OH}/K_{CH_3OH}\right)$$ $$\left\{ K_{MeOH} = \exp\left(\frac{-\Delta G_{MeOH}^\circ}{RT}\right) K_{\gamma,MeOH} \right.$$ $$\Delta G_{MeOH}^\circ = -24306 + 58.57T$$ $$K_{\gamma,MeOH} = 1 - P(6.713 \times 10^{-5})\, \exp\left(1.7308 \times \frac{10^3}{T}\right)$$ $$r_{C_2H5OH} = k_{C_2H5OH}\, P_{CH_3OH}^2$$ $$r_{HA} = k_{HA}\, P_{CH_3\,OH}^{(4-m)} P_{C_2H_5OH}^{(m-2)}$$ $$r_{HC} = k_{Hc}\, P_{CO}^n P_{H_2}^{(2n+1)}$$ $$r_{CO_2} = k_{CO_2}\left(P_{CO}P_{H_2O} - P_{CO_2}P_{H_2}/K_{WS}\right)$$	523–613	4–7	CFCK(2.5)/ CZA(2.5)	Guo et al. [40]
$$r_A = k\, P_{CO}^\beta P_{H_2}^\gamma$$ $$k = k_0 Exp\left(\frac{-E_a}{RT}\right)\left(P_{CO}^A P_{H_2}^B\right)$$	503–563	5	CoCu/SiO_2	Su et al. [41]
$$r_{MeOH} = A_1\, \exp\left(-\frac{E_{a1}}{RT}\right)\left(P_{CO}^A P_{H_2}^B\right)$$ $$r_{CH_4} = A_2\, \exp\left(-\frac{E_{a2}}{RT}\right)\left(P_{H_2}^C P_{MeOH}^D\right)$$	563–583	8–10	alkali-Co doped molybdenum sulfide	Portillo et al. [42]

Rate expressions				
$r_{EtOH} = A_3\,exp\left(-\dfrac{E_{a3}}{RT}\right)\left(P_{H_2}^F P_{MeOH}^G P_{CO}^E\right)$ $r_{PrOH} = A_4\,exp\left(-\dfrac{E_{a4}}{RT}\right)\left(P_{CO}^H P_{H_2}^I P_{EtOH}^J\right)$ $r_{CO_2} = A_5\,exp\left(-\dfrac{E_{a5}}{RT}\right)\left(P_{CO}P_{H_2O} - \dfrac{k_{sc}}{K_{WS}}P_{H_2}P_{CO_2}\right)$	563–623	5.52–9.65	KCoRhMo/CNH	Boahene and Dalai [43]
$r_{MeOH} = A_1\,exp\left(-\dfrac{E_{a,M}}{RT}\right)\left(P_{CO}^a P_{H_2}^b\right)$ $r_{CH_4} = A_2\,exp\left(-\dfrac{E_{a,CH_4}}{RT}\right)\left(P_{MeOH}^c P_{H_2}^d\right)$ $r_{EtOH} = A_3\,exp\left(-\dfrac{E_{a,E}}{RT}\right)\left(P_{MeOH}^e P_{CO}^f P_{H_2}^g\right)$ $r_{PrOH} = A_4\,exp\left(-\dfrac{E_{a,Pr}}{RT}\right)\left(P_{EtOH}^h P_{CO}^i P_{H_2}^j\right)$ $r_{BuOH} = A_5\,exp\left(-\dfrac{E_{a,Bu}}{RT}\right)\left(P_{PrOH}^k P_{CO}^l P_{H_2}^m\right)$ $r_{PeOH} = A_6\,exp\left(-\dfrac{E_{a,Pe}}{RT}\right)\left(P_{BuOH}^n P_{CO}^o P_{H_2}^p\right)$ $r_{CO_2} = A_7\,exp\left(-\dfrac{E_{a,CO_2}}{RT}\right)\left(P_{CO}P_{H_2O} - \dfrac{k_{sc}}{K_{WGS}}P_{H_2}P_{CO_2}\right)$ $r = k P_{CO}^m P_{H_2}^n$	538–568	1.5–6	Co-Cu	Göbel et al. [44]

3.2 Intraparticle diffusion model

One of the key parameters in modeling and simulation of methanol synthesis reactors is effectiveness factor, which indicates the limitations of mass and heat transfer inside the catalyst. This parameter has been investigated in many cases of low-pressure industrial reactors with different approaches. In this regard, Lommerts et al. [45] analyzed the accuracy of mass transfer models with different complexities to describe the intraparticle diffusion limitations in the methanol production reactor. They presented five models as follows:

(a) The dusty-gas model.
(b) The extended or modified Stefan–Maxwell equations.
(c) Classic multicomponent pore diffusion model coupled with convective mass transfer.
(d) Classic multicomponent pore diffusion model.
(e) Using Thiele modulus concept.

In the dusty-gas model (Eq. (24)), both convective and diffusional mass transfers and pressure drop over the catalyst pellet due to the concomitant convective transfer of molecules and the reaction stoichiometry are considered.

$$\frac{N_i}{D_{k,i}^e} + \sum_{\substack{j=1 \\ j \neq i}}^{N} \left(\frac{y_j N_i - y_i N_j}{D_{i,j}^e} \right) = -\frac{P}{RT}\frac{dy_i}{dr} - \frac{y_i}{RT}\left(1 + \frac{B_0 P}{D_{k,i}^e \mu_{gas}}\right)\frac{dp}{dr} \tag{24}$$

In the modified Stefan–Maxwell model (Eq. (25)), isobaric conditions are assumed. So, the dusty-gas model is reduced to the modified Stefan–Maxwell equation by neglecting the pressure drop in the catalyst particle.

$$\frac{N_i}{D_{k,i}^e} + \sum_{\substack{j=1 \\ j \neq i}}^{N} \left(\frac{y_j N_i - y_i N_j}{D_{i,j}^e} \right) = -\frac{dC_i}{dr} \tag{25}$$

In this equation, the net flux ($\sum N_i \neq 0$) is calculated by the nonequimolar stoichiometry of reaction, and the impact of convective transfer of molecules is considered, leading to an unrestricted net flow of gas molecules toward the center of the catalyst pellet during methanol synthesis. Therefore, the total mass balance under isobaric conditions (i.e., $dC_t/dr = 0$) is obeyed throughout the particle.

Assuming that the catalyst particle is spherical, the partial mass balance for component i is as follows:

$$\frac{dN_i}{dr} + \frac{2}{r}N_i - \sum r_i''' = 0 \tag{26}$$

where the diffusional flux J_i and the convective flow Φ form the total molar flux of component i, N_i.

$$N_i = J_i + \Phi \frac{C_i}{4\pi r^2} \tag{27}$$

Substituting Eq. (27) into Eq. (26) results in:

$$\frac{dJ_i}{dr} + \frac{2}{r}J_i + \frac{\Phi}{4\pi r^2}\frac{dC_i}{dr} + \frac{d\Phi}{dr}\frac{C_i}{4\pi r^2} - \sum r_i''' = 0 \tag{28}$$

The multicomponent pore-diffusion models and Thiele modulus model are derived based on Eq. (28). The convective mass transfer expression (Eq. (29)) is determined under the isobaric conditions from the summation of Eq. (28) for all species and taking equimolarity of counter diffusion ($\sum J_i = 0$). After substitution of Eq. (29) in Eq. (28) and considering the Fick's law, differential Eq. (30) is obtained. The multicomponent pore diffusion model coupled with convective mass transfer (Eqs. (29) and (30)) is solved with boundary conditions (31) and (32).

$$\frac{d\Phi}{dr} = -4\pi r^2 \frac{\left(2r_1''' + 2r_2'''\right)}{C_t} \tag{29}$$

$$D_{mi}^e \frac{d^2 C_i}{dr^2} - \left(\frac{2}{r}D_{mi}^e - \frac{\Phi}{4\pi r^2}\right)\frac{dC_i}{dr} + \left(2r_1''' + 2r_2'''\right)\frac{C_i}{C_t} + \sum r_i''' = 0 \tag{30}$$

$$\text{At } r = R_p \ C_i = C_i^0 \tag{31}$$

$$\text{At } r = 0 \ J_i = 0 \text{ or } \Phi = 0 \tag{32}$$

Eq. (30) is reduced to the multicomponent pore diffusion model (Eq. (33)) by neglecting the convective mass transfer. This model could be solved by the same boundary conditions as for the previous model.

$$D_{mi}^e \left(\frac{d^2 C_i}{dr^2} - \frac{2}{r}\frac{dC_i}{dr}\right) + \sum r_i''' = 0 \tag{33}$$

The effectiveness factors for the production of CH_3OH and H_2O are calculated on the basis of total molar flux at the catalyst surface.

$$\eta_i = \frac{3}{R}\frac{(N_i)_{r=R_p}}{\left(\sum r_i'''\right)_{r=R_p}} \tag{34}$$

Eq. (33) can be solved analytically by substituting pseudo-first-order rate expressions to $\sum r_i'''$. The Thiele modulus, M_T, is defined as follows when applying the first-order kinetic rates for the CH_3OH and H_2O production.

$$M_T = \frac{r_p}{3} \sqrt{\frac{k_i''' \left(K_i^{eq} + 1\right)}{D_{mi}^e K_i^{eq}}} \tag{35}$$

where k_i''' and K_i^{eq} are pseudo-first-order rate and pseudo-equilibrium constants, respectively, for the CH_3OH or H_2O production. k_{CH_3OH}''' and k_{H_2O}''' can be obtained from calculating the real kinetic rate expressions (i.e., r_i''') reported by Graaf et al. [22] and linearization of those real expressions according to:

$$r_{CH_3OH}''' = k_{CH_3OH}''' \left(C_{H_2} - C_{CH_3OH}/K_{CH_3OH}^{eq}\right) \tag{36}$$

$$r_{H_2O}''' = k_{H_2O}''' \left(C_{H_2} - C_{H_2O}/K_{H_2O}^{eq}\right) \tag{37}$$

where $K_{CH_3OH}^{eq}$ and $K_{H_2O}^{eq}$, which are dependent on composition, pressure, and temperature, are presented by:

$$K_{CH_3OH}^{eq} = \left(\frac{C_{CH_3OH}}{C_{H_2}}\right)_{eq} \tag{38}$$

$$K_{H_2O}^{eq} = \left(\frac{C_{H_2O}}{C_{H_2}}\right)_{eq} \tag{39}$$

The effectiveness factor can be determined using Thiele modulus concept according to pseudo-first-order kinetic assumption [45].

$$\eta_i = \frac{1}{M_T} \frac{(3M_T)\coth(3M_T) - 1}{3M_T} \tag{40}$$

Lommerts et al. [45] showed that using the Thiele modulus concept to consider intraphase diffusion and surface reaction phenomena could be beneficial as a result of easier application in methanol synthesis reactor modeling, decreasing computation time, and good agreement with other complex models, such as the dusty-gas or the Stefan–Maxwell model. However, it should be noted that the dusty-gas model is widely used to calculate intraparticle diffusion limitation in methanol synthesis reactor simulation and modeling studies [46–58].

For the first time, Graaf et al. [59] adopted dusty-gas model to calculate intraparticle diffusion inside the commercial Cu-Zn-Al methanol synthesis catalyst under the operating conditions of 210–275°C and 10–50 bar, with the inlet feed containing CO, CO_2, and H_2. To calculate the effectiveness factor of methanol synthesis reactions based on the dusty-gas model, the material balances and the stoichiometric relations (Eqs. (41) and (42)) with the assumption of spherical and isothermal particles were presented in detail. The auxiliary parameters and the pressure drop in radial direction are given by Eqs. (43)–(46). In addition, the boundary conditions required to solve these equations are presented in Eq. (47). Finally, effectiveness factor of reaction k is calculated by Eq. (48) [49].

$$\frac{d\Omega_k}{dr} = r^2 r_k \quad k = 1, 2, 3, 4 \tag{41}$$

$$\frac{dy_i}{dr} = -\frac{1}{r^2} \left(\sum_{k=1}^{3} \Omega_k \sum_{j=1}^{5} v_{kj} F_{ij} \right), \quad i = 1, 2, \ldots, N-1 \tag{42}$$

$$F_{ii} = \frac{RT}{P} \left(\frac{1}{D_i^{\text{Eff}}} + \sum_{j=1, j \neq i}^{5} \frac{y_i}{D_{ij}^{\text{Eff}}} - \frac{y_i}{D_i^{\text{Eff}}} \left(1 + \frac{B_0 P}{\mu D_i^{\text{Eff}}} \right) \frac{1}{w} \right) \tag{43}$$

$$F_{ij} = -\frac{RT}{P} \left(\frac{y_i}{D_i^{\text{Eff}}} + \frac{y_i}{D_j^{\text{Eff}}} \left(1 + \frac{B_0 P}{\mu D_i^{\text{Eff}}} \right) \frac{1}{w} \right) \tag{44}$$

$$w = 1 + \frac{B_0 P}{\mu} \sum_{i=1}^{N} \frac{y_i}{D_i^{\text{Eff}}} \tag{45}$$

$$\frac{dP}{dr} = -\frac{1}{r^2} \frac{RT}{w} \sum_{i=1}^{N} \left(\frac{1}{D_i^{\text{Eff}}} \sum_{k} v_{ik} \Omega_k \right) \tag{46}$$

$$\begin{cases} \text{at } r = 0 \quad \Omega_k = 0 \\ \text{at } r = R_p \quad y_i = y_i^s \\ \text{at } r = R_p \quad P = P^s \end{cases} \tag{47}$$

$$\eta_k = \frac{\int_0^{R_p} r_k dr}{R_p r_k^s} \tag{48}$$

Their results showed that the industrial catalyst size has intraparticle diffusion limitation, and the effectiveness factors vary from 0.32 to 0.92 depending on the operating conditions. They mentioned, under low-pressure condition, Knudsen diffusion is the main diffusion mechanism and under high-pressure condition, bulk diffusion is the dominant diffusion mechanism. In addition, although the effect of bulk gas composition and pressure on the effectiveness factor is small, this factor is strongly dependent on temperature and decreases with increasing temperature [59].

3.3 Catalyst deactivation model

The catalyst can be used for several years in an industrial-scale methanol plant. Over this time, the reactor temperature is gradually increased to compensate for the loss of catalyst activity. Catalyst sintering and poisoning are the two phenomena that have a considerable effect on deactivation of catalysts. Catalyst sintering reduces the active site surface area, which causes limitation to the mass transfer phenomena. The proper sulfur concentration in the feed should not exceed 0.5 ppm, but less than 0.1 ppm is preferred to prevent poisoning [60]. The sintering rate depends on the temperature of the reactor, and it is observed at temperatures above 500 K. In addition, ZnO crystallites form at temperatures higher than 573 K, which causes the catalyst to sinter further [61]. It should be noted that the H_2O concentration raises the catalyst sintering, decreases the catalyst active sites, and thereby inactivates the catalyst [62].

Sun et al. [60] investigated the impact of feed composition on the $Cu/ZnO/Al_2O_3$ methanol synthesis catalyst deactivation. Accordingly, they considered two cases of differential and finite conversion reactor. In the differential conversion reactor, the catalyst was not exposed to the product gas, and just the influence of CO/CO_2 ratio on catalyst deactivation was studied. In this case, the results showed that the main reason for catalyst deactivation was CO concentration, and CO_2 has an insignificant influence. In the finite conversion reactor, the catalyst deactivation was investigated in the presence of product gas, especially H_2O. In this case, the CO_2 concentration increases the catalyst deactivation, and the CO has a lower impact on the deactivation of catalyst compared with the differential mode.

Among the deactivation rates of commercial methanol synthesis catalysts, the model proposed by Hanken [63] has a good agreement with the data of industrial methanol synthesis reactors due to the adoption of industrial plant data [64]. Besides, it should be mentioned that this model (Eq. (49)) has been used in many dynamic modelings and simulations of methanol synthesis reactor, considering catalyst deactivation.

$$\frac{da}{dt} = -K_d \exp\left(\frac{-E_d}{R}\left(\frac{1}{T} - \frac{1}{T_R}\right)\right)a^5 \tag{49}$$

where a is the activity of the catalyst, and K_d, E_d, and T_R refer to the deactivation constant, activation energy, and temperature of the catalyst at the reference point, respectively, and the values of these parameters are: $K_d = 0.00439\ \frac{1}{hr}$, $E_d = 91270\ \frac{J}{mol}$, and $T_R = 513\ K$ [63].

Furthermore, Kuechen and Hoffmann [65] reported a deactivation model (Eq. (50)) for Cu/ZnO catalysts under operating conditions of 483–543K and 3–5MPa that depends on the temperature and fugacity of CO and CO_2.

$$\frac{da}{dt} = -K_d(T)\left(\frac{f_{CO}}{f_{CO_2}}\right)^m a^2 \tag{50}$$

Sahibzada et al. [66] investigated the deactivation rate of Pd-promoted $Cu/ZnO/Al_2O_3$ catalysts at 523 K and 5 MPa. They proposed a power-law kinetic model for the deactivation of the catalyst but did not report rate constants. Rahimpour et al. [67] adopted the Langmuir–

Hinshelwood–Hougen–Watson type kinetic model to describe catalyst deactivation rate of Cu/ZnO catalyst. In this model, two types of active sites were assumed, in which some active sites were deactivated by CO and others by CO_2. The kinetic models shown by Eqs. (51) and (52) are related to the catalyst deactivation by CO_2 and CO, respectively.

$$\frac{da_1}{dt} = -a_1^{d_1}\left(\frac{k_{d_1}p_{CO_2}^2}{1 + K_{CO}p_{CO} + K_{CO}K_H^{3/2}K_{CH}p_{CO}p_{H_2}^{3/2}}\right) \tag{51}$$

$$\frac{da_2}{dt} = -a_2^{d_2}\left(\frac{k_{d_2}p_{CO}^2}{1 + K_{CO_2}p_{CO_2} + K_{CO_2}K_H^{1/2}K_{HCO_2}p_{CO_2}p_{H_2}^{1/2}}\right) \tag{52}$$

3.4 Mass and energy conservation laws

To date, many research studies have been theoretically accomplished on modeling of various types of reactor configurations to enhance the methanol productivity, which include the following: fixed-bed reactor, membrane reactor, fluidized-bed reactor, thermally coupled reactor, sorption-enhanced methanol synthesis reactor, slurry-bubble column reactor, moving-bed reactor, monolithic reactor, trickle-bed reactor, and radial- and axial-flow spherical-bed reactor. This section includes reactor modeling based on mass and energy conservation laws as well as model assumptions to describe the concentration and temperature profiles in various reactors.

3.4.1 Fixed-bed reactor

Fixed-bed reactors are the most common reactors in the methanol synthesis process. Although an accurate mathematical model is required to predict the actual behavior of the methanol synthesis reactor, employing complex phenomena in the mathematical model increases the computational time and, in some cases, does not have a significant effect on the reliability of the model. Considering that, many researchers simplified the general mathematical model with appropriate assumptions to enhance simplicity and accuracy at the same time. Two different approaches, namely pseudo-homogeneous and heterogeneous models, have been used in modeling and simulation of methanol synthesis reactors. In the pseudo-homogeneous model, the gradients between the solid (catalysts) and the gas phase is ignored, but in the heterogeneous model, the intraparticle diffusion limitation is considered, and the model includes the mass and energy balances for both solid and gas phases. The general form of one- and two-dimensional heterogeneous and pseudo-homogeneous mathematical models under unsteady-state conditions is presented in Table 4.

Table 4 One- and two-dimensional heterogeneous and pseudo-homogeneous mathematical models under unsteady-state conditions.

Equation	Conservation law	Phase
Dynamic two-dimensional heterogeneous model		
$\varepsilon_s \dfrac{\partial C_{is}}{\partial t} = a_v C_t k_{g,i}\left(y_i - y_s\right) + \eta r_i \rho_B a$	Mass	Solid
$\left((1-\varepsilon_s)\rho_s C_{p,s} + \varepsilon_s \displaystyle\sum_{i=1}^{N_c}\left(\rho_{s,i} C_{p,s,i}\right)\right)\dfrac{\partial T_s}{\partial t} = a_v h_f\left(T - T_s\right) + \rho_B a \displaystyle\sum_{j=1}^{N}\eta r_j\left(-\Delta H_{R,j}\right)$	Energy	Solid
$\varepsilon_B \dfrac{\partial C_i}{\partial t} = -\dfrac{\partial(u C_i)}{\partial z} + a_v C_t k_{g,i}\left(y_{is} - y_i\right) + \varepsilon_B \dfrac{\partial}{\partial z}\left(D_z \dfrac{\partial C_i}{\partial z}\right) + \dfrac{\varepsilon_B}{r}\dfrac{\partial}{\partial r}\left(r D_r \dfrac{\partial C_i}{\partial r}\right)$	Mass	Fluid
$\varepsilon_B C_t c_{pg}\dfrac{\partial T}{\partial t} = -u\,C_t c_{pg}\dfrac{\partial T}{\partial z} + a_v h_f\left(T_s - T\right) + \varepsilon_B \dfrac{\partial}{\partial z}\left(\lambda_z \dfrac{\partial T}{\partial z}\right) + \dfrac{\varepsilon_B}{r}\dfrac{\partial}{\partial r}\left(r\lambda_r \dfrac{\partial T}{\partial r}\right)$	Energy	Fluid
Dynamic one-dimensional heterogeneous model		
$\varepsilon_s \dfrac{\partial C_{is}}{\partial t} = a_v C_t k_{g,i}\left(y_i - y_s\right) + \eta r_i \rho_B a$	Mass	Solid
$\left((1-\varepsilon_s)\rho_s C_{p,s} + \varepsilon_s \displaystyle\sum_{i=1}^{N_c}\left(\rho_{s,i} C_{p,s,i}\right)\right)\dfrac{\partial T_s}{\partial t} = a_v h_f\left(T - T_s\right) + \rho_B a \displaystyle\sum_{j=1}^{N}\eta r_j\left(-\Delta H_{R,j}\right)$	Energy	Solid
$\varepsilon_B \dfrac{\partial C_i}{\partial t} = -\dfrac{\partial(u C_i)}{\partial z} + a_v C_t k_{g,i}\left(y_{i,s} - y_i\right) + \varepsilon_B \dfrac{\partial}{\partial z}\left(D_z \dfrac{\partial C_i}{\partial z}\right)$	Mass	Fluid
$\varepsilon_B C_t c_{pg}\dfrac{\partial T}{\partial t} = -u\,C_t c_{pg}\dfrac{\partial T}{\partial z} + a_v h_f\left(T_s - T\right) + \varepsilon_B \dfrac{\partial}{\partial z}\left(\lambda_z \dfrac{\partial T}{\partial z}\right) + \dfrac{\pi D_o}{A_c} U(T^s - T)$	Energy	Fluid
Dynamic two-dimensional pseudo-homogeneous model		
$\varepsilon_B \dfrac{\partial C_i}{\partial t} = -\dfrac{\partial(u C_i)}{\partial z} + \eta r_i \rho_B a + \varepsilon_B \dfrac{\partial}{\partial z}\left(D_z \dfrac{\partial C_i}{\partial z}\right) + \dfrac{\varepsilon_B}{r}\dfrac{\partial}{\partial r}\left(r D_r \dfrac{\partial C_i}{\partial r}\right)$	Mass	Fluid
$\left(\varepsilon_B \rho_g C_{pg} + (1-\varepsilon_B)\rho_s C_{p,s}\right)\dfrac{\partial T}{\partial t} = -u\,\rho_g C_{pg}\dfrac{\partial T}{\partial z} + \rho_B a \displaystyle\sum_{j=1}^{N}\eta r_j\left(-\Delta H_{R,j}\right) + \varepsilon_B \dfrac{\partial}{\partial z}\left(\lambda_z \dfrac{\partial T}{\partial z}\right) + \dfrac{\varepsilon_B}{r}\dfrac{\partial}{\partial r}\left(r\lambda_r \dfrac{\partial T}{\partial r}\right)$	Energy	Fluid
Dynamic one-dimensional pseudo-homogeneous model		
$\varepsilon_B \dfrac{\partial C_i}{\partial t} = -\dfrac{\partial(u C_i)}{\partial z} + \eta r_i \rho_B a + \varepsilon_B \dfrac{\partial}{\partial z}\left(D_z \dfrac{\partial C_i}{\partial z}\right)$	Mass	Fluid
$\left(\varepsilon_B \rho_g C_{pg} + (1-\varepsilon_B)\rho_s C_{p,s}\right)\dfrac{\partial T}{\partial t} = -u\,\rho_g C_{pg}\dfrac{\partial T}{\partial z} + \rho_B a \displaystyle\sum_{j=1}^{N}\eta r_j\left(-\Delta H_{R,j}\right) + \varepsilon_B \dfrac{\partial}{\partial z}\left(\lambda_z \dfrac{\partial T}{\partial z}\right) + \dfrac{\pi D_o}{A_c} U(T^s - T)$	Energy	Fluid

The most common assumptions applied by researchers to develop mathematical models for methanol synthesis reactor are as follows:

(1) The mass and heat radial diffusion are ignored because of the turbulent flow regime inside the tubes and the relatively small diameter of the tube (e.g., 4 cm) [68].
(2) The mass and heat axial diffusion are ignored.
(3) Catalyst deactivation is ignored.
(4) The effectiveness factor of catalyst is calculated by the dusty-gas model or the Thiele modulus.

In the following, some interesting studies on mathematical modeling and corresponding assumptions are discussed. Løvik [3] proposed pseudo-steady-state two-dimensional pseudo-homogeneous and heterogeneous models. A comparison between the relative activity of catalysts based on pseudo-homogeneous and heterogeneous models is indicated in Fig. 4A. The results showed that to compensate the catalyst deactivation, the coolant side temperature in the pseudo-homogeneous model should increase further over the time compared with that in the heterogeneous model (see Fig. 4B).

Manenti et al. [72] compared three different mathematical models for the typical Lurgi's fixed-bed methanol synthesis reactor, consisting of the pseudo-homogeneous model with or without considering the molar changes along the reactor and the heterogeneous model. They implemented the assumptions (1)–(4) as well as the uniform distribution of concentration and temperature in solid particles in their models. The results of comparing two types of pseudo-homogeneous models showed that the constant molar flow rate assumption can affect the outlet methanol fraction of reactor by 10%, and the hot spot temperature and its position

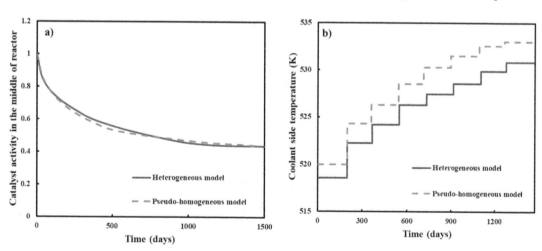

Fig. 4
A comparison between the (A) catalyst activity [69] and (B) coolant side temperature in the pseudo-homogeneous and heterogeneous model [70,71].

have a significant deviation. However, a small deviation between the heterogeneous and pseudo-homogeneous model was observed, when the system was operating under typical conventional conditions. The difference becomes considerable when the reactor operates far from the common condition. Also, Rezaie et al. [57] performed a comparison between the homogeneous and heterogeneous models to simulate the Lurgi-type industrial methanol reactor, considering the same assumptions. A good agreement was found between the two models, and similar results were achieved under typical industrial operating conditions. Jahanmiri and Eslamloueyan [73] used one- and two-dimensional pseudo-homogeneous model to simulate industrial methanol synthesis reactor. They assumed that there were no interfacial mass and heat gradients between the fluid and catalyst, there was only intraparticle mass diffusion limitation, and the catalysts were isothermal. Figs. 5A and B show a comparison of the temperature distribution and molar flow rate of CH_3OH and CO, respectively, which are predicted by one- and two-dimensional models. As it can be seen, an excellent similarity was found between the results of the 1-D and 2-D models, and the maximum absolute error between the predicted temperatures of one- and two-dimensional models was about 4°C. Thus, the one-dimensional model was a reliable approach to predict the performance of methanol synthesis reactor.

Furthermore, there are some interesting studies on the various configurations of fixed-bed methanol synthesis reactors based on mathematical modeling. Rahimpour [53] investigated the performance of dual-catalyst fixed-bed reactor, which is a good alternative to a single catalyst bed. Fig. 6 shows a schematic of this reactor. The dual-catalyst fixed-bed reactor contains conventional water-cooled Lurgi-type and gas-cooled reactors. A 1-D quasi-steady model was

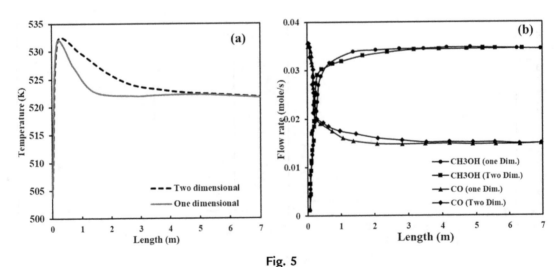

Fig. 5

A comparison of (A) temperature and (B) CH_3OH and CO molar flow rate distributions predicted by one- and two-dimensional models [73].

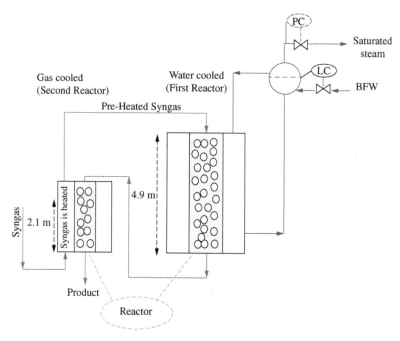

Fig. 6

A schematic diagram of dual-bed type methanol synthesis reactor [53].

used to compare the performance of two types of reactors. The results indicated that the dual-bed reactor has a higher conversion and a longer catalyst lifetime.

Mirvakili et al. [74] used a 1-D heterogeneous model to simulate different configurations of methanol synthesis reactors, including two parallel fixed-bed water-cooled reactors together with a gas-cooled type reactor and three proposed configurations in which the gas-cooled was parallel to the two water-cooled reactors under different feeding strategies. Recently, Nassirpour and Khademi [75] compared the performance of three types of industrial methanol synthesis reactors, including direct-, quench-, and indirect-cooling reactors with regard to energy efficiency and methanol yield terms under optimal conditions, which is achieved by performing an economic optimization. A 1-D heterogeneous model was employed to determine the effect of various parameters on methanol yield. Comparison between these types of reactors in terms of profit value, energy efficiency, and methanol yield under optimal operating conditions (see Fig. 7) indicated that there is no significant difference between energy efficiency of all reactors. They also concluded that the direct-cooling method has the most profit value and the highest methanol yield compared with other reactors, and the minimum profit value and lowest methanol yield were found for the quench-cooling reactor. Other researchers have identified the fixed-bed methanol synthesis reactor with various types of mathematical models and its simplifying assumptions, which are summarized in Table 5. By

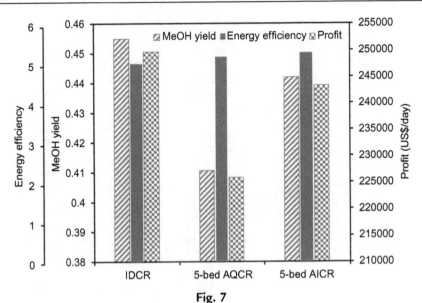

Fig. 7

A comparison between direct-, quench-, and indirect-cooling reactors in terms of profit value, energy efficiency, and methanol yield under optimal operating conditions [75].

Table 5 Various types of mathematical models and its simplifying assumptions to simulate the fixed-bed methanol synthesis reactor.

Author(s)	Assumptions
One-dimensional heterogeneous model	
Velardi and Barresi [76]	Assumption (1) and (3) Dynamic condition Ignoring pressure drop Ideal gas behavior In the solid phase energy balance equation, variations of gas energy in the accumulation term were not considered
Rezaie et al. [57]	Assumption (1), (2), and (4) Dynamic condition In the solid phase energy balance equation, variations of gas energy in the accumulation term were not considered
Shahrokhi and Baghmisheh [77]	Assumption (1), (2) and (3) Dynamic condition Calculating effectiveness factor by considering pseudo steady-state model for catalysts
Kordabadi and Jahanmiri [78]	Assumption (1) and (2) Isotherm catalyst
Kordabadi and Jahanmiri [79]	Ignoring viscous flow on the catalyst pellets In the solid phase energy balance equation, variations of gas energy in the accumulation term were not considered

Table 5 **Various types of mathematical models and its simplifying assumptions to simulate the fixed-bed methanol synthesis reactor—Cont'd**

Author(s)	Assumptions
Rahimpour [53]	Assumption (1), (2) and (4) Steady-state condition
Rahimpour [80] Rahimpour and Lotfinejad [81] Rahimpour and Lotfinejad [55]	Assumption (1), (2) and (4) Dynamic condition Considering linear pressure drop No viscous ideal gas behavior Considering catalyst deactivation In the solid phase energy balance equation, variations of gas energy in the accumulation term were not considered
Manenti et al. [72] Mirvakili and Rahimpour [51] Mirvakili et al. [74] Khanipour et al. [50] Khanipour et al. [82] Keshavarz et al. [83] Nassirpour and Khademi [75]	Assumption (1)–(4) Steady-state condition
De María et al. [84]	Assumption (1)–(3) Considering effectiveness factor equal to one Ideal gas behavior Steady-state condition Constant coolant temperature
Two-dimensional pseudo-homogeneous model	
Jahanmiri and Eslamloueyan [73]	Assumption (2) and (3) Steady-state condition Considering only intra-particle mass diffusion limitation, and assuming the isothermal catalysts
Leonzio and Foscolo [85]	Assumption (2) and (3) Constant pressure and temperature distribution inside the catalyst pellet Ideal gas behavior Considering effectiveness factor equal to one Steady-state condition
One-dimensional pseudo-homogeneous model	
Jahanmiri and Eslamloueyan [73]	Assumption (1)–(3) Steady-state condition Considering only intra-particle mass diffusion limitation, and assuming the isothermal catalysts
Rezaie et al. [57]	Assumption (1), (2), and (4) Dynamic condition In the solid phase energy balance equation, variations of gas energy in the accumulation term were not considered
Abrol and Hilton [86]	Assumption (1)–(4) Dynamic condition

Continued

Table 5 Various types of mathematical models and its simplifying assumptions to simulate the fixed-bed methanol synthesis reactor—Cont'd

Author(s)	Assumptions
Manenti et al. [87]	In the solid phase energy balance equation, variations of gas energy in the accumulation term were not considered Assumption (1), (3) and (4) Dynamic condition
De María et al. [84]	Constant pressure and temperature distribution inside the catalyst pellet Assumption (1)–(3) Considering effectiveness factor equal to one Ideal gas behavior Steady-state condition Constant coolant temperature Constant molar flux along the reactor

applying these corresponding assumptions to the energy and mass conservation equations tabulated in Table 4, the mathematical models developed by the researchers are obtained.

3.4.2 Membrane reactor

Some methods have been suggested by researchers to upgrade methanol synthesis in the conventional reactors. These methods include catalyst improvement, unconverted syngas recycling, and the use of membranes. The use of membrane reactors has some remarkable advantages, as membranes are less energy intensive, do not need solvents or adsorbents, are simple, and have low maintenance operations [88]. Generally, membrane reactors can be used in two different ways to improve methanol production from syngas. One way is to deliver the reactant to the catalyst bed in a controlled way. In this case, it can be beneficial to add H_2 through the membrane to the catalyst zone, which can lead to achieve the desired concentration in the catalyst bed [8].

Pd-alloy membranes along with a stainless-steel support are commonly utilized to separate H_2. This type of membrane has some remarkable properties such as low cost, good permselectivity (100% selective to hydrogen), and acceptable mechanical, long-term, and thermal stability [89]. The hydrogen permeation flux through Pd/Ag and pure-Pd membranes, expressed according to the Sievert's law (half-power pressure law), is given in Table 6. Researchers who have used these permeation rates are also listed in this table.

Rahimpour and Ghader [90] used a palladium–silver membrane reactor for methanol production, in which hydrogen permeated through the membrane into the catalyst bed. They investigated the impact of operating conditions on the performance of reactor and showed that the use of membrane can increase carbon monoxide conversion beyond equilibrium. Rahimpour and Lotfinejad [89] compared the performance of membrane and conventional dual-type methanol synthesis reactors using a dynamic 1-D heterogeneous model and the

Table 6 Two common expressions for hydrogen permeation through Pd/Ag and pure-Pd membranes.

Equation	Membrane type	Researcher who has used this equation
$$\begin{cases} J_{H_2} = \alpha_{H_2}\left(\sqrt{P_{H_2}{}^t} - \sqrt{P_{H_2}{}^s}\right) \\ \alpha_{H_2} = \dfrac{2\pi L\bar{P}}{\ln\left(\dfrac{D_o}{D_i}\right)} \\ \bar{P} = P_0 \exp\left(-\dfrac{E_p}{RT}\right) \end{cases}$$ for $T > 473.15\ K$ $\begin{cases} P_0 = 6.33 \times 10^{-8}\ \text{mol m}^{-2}\ \text{s}^{-1}\ \text{pa}^{-1/2} \\ E_p = 15.7\ \text{kJ kmol}^{-1} \end{cases}$	Pd/Ag	Rahimpour and Ghader [90] Rahimpour and Lotfinejad [89] Rahimpour and Lotfinejad [91] Rahimpour et al. [92] Rahimpour and Behjati [93] Parvasi et al. [94] Khademi, et al. [95] Khademi et al. [47] Rahimpour and Bayat [96] Rahimpour, et al. [97] Rahimpour and Bayat [54] Rahmani et al. [56] Amirabadi et al. [98] Bayat and Rahimpour [99] Farniaei et al. [88] Bayat and Rahimpour [100]
$$\begin{cases} J_{H_2} = \alpha_{H_2}\left(\sqrt{P_{H_2}{}^t} - \sqrt{P_{H_2}{}^s}\right) \\ \alpha_{H_2} = \dfrac{2\pi L}{\ln\left(\dfrac{D_o}{D_i}\right)}\dfrac{DC_0}{\sqrt{P_0}} \end{cases}$$ $\begin{cases} P_0 = 1.013 \times 10^3\ \text{pa} \\ C_0 = 1280\ \text{mol m}^{-3}\ \text{at 473 K} \end{cases}$	Pure-Pd	Rahimpour, et al. [92] Parvasi et al. [94]

presence of catalyst deactivation. The assumptions considered to derive model equations were the plug flow pattern, the ideal gas behavior, and the neglect of the axial heat dispersion compared with the convective flow. Dual-type methanol synthesis reactor is a shell and tube configuration, in which the first and second reactors are cooled by water and syngas, respectively. In the membrane dual-type reactor, the wall of the catalytic bed, which is cooled by the syngas, is covered with a palladium–silver H_2 permselective membrane. So, the H_2 content in the syngas permeates through the membrane into the catalyst bed. Hydrogen side feeding could disturb the thermodynamic equilibrium and shift the reaction toward further production of methanol. The results indicated that the membrane reactor has better performance compared with traditional reactors in terms of the methanol production and catalyst deactivation. Rahimpour and Lotfinejad [91], following their previous research, compared the performance of countercurrent and cocurrent membrane dual-type methanol synthesis reactors. The results showed that the countercurrent mode has a longer catalyst lifetime but lower methanol production and hydrogen permeation rate.

Rahimpour et al. [92] suggested a novel membrane reactor instead of the conventional methanol synthesis reactor. In this novel reactor, the synthesis gas is fed to the shell side (reaction zone), while the recycled product flows in cocurrent mode to the tube side. H_2 passes through the membrane from the tube to the shell side. In the meantime, the reaction zone is cooled by saturated water in the outer shell. They used one-dimensional steady state and quasi-steady state to study the influence of key operating parameters on performance of reactor and indicated that the membrane reactor has higher methanol production rate compared with conventional reactors. A one-dimensional homogeneous model considered by Rahimpour et al. [92] for hydrogen permselective membrane reactor is typically given in Table 7. A schematic diagram of this reactor is demonstrated in Fig. 8A.

Rahimpour and Bayat [96] developed a dynamic 1-D heterogeneous model to simulate a cascade membrane reactor for methanol production while considering catalyst deactivation. Following the previous work, they used genetic algorithm to achieve the optimum conditions for this system [100]. Rahimpour et al. [97] developed a dynamic 1-D heterogeneous mathematical model to investigate the performance of a two-stage methanol synthesis hydrogen-permselective membrane reactor with two different flow modes. In the suggested configurations, the fresh synthesis gas flows in the tube side of the membrane reactor co- and countercurrently with the reacting gas in the shell side. The results showed that the countercurrent mode had higher CO conversion and more H_2 permeation rate, while the cocurrent mode had longer catalyst lifetime.

The second approach for enhancement of methanol production from syngas is to remove products from the catalyst bed [8]. Removing the products from the reaction zone can be useful to overcome the thermodynamic equilibrium limitations in chemical reactions. The membrane changes the reaction equilibrium through the Le Chatelier–Brown principle, so that higher conversion can be achieved beyond the equilibrium [90]. The H_2O permeation flux through the alumina–silica composite and H-SOD membranes is given in Table 8. Researchers who have used these permeation rates are also listed in this table.

Farsi and Jahanmiri [101] investigated the performance of H_2O-permselective membrane methanol synthesis reactor using a dynamic 1-D heterogeneous model in which catalyst deactivation was considered. In this regard, alumina–silica composite membrane layer was used to separate H_2O from the catalyst bed. The results indicated that the use of this type of reactor increases methanol production rate by 4.06% compared with the traditional reactor. Table 7 shows a typical one-dimensional heterogeneous model introduced by Farsi and Jahanmiri [101] for H_2O permselective membrane reactor. A schematic diagram of this reactor is demonstrated in Fig. 8B.

Table 7 Typical one-dimensional models for H_2- and H_2O-permselective membrane reactor.

Equation	Description
A typical 1-D homogeneous model for H_2-permselective membrane reactor	
$\dfrac{F_t^s}{A_c}\dfrac{\partial y_i}{\partial z} = r_i \rho_B a + \dfrac{a_{H_2}}{A_c}\left(\sqrt{p_{H_2}^t} - \sqrt{p_{H_2}^s}\right)$	Fluid-phase mass species conservation equation in the shell-side
$\dfrac{F_t^s}{A_c} c_{p,g}\dfrac{\partial T}{\partial z} = \dfrac{\pi D_{ti}}{A_c} U_{shell}(T^c - T) + \dfrac{\pi D_{to}}{A_c} U_{tube}(T^t - T) + \rho_B a \sum_{j=1}^{N}\eta r_j(-\Delta H_{R,j})$	Fluid-phase energy conservation equation in the shell-side
$\dfrac{\partial F_i^t}{\partial z} = a_{H_2}\left(\sqrt{p_{H_2}^s} - \sqrt{p_{H_2}^t}\right)$	Fluid-phase mass species conservation equation in the tube-side
$F_t^t c_{p,g}\dfrac{\partial T^t}{\partial z} = \pi D_{to} U_{tube}(T^t - T)$	Fluid-phase energy conservation equation in the tube-side
A typical 1-D heterogeneous model for H_2O-permselective membrane reactor	
$\varepsilon_s \dfrac{\partial(C_t y_{is})}{\partial t} = C_t a_v k_{g,i}\left(y_{i,2} - y_s\right) + \eta r_i \rho_B a$	Solid-phase mass species conservation equation in the shell-side
$\rho_B C_{p,s}\dfrac{\partial T_s}{\partial t} = a_v h_f(T_2 - T_s) + \rho_B a \sum_{i=1}^{N}\eta r_i(-\Delta H_{f,i})$	Solid-phase energy conservation equation in the shell-side
$\varepsilon_B \dfrac{\partial(C_t y_{i,2})}{\partial t} = -\dfrac{1}{A_c}\dfrac{\partial(F_{t,2} y_{i,2})}{\partial z} + a_v C_t k_{g,i}\left(y_{is} - y_{i,2}\right) - \dfrac{\pi D_1}{A_c}\alpha_{H_2O}\left(p_{H_2O,2} - p_{H_2O,1}\right)$	Fluid-phase mass species conservation equation in the shell-side
$\rho c_{p,g}\dfrac{\partial T_2}{\partial t} = -\dfrac{c_{p,g}}{A_c}\dfrac{\partial(F_{t,2}T_2)}{\partial z} + a_v h_f(T_s - T_2) - \dfrac{\pi D_1}{A_c}\int_{T_0}^{T_1} j_{H_2O} c_p dT - \dfrac{\pi D_1}{A_c}U_{1-2}(T_2 - T_1) - \dfrac{\pi D_2}{A_c}U_{2-3}(T_2 - T^s)$	Fluid-phase energy conservation equation in the shell-side
$\varepsilon_B\dfrac{\partial(C_t y_{i,1})}{\partial t} = -\dfrac{1}{A_c}\dfrac{\partial(F_{t,1} y_{i,1})}{\partial z} + \dfrac{\pi D_1}{A_c}\alpha_{H_2O}\left(p_{H_2O,2} - p_{H_2O,1}\right)$	Fluid-phase mass species conservation equation in the tube-side
$\rho C_{p,g}\dfrac{\partial T_1}{\partial t} = -\dfrac{c_{p,g}}{A_c}\dfrac{\partial(F_{t,1}T_1)}{\partial z} + \dfrac{\pi D_1}{A_c}U_{1-2}(T_2 - T_1) + \dfrac{\pi D_1}{A_c}\int_{T_0}^{T_2} j_{H_2O} c_p dT$	Fluid-phase energy conservation equation in the tube-side

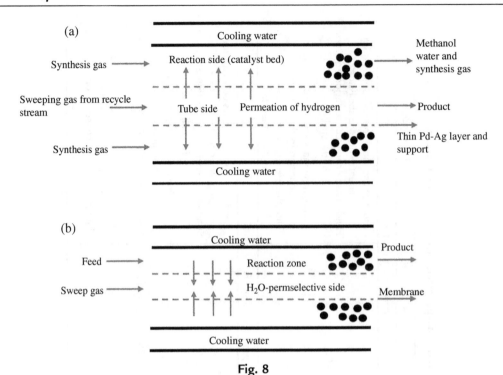

Fig. 8

A schematic diagram of membrane reactors, considering (A) hydrogen permselective and (B) steam perm-selective membrane [92,101].

Table 8 Two common expressions for H_2O permeation through H-SOD and alumina–silica composite membranes.

Equation	Membrane type	Researcher who has used this equation
$J_{H_2O} = \alpha_{H_2O}(P^t_{H_2O} - P^s_{H_2O})$ $\alpha_{H_2O} = 1.14 \times 10^{-7}$ mol m^{-2} s^{-1} pa^{-1} at 523 K	Alumina–silica composite	Farsi and Jahanmiri [102] Farsi and Jahanmiri [103] Farsi and Jahanmiri [101] Farsi and Jahanmiri [104]
$J_{H_2O} = \alpha_{H_2O}(P_{H_2O}{}^t - P_{H_2O}{}^s)$ $\alpha_{H_2O} = 10^{-6} - 10^{-7}$ mol m^{-2} s^{-1} pa^{-1}	H_2O-selective H-SOD	Rahmani et al. [56] Mirvakili et al. [52]

Following the pervious study, this group of authors suggested a new methanol synthesis reactor consisting of three concentric cylinders in which H_2 passes through a Pd/Ag membrane to the reaction zone (middle cylinder), and H_2O removes from the reaction zone into the inner cylinder through alumina–silica composite membrane layer. A steady-state [102] and dynamic [104] 1-D heterogeneous model was applied to simulate this reactor. The numerical results indicated that the use of two membranes in the methanol synthesis reactor can increase methanol yield by 10.02% [102]. They also optimized this reactor using genetic algorithm, considering maximization of methanol production as an objective function [103].

Regarding the optimization of methanol synthesis membrane reactor, Rahimpour and Behjati [93] optimized a membrane dual-type methanol synthesis reactor and compared the results with the traditional reactor. The optimized membrane reactor could have increased the methanol production rate by 5.95%. Parvasi et al. [94] optimized a novel methanol synthesis loop consisting of a membrane reactor using the Differential Evolution (DE) algorithm and compared its performance with the conventional loop. Although a reduction in methanol production was observed in the suggested loop, it was compensated by improvement in catalyst deactivation. Bayat et al. [105] simulated and optimized a methanol synthesis membrane reactor with in situ water adsorption using a dynamic 1-D heterogeneous model in the presence of deactivation of catalyst. Adsorbents are fed into the outer tube, where the methanol synthesis reactions take place. Fine particles adsorb water during the reaction, which increases methanol synthesis rate. The recycle gas was fed to the reactor inner tube, which is covered by a Pd/Ag membrane. The H_2 permeation from the inner tube to the outer tube leads to more methanol production.

3.4.3 Thermally coupled reactor

Chemical engineering designers are interested in increasing the efficiency and decreasing the total costs and environmental side effects of processes. One of the promising solutions is the coupling of endothermic and exothermic reactions, in which the heat generated in exothermic reactions is used to drive endothermic reactions [88]. Several researchers (see Table 9) focused on modeling, optimizing, and investigating the concept of coupling methanol production from

Table 9 Coupling of the methanol synthesis reaction with other endothermic reactions in various configurations.

Endothermic reaction	Description	Author
Cyclohexane dehydrogenation to benzene (CDB)	• The reactor consists of three concentric cylinders, • The endothermic-side is covered with Pd/Ag membrane to extract H_2 from this side, • Pure hydrogen, benzene, and methanol are produced simultaneously.	Khademi et al. [95]
CDB	➤ The DE algorithm was used to find the optimal conditions in which benzene and methanol mole fractions are the main objectives.	Khademi et al. [49]
CDB	✓ The thermally coupled membrane reactor proposed by Khademi et al. [95] was optimized using the DE method.	Khademi et al. [47]

Continued

Table 9 **Coupling of the methanol synthesis reaction with other endothermic reactions in various configurations—Cont'd**

Endothermic reaction	Description	Author
CDB	• A fluidized-bed thermally coupled membrane reactor was proposed, • A Steady-state 1-D heterogeneous model and two-phase theory in bubbling regime of fluidization were used for cyclohexane dehydrogenation and methanol production sides, respectively, • This reactor solves some issues of thermally coupled membrane reactor suggested by Khademi et al. [95], such as radial gradient of temperature and concentration in both sides, pressure drop and catalyst internal mass transfer limitations.	Rahimpour and Bayat [54]
Dehydrogenation of methyl-cyclohexane to toluene	➢ The temperature profile of this reactor was significantly lower than that of the traditional methanol synthesis reactor, which reduces the rate of deactivation of catalyst.	Rahimpour et al. [106]
CDB	✓ The startup and transient response of the system were evaluated using a dynamic heterogeneous model.	Khademi et al. [48]
CDB	• The performance of a thermally coupled dual methanol production process consisting of two reactors was investigated, • In the first reactor, methanol synthesis is coupled with cyclohexane dehydrogenation, and in the second one, cyclohexane is produced due to the transport of all hydrogen produced from the first reactor to the second one for benzene hydrogenation reaction. • The two advantages of this process are the need for a small amount of exterior cyclohexane injection rate and the solution of the hydrogen storage problem by the simultaneous production and usage of H_2.	Mirvakili et al. [52]
Methyl formate synthesis reaction	➢ The methanol produced in the exothermic-side was recycled and utilized as the feed of the endothermic-side to produce methyl formate.	Goosheneshin et al. [107]
CDB	✓ A two-zone thermally coupled membrane reactor was developed to enhance the simultaneous synthesis of H_2, benzene and methanol, ✓ In the first part, the syngas is partially converted to methanol, where the gas is cooled with boiling water,	Bayat and Rahimpour [108]

Table 9 Coupling of the methanol synthesis reaction with other endothermic reactions in various configurations—Cont'd

Endothermic reaction	Description	Author
CDB	✓ In the second part, the heat required to conduct the cyclohexane dehydrogenation is provided by methanol synthesis, ✓ Pd—Ag membrane was used to extract ultrapure hydrogen from cyclohexane dehydrogenation side, ✓ A 13.14% enhancement in the pure hydrogen production was observed compared to the configuration presented by Khademi et al. [47]. • The performance of two different auto-thermal membrane configurations, containing in situ H_2 addition and in situ H_2O removal was compared to the nonmembrane one, • In situ H_2O removal configuration improves methanol yield and more CO_2 removal which leads to a lower environmental issues and longer exothermic-side catalyst lifetime due to the more favorable temperature profile and reduced H_2O promoted catalyst deactivation.	Rahmani et al. [56]
CDB	➢ The results showed that the coupling of these reactions in a recuperative reactor is possible provided that the shell diameter, inlet temperatures, and flow rates are properly designed.	Khademi et al. [46]
CDB	✓ A thermally double-coupled two-membrane reactor was simulated, ✓ Cyclohexane dehydrogenation was coupled with two exothermic reactions including direct dimethyl ether and methanol synthesis from synthesis gas, ✓ Two Pd/Ag membranes were utilized to separate the H_2 from unconverted products in exothermic sides, ✓ This reactor can increase methanol production rate by 3.39%, 7.03%, and 10.94% compared to the traditional methanol synthesis reactor, methanol synthesis and cyclohexane dehydrogenation coupled reactor, and thermally double-coupled reactor, respectively.	Farniaei et al. [88]
Dehydrogenation of methyl-cyclohexane to toluene	• A thermally coupled double-membrane reactor was optimized using the DE algorithm,	Amirabadi et al. [98]

Continued

Table 9   Coupling of the methanol synthesis reaction with other endothermic reactions in various configurations—Cont'd

Endothermic reaction	Description	Author
Dehydrogenation of decline	• Pd/Ag and hydroxy sodalite membranes were used to extract the hydrogen from the endothermic-side and remove the water from the exothermic-side, respectively, • Water removal enhances the selectivity and activity of the methanol production and prevents catalyst recrystallization, • This reactor increases the methanol yield by 13.8% compared to the traditional reactor. ➤ The performance of a thermally double-coupled reactor was investigated, in which methanol and Fischer-Tropsch synthesis were the exothermic reactions, ➤ In addition to methanol, H_2, gasoline, and naphthalene are additionally produced, ➤ The DE algorithm was applied to find the maximum methanol mole fraction and gasoline yield, ➤ This configuration increases methanol production by 10.52% compared to the conventional reactor.	Samimi et al. [58]
CDB	✓ A thermally coupled two-membrane reactor was evaluated with the goal of maximizing the methanol and H_2 production, ✓ Two membranes were used to separate H_2 and H_2O from cyclohexane dehydrogenation and methanol production sides, respectively, ✓ A 10.51% improvement in methanol synthesis was found compared to the conventional reactor.	Bayat and Rahimpour [99]
CDB	• A thermally-coupled double-membrane reactor consisting of four concentered tubes was optimized with the aim of maximizing methanol, benzene and hydrogen, • The syngas was fed to the exothermic-side and the high-pressure effluent gas was recycled to the recycling-side; at the same time allowing the excess H_2 to permeate through the membrane into the exothermic-side for facilitating the methanol synthesis reaction, • The endothermic-side is covered with Pd/Ag membrane to recover H_2 from this side.	Rahmanifard et al. [109]

syngas as an exothermic reaction with other endothermic reactions because of the following benefits:

- Energy saving
- Attaining a multiple products and reactants configuration
- Reducing the capital and operational costs

All the mathematical modeling related to the research mentioned in Table 9 is based on a one-dimensional heterogeneous model. However, Khademi et al. [49], for the first time, presented a thermally coupled reactor, in which the methanol production reaction provides the heat required for cyclohexane dehydrogenation to benzene (CDB). Fig. 9 shows a schematic diagram of this configuration. Some of the assumptions considered for this model in both catalytic sides are as follows:

- Ideal gas behavior,
- Steady-state condition,
- Ignoring the radial gradient of temperature and concentration because of small ratio of radius to tube length,
- Not considering axial diffusion of mass and heat compared with the bulk movement (plug flow pattern),
- Uniform bed porosity in radial and axial directions,
- Uniform temperature inside the catalyst pellets,
- No heat loss to the surroundings,
- Calculating the effectiveness factor based on the dusty-gas model.

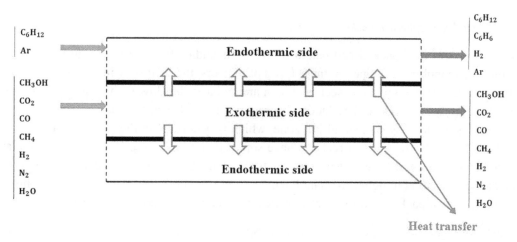

Fig. 9

A schematic diagram of methanol and benzene synthesis coupled reactor [49].

Table 10 The conservation laws for a typical heat-exchanger reactor.

Equation	No.
Catalyst-phase mass and energy conservation equations in both sides	
$a_v C_i k_{g,i}(y_i - y_s) + \eta r_i \rho_B = 0$ $a_v h_f(T - T_s) + \rho_B \sum_{j=1}^{N} \eta r_j (-\Delta H_{R_j}) = 0$	(53) (54)
Gas-phase mass and energy conservation equations in both sides	
$-\frac{F_i}{A_c}\frac{\partial y_i}{\partial z} + a_v C_i k_{g,i}(y_s - y_i) = 0$ $-\frac{F_i}{A_c} c_{p,g}\frac{\partial T}{\partial z} + a_v h_f(T_s - T) \pm \frac{\pi D_i}{A_c} U_{1-2}(T_2 - T_1) = 0$	(55) (56)
Pressure drop	
$\frac{dp}{dz} = \frac{150\mu u_g}{d_p^2}\frac{(1-\varepsilon)^2}{\varepsilon^3} + \frac{1.75\rho u_g^2}{d_p}\frac{(1-\varepsilon)}{\varepsilon^3}$	(57)

The mass and energy conservation equations for the catalyst and gas phases as well as pressure drop in both sides are presented in Table 10. The third term of Eq. (56) refers to the heat transfer between the exothermic and endothermic sections, in which the positive sign is applied for methanol synthesis and the negative sign for the dehydrogenation reaction. In this equation, subscripts of 1 and 2 stand for the exothermic- and endothermic side, respectively. The results indicated that this recuperative reactor has considerable benefits compared with the traditional methanol synthesis reactor such as: enhancement in equilibrium conversion due to the lower outlet temperature of product stream; reducing the size of the reactors; achieving the autothermality within the reactor; and producing benzene as an additional valuable product. Therefore, this configuration can be a feasible and beneficial choice.

3.4.4 Radial-flow spherical packed-bed reactor

Radial-flow spherical packed-bed reactors have been studied from more than a decade ago to produce methanol from syngas. In these types of reactors, the feed flow in the radial direction provides a larger average cross-sectional area and reduces the flow path compared with the vertical columns, which leads a reduction in pressure drop. Also, another advantage of this type of reactor is the making of a small hot zone, which is beneficial in the reversible exothermic reactions [110]. These types of reactors have some advantages compared with multitubular packed-bed reactors, such as lower pressure drop and manufacturing cost due to smaller wall thickness and getting a higher methanol synthesis rate. In addition, it is possible to connect several spherical packed-bed reactors in series with an external cooler between them [64].

Rahimpour et al. [110] proposed a radial-flow, spherical-bed methanol synthesis reactor, and a schematic diagram of that is shown in Fig. 10. The catalyst is placed between two perforated spherical shells, the feed enters the catalyst bed from the outer sphere and the products finally

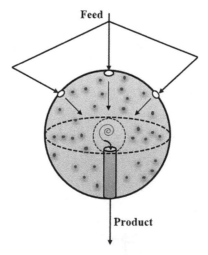

Fig. 10

A schematic diagram of a radial-flow, spherical packed-bed methanol synthesis reactor [111].

exit the inner sphere. A dynamic one-dimensional pseudo-homogeneous model, considering catalyst deactivation, was developed with the assumption of ignoring concentration and temperature gradients between catalyst and gas phases and neglecting the axial heat dispersion.

$$\varepsilon \frac{\partial C_i}{\partial t} = -\frac{1}{r^2}\frac{\partial}{\partial r}\left(r^2 u_r C_i\right) + \varepsilon_s(1-\varepsilon)a\rho_B \sum_{i=1}^{n} v_{i,j} r_i \tag{58}$$

$$(1-\varepsilon_s)\rho C_{p,s}\frac{\partial T}{\partial t} = -\frac{1}{r^2}\frac{\partial}{\partial r}\left(\rho u_r r^2 C_{p,g}\left(T - T_{ref}\right)\right) + \varepsilon_s(1-\varepsilon)a\rho_B \sum_{j=1}^{N} \Delta H_{R,j} r_j \tag{59}$$

Rahimpour et al. [64] studied the influence of operating conditions and design parameters on the performance of the reactor; and better performance was found in the two-stage spherical-bed reactor compared with the conventional tubular reactors, and single- and three-stage spherical-bed reactors, under the same conditions. After that, a theoretical optimization was conducted by Rahimpour et al. [110] to find the optimal operating conditions of the spherical-bed reactor to improve the methanol production rate. The optimization results showed that the catalyst lifetime of this type of reactor is 4 years.

3.4.5 Axial-flow spherical packed-bed reactor

Rahimpour et al. [112] suggested a new configuration of spherical packed-bed methanol synthesis reactor, considering the axial flow. In this configuration, the feed enters the reactor form one side, passes through the catalyst bed placed inside the dome, and then exits from the opposite side of the reactor (see Fig. 11). Due to the small area at the inlet and outlet of the

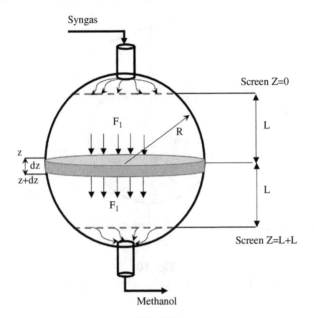

Fig. 11

A schematic diagram of an axial-flow spherical packed-bed reactor suggested by Rahimpour et al. [112].

reactor, the pressure drop can be increased, so to solve this problem, the catalysts in these areas are held by screens. Also, since the spherical packed-bed reactor has a lower contact surface with the cooling water for heat transfer compared with conventional reactors, the use of cooling water has little effect and several adiabatic reactors in series with external cooler between them were considered. They used a dynamic one-dimensional pseudo-homogeneous model to investigate concentration and temperature changes and to compare the performance of this configuration with a multitubular packed-bed reactor.

$$D_z \frac{1}{A_C} \frac{\partial}{\partial z}\left(A_c \frac{\partial C_j}{\partial z}\right) - \frac{1}{A_C} \frac{\partial}{\partial z}(A_c U_z C_i) + \rho_B a \sum_{i=1}^{m} v_{i,j} r_i = \varepsilon \frac{\partial C_i}{\partial t} \tag{60}$$

$$\lambda_r \frac{1}{A_C} \frac{\partial}{\partial z}\left(A_c \frac{\partial T}{\partial z}\right) - \frac{1}{A_C} \frac{\partial}{\partial z}\left(\rho_g A_c U_z C_p (T - T_{ref})\right) + \rho_B a \sum_{i=1}^{m} \Delta H_i r_i = \varepsilon \frac{\partial(\rho C_{p,s}(T - T_{ref}))}{\partial t}$$

$$\tag{61}$$

Eqs. (62) and (63) should be added to the conservation laws to complete the model, where z, R, and L are axial coordinate, radius, and the length of the reactor in axial direction, respectively.

$$A_c = \pi\left(R^2 - (Z - L)^2\right) \tag{62}$$

$$R = \left(\frac{\frac{m_{cat}}{\rho_B \pi}}{2\frac{L}{R} - \frac{2}{3}\left(\frac{L}{R}\right)^3} \right)^{1/3} \tag{63}$$

Ergun equation (Eq. (64)) was used to determine the pressure drop:

$$\frac{dp}{dz} = \frac{150\mu}{\varphi_s^2 d_p^2} \frac{(1-\varepsilon)^2}{\varepsilon^3} \frac{Q}{A_c} + \frac{1.75\rho}{\varphi_s d_p} \frac{(1-\varepsilon)}{\varepsilon^3} \frac{Q^2}{A_c^2} \tag{64}$$

The modeling results showed that the use of three- and four-stages spherical packed-bed reactors improves the methanol synthesis rate by 4.4% and 7.7%, respectively, under the steady-state condition.

3.4.6 Radial-flow packed-bed reactor

The radial-flow packed-bed reactor has been applied to a variety of catalytic petrochemical industrial processes. The main advantages of these reactors compared with the axial-flow fixed-bed reactors are as follows: (1) high flux capacity, (2) use of small catalysts, (3) high flow surface area per volume of catalyst, and (4) low pressure drop. In this reactor, the catalysts are located between two coaxial perforated cylinders and the gas phase flows radially across the catalyst bed.

The Toyo Engineering Corporation (TEC) was the first and only company that utilized a multistage intermediate cooling radial-flow reactor, namely MFR-Z$^@$ for methanol synthesis, and its performance has already been proven in a commercial scale of 5000t/d for more than 7 years [113]. Hirotani et al. [113] compared the mechanical design and process performance of the MRF-Z$^®$ reactor with a quench cooling reactor and reported the inlet and outlet of gas composition, operating conditions, and dimensions of the MRF-Z$^@$ reactor.

Three years ago, Farsi [114] focused on the modeling and optimization of a radial-flow tubular reactor to maximize the methanol productivity by considering the inlet temperature and pressure and cooling side temperature as the decision variables. In this regard, a one-dimensional heterogeneous model (see Table 11) under steady-state condition was considered to study the performance of this type of reactor. As shown in Fig. 12, some tubes are located in the space between two concentric perforated cylindrical vessels, where methanol production reaction takes place over the $CuO/ZnO/Al_2O_3$ catalyst, and the released heat from the reaction is removed by boiling water circulated in the tubes. The results showed that the proposed reactor, in terms of methanol production and pressure drop, has a better performance compared with conventional axial-flow reactors under optimal conditions.

3.4.7 Fluidized-bed reactor

The high pressure drop along the reactor axis, low heat transfer rate, low effectiveness factor due to catalyst size, and low production capacity are the potential drawbacks of fixed-bed reactors. Fluidized-bed reactors are a proper candidate to solve these issues [116]. The

Table 11   1-D heterogeneous model for the radial-flow packed-bed methanol synthesis reactor under the steady-state condition.

Equation	No.
Fluid-phase mass and energy conservation equations	
$\left(-u_g \frac{dC_i}{dr} - C_i \frac{du_g}{dr} - \frac{u_g C_i}{r}\right) - a_v k_{gi}(C_i - C_s) = 0$	(65)
$-c_{p,g}\left(u_g C_t \frac{dT^s}{dr} - T u_g \frac{dC_t}{dr} - T C_i \frac{du_g}{dr} - \frac{T u_g C_t}{r}\right) - a_v h_f(T - T_s) + \frac{UA_h}{2\pi r L}(T^s - T) = 0$	(66)
Solid-phase mass and energy conservation equations	
$a_v k_{gi}(C_i - C_s) + \rho_B \sum_{i=1}^{N} \alpha \eta r_i = 0$	(67)
$a_v h_f(T - T_s) + \rho_B \sum_{i=1}^{N} \alpha \eta r_i (\Delta H_i) = 0$	(68)
Total mole balance	
$\left(-u_g \frac{dC_t}{dr} - C_t \frac{du_g}{dr} - \frac{u_g C_t}{r}\right) + \rho_B \sum_{j=1}^{3}\sum_{i=1}^{N} \alpha \eta_i \, r_{i,j} = 0$	(69)
Pressure drop	
$\frac{dP}{dr} = f \frac{\rho u_g^2}{D_p}$	(70)
$f = \frac{150}{Re}\frac{(1-\varepsilon)^2}{\varepsilon^3} + \frac{4.2}{Re^{1/6}}\frac{(1-\varepsilon)^{1.166}}{\varepsilon^3}$	(71)

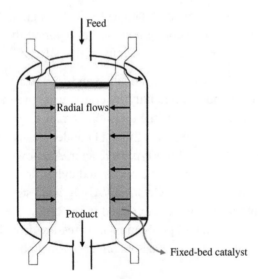

Fig. 12

A schematic diagram of the radial-flow packed-bed methanol synthesis reactor [115].

fluidized-bed reactor can operate at a lower pressure drop compared with fixed-bed reactors, which results in employing smaller catalysts and reducing the intraparticle limitations [117]. Additionally, the use of fluidized-bed reactor can provide a uniform temperature distribution throughout the reactor, which improves catalyst lifetime [118]. However, difficulties in reactor construction, catalyst attrition, and erosion of reactor are the disadvantages of this type of reactor [119].

Numerous researchers have conducted comparisons between fixed- and fluidized-bed methanol synthesis reactors. In this regard, Rahimpour and Alizadehhesari [117] used the two-phase theory of fluidization to model and determine the performance of fluidized-bed membrane dual-type methanol reactors. Following the previous study, Rahimpour and Elekaei [116] developed a dynamic model in the presence of deactivation of catalyst to investigate the performance of bubbling fluidized-bed membrane dual-type methanol synthesis reactor and compared the results with membrane dual-type and conventional dual-type methanol synthesis reactors. The suggested reactor improved some issues of conventional type such as internal mass transfer rate, pressure drop, and radial temperature and concentration gradient in gas-cooled reactor.

Rahimpour et al. [119] proposed a cascading fluidized-bed hydrogen permselective dual-type membrane reactor to produce methanol. To develop a mathematical model for the fluidized-bed methanol synthesis reactor cooled by water, the following assumptions are considered:

- Ideal gas behavior.
- Reactions take place mostly in the emulsion phase.
- The intraparticle diffusional resistance is ignored due to small particle size.
- The dense catalyst bed consists of the bubble and emulsion phases.
- The shape of the bubbles is assumed to be spherical.
- Even though some catalyst particles exist in the bubble phase which are involved in the reactions but the emulsion phase has higher extend of reaction compared with the bubble phase.
- The reactor operates under isothermal conditions due to rapid mixing, meaning that the temperatures of emulsion and bubble phases are the same.
- Three meters from the top of the reactor is assumed to be a freeboard region, where no catalyst is present.

According to these assumptions, a one-dimensional model including the differential mass balance equations for the bubble and emulsion phases, respectively, as well as the energy balance equation, is presented typically as follows:

$$\delta K_{bei} C_t a_b \left(y_{ie} - y_{ib} \right) - \frac{\delta}{A_c} \frac{dF_{i,b}}{dz} + \delta . \gamma . \rho_s \sum_{j=1}^{3} r_{bij} = 0 \qquad (72)$$

$$\delta K_{bei} C_t a_b (y_{ib} - y_{ie}) - \frac{(1-\delta)}{A_c} \frac{F_{i,e}}{dz} + (1-\delta)\rho_e . \eta . \sum_{j=1}^{3} r_{ij} = 0 \tag{73}$$

$$\frac{\pi D_i}{A_c} U_{tube}(T^s - T) + (1-\delta)\rho_e . \eta . a . \sum_{i=1}^{3} r_j(-\Delta H_{f,j}) + \delta . \gamma . \rho_B . \eta . a . \sum_{j=1}^{3} r_{bj}(-\Delta H_{f,j}) = 0 \tag{74}$$

The numerical results illustrated that the proposed reactor enhances the methanol production yield by 9.53% in comparison with the conventional dual-type methanol synthesis reactor [119]. They also proposed a dynamic mathematical model to simulate that reactor, considering the long-term catalyst deactivation [118]. After that, this group of authors evaluated a fluidized-bed membrane dual-type reactor for CO_2 removal. Based on a dynamic mathematical model, considerable CO_2 conversion was observed compared with the results of conventional and membrane dual-type methanol synthesis reactors [120].

Rahimpour and Bayat [121] compared the performance of countercurrent and cocurrent fluidized-bed membrane methanol synthesis reactors. It was found that the cocurrent mode has a higher CO removal and H_2 permeation rate as well as lower water production, whereas the countercurrent mode has a higher methanol conversion and CO_2 removal as well as longer catalyst lifetime.

3.4.8 Slurry bubble column reactor

Generally, slurry bubble column reactors are used for highly exothermic and rapid reactions, in which the slurry phase consists of an inert liquid such as paraffin wax, paraffin oil, decahydronaphthalene, tetrahydronaphthalene, and the like, to absorb the heat released by the exothermic reactions and transfer the catalyst dispersed in the liquid phase. Gas bubbles flow upward through the liquid phase, reactions take place on the catalyst surface, and mass transfer occurs between the gas, liquid, and solid phases. A schematic diagram of the slurry bubble column reactor is shown in Fig. 13. These types of reactors offer some remarkable advantages, such as better use of the catalyst (varying catalyst size from 5 to 150μm and catalyst loading up to 50% volume), low pressure drop, approximately isothermal condition, low maintenance and operating costs, ease of operation, and good mass and heat transfer characteristics due to the high level of back mixing and high contact interface [122].

Due to the advantages of the slurry bubble column reactor, several researchers have studied the simulation and modeling of this type of reactor. Wu and Gidaspow [123] applied a 2-D hydrodynamic transient model to simulate the slurry bubble column reactor for methanol synthesis from synthesis gas. Gamwo et al. [124] developed a transient, reactive, 2-D multiphase-flow model for the methanol production from syngas, considering CO hydrogenation as the only reaction occurred in the simulation. Wang et al. [125] compared the performance of slurry bubble column and trickle-bed reactors for the synthesis of methanol from syngas using two- and three-phase models. They concluded that in pilot scale, the trickle-bed reactor has a better performance compared with the slurry bubble column reactor. Salehi et al. [122] suggested a commercial-scale slurry bubble column methanol synthesis reactor instead of the conventional gas-phase reactor and simulated it based on a one-dimensional model (see Table 12) and the following assumptions:

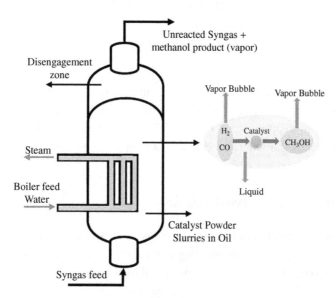

Fig. 13
A schematic diagram of the slurry bubble column reactor [122].

Table 12 One-dimensional model for slurry bubble column methanol synthesis reactor under the steady-state condition.

Equation	No.
Gas-phase mass balance	
$\dfrac{d\left(u_g C_{i,g}\right)}{dz} - (k_l a)_i \left(C_{i,l}^* - C_{i,l}\right) = 0$	(75)
Liquid-phase mass balance	
$\varepsilon_l E_l \dfrac{d^2 C_{i,l}}{dz^2} - \dfrac{d\left(u_{ss} C_{i,l}\right)}{dz} + (k_l a)_i \left(C_{i,l}^* - C_{i,l}\right) - a\eta r_i C_s = 0$	(76)
Henry's law to calculate the solubility of each species in the liquid phase	
$C_{i,l}^* = \dfrac{P_i}{He_i}$	(77)
Energy balance	
$\varepsilon_{sl} \lambda_{ax} \dfrac{\partial^2 T}{\partial z^2} + \varepsilon_s \rho_s \sum a\eta r_i \left(-\Delta H_{f,i}\right) - \alpha_{eff} \dfrac{A_{heat}}{A_{CS}} \left(T - T_{sat.water}\right) = 0$	(78)
Pressure drop	
$\dfrac{dP}{dz} = -\left(\rho_g \varepsilon_g + \rho_l \varepsilon_l + \rho_s \varepsilon_s\right) g$	(79)

- Plug flow pattern is considered for the gas phase.
- Axial dispersion model is applied for the slurry phase.
- The effectiveness factor is assumed to be 1 because of small size of the catalyst.
- Mass transfer resistance between solid and liquid phases is ignored. Thus, solid and liquid are a homogeneous phase in which reactions take place.
- Isothermal condition is assumed.
- Methanol is produced only by CO hydrogenation and the WGS reaction is in dynamic equilibrium.

Although lower methanol production rate was found in the slurry bubble column reactor compared with the traditional reactor, the methanol synthesis deficiency was compensated by modifying the feed composition as 5% CO and 95% conventional feed. Moreover, the addition of excess CO diminishes the water production, which reduces the catalyst deactivation rate.

4. Modeling of methanol synthesis loop

This section describes a mathematical model for typical methanol production loop. At first stage, methanol is produced by passing syngas through the reactor. The product stream temperature is reduced to the dew point of methanol by a heat exchanger, and then, methanol is separated in the liquid phase in a separator. The unconverted reactant is recycled to the reactor, so this process operates as a loop statement. Besides, to avoid the accumulation of inert gases in the synthesis loop, a purge gas stream is taken from the separator. Fig. 14 shows a schematic diagram of the methanol production loop consisting of four parts [94]. In the following, the mathematical model for each part of this loop is presented.

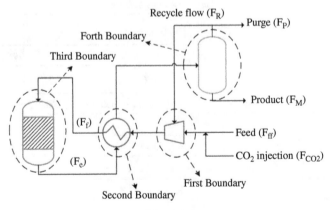

Fig. 14
A schematic diagram of the methanol synthesis loop [94].

4.1 Mixer

The partial mass balance, total mass balance, and enthalpy balances around the first boundary (mixer) are given by Eqs. (80)–(82), respectively.

$$F_{out}x_{out}^i = \sum_{k=1}^{n_{is}} F_{in}^k x_{in}^i, k \tag{80}$$

$$F_{out} = \sum_{k=1}^{n_{is}} F_{in}^k \tag{81}$$

$$F_{out}H_{out} = \sum_{k=1}^{n_{is}} F_{in}^k H_{in}^k \tag{82}$$

Eq. (83) is also used to calculate the pressure of outlet stream from the mixer [3].

$$P_{out} = \frac{\sum_{k=1}^{n_{is}} P_{in}^k}{n_{is}} \tag{83}$$

4.2 Heat exchanger

The second boundary contains the heat exchanger in which no phase change occurs during the process. This unit is smaller than the reactor unit, so a lump model is considered to be able to predict the gas temperature. Eqs. (84) and (85) describe the heat transfer from the tube wall to the cold gas and from the hot gas to the tube wall, in which T_c, T_H, and T_M are specified as the temperature of the cold side, hot side, and tube wall, respectively. Finally, the governing energy equations (Eqs. (86) and (87)), which are time-dependent, are presented to estimate the outlet hot and cold gas temperatures [94].

$$Q_C = U_c A_c (T_M - T_c) \tag{84}$$

$$Q_H = U_H A_H (T_H - T_M) \tag{85}$$

$$C_p \rho_{av} V_H \frac{dT_H}{dt} = \left(F_H C_{p_H} T_H\right)_L - \left(F_H C_{p_H} T_H\right)_o - Q_H \tag{86}$$

$$C_p \rho_{av} V_c \frac{dT_c}{dt} = \left(F_c(t) C_{p_c} T_c\right)_L - \left(F_c(t) C_{p_c} T_c\right)_o - Q_c \tag{87}$$

4.3 Methanol synthesis reactor

Mathematical modeling of the fixed-bed reactor as the heart of the methanol synthesis loop, based on pseudo-homogeneous and heterogeneous models, is described in detail in Section 3.

4.4 Separator

The outlet stream from the methanol synthesis reactor at thermodynamic equilibrium is fed to a flash drum to separate the gas and liquid phases. This thermodynamic equilibrium is achieved by passing the feed stream through an expansion valve, which is placed before the separator. The liquid stream is removed from the bottom and the vapor phase goes out from the top. Thermodynamic models are applied to calculate the amount of methanol in the liquid phase. Equations of state and liquid activity coefficient models are two conventional classes of thermodynamic models for phase equilibrium calculations. Cubic equations of state are widely used to predict phase behavior in refineries and petroleum equipment owing to the accuracy and simplicity of the predicted K-values. The Peng–Robinson equation of state predicts liquid densities better than other two-parameter equation of state such as Soave–Redlich–Kwong (SRK). The Peng–Robinson (PR) equation of state formula is shown in Eqs. (88)–(90).

$$P = \frac{RT}{v - b} - \frac{a}{v(v + b) + b(v - b)} \tag{88}$$

$$b = \sum_i y_i b_i \tag{89}$$

$$a = \sum_i \sum_j y_i y_j \sqrt{a_i a_j} (1 - k_{ij}) \tag{90}$$

The vapor and liquid phases, compressibility factor expression, and fugacity coefficient of component *i* derived from the PR equation of state are given in Eqs. (91) and (92), respectively. Binary interaction coefficients are evaluated by Eq. (94), which is used for the calculation of multicomponent phase equilibrium, in which the system consists of CH_4, CO_2, and N_2. This relation is a function of acentric factor, temperature, and pressure. Due to the high-pressure conditions in the separator, Eqs. (95) and (96) are applied to consider the effect of pressure, where the pressure is in psi [94].

$$Z^3 + (B - 1)Z^2 + (A - 2B - 3B^3)Z + (B^2 + B^3 - AB) = 0 \tag{91}$$

$$RT \ln \phi_i = \frac{b_i}{b}(Z - 1) - \ln(Z - B) + \frac{A}{2\sqrt{2}B}\left(\frac{2a'}{a} - \frac{b_i}{b}\right)\left[\ln \frac{Z + (1 - \sqrt{2})B}{Z + (1 + \sqrt{2})B}\right] \tag{92}$$

where

$$a' = \sum_j y_i y_j \sqrt{a_i a_j}(1 - k_{ij}) \tag{93}$$

$$k_{ij} = \delta_2 T_{rj}^2 + \delta_1 T_{rj} + \delta_0 \tag{94}$$

$$k'_{rj} = k_{ij}(1.04 - 4.2 \times 10^{-5}P) \text{ for } N_2 \text{ interaction parameters} \tag{95}$$

$$k'_{rj} = k_{ij}(1.044269 - 4.375 \times 10^{-5}P) \text{ for } CO_2 \text{ interaction parameters} \tag{96}$$

5. *Simulation of the methanol plant using simulators*

Simulators are attractive tools for simulation, design, and optimization of complicated chemical engineering processes. Aspen Plus and Aspen HYSYS are the two simulators that have been widely used to study chemical processes. According to the capacity of simulator, numerous researchers have investigated methanol production from syngas using these software packages. Table 13 indicates a summary of studies performed on the methanol plants, which is classified according to the type of simulator, methanol synthesis reactor model, and the equation of state used to calculate the thermodynamic properties. RPLUG, RSTOICH, REQUIL, and RGIBBS are the common models used to simulate the reactors. RPLUG is a rigorous model for plug flow; RSTOICH is used in reactor modeling where the stoichiometry of the reaction is known but its kinetics is either negligible or unknown. REQUIL bases its calculations on simultaneous solution of phase equilibrium and stoichiometric chemical computations, whereas RGIBBS calculates the equilibrium compositions by minimizing Gibbs free energy. As shown in Table 13, SRK and PR are the most popular equations of state for predicting the thermodynamic properties in the modeling of reactors.

Table 13 A summary of the studies performed on the methanol plants using simulators.

Simulator	Reactor model	Equation of state	Description	Author
Aspen Plus	RPLUG	–	• Simulation of a methanol plant was carried out via H_2-rich biomass-derived syngas from biomass gasification.	Zhang et al. [126]
Aspen HYSYS	REQUIL	PR-SV	➤ The large-scale integrated methanol and electricity production from natural gas was investigated from an economic viewpoint, ➤ Both operating and capital costs were considered, ➤ The process comprised steam reforming of methane and methanol synthesis units integrated with a power plant in which the unreacted gases are burned.	Pellegrini et al. [127]
Aspen Plus	REQUIL	PR	✓ The methanol production plant from tri-reforming process was simulated, ✓ An optimum heat exchange network was achieved with the aim of minimizing both capital and utility costs,	Zhang et al. [128]

Continued

Table 13   A summary of the studies performed on the methanol plants using simulators—Cont'd

Simulator	Reactor model	Equation of state	Description	Author
Aspen Plus	RPLUG	–	✓ An economic evaluation was performed to prove the potential profits based on the conceptual results from heat integration, ✓ Results showed that 54.9% of energy is saved through heat integration. • The synthesis gas produced in the reformer was mixed with the CO_2 stream exiting from the power-plant carbon capture process, • Co-feeding of CO_2 with syngas increased the methanol to methane ratio from 1.69 to 2.27 compared to the conventional unit, • This integration contributed to reduce the CO_2 emissions by 21.9% and to decrease the methane uptake by 25.6%.	Milani et al. [129]
Aspen HYSYS and PILOT	RSTOICH	PR-B	➤ A turbine expander was integrated with an industrial methanol synthesis plant, ➤ Pinch theory was combined with exergy analysis to achieve optimal conditions of the process, ➤ The power generation led to profitability of the gas turbine integrated plant.	Azadi et al. [130]
Aspen Plus	RPLUG	SRK	✓ The methanol production plant using natural gas with CO_2 utilization was evaluated, ✓ Energy and exergy analysis of a medium-capacity methanol synthesis plant was performed, ✓ The use of CO_2 led to an increase in the methanol production capacity by 20%.	Blumberg et al. [131]
Aspen Plus	RGIBBS	–	• The methanol synthesis from air–steam gasification of biomass, coal, and coal-biomass was simulated,	Liu [132]

Table 13 A summary of the studies performed on the methanol plants using simulators—Cont'd

Simulator	Reactor model	Equation of state	Description	Author
Aspen Plus	REQUIL	PR	• The effect of plant size and reaction temperature on methanol selling price and methanol yield was investigated, • Increasing the plant size decreases the methanol selling price, • Methanol synthesis from coal gasification was more affordable than biomass-coal and biomass. ➢ The methanol production from synthesis gas derived from pine biomass gasification was analyzed, ➢ Three integrated processes, including pine gasification, syngas cleaning using a PSA, and methanol production were simulated.	Puig-Gamero et al. [133]
Aspen Plus	RPLUG	SRK	✓ The CO_2-utilization in the methanol production from natural gas was investigated in terms of exergy analysis. ✓ The CO_2-integration potential by dry reforming, steam reforming (SR) with direct CO_2 hydrogenation, and combined reforming was studied.	Blumberg et al. [134]
Aspen Plus	RPLUG	SRK	• A comparison between three different approaches for mitigating CO_2, containing hydrogenation, bi- and tri-reforming, in terms of environmental and economical factors, were carried out to produce methanol, • Hydrogenation process with hydrogen from renewable sources is superior to other methods due to the achievement of near-zero CO_2 emission.	Nguyen and Zondervan [135]
Aspen Plus	RGIBBS	SRK	➢ The simultaneous production of methanol and power was proposed from coal gasification and methane steam reforming, ➢ The total efficiency of plant increases with the methane/coal ratio.	Lv et al. [136]

Continued

Table 13 A summary of the studies performed on the methanol plants using simulators—Cont'd

Simulator	Reactor model	Equation of state	Description	Author
Aspen Plus	RPLUG	SRK	✓ The economics of methanol production from natural gas by different single reforming technologies, including auto-thermal reforming, steam reforming, and dry reforming of methane was investigated, ✓ Three advanced synthesis routes with different combinations of reforming and synthesis gas conditioning technologies were also studied, ✓ It was found that advanced synthesis technologies have higher methanol yield (i.e., 10–15 mol%) than single reforming technologies (i.e., 5–8 mol%).	Blumberg et al. [137]
Aspen Plus	RPLUG and REQUIL	PR	• A techno-economic assessment of a methanol synthesis plant was conducted, • In order to achieve more degrees of freedom to control the syngas ratio, dry reforming, SR, and partial oxidation of methane in parallel arrangement were suggested to produce syngas for the methanol synthesis reactor.	Alsuhaibani et al. [138]
Aspen Plus	RGIBBS	–	➢ An innovative methanol synthesis process was proposed and simulated using dry and steam reforming of methane by adding triple CO_2 feeds, ➢ Genetic algorithm was used to optimize the process with the aim of maximizing CO_2 conversion and minimizing energy consumption per unit of product.	Wang et al. [139]
Aspen Plus	REQUIL	PR-BM	✓ The performance of the methanol unit was evaluated with the aim of increasing the methanol production capacity, reducing energy requirements, and improving process economics. ✓ Coal gasification process was integrated in the parallel design configuration with the methane steam reforming to produce synthesis gas.	Ahmed [140]

Table 13   **A summary of the studies performed on the methanol plants using simulators—Cont'd**

Simulator	Reactor model	Equation of state	Description	Author
Aspen HYSYS	REQUIL	–	✓ The results showed that the integration of coal gasification with the reforming technology has a higher methanol production capacity as well as lower energy consumption (~10.1%) compared to the coal to methanol process. • The methanol production unit integrated with methane tri-reforming was simulated and its economic feasibility was evaluated.	Borreguero et al. [141]
Aspen HYSYS	RPLUG	PR	➤ A technical-economic analysis was carried out to produce methanol from water electrolysis and methane tri-reforming, ➤ H_2 and O_2 were produced from H_2O electrolysis; O_2 was utilized in the tri-reformer and H_2 was gathered to provide syngas with an optimum stoichiometric ratio of 2 for methanol production.	Shi et al. [142]
Aspen HYSYS and MATLAB	RPLUG	PR	✓ The modeling, simulation, and optimization of methanol production from syngas obtained from the methane tri-reforming process were studied, ✓ Genetic algorithm was applied to gain the maximum methanol production rate.	Farsi and Lari [7]

6. Conclusion

Due to the wide range of applications of alcohols in chemical processes, from clean fuel, hydrogen carriers, and solvent to pharmaceutical usages, a better understanding of alcohols synthesis from synthesis gas is essential. A comprehensive overview of alcohols synthesis plant simulation was conducted in this chapter focusing on modeling different equipment and describing various processes.

Given the importance of producing a variety of alcohols, many researchers have studied the synthesis of methanol and HAs from syngas. However, the lack of studies in the field of synthesis of HAs from natural-gas-based syngas on an industrial scale could be felt in the literature. One of the main fields in the literature was choosing the appropriate reactor configuration to increase the process efficiency. Therefore, different reactor configurations and corresponding mathematical models were investigated, and the benefits and drawbacks of these reactors were also discussed. Additionally, according to the importance of kinetic models in reactor modeling, the kinetic models of methanol and HAs synthesis reported in the literature were mentioned in detail. Aspen Plus and Aspen HYSYS were two attractive tools for simulation, design, and optimization of the methanol synthesis processes, so that the use of these software was reviewed in terms of simulator type, methanol synthesis reactor model, and the equation of state used to calculate the thermodynamic properties.

Finally, during the past decades, researchers have recommended major changes in methanol and HAs synthesis processes from syngas and have evaluated the performance of their ideas with mathematical modeling and simulation tools. This indicates that the simulation of methanol and HAs production processes from syngas can make significant progress in this particular process of interest

Abbreviations and symbols

A_c	cross-sectional area (m^2)
a	catalyst activity
a_b	bubble specific surface area ($m^2\ m^{-3}$)
a_v	catalyst pellet specific surface area ($m^2\ m^{-3}$)
c_p	specific heat capacity ($J\ mol^{-1}\ K^{-1}$)
C_p	specific heat capacity ($J\ Kg^{-1}\ K^{-1}$)
C	concentration ($mol\ m^{-3}$)
D_o	outside diameter of the tube (m)
D_i^e	diffusion coefficient of component i ($m^2\ s^{-1}$)
$D_{k,i}^e$	Knudsen diffusion coefficient of component i ($m^2\ s^{-1}$)
D_z	axial dispersion coefficient of species ($m^2\ s^{-1}$)
D_r	radial dispersion coefficient of species ($m^2\ s^{-1}$)
E_l	Liquid-phase dispersion coefficient ($m^2\ s^{-1}$)
F	molar flow rate ($mol\ s^{-1}$)
h_f	gas-catalyst heat transfer coefficient ($W\ m^{-2}\ K^{-1}$)
J	membrane permeation rate
K_{bei}	mass transfer coefficient for component i in fluidized-bed ($m\ s^{-1}$)
k_g	gas–solid mass transfer coefficient ($m\ s^{-1}$)
P	pressure (bar)
r_j	rate of reaction j ($mol\ kg^{-1}\ s^{-1}$)
R	universal gas constant ($J\ mol^{-1}\ K^{-1}$)
T	temperature (K)

u fluid-phase velocity (m s^{-1})

U overall heat transfer coefficient (W m^{-2} K^{-1})

$\nu_{i,j}$ stoichiometric coefficient of component i in reaction j

y mole fraction (mol mol^{-1})

z axial coordinate (m)

r radial coordinate (m)

α_{H_2} hydrogen permeation rate constant (mol m^{-1}s^{-1} Pa$^{-0.5}$)

α_{H_2O} H$_2$O permeation rate constant (mol m^{-2}s^{-1} Pa^{-1})

α_{eff} slurry-phase overall heat transfer coefficient between process streams and coolant (W m^{-2} K^{-1})

ΔH_R enthalpy of reaction (J mol^{-1})

ρ density (kg m^{-3})

δ bubble-phase volume as a fraction of total bed volume

ε porosity

η effectiveness factor

γ volume fraction of catalyst bed occupied by solid particles in bubble phase

λ_z axial heat dispersion coefficient (W m^{-1} K^{-1})

λ_r radial heat dispersion coefficient (W m^{-1} K^{-1})

μ viscosity (kg m^{-1} s^{-1})

t tube side

s shell side

b bubble phase

B catalyst bed

e emulsion phase

i chemical species

j reaction number

g gas phase

s solid phase

sl liquid solid suspension

References

[1] M.H. Khademi, A. Alipour-Dehkordi, M. Tabesh, Optimal design of methane tri-reforming reactor to produce proper syngas for Fischer-Tropsch and methanol synthesis processes: a comparative analysis between different side-feeding strategies, Int. J. Hydrogen Energy 46 (27) (2021) 14441–14454.

[2] P. Liao, C. Zhang, L. Zhang, Y. Yang, L. Zhong, H. Wang, Y. Sun, Higher alcohol synthesis via syngas over CoMn catalysts derived from hydrotalcite-like precursors, Catal. Today 311 (2018) 56–64.

[3] I. Løvik, Modelling, Estimation and Optimization of the Methanol Synthesis with Catalyst Deactivation, Doctoral Dissertation, Norwegian University of Science and Technology, 2001.

[4] G. Graaf, A. Beenackers, Comparison of two-phase and three-phase methanol synthesis processes, Chem. Eng. Process. Process Intensif. 35 (6) (1996) 413–427.

[5] P. Tijm, F. Waller, D. Brown, Methanol technology developments for the new millennium, Appl. Catal. A. Gen. 221 (1–2) (2001) 275–282.

[6] M. Ao, G.H. Pham, J. Sunarso, M.O. Tade, S. Liu, Active centers of catalysts for higher alcohol synthesis from syngas: a review, ACS Catal. 8 (8) (2018) 7025–7050.

[7] M. Farsi, M.F. Lari, Methanol production based on methane tri-reforming: process modeling and optimization, Process Saf. Environ. Prot. 138 (2020) 269–278.

[8] F. Gallucci, A. Basile, A theoretical analysis of methanol synthesis from CO_2 and H_2 in a ceramic membrane reactor, Int. J. Hydrogen Energy 32 (18) (2007) 5050–5058.

[9] J.-P. Lange, Methanol synthesis: a short review of technology improvements, Catal. Today 64 (1–2) (2001) 3–8.

[10] E. Fiedler, G. Grossmann, D.B. Kersebohm, G. Weiss, C. Witte, Methanol, in: Ullmann's Encyclopedia of Industrial Chemistry, Wiley-VCH, Weinheim, 2000.

[11] G. Chinchen, P. Denny, J. Jennings, M. Spencer, K. Waugh, Synthesis of methanol: part 1. Catalysts and kinetics, Appl. Catal. 36 (1988) 1–65.

[12] R.J. Dry, Possibilities for the development of large-capacity methanol synthesis reactors for synfuel production, Ind. Eng. Chem. Res. 27 (4) (1988) 616–624.

[13] P. Parvasi, M.R. Rahimpour, A. Jahanmiri, Incorporation of dynamic flexibility in the design of a methanol synthesis loop in the presence of catalyst deactivation, Chem. Eng. Technol. 31 (1) (2008) 116–132.

[14] J. Pang, M. Zheng, T. Zhang, Synthesis of ethanol and its catalytic conversion, Adv. Catal. 64 (2019) 89–191.

[15] V. Subramani, S.K. Gangwal, A review of recent literature to search for an efficient catalytic process for the conversion of syngas to ethanol, Energy Fuel 22 (2) (2008) 814–839.

[16] O. Pardo-Planas, H.K. Atiyeh, J.R. Phillips, C.P. Aichele, S. Mohammad, Process simulation of ethanol production from biomass gasification and syngas fermentation, Bioresour. Technol. 245 (2017) 925–932.

[17] V.E. Leonov, M.M. Karavaev, E.N. Tsybina, G.S. Petrishcheva, Kinetics of methanol synthesis on a low-temperature catalyst, Kinet. Katal. 14 (1973) 970–975.

[18] K. Klier, V. Chatikavanij, R. Herman, G. Simmons, Catalytic synthesis of methanol from COH_2: IV. The effects of carbon dioxide, J. Catal. 74 (2) (1982) 343–360.

[19] P. Villa, P. Forzatti, G. Buzzi-Ferraris, G. Garone, I. Pasquon, Synthesis of alcohols from carbon oxides and hydrogen. 1. Kinetics of the low-pressure methanol synthesis, Ind. Eng. Chem. Process. Des. Dev. 24 (1) (1985) 12–19.

[20] R.M. Agny, C.G. Takoudis, Synthesis of methanol from carbon monoxide and hydrogen over a copper-zinc oxide-alumina catalyst, Ind. Eng. Chem. Prod. Res. Dev. 24 (1) (1985) 50–55.

[21] M. Takagawa, M. Ohsugi, Study on reaction rates for methanol synthesis from carbon monoxide, carbon dioxide, and hydrogen, J. Catal. 107 (1) (1987) 161–172.

[22] G. Graaf, E. Stamhuis, A. Beenackers, Kinetics of low-pressure methanol synthesis, Chem. Eng. Sci. 43 (12) (1988) 3185–3195.

[23] M.A. McNeil, C.J. Schack, R.G. Rinker, Methanol synthesis from hydrogen, carbon monoxide and carbon dioxide over a $CuO/ZnO/Al_2O_3$ catalyst: II. Development of a phenomenological rate expression, Appl. Catal. 50 (1) (1989) 265–285.

[24] J. Skrzypek, M. Lachowska, H. Moroz, Kinetics of methanol synthesis over commercial copper/zinc oxide/alumina catalysts, Chem. Eng. Sci. 46 (11) (1991) 2809–2813.

[25] K.V. Bussche, G. Froment, A steady-state kinetic model for methanol synthesis and the water gas shift reaction on a commercial $Cu/ZnO/Al_2O_3$ catalyst, J. Catal. 161 (1) (1996) 1–10.

[26] T. Kubota, I. Hayakawa, H. Mabuse, K. Mori, K. Ushikoshi, T. Watanabe, M. Saito, Kinetic study of methanol synthesis from carbon dioxide and hydrogen, Appl. Organomet. Chem. 15 (2) (2001) 121–126.

[27] A.Y. Rozovskii, G.I. Lin, Fundamentals of methanol synthesis and decomposition, Top. Catal. 22 (3) (2003) 137–150.

[28] H.-W. Lim, M.-J. Park, S.-H. Kang, H.-J. Chae, J.W. Bae, K.-W. Jun, Modeling of the kinetics for methanol synthesis using $Cu/ZnO/Al_2O_3/ZrO_2$ catalyst: influence of carbon dioxide during hydrogenation, Ind. Eng. Chem. Res. 48 (23) (2009) 10448–10455.

[29] N. Park, M.-J. Park, Y.-J. Lee, K.-S. Ha, K.-W. Jun, Kinetic modeling of methanol synthesis over commercial catalysts based on three-site adsorption, Fuel Process. Technol. 125 (2014) 139–147.

[30] G. Natta, Synthesis of methanol, Catalysis 3 (1955). Reinhold, New York.

[31] L. Grabow, M. Mavrikakis, Mechanism of methanol synthesis on Cu through CO_2 and CO hydrogenation, ACS Catal. 1 (4) (2011) 365–384.

[32] G. Leonzio, Mathematical modeling of a methanol reactor by using different kinetic models, J. Ind. Eng. Chem. 85 (2020) 130–140.

[33] T.Y. Park, I.-S. Nam, Y.G. Kim, Kinetic analysis of mixed alcohol synthesis from syngas over K/MoS_2 catalyst, Ind. Eng. Chem. Res. 36 (12) (1997) 5246–5257.

[34] A. Beretta, E. Micheli, L. Tagliabue, E. Tronconi, Development of a process for higher alcohol production via synthesis gas, Ind. Eng. Chem. Res. 37 (10) (1998) 3896–3908.

[35] A.K. Gunturu, E.L. Kugler, J.B. Cropley, D.B. Dadyburjor, A kinetic model for the synthesis of high-molecular-weight alcohols over a sulfided Co-K-Mo/C catalyst, Ind. Eng. Chem. Res. 37 (6) (1998) 2107–2115.

[36] M. Kulawska, J. Skrzypek, Kinetics of the synthesis of higher aliphatic alcohols from syngas over a modified methanol synthesis catalyst, Chem. Eng. Process. Process Intensif. 40 (1) (2001) 33–40.

[37] J.M. Christensen, P.M. Mortensen, R. Trane, P.A. Jensen, A.D. Jensen, Effects of H_2S and process conditions in the synthesis of mixed alcohols from syngas over alkali promoted cobalt-molybdenum sulfide, Appl. Catal. A. Gen. 366 (1) (2009) 29–43.

[38] V.R. Surisetty, A.K. Dalai, J. Kozinski, Intrinsic reaction kinetics of higher alcohol synthesis from synthesis gas over a sulfided alkali-promoted Co − Rh − Mo trimetallic catalyst supported on multiwalled carbon nanotubes (MWCNTs), Energy Fuel 24 (8) (2010) 4130–4137.

[39] S.F. Zaman, K.J. Smith, Synthesis gas conversion over a Rh–K–MoP/SiO_2 catalyst, Catal. Today 171 (1) (2011) 266–274.

[40] H. Guo, S. Li, H. Zhang, F. Peng, L. Xiong, J. Yang, et al., Reaction condition optimization and lumped kinetics study for lower alcohols synthesis from syngas using a two-stage bed catalyst combination system, Ind. Eng. Chem. Res. 53 (1) (2014) 123–131.

[41] J. Su, W. Mao, X.C. Xu, Z. Yang, H. Li, J. Xu, Y.F. Han, Kinetic study of higher alcohol synthesis directly from syngas over CoCu/SiO_2 catalysts, AICHE J. 60 (5) (2014) 1797–1809.

[42] M. Portillo, A.V. Perales, F. Vidal-Barrero, M. Campoy, A kinetic model for the synthesis of ethanol from syngas and methanol over an alkali-Co doped molybdenum sulfide catalyst: model building and validation at bench scale, Fuel Process. Technol. 151 (2016) 19–30.

[43] P.E. Boahene, A.K. Dalai, Higher alcohols synthesis over carbon nanohorn-supported KCoRhMo catalyst: pelletization and kinetic modeling, Ind. Eng. Chem. Res. 57 (16) (2018) 5517–5528.

[44] C. Göbel, S. Schmidt, C. Froese, T. Bujara, V. Scherer, M. Muhler, The steady-state kinetics of CO hydrogenation to higher alcohols over a bulk Co-Cu catalyst, J. Catal. 394 (2021) 465–475.

[45] B. Lommerts, G. Graaf, A. Beenackers, Mathematical modeling of internal mass transport limitations in methanol synthesis, Chem. Eng. Sci. 55 (23) (2000) 5589–5598.

[46] M.H. Khademi, M. Rahimpour, A. Jahanmiri, A comparison of a novel recuperative configuration and conventional methanol synthesis reactor, Chem. Eng. Commun. 199 (7) (2012) 889–911.

[47] M.H. Khademi, M.R. Rahimpour, A. Jahanmiri, Differential evolution (DE) strategy for optimization of hydrogen production, cyclohexane dehydrogenation and methanol synthesis in a hydrogen-permselective membrane thermally coupled reactor, Int. J. Hydrogen Energy 35 (5) (2010) 1936–1950.

[48] M.H. Khademi, M.R. Rahimpour, A. Jahanmiri, Start-up and dynamic analysis of a novel thermally coupled reactor for the simultaneous production of methanol and benzene, Ind. Eng. Chem. Res. 50 (21) (2011) 12092–12102.

[49] M.H. Khademi, P. Setoodeh, M.R. Rahimpour, A. Jahanmiri, Optimization of methanol synthesis and cyclohexane dehydrogenation in a thermally coupled reactor using differential evolution (DE) method, Int. J. Hydrogen Energy 34 (16) (2009) 6930–6944.

[50] M. Khanipour, A. Mirvakili, A. Bakhtyari, M. Farniaei, M.R. Rahimpour, Enhancement of synthesis gas and methanol production by flare gas recovery utilizing a membrane based separation process, Fuel Process. Technol. 166 (2017) 186–201.

[51] A. Mirvakili, M.R. Rahimpour, Mal-distribution of temperature in an industrial dual-bed reactor for conversion of CO_2 to methanol, Appl. Therm. Eng. 91 (2015) 1059–1070.

[52] A. Mirvakili, M. Rostami, K. Paymooni, M.R. Rahimpour, B. Moghtaderi, Hydrogen looping approach in optimized methanol thermally coupled membrane reactor, Int. J. Hydrogen Energy 37 (1) (2012) 235–249.

[53] M.R. Rahimpour, A dual-catalyst bed concept for industrial methanol synthesis, Chem. Eng. Commun. 194 (12) (2007) 1638–1653.

[54] M.R. Rahimpour, M. Bayat, Production of ultrapure hydrogen via utilizing fluidization concept from coupling of methanol and benzene synthesis in a hydrogen-permselective membrane reactor, Int. J. Hydrogen Energy 36 (11) (2011) 6616–6627.

[55] M.R. Rahimpour, M. Lotfinejad, A comparison of co-current and counter-current modes of operation for a dual-type industrial methanol reactor, Chem. Eng. Process. Process Intensif. 47 (9–10) (2008) 1819–1830.

[56] F. Rahmani, M. Haghighi, P. Estifaee, M.R. Rahimpour, A comparative study of two different membranes applied for auto-thermal methanol synthesis process, J. Nat. Gas Sci. Eng. 7 (2012) 60–74.

[57] N. Rezaie, A. Jahanmiri, B. Moghtaderi, M.R. Rahimpour, A comparison of homogeneous and heterogeneous dynamic models for industrial methanol reactors in the presence of catalyst deactivation, Chem. Eng. Process. Process Intensif. 44 (8) (2005) 911–921.

[58] F. Samimi, S. Kabiri, A. Mirvakili, M.R. Rahimpour, The concept of integrated thermally double coupled reactor for simultaneous production of methanol, hydrogen and gasoline via differential evolution method, J. Nat. Gas Sci. Eng. 14 (2013) 144–157.

[59] G. Graaf, H. Scholtens, E. Stamhuis, A. Beenackers, Intra-particle diffusion limitations in low-pressure methanol synthesis, Chem. Eng. Sci. 45 (4) (1990) 773–783.

[60] J.T. Sun, I.S. Metcalfe, M. Sahibzada, Deactivation of $Cu/ZnO/Al_2O_3$ methanol synthesis catalyst by sintering, Ind. Eng. Chem. Res. 38 (10) (1999) 3868–3872.

[61] H.H. Kung, Deactivation of methanol synthesis catalysts-a review, Catal. Today 11 (4) (1992) 443–453.

[62] A. Prašnikar, A. Pavlišič, F. Ruiz-Zepeda, J. Kovač, B. Likozar, Mechanisms of copper-based catalyst deactivation during CO_2 reduction to methanol, Ind. Eng. Chem. Res. 58 (29) (2019) 13021–13029.

[63] L. Hanken, Optimization of Methanol Reactor, Master's thesis, The Norwegian University of Science and Technology, 1995.

[64] M.R. Rahimpour, A. Abbasloo, J. Sayyad Amin, A novel radial-flow, spherical-bed reactor concept for methanol synthesis in the presence of catalyst deactivation, Chem. Eng. Technol. 31 (11) (2008) 1615–1629.

[65] C. Kuechen, U. Hoffmann, Investigation of simultaneous reaction of carbon monoxide and carbon dioxide with hydrogen on a commercial copper/zinc oxide catalyst, Chem. Eng. Sci. 48 (22) (1993) 3767–3776.

[66] M. Sahibzada, D. Chadwick, I. Metcalfe, Methanol synthesis from CO_2/H_2 over Pd-promoted $Cu/ZnO/Al_2O_3$ catalysts: kinetics and deactivation, Stud. Surf. Sci. Catal. 107 (1997) 29–34. Elsevier.

[67] M.R. Rahimpour, J. Fathikalajahi, A. Jahanmiri, Selective kinetic deactivation model for methanol synthesis from simultaneous reaction of CO_2 and CO with H_2 on a commercial copper/zinc oxide catalyst, Can. J. Chem. Eng. 76 (4) (1998) 753–761.

[68] G. Bozzano, F. Manenti, Efficient methanol synthesis: perspectives, technologies and optimization strategies, Prog. Energy Combust. Sci. 56 (2016) 71–105.

[69] I. Løvik, M. Hillestad, T. Hertzberg, Long term dynamic optimization of a catalytic reactor system, Comput. Chem. Eng. 22 (1998) S707–S710.

[70] I. Løvik, M. Hillestad, T. Hertzberg, Modeling and optimization of a reactor system with deactivating catalyst, Comput. Chem. Eng. 23 (1999) S839–S842.

[71] I. Løvik, M. Hillestad, T. Hertzberg, Sensitivity in optimization of a reactor system with deactivating catalyst, Comput. Aided Chem. Eng. 8 (2000) 517–522. Elsevier.

[72] F. Manenti, S. Cieri, M. Restelli, Considerations on the steady-state modeling of methanol synthesis fixed-bed reactor, Chem. Eng. Sci. 66 (2) (2011) 152–162.

[73] A. Jahanmiri, R. Eslamloueyan, Optimal temperature profile in methanol synthesis reactor, Chem. Eng. Commun. 189 (6) (2002) 713–741.

[74] A. Mirvakili, S. Chahibakhsh, M. Ebrahimzadehsarvestani, E. Soroush, M.R. Rahimpour, Modeling and assessment of novel configurations to enhance methanol production in industrial mega-methanol synthesis plant, J. Taiwan Inst. Chem. Eng. 104 (2019) 40–53.

[75] M. Nassirpour, M.H. Khademi, Evaluation of different cooling technologies for industrial methanol synthesis reactor in terms of energy efficiency and methanol yield: an economic-optimization, J. Taiwan Inst. Chem. Eng. 113 (2020) 302–314.

[76] S.A. Velardi, A.A. Barresi, Methanol synthesis in a forced unsteady-state reactor network, Chem. Eng. Sci. 57 (15) (2002) 2995–3004.

[77] M. Shahrokhi, G. Baghmisheh, Modeling, simulation and control of a methanol synthesis fixed-bed reactor, Chem. Eng. Sci. 60 (15) (2005) 4275–4286.

[78] H. Kordabadi, A. Jahanmiri, Optimization of methanol synthesis reactor using genetic algorithms, Chem. Eng. J. 108 (3) (2005) 249–255.

[79] H. Kordabadi, A. Jahanmiri, A pseudo-dynamic optimization of a dual-stage methanol synthesis reactor in the face of catalyst deactivation, Chem. Eng. Process. Process Intensif. 46 (12) (2007) 1299–1309.

[80] M.R. Rahimpour, A two-stage catalyst bed concept for conversion of carbon dioxide into methanol, Fuel Process. Technol. 89 (5) (2008) 556–566.

[81] M.R. Rahimpour, M. Lotfinejad, A comparison of auto-thermal and conventional methanol synthesis reactor in the presence of catalyst deactivation, Chem. Eng. Process. Process Intensif. 47 (12) (2008) 2121–2130.

[82] M. Khanipour, A. Mirvakili, A. Bakhtyari, M. Farniaei, M.R. Rahimpour, A membrane-assisted hydrogen and carbon oxides separation from flare gas and recovery to a commercial methanol reactor, Int. J. Hydrogen Energy 45 (12) (2020) 7386–7400.

[83] A. Keshavarz, A. Mirvakili, S. Chahibakhsh, A. Shariati, M.R. Rahimpour, Simultaneous methanol production and separation in the methanol synthesis reactor to increase methanol production, Chem. Eng. Process. Process Intensif. 158 (2020), 108176.

[84] R. De María, I. Díaz, M. Rodríguez, A. Sáiz, Industrial methanol from syngas: kinetic study and process simulation, Int. J. Chem. React. Eng. 11 (1) (2013) 469–477.

[85] G. Leonzio, P.U. Foscolo, Analysis of a 2-D model of a packed bed reactor for methanol production by means of CO_2 hydrogenation, Int. J. Hydrogen Energy 45 (18) (2020) 10648–10663.

[86] S. Abrol, C.M. Hilton, Modeling, simulation and advanced control of methanol production from variable synthesis gas feed, Comput. Chem. Eng. 40 (2012) 117–131.

[87] F. Manenti, S. Cieri, M. Restelli, G. Bozzano, Dynamic modeling of the methanol synthesis fixed-bed reactor, Comput. Chem. Eng. 48 (2013) 325–334.

[88] M. Farniaei, M. Abbasi, A. Rasoolzadeh, M.R. Rahimpour, Enhancement of methanol, DME and hydrogen production via employing hydrogen permselective membranes in a novel integrated thermally double-coupled two-membrane reactor, J. Nat. Gas Sci. Eng. 14 (2013) 158–173.

[89] M.R. Rahimpour, M. Lotfinejad, Enhancement of methanol production in a membrane dual-type reactor, Chem. Eng. Technol. 30 (8) (2007) 1062–1076.

[90] M.R. Rahimpour, S. Ghader, Enhancement of CO conversion in a novel Pd–Ag membrane reactor for methanol synthesis, Chem. Eng. Process. Process Intensif. 43 (9) (2004) 1181–1188.

[91] M.R. Rahimpour, M. Lotfinejad, Co-current and countercurrent configurations for a membrane dual type methanol reactor, Chem. Eng. Technol. 31 (1) (2008) 38–57.

[92] M.R. Rahimpour, A.K. Mostafazadeh, M. Barmaki, Application of hydrogen-permselective Pd-based membrane in an industrial single-type methanol reactor in the presence of catalyst deactivation, Fuel Process. Technol. 89 (12) (2008) 1396–1408.

[93] M.R. Rahimpour, H.E. Behjati, Dynamic optimization of membrane dual-type methanol reactor in the presence of catalyst deactivation using genetic algorithm, Fuel Process. Technol. 90 (2) (2009) 279–291.

[94] P. Parvasi, A.K. Mostafazadeh, M. Rahimpour, Dynamic modeling and optimization of a novel methanol synthesis loop with hydrogen-permselective membrane reactor, Int. J. Hydrogen Energy 34 (9) (2009) 3717–3733.

[95] M.H. Khademi, A. Jahanmiri, M. Rahimpour, A novel configuration for hydrogen production from coupling of methanol and benzene synthesis in a hydrogen-permselective membrane reactor, Int. J. Hydrogen Energy 34 (12) (2009) 5091–5107.

[96] M.R. Rahimpour, M. Bayat, A novel cascade membrane reactor concept for methanol synthesis in the presence of long-term catalyst deactivation, Int. J. Energy Res. 34 (15) (2010) 1356–1371.

[97] M.R. Rahimpour, S. Mazinani, B. Vaferi, M. Baktash, Comparison of two different flow types on CO removal along a two-stage hydrogen permselective membrane reactor for methanol synthesis, Appl. Energy 88 (1) (2011) 41–51.

[98] S. Amirabadi, S. Kabiri, R. Vakili, D. Iranshahi, M.R. Rahimpour, Differential evolution strategy for optimization of hydrogen production via coupling of methylcyclohexane dehydrogenation reaction and methanol synthesis process in a thermally coupled double membrane reactor, Ind. Eng. Chem. Res. 52 (4) (2013) 1508–1522.

[99] M. Bayat, M.R. Rahimpour, Production of hydrogen and methanol enhancement via a novel optimized thermally coupled two-membrane reactor, Int. J. Energy Res. 37 (2) (2013) 105–120.

[100] M. Bayat, M.R. Rahimpour, Dynamic optimal analysis of a novel cascade membrane methanol reactor by using genetic algorithm (GA) method, Energy Syst. 4 (2) (2013) 137–164.

[101] M. Farsi, A. Jahanmiri, Dynamic modeling of a H_2O-permselective membrane reactor to enhance methanol synthesis from syngas considering catalyst deactivation, J. Nat. Gas Chem. 21 (4) (2012) 407–414.

[102] M. Farsi, A. Jahanmiri, Application of water vapor and hydrogen-permselective membranes in an industrial fixed-bed reactor for large scale methanol production, Chem. Eng. Res. Des. 89 (12) (2011) 2728–2735.

[103] M. Farsi, A. Jahanmiri, Methanol production in an optimized dual-membrane fixed-bed reactor, Chem. Eng. Process. Process Intensif. 50 (11 – 12) (2011) 1177–1185.

[104] M. Farsi, A. Jahanmiri, Dynamic modeling and operability analysis of a dual-membrane fixed bed reactor to produce methanol considering catalyst deactivation, J. Ind. Eng. Chem. 20 (5) (2014) 2927–2933.

[105] M. Bayat, Z. Dehghani, M. Rahimpour, Membrane/sorption-enhanced methanol synthesis process: dynamic simulation and optimization, J. Ind. Eng. Chem. 20 (5) (2014) 3256–3269.

[106] M.R. Rahimpour, R. Vakili, E. Pourazadi, A. Bahmanpour, D. Iranshahi, Enhancement of hydrogen production via coupling of MCH dehydrogenation reaction and methanol synthesis process by using thermally coupled heat exchanger reactor, Int. J. Hydrogen Energy 36 (5) (2011) 3371–3383.

[107] A. Goosheneshin, R. Maleki, D. Iranshahi, M.R. Rahimpour, A. Jahanmiri, Simultaneous production and utilization of methanol for methyl formate synthesis in a looped heat exchanger reactor configuration, J. Nat. Gas Chem. 21 (6) (2012) 661–672.

[108] M. Bayat, M. Rahimpour, Simultaneous hydrogen and methanol enhancement through a recuperative two-zone thermally coupled membrane reactor, Energy Syst. 3 (4) (2012) 401–420.

[109] H. Rahmanifard, R. Vakili, T. Plaksina, M.R. Rahimpour, M. Babaei, X. Fan, On improving the hydrogen and methanol production using an auto-thermal double-membrane reactor: model prediction and optimisation, Comput. Chem. Eng. 119 (2018) 258–269.

[110] M.R. Rahimpour, P. Parvasi, P. Setoodeh, Dynamic optimization of a novel radial-flow, spherical-bed methanol synthesis reactor in the presence of catalyst deactivation using differential evolution (DE) algorithm, Int. J. Hydrogen Energy 34 (15) (2009) 6221–6230.

[111] D. Iranshahi, A. Golrokh, E. Pourazadi, S. Saeidi, F. Gallucci, Progress in spherical packed-bed reactors: opportunities for refineries and chemical industries, Chem. Eng. Process. Process Intensif. 132 (2018) 16–24.

[112] M.R. Rahimpour, E. Pourazadi, D. Iranshahi, A. Bahmanpour, Methanol synthesis in a novel axial-flow, spherical packed bed reactor in the presence of catalyst deactivation, Chem. Eng. Res. Des. 89 (11) (2011) 2457–2469.

[113] K. Hirotani, H. Nakamura, K. Shoji, Optimum catalytic reactor design for methanol synthesis with TEC MRF-Z reactor, Catal. Surv. Jpn. 2 (1998) 99–106.

[114] M. Farsi, Mathematical modeling and optimization of a radial flow tubular reactor to produce methanol from syngas, Pet. Chem. 58 (12) (2018) 1091–1098.

[115] M. Bayat, Z. Dehghani, M. Rahimpour, Dynamic multi-objective optimization of industrial radial-flow fixed-bed reactor of heavy paraffin dehydrogenation in LAB plant using NSGA-II method, J. Taiwan Inst. Chem. Eng. 45 (4) (2014) 1474–1484.

[116] M.R. Rahimpour, H. Elekaei, Enhancement of methanol production in a novel fluidized-bed hydrogen-permselective membrane reactor in the presence of catalyst deactivation, Int. J. Hydrogen Energy 34 (5) (2009) 2208–2223.

[117] M.R. Rahimpour, K. Alizadehhesari, A novel fluidized-bed membrane dual-type reactor concept for methanol synthesis, Chem. Eng. Technol. 31 (12) (2008) 1775–1789.

[118] M.R. Rahimpour, M. Bayat, F. Rahmani, Dynamic simulation of a cascade fluidized-bed membrane reactor in the presence of long-term catalyst deactivation for methanol synthesis, Chem. Eng. Sci. 65 (14) (2010) 4239–4249.

[119] M.R. Rahimpour, M. Bayat, F. Rahmani, Enhancement of methanol production in a novel cascading fluidized-bed hydrogen permselective membrane methanol reactor, Chem. Eng. J. 157 (2–3) (2010) 520–529.

[120] M.R. Rahimpour, F. Rahmani, M. Bayat, Contribution to emission reduction of CO_2 by a fluidized-bed membrane dual-type reactor in methanol synthesis process, Chem. Eng. Process. Process Intensif. 49 (6) (2010) 589–598.

[121] M.R. Rahimpour, M. Bayat, Comparative study of two different hydrogen redistribution strategies along a fluidized-bed hydrogen permselective membrane reactor for methanol synthesis, Ind. Eng. Chem. Res. 49 (2) (2010) 472–480.

[122] K. Salehi, S. Jokar, J. Shariati, M. Bahmani, M. Sedghamiz, M.R. Rahimpour, Enhancement of CO conversion in a novel slurry bubble column reactor for methanol synthesis, J. Nat. Gas Sci. Eng. 21 (2014) 170–183.

[123] Y. Wu, D. Gidaspow, Hydrodynamic simulation of methanol synthesis in gas–liquid slurry bubble column reactors, Chem. Eng. Sci. 55 (3) (2000) 573–587.

[124] I.K. Gamwo, J.S. Halow, D. Gidaspow, R. Mostofi, CFD models for methanol synthesis three-phase reactors: reactor optimization, Chem. Eng. J. 93 (2) (2003) 103–112.

[125] J. Wang, R.G. Anthony, A. Akgerman, Mathematical simulations of the performance of trickle bed and slurry reactors for methanol synthesis, Comput. Chem. Eng. 29 (11–12) (2005) 2474–2484.

[126] Y. Zhang, J. Xiao, L. Shen, Simulation of methanol production from biomass gasification in interconnected fluidized beds, Ind. Eng. Chem. Res. 48 (11) (2009) 5351–5359.

[127] L.A. Pellegrini, G. Soave, S. Gamba, S. Langè, Economic analysis of a combined energy–methanol production plant, Appl. Energy 88 (12) (2011) 4891–4897.

[128] Y. Zhang, J. Cruz, S. Zhang, H.H. Lou, T.J. Benson, Process simulation and optimization of methanol production coupled to tri-reforming process, Int. J. Hydrogen Energy 38 (31) (2013) 13617–13630.

[129] D. Milani, R. Khalilpour, G. Zahedi, A. Abbas, A model-based analysis of CO_2 utilization in methanol synthesis plant, J. CO_2 Util. 10 (2015) 12–22.

[130] M. Azadi, N. Tahouni, M.H. Panjeshahi, Energy conservation in methanol plant using CHP system, Appl. Therm. Eng. 107 (2016) 1324–1333.

[131] T. Blumberg, T. Morosuk, G. Tsatsaronis, Exergy-based evaluation of methanol production from natural gas with CO_2 utilization, Energy 141 (2017) 2528–2539.

[132] Z. Liu, Economic analysis of methanol production from coal/biomass upgrading, Energy Sources Part B Econ. Plann. Policy 13 (1) (2018) 66–71.

[133] M. Puig-Gamero, J. Argudo-Santamaria, J. Valverde, P. Sánchez, L. Sanchez-Silva, Three integrated process simulation using aspen plus®: pine gasification, syngas cleaning and methanol synthesis, Energ. Conver. Manage. 177 (2018) 416–427.

[134] T. Blumberg, T. Morosuk, G. Tsatsaronis, CO_2-utilization in the synthesis of methanol: potential analysis and exergetic assessment, Energy 175 (2019) 730–744.

[135] T.B. Nguyen, E. Zondervan, Methanol production from captured CO_2 using hydrogenation and reforming technologies_environmental and economic evaluation, J. CO_2 Util. 34 (2019) 1–11.

[136] L. Lv, L. Zhu, H. Li, B. Li, Methanol-power production using coal and methane as materials integrated with a two-level adjustment system, J. Taiwan Inst. Chem. Eng. 97 (2019) 346–355.

[137] T. Blumberg, G. Tsatsaronis, T. Morosuk, On the economics of methanol production from natural gas, Fuel 256 (2019), 115824.

[138] A.S. Alsuhaibani, S. Afzal, M. Challiwala, N.O. Elbashir, M.M. El-Halwagi, The impact of the development of catalyst and reaction system of the methanol synthesis stage on the overall profitability of the entire plant: a techno-economic study, Catal. Today 343 (2020) 191–198.

[139] H. Wang, Y. Su, D. Wang, S. Jin, S.A. Wei, W. Shen, Optimal design and energy-saving investigation of the triple CO_2 feeds for methanol production system by combining steam and dry methane reforming, Ind. Eng. Chem. Res. 59 (4) (2020) 1596–1606.

[140] U. Ahmed, Techno-economic feasibility of methanol synthesis using dual fuel system in a parallel process design configuration with control on green house gas emissions, Int. J. Hydrogen Energy 45 (11) (2020) 6278–6290.

[141] A. Borreguero, F. Dorado, M. Capuchino-Biezma, L. Sánchez-Silva, J. García-Vargas, Process simulation and economic feasibility assessment of the methanol production via tri-reforming using experimental kinetic equations, Int. J. Hydrogen Energy 45 (2020) 26623–26636.

[142] C. Shi, B. Labbaf, E. Mostafavi, N. Mahinpey, Methanol production from water electrolysis and tri-reforming: process design and technical-economic analysis, J. CO_2 Util. 38 (2020) 241–251.

CHAPTER 15

Modeling, simulation, and optimization of combined heat and power generation from produced syngas

Ilenia Rossetti

Chemical Plants and Industrial Chemistry Group, Dipartimento di Chimica, Università degli Studi di Milano, INSTM Milano Università-Unit, CNR-SCITEC, Milano, Italy

1. Introduction

Combined heat and power (CHP) cogeneration units are systems for the simultaneous conversion of the chemical energy of a fuel into electricity and heat. Possibilities are available also for cooling options, and such systems are known as combined cooling, heat and power systems (CCHP). The panorama is very wide and includes plants of different scale, forms microgeneration systems, suitable for single houses or small districts, to large facilities, which exploit distributed heating networks and connection with the grid. The primary source can be fossil (e.g., natural gas or coal) or renewable. In this case, solid biomass of different origin (agricultural residua, energy crops, wastes) can be indirectly used to produce steam after burning in furnaces, or it can be transformed into syngas through gasification. Alternatively, biomass-derived compounds, e.g., biogas, obtained by anaerobic digestion of biomass, or bioethanol, obtained by biomass fermentation, can be transformed into syngas by reforming processes.

The produced syngas can be burned in internal combustion engines (ICEs) coupled with electrical generators (EG) or in alternative systems suitable for small-size cogeneration such as Stirling engines (SE) or organic Rankine cycles (ORC). For large-scale applications, steam or gas turbine (ST, GT) cycles ensure higher electric efficiencies (η_{el}). In all these cases, however, the efficiency of transformation of the power content of the primary fuel to electric power (the most valuable output) is quite low, usually below 30% (far lower for small-power systems). A solution to improve the efficiency to electric power is to use fuel cells (FCs), electrochemical devices that are not limited by Carnot cycle efficiency, and whose η_{el} ranges between 35% and

Advances in Synthesis Gas: Methods, Technologies and Applications. https://doi.org/10.1016/B978-0-323-91879-4.00016-3

60%. FCs can be fed with more or less pure hydrogen, depending on the device type, or in some cases, internal reforming is also possible, in case of FCs operating at high temperature.

In the following some examples of CHP units are reported, with particular attention on the modeling methods. Only the main equations are reported for the sake of brevity, and the reader is recommended to refer to the cited sources for the full model details. In particular, examples of CHP, in particular fed with biomass, are first reported, followed by the integration with FCs. At last, the models most frequently used for optimization of these energy systems are summarized.

2. CHP systems based on biomass

An example of valorization of agricultural residua through gasification is reported by Almpantis et al. [1]. A case history has been developed assessing the possibility of delocalized CHP using rice husk as waste biomass. A possible plant layout has been simulated using Aspen Plus© and includes a fluidized-bed reactor, gas cleaning units, and condensation. Different gasification options have been considered, using air, steam, or steam+oxygen, with a plant capacity of 25,000t/y of rice husk. The produced electricity was either used on-site or sold to the grid, while the heat generated was used internally. The simulation results were compared with experimental data. The simulation included important simplifications, considering ideal reactors and fluids, neglecting the reactions involving heteroatoms (S, N), approximating tar as benzene and char as pure graphite carbon. The most important simplifying assumption was to simulate equilibrium (Gibbs type) reactors, which prevented accurate prediction of the concentration of most of the components, except H_2 [1]. Despite the evident complexity to account for kinetics of transformation of such complex raw materials and wide array of reactions, a more reliable approach is the one allowing a reasonable definition of costs. However, even according to this simplified thermodynamic elaboration, the advantage to gasify the biomass using steam is evident, since in spite of higher energy consumption and cost for steam generation, much higher H_2 yield is achieved [1].

The gasification of biomass to syngas is also often presented as an option to feed CHP units, in the simplest cases selecting SE or simple ICE for power generation. As said, the modeling of biomass conversion to syngas is very complex, as it depends on the type of biomass, and even the same source is difficult to standardize. Also, the reaction itself is hard to model being composed of a sequence of consecutive steps that progressively decompose the starting material. In addition, the inorganic content of the starting biomass is transformed into ash. The latter, besides constituting a practical challenge for management of real plants, due to possible melting and clogging in the apparatus, is often neglected in models, due to its undefined properties and fate. Furthermore, tar is often formed and removed from syngas by condensation. Many reports do not account for these partially decomposed compounds, others try to incorporate them simplifying their composition (which is indeed difficult to define and standardize). When it is considered, very often tar is modeled as benzene. Different examples of application of biomass gasification for the production of syngas in CHP applications are discussed in the following.

Two different approaches can be followed when dealing with biomass as raw material, i.e., using energy crops or possibly agricultural residua or using different types of wastes [2]. The latter pose even bigger challenges for modeling because, in addition to the difficult standardization of their composition, they are often available in physical states incompatible with direct gasification. For instance, digestate and manure have too high moisture content to be treated as such. This imposes the design of a more complex integrated system including a drying stage, which ideally makes use of the same heat cogenerated to accomplish the removal of excess water [3,4]. Of course, this limits the energy output and thus the overall efficiency and performance of the system. Possible improvements have been suggested by coupling different renewable heating systems such as solar heating [3].

Modeling is mostly based on equilibrium reactions, where the gas products derive from a selected stoichiometry of an apparent lumped reaction of the type [5]:

$$CH_xO_yN_zS_u + m\,H_2O + n_1\,(O_2 + 3.76\,N_2) \rightarrow n_2\,CO + n_3\,CO_2 + n_4\,H_2 + n_5\,CH_4 + n_6\,H_2O$$
$$+ n_7\,H_2S + n_8\,N_2 + n_{tar}\,C_6H_6 + n_{char}\,C \quad (1)$$

In which n_i are coefficients retrieved by comparison with experiments. Often the S and N contents are considered negligible and not included in the mass balance, causing missing convergence for balances if the ultimate and proximate composition of the starting biomass is not precisely rescaled withdrawing these components. Tar formation may be computed semiempirically through a tar coefficient, again defined based on experiments, e.g., [3]:

$$\eta_{tar} = 35.98 \, \exp\left(-0.0029^* T\right) \quad (2)$$

The performance of the system is then evaluated against experimental data optimizing the apparent stoichiometric coefficients of the stoichiometric equation and, according to such calculated products distribution, computing the lower heating value (LHV) of the gas mixture. Quantification of efficiency follows, such as in form of cold gas efficiency (CGE). For instance [3]:

$$LHV_{gas} = 10.8 \, x_{H2} + 12.64 \, x_{CO} + 35.82 \, x_{CH4} \, \left(MJ/Nm^3\right) \quad (3)$$

$$CGE \, (\%) = \frac{LHV_{gas}G_{gas}}{LHV_{biomass}G_{biomass}} \times 100 \quad (4)$$

Furthermore, in many studies, the addition of detailed models for the downstream CHP unit is not included, relying only on the literature performance, e.g., 30% electrical efficiency and 52% for heat recovery for a CHP configuration based on ICE [6]. Downdraft gasifiers are considered valid options for small-scale biomass conversion, limiting tar yield [7].

A drying model has been included in [3], representing a belt drier fed with hot air and included in the overall energy balance:

$$G_{DA} \cdot \left(\Delta h_{air,pre-heater} - \Delta h_{air,ext}\right) = G_{DB} \cdot \left(\Delta h_{biomass,in} - \Delta h_{biomass,ext}\right) + Q_{loss} \tag{5}$$

$$\Delta h_{air,pre-heater} = (1-R) \cdot \Delta h_{air,in} + R \cdot \Delta h_{air,exit} \tag{6}$$

$$G_{DA} \cdot \left(Hum_{air,ext} - Hum_{air,in}\right) = G_{DB} \cdot \left(MC_{out} - MC_{in}\right) \tag{7}$$

$$Hum_{air,pre-heat} = (1-R) \cdot Hum_{air,in} + R \cdot Hum_{air,ext} \tag{8}$$

$$Q_{in} = G_{DA} \cdot \left(\Delta h_{air,mix} - \Delta h_{air,pre-heater}\right) \tag{9}$$

$$\Delta h_{air,j} = C_{p,air} \cdot T_{air,j} + Hum_{air,j} \cdot C_{p,vap} \cdot T_{air,j} \tag{10}$$

$$\Delta h_{biomass,j} = C_{p,biomass} \cdot T_{biomass} + MC_j \cdot C_{p,water} \cdot T_{biomass,j} + MC_j \cdot \lambda_{vap} \tag{11}$$

where $T_{air,\ ini}$ $T_{air,\ pre-heat}$ $T_{air,\ exit}$ are air temperature inlet, preheated and at outlet, $T_{biomass}$ is biomass temperature in and out, G is the flow rate in kg/h of dry biomass (DB) or dry air (DA), Q_{loss} is the heat lost due to dissipation, R is the air recycle ratio, Hum_{air} is the humidity of air, λ_{vap} is the water latent heat of vaporization, C_p is the heat capacity, and MC is the moisture content.

The simulation of different biowastes has been compared [3] observing that the computation of ashes and by-products is important to avoid severe overestimation of syngas yield. Besides the intrinsic error introduced by considering equilibrium conversion, for instance, neglecting ashes avoids consideration of a thermal sink that decreases the temperature of the gasifier, limiting the syngas production. This is particularly relevant for some types of wastes, such as poultry litter. Therefore, a significant challenge in this field is the progressive turn into more detailed models that can more effectively compute also kinetics and accurate products distributions. Furthermore, the use of high-moisture wastes poses a big challenge for the overall efficiency of the system.

Other important issues in modeling are related to the prediction of the main economic assessment parameters, starting from the evaluation of the main cost items, through the definition of indicators of the economic performance. In particular, stochastic methods such as the Monte Carlo simulation help in assessing the risk for evaluation of the dependence on some economic parameters. This is particularly important when investments should base on emerging technologies for which uncertainties are very large.

Examples are reported for a 11-MW (and bigger) gasification power plant [8,9], for a bagasse sugarcane biorefinery [10], and for micro-CHP systems [11,12]. Colantoni et al. [13] focused on biomass gasification for CHP focusing on the specific location in Italy, considering small, medium, and large-size plants, i.e., $100\,kW_{th}$, $1\,MW_{th}$, and $10\,MW_{th}$. The plant was modeled considering a bubbling fluidized bed (BFB) gasifier coupled with an ICE and an EG. The cash flow (CF), net present value (NPV), and pay-back time (PBT) were calculated. Then, Monte Carlo simulation was used to calculate the probability of a positive NPV with variable uncertain parameters [11,14–16]. The authors specifically selected biomass pruning residua from olives and wine industry, but the concept was also generalized to straw, to extend the analysis to non-Mediterranean areas. The scheme of the plant is reported in Fig. 1, which includes a hopper, a screw, the gasifier, gas cleaning units, and the ICE-EG unit [13].

Green certificates or other incentives were not considered in cost estimation. A sensitivity analysis allowed to identify the most critical items for each scale of production, but in every case a key factor was the possibility to valorize heat, not only electrical power. Based on the Monte Carlo uncertainty analysis, the most sensitive point influencing NPV was biomass cost and availability, irrespectively from the scale, while for small-scale plants, also electricity price was determinant.

Computational fluid dynamics (CFD) may provide a good insight into heat distribution and performance of CHP systems. For instance, furnaces in heavy industries, such as cement and steel are very energy-intensive and account for huge consumption of natural gas and CO_2 emissions. Syngas obtained either from biomass gasification or reforming can significantly improve the sustainability of such processes. The URANS (Unsteady Reynolds-Averaged Navier-Stokes) model was used to average out turbulent fluctuations,

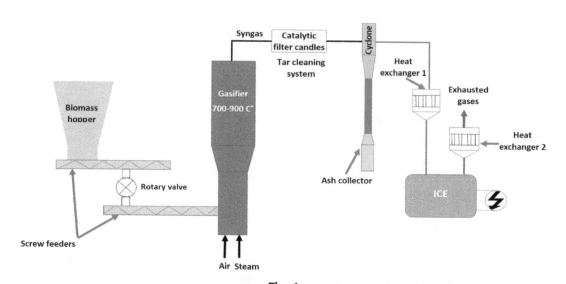

Fig. 1

Biomass gasifier-ICE-EG plant layout. *Reproduced from Colantoni A, Villarini M, Monarca D, Carlini M, Mosconi EM, Bocci E, et al. Economic analysis and risk assessment of biomass gasification CHP systems of different sizes through Monte Carlo simulation. Energy Rep. 2021; 7: 1954–61. https://doi.org/10.1016/j.egyr.2021.03.028 under the Creative Commons CC-BY-NC-ND license (Elsevier Ltd.).*

overperforming for efficiency the LES approach (large eddy simulation), either with single type fuel or for cofiring [17].

Micro-CHP plants using syngas from biomass gasification have been designed also with ICEs as a way to decrease particulate emissions with respect to direct biomass combustion. A 3D CFD model has been developed to describe a unit with 20+40 kW capacity, electrical and thermal power, respectively. An estimation of ca. 32% indicated electrical efficiency was calculated [18].

3. CHP using fuel cells

Among the options for the exploitation of syngas, its conversion to electrical power and heat through FCs has several advantages. FCs are electrochemical devices where the anode can elaborate the fuel, usually hydrogen or its mixtures, but options exist to directly manage more complex fuels, such as CH_4 or CH_3OH, whereas the cathode is fed with air. Formally, full fuel oxidation to H_2O (or for hydrocarbons to CO_2+H_2O) is accomplished, with negligible local emission. Overall emission estimation must, however, include those for the production and purification of the fuel. With respect to conventional power systems, FCs exploit higher efficiency (net electrical efficiency at least 35% for FCs vs. 10%–25% for heat engines; overall system efficiency up to 75%–90%), better ratio of electrical power vs. heat, absence of noise and vibrations [19].

Different FC types exist, some of which have demonstrated higher maturity and are near to commercialization, and the following examples will mainly deal with these.

Proton-exchange membrane FCs (PEMFCs) are suitable from small to medium-scale power, offer rapid start-up and high power density. They may operate at low temperature (60–80°C) using a Nafion membrane as proton exchange electrolyte, but in this case water management is very critical. More recently, high-temperature devices, operating between 140°C and 180°C, have been developed, using a polybenzimidazole membrane, which allows to keep water in the gas phase all over the cell, and thanks to the higher operating temperature, it is also more resistant to poisons, such as CO. Indeed, operation at higher temperature is characterized by higher efficiency and allows lower adsorption of CO over the electrode catalyst. This is an important advantage for the design of the plant, since the H_2 purification requests are much less demanding, and simpler fuel processors are admissible [19].

Different types of FCs use solid oxides (SOs) as electrolyte, thus the ionic mobility involves O^{2-} transfer. SOFCs operate at higher temperature (up to 1000°C, though for small-scale applications more manageable temperature is indicated, i.e., 500–700°C) and are characterized by higher efficiency than PEMFCs and higher tolerance toward CO in virtue of the high-temperature operation. Y-stabilized zirconia (YSZ), Gd-doped ceria (GDC), and SrO- and MgO-doped lanthanum gallates (LSGM) are mainly exploited as electrolyte. Some proton or mixed conducting materials are also explored and even electrolyte-free systems [20]. YSZ becomes an O^{2-} conductor at temperature higher than 800°C, and the typical operating range is between 800°C and 1100°C, with ionic conductivity of ca. $0.02\,S\,m^{-1}$ and $0.1\,S\,cm^{-1}$ at the two temperature limits, respectively [21]. The theoretical efficiency depends on temperature and may reach ca. 60%. The real efficiency tightly depends on operating conditions (temperature, pressure) and fuel utilization factor, but it is commonly ca. 50% for electrical power production and between 25% and 35% for thermal power [20]. SOFCs also offer interesting options for heat integration and optimization of heat management, thanks to high-temperature operation.

FCs allow the cogeneration of heat and power, but the maximization of efficiency needs the concomitant consumption of all the energy produced. The release of excess heat to a heat sink and its dispersion indeed would decrease considerably the efficiency and limit the advantages of the technology. Therefore, in case of mismatch between the electrical and thermal demands, it is compulsory to decide whether operating the system as electrically driven, thermally driven, or combined, i.e., deciding if sizing the CHP plant based on the electrical or thermal demand, supplying the remaining needs through auxiliary units. For instance, when the system is in any case connected to the electric grid, it may be convenient to operate the system exploiting the maximum thermal power, selling or buying the electrical mismatch through the grid. Another option is the coupling with energy storage systems (either thermal, such as a phase change material, or electrical, such as a rechargeable battery), to accumulate peak energy production.

CHP based on FCs has reached a semi-commercial development, though the capital costs are still prohibitive in the absence of public support. For this reason, the interest is focused on small delocalized CHP systems, also defined micro-CHP, intended for residential purposes as substitutes for domestic heating and boilers and supplying at least part of the power demand. Large CHP installations based on FCs still require huge investments and are currently at demonstrative stage. A very recent review summarizes the state of the art on micro-CHP based on fuel cells [22] and exemplifies some applications, e.g., 70 household systems by Bosch in Germany, United Kingdom, and Nederland in 2013 [23], or a PEMFC developed by Panasonic and Tokio Gas ($0.7\,kW_{el}$, with bare module cost of 22,500 US$) [24].

Modeling of micro-CHP units requires advanced mathematical tools, typically user-defined models written in MATLAB©, Aspen Plus© user routines, Fortran, etc. [22]. The models may be steady state, for the estimation of the parametric dependence of performance, or dynamic, to account for transient operation and control purposes. Specific models have been developed to account for aging and deactivation.

The sequence to build a steady-state model is summarized in Fig. 2 [22]: constant parameters are set, the model and submodels are built and integrated in the overall system model and then validated against experimental data. The required load data are then introduced, calculating the consumption expected.

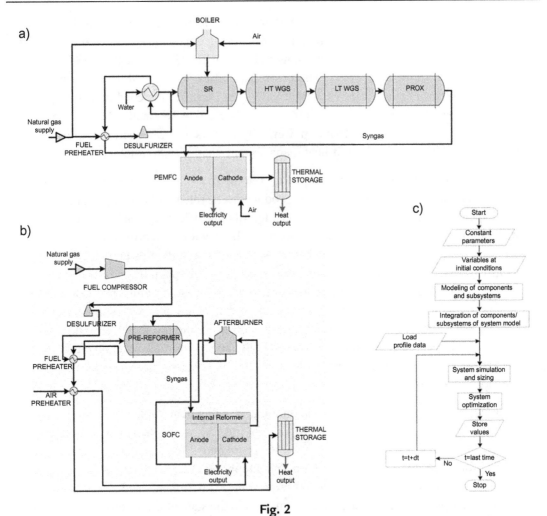

Fig. 2

Typical scheme for fuel processing and FC integration in the case of (A) PEMFC, (B) SOFC. (C) Conceptual flow diagram of model building for a micro-CHP system. *Reproduced from Arsalis A. A comprehensive review of fuel cell-based micro-combined-heat-and-power systems. Renew. Sustain. Energy Rev. 2019; 105: 391–414. https://doi.org/10.1016/J.RSER.2019.02.013 by kind permission of Elsevier.*

The CHP unit must include not only the model for the fuel cell, which is of course one of the less conventional, but should also include the fuel processor subunits. A straightforward computing should account for kinetic models to describe the performance of each reactor, but to simplify the model, also equilibrium stages for some noncritical units can be considered if sizing is not required. This is to avoid dealing with a strongly nonlinear behavior of many units. For instance, a preliminary exploratory investigation on the steam reforming of bioethanol to feed a 5-kW$_{el}$ PEMFC system was developed [25]. The system was based on a steam reforming unit, followed by high- and low-temperature water gas shift (HT- and LT-WGS) reactors and two methanators, to comply with the strict CO limitations of the PEMFC used (<20ppm CO). The system was represented in Aspen Plus© to optimize the system performance according to a parametric sensitivity analysis. The really unexplored part of the system was the steam reformer, which was sized according to a detailed kinetic model [26–29]. The following units, more conventional, were considered in equilibrium to avoid convergence problems with the solution algorithm [30,31]. Later, when economic assessment was needed, the detailed kinetics introduced also in the other stages of the fuel processor allowed the overall feasibility and economic assessment [32–34].

Examples of flow diagrams for micro CHP systems based on PEMFC and SOFC are sketched in Fig. 2A and B. For each subsystem, mass and energy balances under steady state are computed and a general overview of the optimization algorithm to search for suitable operation conditions is proposed in Fig. 2C:

$$\sum \dot{m}_{i,in} = \sum \dot{m}_{i,out} \tag{12}$$

$$\sum H_{i,in} = \sum H_{i,out} + \dot{W}_{el} \tag{13}$$

A 0-D model was developed in Aspen Plus© allowing to compute the exergetic efficiency under different operating conditions and comparing an SOFC and a PEMFC [35]. The latter was more efficient for operation under atmospheric conditions. A quasi 2-D model was instead developed to compute nonuniformity in the distribution of reactants, local current densities, and temperature, with the aim to predict thermal stresses [36].

An energetic/exergetic analysis of a CHP system based on PEMFC has been carried out by Barelli et al. [35]. They implemented the electrochemical model in Aspen Plus© through a Fortran routine, adapting semiempirical correlations between voltage and current density at variable power demand. Different efficiency quantifiers for the production of electrical power only or for cogeneration (cog) have been identified as first law (of thermodynamic) efficiency, with W_e and Q the electric and heat powers produced, $W_{fuel}=HHV_{H2}\,n_{H2}$ (higher heating value and molar flow rate of H_2) and Q_i being the thermal power to ensure the required temperature and humidity of such stream:

$$\eta(I) = \frac{W_e}{W_{fuel} + Q_{air} + Q_{fuel}} \tag{14}$$

$$\eta(I)_{cog} = \frac{W_e + Q}{W_{fuel} + Q_{air} + Q_{fuel}} \tag{15}$$

According to the second law of thermodynamic, the following efficiency was also drawn:

$$\eta(II) = \frac{W_e}{Ex_{in}} \tag{16}$$

Ex being the exergy input calculated from the feed flowrate and the preheating heat flow. $E_x = ne$, with $e = e^{ch} + e^{ph}$, the sum of physical and chemical contributions to exergy.

$$e^{ph} = (h - h_0) - T_0(s - s_0) = C_p T_0 \left[\frac{T}{T_0} - 1 - ln\left(\frac{T}{T_0}\right) + ln\left(\frac{P}{P_0}\right)^{\frac{k-1}{k}} \right] \tag{17}$$

$$e^{ch} = \sum_n x_n e_n^{ch} + RT_0 \sum_n x_n \, ln \, x_n \tag{18}$$

with x_n the molar fraction and e_n^{ch} the specific chemical exergy.

$$\eta(II)_{cog} = \frac{W_e + Ex_Q}{Ex_{in}} \tag{19}$$

$$Ex_Q = \sum Ex_{QCool} \tag{20}$$

Ex_Q is the exergy related to cooling heat exchange.

An example of efficiency trend depending on the cell temperature is reported in Fig. 3 [35]. Optimal operating conditions have been found as 2atm, 378.15K, and 58% relative humidity of feeding air.

The specific cell design for a SOFC has been evaluated based on a Fortran routine incorporating semiempirical models in physical-based equations, to forecast the main cathode and anode

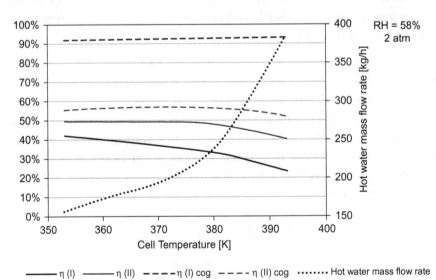

Fig. 3

Efficiency and hot water mass flow trends at RH=58% and P=2atm as a function of cell temperature. *Reproduced from Barelli L, Bidini G, Gallorini F, Ottaviano A. An energetic-exergetic analysis of a residential CHP system based on PEM fuel cell. Appl. Energy 2011; 88: 4334–42. https://doi.org/10.1016/J.APENERGY.2011. 04.059 by kind permission of Elsevier.*

features. A 2D model was developed for a planar SOFC with a suitable mapping of gas composition, temperature, current density, and polarization losses on each plane [37]. The simulation is based on a previously validated code for an MCFC, solving mass, energy, and momentum balances under steady state. As a general approach, the actual voltage is computed from the thermodynamic (Nernstian) potential, subtracting all the losses, ohmic, activation, and concentration.

$$V = E - \eta_{ohm} - \eta_{act} - \eta_{conc} \tag{21}$$

$$E = E_0 + \frac{RT}{zF} \ln \left(\frac{p_{H2,an} p_{O2,cat}^{0.5}}{p_{H2O,an}} \right) - \eta_{leak} \tag{22}$$

$$E_0 = 1.253 - 2.4516 \times 10^{-4} T \tag{23}$$

The model can be validated against impedance and potential curves (maximum error 2%), allowing to assess the best operating conditions in a sensitivity analysis.

An example of integrated straw gasification and CHP based on SOFC is reported in [38] with 100kW size. Biomass decomposition and gasification, supported by burning part of the syngas have been modeled as Gibbs or yield reactors. Instead the SOFC model included the definition of the Nernst potential and the rates of steam reforming and water gas shift reactions. Equations for ohmic, concentration, and activation losses were added as well. Mass and energy balances allowed the evaluation of system performance under steady-state conditions. A more general simulation of the system was done considering the fluctuating demand during time. For periods of higher heat demand than that produced by the CHP unit, a heat storage tank supports the need. The capacity of the tank is sized based on the excursion between the maximum heat load and the average (used for sizing the CHP unit). Possible excess power generation is sold to the grid. This configuration meets the average heat demand (responding to possible peaks through a storage system) and is the most energy saving with a better economy.

Microgeneration from natural gas using PEMFC was explored comparing a conventional fuel processing scheme, a simplified unit coupled with HT-PEMFC (allowing lower purification of the fuel), and a membrane reformer that feeds pure H_2 to the PEMFC and allows possible CO_2 storage [39]. The economics of the proposed solutions have been considered in the Italian and European scenario, concluding a maximum investment cost of 3000 €/kW$_{el}$. The CHP unit was sized to achieve 11 kW$_{el}$ and 6 kW$_{th}$, with overall efficiency reaching 86%. Economic assessment returned a 2.2-year payback time [40].

Wood gasification has been proposed for syngas production, later separated by pressure swing adsorption (PSA) to achieve 99.99% H_2 purity. The latter feeds a tubular solid oxide fuel cell (TSOFC) with ca. 200kW capacity (starting from ca. 13kg/h of biomass). Heat from the fuel cell is internally used to produce 60kg/h of steam for the gasifier. The net efficiency of the integrated system is 27% (only electrical), which is lower than that of a gas turbine (ca. 33%) [41].

A residential-size CHP system has been modeled based on PEMFC, and the key performance indicators have been compared with rival options. As for costs, the possibility to produce H_2 through a centralized steam reforming plant and distribute it (e.g., through pipeline as in the business case considered) allows the lowest cost. H_2 production through photovoltaic sustained electrolyzers is still economically not competitive. Overall 95% efficiency has been estimated for this CHP option, with 50% electrical efficiency [42].

Sorption-enhanced steam reforming of natural gas is another option to maximize the hydrogen yield by CO_2 adsorption with CaO. The reformer was coupled with an SOFC, and a 1D numerical model has been developed.

As for dynamic simulation, a short review is proposed in [43], and Matlab© and Simulink© tools are mainly adopted for this [44].

In addition to all the variables accounted for in steady state, dynamic modeling adds a dynamic computation of the activation losses and of cell thermodynamics due to the formation of a charge double layer (CDL). The following time-dependent equations are used:

$$\frac{d\eta_{Act}}{dt} = \frac{i}{C} - \frac{\eta_{Act}}{R_c C} \tag{24}$$

$$R_c = \frac{\eta_{Act,0} + \eta_{Conc}}{i} \tag{25}$$

where R_c sums the steady-state losses for activation and concentration, and C is the double layer charge, while i is the current density [44]. The temperature also changes with time, affecting in turn the main model variables. It changes in time as:

$$C_T \frac{dT}{dt} = i(E_{Nernst} - E_{cell}) - H(T_{cell} - T_f) \tag{26}$$

Where C_T is the thermal capacitance, H the global heat transfer coefficient, T_f the reference temperature, and T_{cell} the lumped temperature of the FC [44]. Some examples are reported in the literature, e.g., [45–51].

As for deactivation modeling, Costamagna et al. cumulated extensive experience in modeling SOFCs, from specific models for stacks and electrodes performance, to integrated fuel processors and more recently to interpret and predict failure modes. Two classical failure detection approaches can be used, one based on fault signature matrix, the other on data with statistical classification, to be used singly or simultaneously [52–55]. Furthermore, a degradation model for SOFCs has been elaborated to account for coking, sulfur poisoning, local hot spots attributed to peaking current densities, and frequent power cycling [56]. The empirical deactivation rate $(\dot{r}_{deg})$ is expressed as follows and decreases the cell potential (E) during time:

$$\dot{r}_{deg} = f(i, T) \tag{27}$$

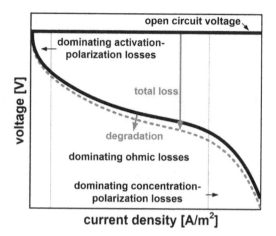

Fig. 4

Principal current density-voltage relationship for an SOFC, including deactivation. *Reproduced from Eichhorn Colombo KW, Kharton V V. Reliability analysis of a multi-stack solid oxide fuel cell from a systems engineering perspective. Chem. Eng. Sci. 2021; 238: 116571. https://doi.org/10.1016/J.CES.2021.116571 by kind permission of Elsevier.*

$$E = E_{Nernst} - \eta_{total}\left(1 + \dot{r}_{deg}dt\right) \tag{28}$$

The integration with a nonlinear solver allows to predict the real potential while aging the device in form of polarization curve including degradation of the cell (Fig. 4).

Optimization models should account for a specific objective function, e.g., maximum energy efficiency, minimum unit cost of electricity or life cycle cost, etc. Often, multiple objective functions coexist, with the need of multiobjective optimization methods. Genetic algorithm optimization methods are useful with strongly nonlinear systems and will be deepened in the last paragraph. When heat integration is used, pinch analysis is also added to the model [57].

4. Hybrid CHP systems

In order to improve the footprint of heat and power production for large-scale applications, hybrid systems have been designed and simulated. An example is the 10th MW power production in petrochemical plants, where electricity needs are coupled with steam generation. A complex and integrated system has been proposed [58], where pine sawdust is exploited as widely available biomass, transformed into syngas in a gasifier. Steam and compressed air are used in the same reactor. Syngas is then used both to feed a gas turbine (GT) and an SOFC. This latter unit accomplishes in part the internal methane reforming, since natural gas is supplied as auxiliary fuel to sustain the CHP process, especially for peak production. All the flue gases and the heat released by the SOFC are recovered to produce steam for the gasifier and for site heating. A typical steam generation and distribution network is used, where very high or high, medium, and low-pressure steam is produced, expanded in steam turbines for power generation, and used for heating. A scheme of the process is reported in Fig. 5 [58].

The integrated system has been modeled using Engineering Equation Solver, under steady-state conditions, considering thermodynamic equilibrium for the reacting systems. Mass and heat balances are detailed for each unit in the original reference [58], where model validation was accomplished based on experimental data collected separately for the gasification of municipal wastes and for the performance of an SOFC. A parametric study on efficiency and the ratio of cogeneration, i.e., the ratio between electric power and thermal output, has been tested against the main operating parameters. Among these, an increase of the biomass flowrate decreased the energy efficiency but increased the cogeneration ratio, while efficiency slightly increased with the temperature of the SOFC, with a cogeneration ratio passing through a maximum (Fig. 6).

5. Optimization of CHP solutions

Integrated CHP systems often need optimization to select the best operating modes of most units according to a given criterion or multiple criteria (e.g. maximize efficiency and minimize CO_2 emission, minimize installation cost and maximize power output, etc.). Various algorithms for multivariable optimization are available and have been recently reviewed for application in CHP systems, especially based on renewable sources [59].

Search-based [60] (using a random or organized sequence to reach the optimum for the desired objective function) or calculus-based algorithms (using first or second derivatives to reach the optimum) can be selected depending on the case. Open-source codes are also available for this purpose, critically assessed elsewhere for their significance and reliability [61].

Different algorithms are available for unconstrained optimization, which consist of a line search method to find the optimum, starting from a guess of the initial point and according to a given step. An example is the Nelder Mead method (Fig. 7), applicable to multidimensional problems. It is based on the Simplex, which is shaped with $n+1$ vertexes for a n-D space. During each iteration, the worst vertex is replaced by basic operations of expansion, reduction, reflection, or contraction, each having a specific coefficient [62]. Further evolutions of this method have been developed [63], and application to the optimization of CHP systems is widely reported (see, e.g., [64,65]). In some cases it is nested as suboptimization algorithm in the case of multiobjective optimization [65].

On the other hand, the Golden Section and Fibonacci algorithms are sequential gradientless methods that use an interval (constant for the Golden Section, variable according to consecutive Fibonacci's numbers in the latter) to search for the optimum. The Fibonacci method returns the smallest interval of uncertainty to find the optimum, but needs a definition of a preliminary range where the optimum is found for sure. In turn, the GS method is simpler and easier to implement for high number of iterations, but both are time-consuming, as they are linear search

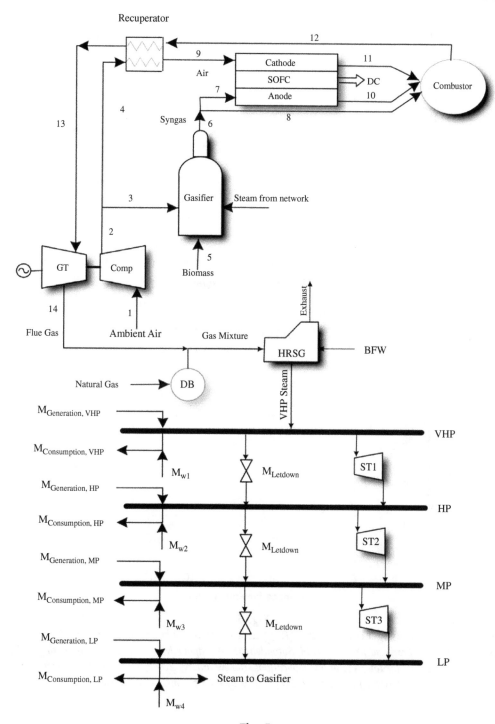

Fig. 5

Conceptual layout of an integrated GT-SOFC cogeneration system, integrated with a steam production and distribution network on an MW scale. *Reproduced from Amiri H, Sotoodeh AF, Amidpour M. A new combined heating and power system driven by biomass for total-site utility applications. Renew. Energy 2021; 163: 1138–52. https://doi.org/10.1016/J.RENENE.2020.09.039 by kind permission of Elsevier.*

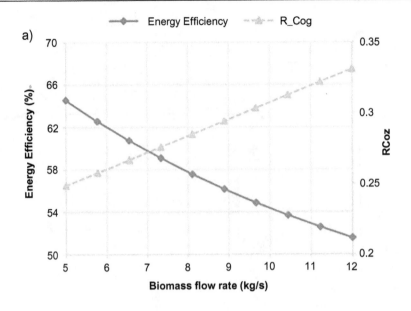

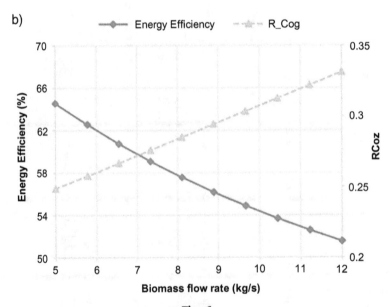

Fig. 6

Energy efficiency and cogeneration ratio (R_{Cog}) as a function of biomass flow rate (A) and SOFC temperature (B). *Reproduced from Amiri H, Sotoodeh AF, Amidpour M. A new combined heating and power system driven by biomass for total-site utility applications. Renew. Energy 2021; 163: 1138–52. https://doi.org/ 10.1016/J.RENENE.2020.09.039 by kind permission of Elsevier.*

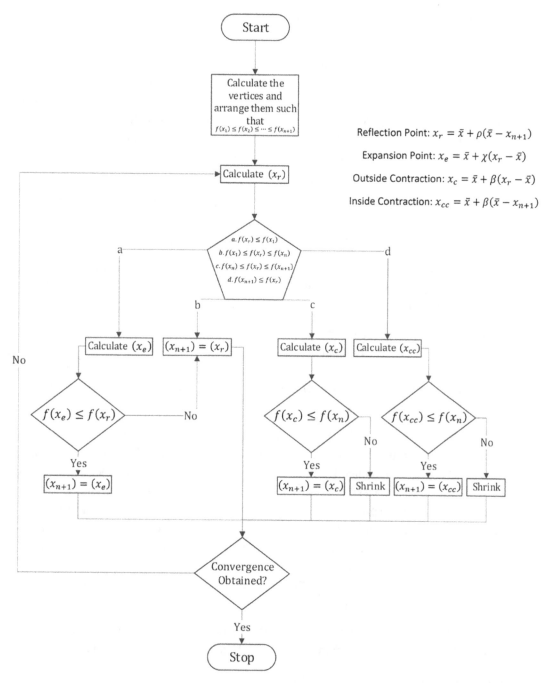

Fig. 7
Example of logical scheme for the application of a Nelder Mead optimization algorithm. *Reproduced from Bagherian MA, Mehranzamir K, Pour AB, Rezania S, Taghavi E, Nabipour-Afrouzi H, et al. Classification and analysis of optimization techniques for integrated energy systems utilizing renewable energy sources: a review for CHP and CCHP systems. Processes 2021; 9: 339. https://doi.org/10.3390/pr9020339 under the Creative Commons License (CC BY, by MDPI, Basel, Switzerland).*

algorithms [66]. An example of application of GS to the CHP based on mixed biomass and natural gas feed is reported in [67].

Sequential techniques search for neighboring values of the object function during subsequent iterations. The convergence of these methods, such as the Hooke-Jeeves one, is quite slow, and they are very sensitive to initial conditions and selected parameters. An improvement in search efficiency was introduced by Powell using conjugate directions to orient the search. However, the global optimum may not be found, so it is primarily used for searching local minima. Examples of application to CHP systems modeling are reported in [68,69].

Gradient-based methods make use of the first-order derivative for optimization, searching the minimum or maximum along the direction of highest slope of the objective function. Examples are the gradient descent or coordinate descent methods, the latter implemented when computation of the gradient is not feasible, and appropriate for local minimization [70].

Other problems involve constrained optimization, and the biggest contribution is given by evolutionary techniques. They offer flexible, simple, and robust methods widely applicable to many problems of CHP optimization.

A $(\mu + \lambda)$-Evolutionary Strategy optimization algorithm is able to work both as a local optimization algorithm and as a global one. The basic principle of this type of algorithm is the capability of survival of the best individuals from one generation to another. This is represented in nature as the Darwin's principle of evolution. For this reason, this kind of algorithms are called Evolutionary Strategy [71].

One of the input arguments for this kind of algorithm is the size of the population at each generation. This value represents how many individuals are evaluated at each generation. In this case, a single value of the objective function is used to evaluate the "goodness of the fitting" for each individual, but it is possible to solve also multiobjective minimization problems. Once all the individuals have been processed in an inner shell, associating to them a fitness value, it is possible to sort them out by putting first the individuals with the best fitness value and at last the ones with the worst fitness value. Then the algorithm creates a new set of individuals from the best individuals of the last generation. A $(\mu + \lambda)$-Evolutionary Strategy selects the new generation from all the individuals of both the children and the parents population, a (μ,λ)-Evolutionary Strategy just from the offspring.

An Evolutionary Strategy algorithm works iteratively passing from generation to generation. Discarding the first iteration, the main steps for each generation can be summarized as:

- Receival as input of the new population;
- Calculation of the fitness values for each individual;
- Sorting of the population by its fitness value;
- *Selection* of individuals to be used as parents for the next population;

- *Reproduction of the parents* to create a new offspring;
- Creation of a new population.

The way the algorithm moves from generation to generation is a determinant step for the effectiveness of the algorithm [71]. So, the selection and reproduction steps are crucial for the effectiveness of the algorithm. More specifically, these steps influence the biodiversity between the individuals of the same generation. If it is too low, the algorithm works as a local optimization algorithm with the risk of converging to a local minimum (maximum). If it is too wide, the algorithm may find the global minimum (maximum), but still failing to converge. Other than the selection and reproduction function, there are other parameters that need to be decided. Their tuning is important, and it has to be done specifically for each kind of regression and if necessary for each regression case.

After the calculation of the fitness values of all the individuals, those are sorted by the fitness function in ascending order. At that point, the algorithm needs to select the best individuals to pass on to the next generation as parents. In order to do that, it assigns a probability to all the individuals to be picked as parents for the next generation. The outcome of a selection function then is a density of probability function, which associates a probability to each of the parents selected from the last generation to be picked as parent for the new child.

A possible selection relates the probability to be picked to the position of the parent in the sorted list, e.g.,

$$p_{(i)} = \frac{1}{\sqrt{k_i}} \tag{29}$$

where $p_{(i)}$ is the probability of the individual i to be picked as the parent, and k_i is the k position of the component i in the sorted list of the parents from the previous generation. Where position 1 represents the individual with the best fitness value, in the last position the individual with the worst fitness value.

It is possible as well to use a selection function, which correlates the outcome directly to the fitness value, where the probability is inversely proportional to the fitness value of the individual. This function could be used to perform different uses of the evolutionary strategy.

The reproduction of the parents is one of the most important parts of an Evolutionary Strategy algorithm. It can be performed in different ways. The most common techniques are as follows:

- Elites. Those are the individuals from the last generation with the best fitness value. In order not to lose them, they are passed without change to the next generation. The number of individuals passed as elites is called *Elite count*.
- Crossover. It takes two or more parents from the parents population to create the new child. The way it combines the parents to create the new child is defined in the crossover function.

- Mutation. This function takes the initial parent and mutates each of its genes along a Gaussian distribution centered at the initial individual.

The elite number of individuals influences on the diversity of the population. If this value is too high, it could make the population stall on certain values for a lot of generations, making the algorithm slower and failing to converge to a global optimum. Otherwise, if this value is too low, the genetic algorithm fails to move from a global optimization to a local optimization. Also, in this case, the algorithm becomes slower and fails to converge. As a rule of thumb, the number of elites has been always set equal to 5% of the population size [72].

The ratio between the individuals generated by crossover and the ones generated by mutation will be called "Crossover Fraction".

$$Crossover\ Fraction = \frac{number\ of\ children\ generated\ by\ crossover}{total\ number\ of\ children\ generated} \tag{30}$$

If the "Crossover Fraction" is 1, all the children are generated by crossover. Otherwise, the children will be generated all by mutation. The reproduction by crossover helps the algorithm to look into new points in the phase space, thus working as a global optimization algorithm. The mutation instead looks into the local area around a single individual, thus working as a local optimization algorithm. The "Crossover Fraction" helps to guide the algorithm between these two kinds of optimization search algorithm.

Crossover is a technique used to create a child from two or more parents. The number of parents used to create a child is an important decision variable. In lexicon of Evolutionary Strategy algorithm, this number is called ρ. For instance, $\rho=2$ means that two parents are used to create a child. There are different possible Crossover Functions and ideally everyone could come up with a new one. "Crossover Scattered" is a technique implemented in Matlab as default option for their Evolutionary algorithm code. It takes two parents and for each gene choses randomly the gene of one of two parents.

"Crossover Intermediate" is another possible function for the crossover. In this case, a random value between the genes of the parents is taken for each gene and the new point always stays within the parents' phase space. Then this type of crossover function introduces the Genetic Repair hypothesis [73], which states that the recombination of different parents enhances the similar genes. In fact, the final child in that genes will have a similar value of both of their parents. Instead, in the other cases, the value will be new and different from the initial ones. Otherwise, building block hypothesis is still object of debate so far. In this case, it is stated that the combination of two good parents will give rise of a new good individual. So far though, a little evidence has been found in favor of this, and only in 2001, Jansen and Wegener [74] were able to build a test function that supported this thesis.

The mutation could be described as a noise added to the initial individual. This helps the algorithm to search the optimal solution as a local minimization algorithm. It should follow the principles below [73]:

- Richness: This principle states that the mutation by itself should be able to explore all the phase space in the generations used by the algorithm.
- Unbiasedness: The mutation should be unbiased. This means that all the directions from the initial point in the phase space should be equally likely.
- Scalability: This means that the average mutation length or *strength* of the mutation should be tuneable to adapt itself to the fitness landscape.

In this case, the mutation function is a Gaussian function centered in the value of the gene. This is made for each gene of the individual, ensuring the unbiasedness of the mutation. The standard deviation is defined as:

$$\sigma_k = \sigma_{k-1}*\left(1 - \frac{k1}{k}\right) \tag{31}$$

where σ_k is the standard deviation at the kth generation, σ_{k-1} is the standard deviation at the $k-1$th generation, k is the total number of generations so far, and k_1 is a constant.

If it is assumed that as the regression proceeds the algorithm tends to convergence, then the probability of finding the individuals in the same region of the state space around the initial individual as the iterations go further is higher. In this sense, it is reasonable to reduce the standard deviation as the number of iterations increases.

The main families of evolutionary strategies used to optimize CHP systems are Genetic Algorithms (GA) [75], Mixed-Integral Linear Programming (MILP), Simulated Annealing (SA), and Particle Swarm Optimization (PSO). The former two are by far the most used for CHP modules.

GA implies that an initial set of values are selected in a heuristic or random way, including the final solution through boundary conditions. The solutions are called Chromosomes and include a set of binary [01] values, which are muted through the following populations to improve the fitness of the guess (Fig. 8).

An example of application of GA to the optimization of a combined SOFC-Gas Turbine cogeneration system is discussed in [76], while an example of HT-PEMFC in [77]. GA is very widely applied to CHP problems, but its application becomes hard with very large datasets.

MIPL algorithms are also widely used in CHP optimization and are suitable to deal with mixed integer and noninteger variables. The problem is branched in nodes searching for the defined tolerance between the Linear Programming solution and the integer solution [77].

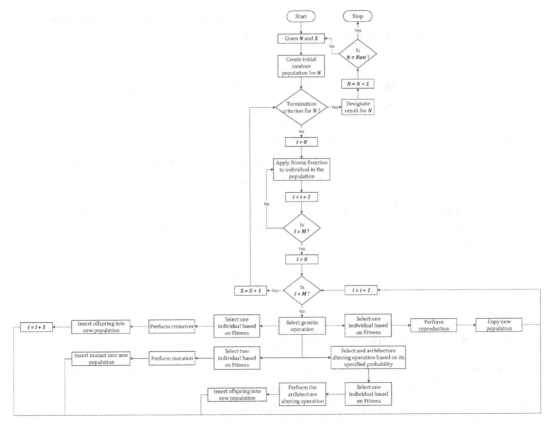

Fig. 8

Example of GA algorithm. *Reproduced from Bagherian MA, Mehranzamir K, Pour AB, Rezania S, Taghavi E, Nabipour-Afrouzi H, et al. Classification and analysis of optimization techniques for integrated energy systems utilizing renewable energy sources: a review for CHP and CCHP systems. Processes 2021; 9: 339. https://doi.org/ 10.3390/pr9020339 under the Creative Commons License (CC BY, by MDPI, Basel, Switzerland).*

SA is the most flexible and suitable for complex combinatorial problems. It is inspired by the crystallization through slow cooling following a minimum free energy pattern [78]. A perfect crystal is considered the global optimum, while the presence of some defects represents a local optimum. The search is iterated in neighboring points that may generate a better solution avoiding further exploration in further iterations in that direction. Two sequences are used: controlled cooling and the Metropolis criterion, which allows random search in the neighboring.

PSO is a stochastic global optimization method that does not require the generation of populations, because it compares individuals. Each element of the population is a dot with a velocity, that changes position depending on its own experience, which constitutes a local best, and that of the other individuals, which is the global best, with its adaptations being one of the best fitted methods for MIL problems. Examples of application in the field of CHP with solar, wind, and FC are reported in [79,80].

6. Conclusion

The development of reliable models to interpret integrated CHP systems is a challenging task. They are needed in order to perform the in silico design of plants, with considerable advantages over extensive experimental investigations and directing in a more efficient way testing of experimental plants in well-defined conditions. The final aim is the saving of money and efforts in the development of CHP commercial units and the considerable decrease of the time to market. However, models should be developed to reproduce in realistic and reliable way very complex systems, where the intrinsic complexity of the feedstock, its variability, and the huge number of reactions taking place simultaneously jeopardize the picture. A sound compromise should be reached between accuracy of the representation and flexibility/versatility of the model. This compromise is not yet found. Quite accurate representation of specific systems is available in various literature reports. Nevertheless, the models are often strictly dependent on semiempirical adjustment of parameters, scaling factors, and approximate/lumped stoichiometries.

All these limitations must be overcome in the future, trying to get better insight in the fundamentals of the reactions, e.g., of biomass gasification, using kinetic models to correctly size reactors and related equipment. In this way, costing and sensitivity analysis can be reliable.

Other challenges come from optimization problems. The parametric dependence of performance indicators on operating parameters should be understood in more detail. This can lead to the selection of the most appropriate tool for optimization, among which GAs are good candidates.

References

[1] D. Almpantis, A. Zabaniotou, Technological solutions and tools for circular bioeconomy in low-carbon transition: simulation modeling of rice husks gasification for CHP by aspen plus V9 and feasibility study by aspen process economic analyzer, Energies 14 (2021), https://doi.org/10.3390/en14072006.
[2] A. Ramos, E. Monteiro, A. Rouboa, Numerical approaches and comprehensive models for gasification process: a review, Renew. Sustain. Energy Rev. 110 (2019) 188–206, https://doi.org/10.1016/j.rser.2019.04.048.
[3] O. de Priall, V. Gogulancea, C. Brandoni, N. Hewitt, C. Johnston, G. Onofrei, et al., Modelling and experimental investigation of small-scale gasification CHP units for enhancing the use of local biowaste, Waste Manag. 136 (2021) 174–183, https://doi.org/10.1016/J.WASMAN.2021.10.012.
[4] X. Zhuang, Y. Song, H. Zhan, X. Yin, C. Wu, Gasification performance of biowaste-derived hydrochar: the properties of products and the conversion processes, Fuel 260 (2020), https://doi.org/10.1016/j.fuel.2019.116320, 116320.
[5] Z.A. Zainal, R. Ali, C.H. Lean, K.N. Seetharamu, Prediction of performance of a downdraft gasifier using equilibrium modeling for different biomass materials, Energ. Conver. Manage. 42 (2001) 1499–1515, https://doi.org/10.1016/S0196-8904(00)00078-9.

[6] F. Patuzzi, D. Prando, S. Vakalis, A.M. Rizzo, D. Chiaramonti, W. Tirler, et al., Small-scale biomass gasification CHP systems: comparative performance assessment and monitoring experiences in South Tyrol (Italy), Energy 112 (2016) 285–293, https://doi.org/10.1016/j.energy.2016.06.077.

[7] W. Elsner, M. Wysocki, P. Niegodajew, R. Borecki, Experimental and economic study of small-scale CHP installation equipped with downdraft gasifier and internal combustion engine, Appl. Energy 202 (2017) 213–227, https://doi.org/10.1016/j.apenergy.2017.05.148.

[8] J. Cardoso, V. Silva, D. Eusébio, Techno-economic analysis of a biomass gasification power plant dealing with forestry residues blends for electricity production in Portugal, J. Clean. Prod. 212 (2019) 741–753, https://doi.org/10.1016/j.jclepro.2018.12.054.

[9] J. Sousa Cardoso, V. Silva, D. Eusébio, I. Lima Azevedo, L.A.C. Tarelho, Techno-economic analysis of forest biomass blends gasification for small-scale power production facilities in the Azores, Fuel 279 (2020), https://doi.org/10.1016/j.fuel.2020.118552, 118552.

[10] M. Mandegari, S. Farzad, J.F. Görgens, A new insight into sugarcane biorefineries with fossil fuel co-combustion: techno-economic analysis and life cycle assessment, Energ. Conver. Manage. 165 (2018) 76–91, https://doi.org/10.1016/j.enconman.2018.03.057.

[11] M.R. Osaki, P. Seleghim, Bioethanol and power from integrated second generation biomass: a Monte Carlo simulation, Energ. Conver. Manage. 141 (2017) 274–284, https://doi.org/10.1016/j.enconman.2016.08.076.

[12] N.J. Williams, P. Jaramillo, J. Taneja, An investment risk assessment of microgrid utilities for rural electrification using the stochastic techno-economic microgrid model: a case study in Rwanda, Energy Sustain. Dev. 42 (2018) 87–96, https://doi.org/10.1016/j.esd.2017.09.012.

[13] A. Colantoni, M. Villarini, D. Monarca, M. Carlini, E.M. Mosconi, E. Bocci, et al., Economic analysis and risk assessment of biomass gasification CHP systems of different sizes through Monte Carlo simulation, Energy Rep. 7 (2021) 1954–1961, https://doi.org/10.1016/j.egyr.2021.03.028.

[14] Y. Degeilh, G. Gross, Stochastic simulation of power systems with integrated intermittent renewable resources, Int. J. Electr. Power Energy Syst. 64 (2015) 542–550, https://doi.org/10.1016/j.ijepes.2014.07.049.

[15] G. Li, P. Zhang, P.B. Luh, W. Li, Z. Bie, C. Serna, et al., Risk analysis for distribution systems in the Northeast U.S. under wind storms, IEEE Trans. Power Syst. 29 (2014) 889–898, https://doi.org/10.1109/TPWRS.2013.2286171.

[16] E. Zio, M. Delfanti, L. Giorgi, V. Olivieri, G. Sansavini, Monte Carlo simulation-based probabilistic assessment of DG penetration in medium voltage distribution networks, Int. J. Electr. Power Energy Syst. 64 (2015) 852–860, https://doi.org/10.1016/j.ijepes.2014.08.004.

[17] P. Jóźwiak, J. Hercog, A. Kiedrzyńska, K. Badyda, CFD analysis of natural gas substitution with syngas in the industrial furnaces, Energy 179 (2019) 593–602, https://doi.org/10.1016/j.energy.2019.04.179.

[18] D. Piazzullo, M. Costa, Z. Petranovic, M. Vujanovic, M. La Villetta, C. Caputo, et al., CFD modelling of a spark ignition internal combustion engine fuelled with syngas for a mCHP system, Chem. Eng. Trans. 65 (2018) 13–18, https://doi.org/10.3303/CET1865003.

[19] A. Albarbar, M. Alrweq, Proton Exchange Membrane Fuel Cells Design, Modelling and Performance Assessment Techniques, Springer, 2018.

[20] B. Zhu, R. Raza, L. Fan, C. Sun, Solid Oxide Fuel Cells. From Electrolyte-Based to Electrolyte-Free Devices, Wiley-VCH, 2020.

[21] A. Tesfai, J.T.S. Irvine, Fuel Cells and Hydrogen Technology in Comprehensive Renewable Energy, Elsevier, 2012.

[22] A. Arsalis, A comprehensive review of fuel cell-based micro-combined-heat-and-power systems, Renew. Sustain. Energy Rev. 105 (2019) 391–414, https://doi.org/10.1016/J.RSER.2019.02.013.

[23] Bosch plans residential CHP field trial for 2014, Fuel Cells Bull. 2013 (2013) 1, https://doi.org/10.1016/S1464-2859(13)70060-9.

[24] Panasonic, Tokyo Gas update Ene-Farm product, Fuel Cells Bull. 2013 (2013) 1, https://doi.org/10.1016/S1464-2859(13)70001-4.

[25] I. Rossetti, C. Biffi, G.F. Tantardini, M. Raimondi, E. Vitto, D. Alberti, 5 kW e + 5 kW t reformer-PEMFC energy generator from bioethanol first data on the fuel processor from a demonstrative project, Int. J. Hydrogen Energy 37 (2012) 8499–8504, https://doi.org/10.1016/j.ijhydene.2012.02.095.

[26] I. Rossetti, M. Compagnoni, M. Torli, Process simulation and optimisation of H2 production from ethanol steam reforming and its use in fuel cells. 1. Thermodynamic and kinetic analysis, Chem. Eng. J. 281 (2015) 1024–1035, https://doi.org/10.1016/j.cej.2015.08.025.

[27] I. Rossetti, M. Compagnoni, M. Torli, Process simulation and optimization of H2 production from ethanol steam reforming and its use in fuel cells. 2. Process analysis and optimization, Chem. Eng. J. 281 (2015) 1036–1044, https://doi.org/10.1016/j.cej.2015.08.045.

[28] A. Tripodi, M. Compagnoni, I. Rossetti, Kinetic modelling and reactor simulation for ethanol steam reforming, ChemCatChem 8 (2016) 3804–3813.

[29] M. Compagnoni, A. Tripodi, I. Rossetti, Parametric study and kinetic testing for ethanol steam refroming, Appl. Catal. Environ. 203 (2017) 899–909.

[30] A. Tripodi, M. Compagnoni, G. Ramis, I. Rossetti, Process simulation of hydrogen production by steam reforming of diluted bioethanol solutions: effect of operating parameters on electrical and thermal cogeneration by using fuel cells, Int. J. Hydrogen Energy 42 (2017), https://doi.org/10.1016/j.ijhydene.2017.04.056.

[31] A. Tripodi, M. Compagnoni, E. Bahadori, I. Rossetti, G. Ramis, Process intensification by exploiting diluted 2nd generation bio- ethanol in the low-temperature steam reforming process, Top. Catal. 61 (2018) 1832, https://doi.org/10.1007/s11244-018-1002-6.

[32] M. Compagnoni, E. Mostafavi, A. Tripodi, N. Mahinpey, I. Rossetti, Techno-economic analysis of a bioethanol to hydrogen centralized plant, Energy Fuel 31 (2017) 12988–12996, https://doi.org/10.1021/acs.energyfuels.7b02434.

[33] M. Compagnoni, A. Tripodi, E. Mostafavi, N. Mahinpey, I. Rossetti, Hydrogen production by steam reforming of bio-ethanol: process design and economic assessment, in: DGMK Tagungsbericht, vol. 2017, 2017.

[34] A. Tripodi, A. Pizzonia, E. Bahadori, I. Rossetti, Integrated plant layout for heat and power cogeneration from diluted bioethanol, ACS Sustain. Chem. Eng. 6 (2018) 5358–5369, https://doi.org/10.1021/acssuschemeng.8b00144.

[35] L. Barelli, G. Bidini, F. Gallorini, A. Ottaviano, An energetic-exergetic analysis of a residential CHP system based on PEM fuel cell, Appl. Energy 88 (2011) 4334–4342, https://doi.org/10.1016/J.APENERGY.2011.04.059.

[36] H. Xu, Z. Dang, B.F. Bai, Analysis of a 1 kW residential combined heating and power system based on solid oxide fuel cell, Appl. Therm. Eng. 50 (2013) 1101–1110, https://doi.org/10.1016/J.APPLTHERMALENG.2012.07.004.

[37] F.R. Bianchi, R. Spotorno, P. Piccardo, B. Bosio, Solid oxide fuel cell performance analysis through local modelling, Catalysts 10 (2020) 519, https://doi.org/10.3390/catal10050519.

[38] Z. Dang, Z. Jiang, J. Ma, X. Shen, G. Xi, Numerical study on the performance of a cogeneration system of solid oxide fuel cell based on biomass gasification, Front. Energy Res. 9 (2021) 1–14, https://doi.org/10.3389/fenrg.2021.609534.

[39] G. Di Marcoberardino, L. Chiarabaglio, G. Manzolini, S. Campanari, A techno-economic comparison of micro-cogeneration systems based on polymer electrolyte membrane fuel cell for residential applications, Appl. Energy 239 (2019) 692–705, https://doi.org/10.1016/J.APENERGY.2019.01.171.

[40] G. Diglio, P. Bareschino, E. Mancusi, F. Pepe, F. Montagnaro, D.P. Hanak, et al., Feasibility of CaO/CuO/NiO sorption-enhanced steam methane reforming integrated with solid-oxide fuel cell for near-zero-CO2 emissions cogeneration system, Appl. Energy 230 (2018) 241–256, https://doi.org/10.1016/J.APENERGY.2018.08.118.

[41] A. Gonzalez-Diaz, S. Carlos Juan Ladrón De Guevara, L. Jiang, M. Ortencia Gonzalez-Diaz, P. Díaz-Herrera, et al., Techno-environmental analysis of the use of green hydrogen for cogeneration from the gasification of wood and fuel cell, Sustainability (2021), https://doi.org/10.3390/su13063232.

[42] A. Herrmann, A. Mädlow, H. Krause, Key performance indicators evaluation of a domestic hydrogen fuel cell CHP, Int. J. Hydrogen Energy 44 (2019) 19061–19066, https://doi.org/10.1016/J.IJHYDENE.2018.06.014.

[43] A. Rubio, W. Agila, Dynamic model of proton exchane membrane fuel cells: a critical review and a novel model, in: 8th Int Conf Renew Energy Res Appl Brasov, Rom, 2019, pp. 353–358.

[44] S.A. Ansari, M. Khalid, K. Kamal, T. Abdul Hussain Ratlamwala, G. Hussain, M. Alkahtani, Modeling and simulation of a proton exchange membrane fuel cell alongside a waste heat recovery system based on the

organic rankine cycle in MATLAB/SIMULINK environment, Sustainability 13 (2021) 1–21, https://doi.org/10.3390/su13031218.

[45] M. Azri, A.N.A. Mubin, Z. Ibrahim, N.A. Rahim, S.R.S. Raihan, Mathematical modelling for proton exchange membrane fuel cell (PEMFC), J. Theor. Appl. Inf. Technol. 86 (2016) 409–419.

[46] N. Benchouia, A. Hadjadj, A. Derghal, L. Khochemane, B. Mahmah, Modeling and validation of fuel cell PEMFC-accepté le 30 juin 2013, Rev. Energ. Renouv. 16 (2013) 365–377.

[47] J. Acedo-Valencia, J. Sierra, S. Figueroa-Ramírez, H. Mandujano, M. Meza, J. Tadeo, et al., Numerical study of heat transfer in a PEM fuel cell with different flow-fields, in: Proc. XV Int. Congr. Mex. Hydrog, 2015.

[48] X.D. Xue, K.W.E. Cheng, D. Sutanto, Unified mathematical modelling of steady-state and dynamic voltage-current characteristics for PEM fuel cells, Electrochim. Acta 52 (2006) 1135–1144, https://doi.org/10.1016/j.electacta.2006.07.011.

[49] J. Zhao, Q. Jian, L. Luo, B. Huang, S. Cao, Z. Huang, Dynamic behavior study on voltage and temperature of proton exchange membrane fuel cells, Appl. Therm. Eng. 145 (2018) 343–351, https://doi.org/10.1016/j.applthermaleng.2018.09.030.

[50] X. Guo, H. Zhang, J. Zhao, F. Wang, J. Wang, H. Miao, et al., Performance evaluation of an integrated high-temperature proton exchange membrane fuel cell and absorption cycle system for power and heating/cooling cogeneration, Energ. Conver. Manage. 181 (2019) 292–301, https://doi.org/10.1016/j.enconman.2018.12.024.

[51] A. Tripodi, E. Bahadori, G. Ramis, I. Rossetti, Feasibility assessment, process design and dynamic simulation for cogeneration of heat and power by steam reforming of diluted bioethanol, Int. J. Hydrogen Energy 44 (2019) 2–22, https://doi.org/10.1016/j.ijhydene.2018.02.122.

[52] P. Costamagna, A. De Giorgi, A. Gotelli, L. Magistri, G. Moser, E. Sciaccaluga, et al., Fault diagnosis strategies for SOFC-based power generation plants, Sensors (Switzerland) 16 (2016) 1–17, https://doi.org/10.3390/s16081336.

[53] P. Costamagna, A. De Giorgi, G. Moser, S.B. Serpico, A. Trucco, Data-driven techniques for fault diagnosis in power generation plants based on solid oxide fuel cells, Energ. Conver. Manage. 180 (2019) 281–291, https://doi.org/10.1016/J.ENCONMAN.2018.10.107.

[54] P. Costamagna, A. De Giorgi, G. Moser, L. Pellaco, A. Trucco, Data-driven fault diagnosis in SOFC-based power plants under off-design operating conditions, Int. J. Hydrogen Energy 44 (2019) 29002–29006, https://doi.org/10.1016/J.IJHYDENE.2019.09.128.

[55] P. Costamagna, A. De Giorgi, L. Magistri, G. Moser, L. Pellaco, A. Trucco, A classification approach for model-based fault diagnosis in power generation systems based on solid oxide fuel cells, IEEE Trans. Energy Convers. 31 (2016) 676–687, https://doi.org/10.1109/TEC.2015.2492938.

[56] K.W. Eichhorn Colombo, V.V. Kharton, Reliability analysis of a multi-stack solid oxide fuel cell from a systems engineering perspective, Chem. Eng. Sci. 238 (2021), https://doi.org/10.1016/J.CES.2021.116571, 116571.

[57] A.M. Rabiu, N. Dlangamandla, Ø. Ulleberg, Novel heat integration in a methane reformer and high temperature PEM fuel cell-based mCHP system, APCBEE Proc. 3 (2012) 17–22, https://doi.org/10.1016/J.APCBEE.2012.06.039.

[58] H. Amiri, A.F. Sotoodeh, M. Amidpour, A new combined heating and power system driven by biomass for total-site utility applications, Renew. Energy 163 (2021) 1138–1152, https://doi.org/10.1016/J.RENENE.2020.09.039.

[59] M.A. Bagherian, K. Mehranzamir, A.B. Pour, S. Rezania, E. Taghavi, H. Nabipour-Afrouzi, et al., Classification and analysis of optimization techniques for integrated energy systems utilizing renewable energy sources: a review for CHP and CCHP systems, Processes 9 (2021) 339, https://doi.org/10.3390/pr9020339.

[60] A. Zakaria, F.B. Ismail, M.S.H. Lipu, M.A. Hannan, Uncertainty models for stochastic optimization in renewable energy applications, Renew. Energy 145 (2020) 1543–1571, https://doi.org/10.1016/J.RENENE.2019.07.081.

[61] M. Groissböck, Are open source energy system optimization tools mature enough for serious use? Renew. Sustain. Energy Rev. 102 (2019) 234–248, https://doi.org/10.1016/J.RSER.2018.11.020.

[62] J.A. Nelder, R. Mead, A simplex method for function minimization, Comput. J. 7 (1965) 308–313, https://doi.org/10.1093/comjnl/7.4.308.

[63] M. Gilli, D. Maringer, E. Schumann, Chapter 12. Heuristic methods a nutshell, in: M. Gilli, D. Mar, E. Schumann (Eds.), Numerical Methods and Optimisation in Finance, second ed., Academic Press, London, UK, 2019, pp. 273–318.

[64] S. Lecompte, H. Huisseune, M. van den Broek, S. De Schampheleire, M. De Paepe, Part load based thermo-economic optimization of the organic rankine cycle (ORC) applied to a combined heat and power (CHP) system, Appl. Energy 111 (2013) 871–881, https://doi.org/10.1016/j.apenergy.2013.06.043.

[65] A. Datas, A. Ramos, C. del Cañizo, Techno-economic analysis of solar PV power-to-heat-to-power storage and trigeneration in the residential sector, Appl. Energy 256 (2019), https://doi.org/10.1016/j.apenergy.2019.113935, 113935.

[66] J.S. Rustagi, Numerical methods of optimization, in: Optimization Techniques in Statistics, Elsevier, 1994, pp. 53–88, https://doi.org/10.1016/B978-0-12-604555-0.50010-8.

[67] J. Wang, T. Mao, J. Sui, H. Jin, Modeling and performance analysis of CCHP (combined cooling, heating and power) system based on co-firing of natural gas and biomass gasification gas, Energy 93 (2015) 801–815, https://doi.org/10.1016/j.energy.2015.09.091.

[68] N. Shokati, F. Ranjbar, M. Yari, A comprehensive exergoeconomic analysis of absorption power and cooling cogeneration cycles based on Kalina, part 2: parametric study and optimization, Energ. Conver. Manage. 161 (2018) 74–103, https://doi.org/10.1016/j.enconman.2018.01.080.

[69] E. Bellos, C. Tzivanidis, Parametric analysis and optimization of an organic rankine cycle with nanofluid based solar parabolic trough collectors, Renew. Energy 114 (2017) 1376–1393, https://doi.org/10.1016/j.renene.2017.06.055.

[70] Y. Nesterov, Efficiency of coordinate descent methods on huge-scale optimization problems, SIAM J. Optim. 22 (2012) 341–362, https://doi.org/10.1137/100802001.

[71] H. Beyer, H. Schwefel, Evolution strategies—a comprehensive introduction, Nat. Comput. 1 (2002) 3–52, https://doi.org/10.1023/A:1015059928466.

[72] G.M. Bollas, P.I. Barton, A. Mitsos, Bilevel optimization formulation for parameter estimation in vapor-liquid(-liquid) phase equilibrium problems, Chem. Eng. Sci. 64 (2009) 1768–1783, https://doi.org/10.1016/j.ces.2009.01.003.

[73] H.-G. Beyer, An alternative explanation for the manner in which genetic algorithms operate, Biosystems 41 (1997) 1–15, https://doi.org/10.1016/S0303-2647(96)01657-7.

[74] T. Jansen, I. Wegener, Real royal road functions—where crossover provably is essential, in: Spectre Led (Ed.), GECCO'01: Proceedings of the Genetic and Evolutionary Computation Conference. Morgan Kaufmann, San Francisco, CA, 2001.

[75] J.H. Holland, Adaptation in Natural and Artificial Systems: An Introductory Analysis with Applications to Biology, Control and Artificial Intelligence, MIT Press, Cambridge, MA, USA, 1992.

[76] B. Fredriksson Möller, J. Arriagada, M. Assadi, I. Potts, Optimisation of an SOFC/GT system with CO2-capture, J. Power Sources 131 (2004) 320–326, https://doi.org/10.1016/j.jpowsour.2003.11.090.

[77] A. Haghighat Mamaghani, B. Najafi, A. Casalegno, F. Rinaldi, Optimization of an HT-PEM fuel cell based residential micro combined heat and power system: a multi-objective approach, J. Clean. Prod. 180 (2018) 126–138, https://doi.org/10.1016/j.jclepro.2018.01.124.

[78] J.C. Spall, Introduction to Stochastic Search and Optimization: Estimation, Simulation, and Control, Wiley, Hoboken, NJ, USA, 2005.

[79] M. Bornapour, R.-A. Hooshmand, A. Khodabakhshian, M. Parastegari, Optimal coordinated scheduling of combined heat and power fuel cell, wind, and photovoltaic units in micro grids considering uncertainties, Energy 117 (2016) 176–189, https://doi.org/10.1016/j.energy.2016.10.072.

[80] S. Soheyli, M.H. Shafiei Mayam, M. Mehrjoo, Modeling a novel CCHP system including solar and wind renewable energy resources and sizing by a CC-MOPSO algorithm, Appl. Energy 184 (2016) 375–395, https://doi.org/10.1016/j.apenergy.2016.09.110.

CHAPTER 16

Study of syngas-powered fuel cell, simulation, modeling, and optimization

Mohammad Rahmani and Hadis Najafi Maharluie

Department of Chemical Engineering, Amirkabir University of Technology, Tehran, Iran

1. Introduction

Because of the energy crisis, restricted fuels, and pollution issues caused by fossil fuels, the utilization of current energy sources has received additional attention than before. These reasons have caused many studies into using new energy conversion devices that are green and environmentally friendly. Systems can convert the potential chemical energy of fuels into electrical energy with minimum waste and minimizing pollutants. Among the new energy conversion devices, the fuel cell has received much attention as a green example, and considerable efforts are being made to advance and expand this clean technology in many countries, including the United States, Japan, India, China, Norway, and Germany. A fuel cell is a new energy generation system that produces high-efficiency electrical energy with water and heat from a direct combination of fuel and oxygen without causing environmental or noise pollution. Fuel cell can use hydrogen-containing fuels such as methanol, ethanol, natural gas, and even gasoline and diesel. The components of a fuel cell are a positive electrode (anode) and a negative (cathode), which are located on either side of an electrolyte (solid or liquid). The movement of electrons from an electrochemical reaction in an external circuit that connects two electrodes generates electricity.

Fuel cells are divided based on various parameters such as type of electrolyte, application, efficiency, temperature range, and the type of carrier ions in the electrolyte. In Table 1, [1,2] the most common types of fuel cells are listed such as phosphoric acid fuel cell (PAFC), polymer electrolyte fuel cell (PEFC), molten carbonate fuel cell (MCFC), alkaline fuel cell (AFC), and solid oxide fuel cell (SOFC). The SOFC operates at a high operating temperature of $500–1000^{\circ}C$ [3]. Each cell consists of two positive and negative electrodes and an electrolyte layer that is entirely ceramic.

Advances in Synthesis Gas: Methods, Technologies and Applications. https://doi.org/10.1016/B978-0-323-91879-4.00014-X

Table 1 Review and comparison of common types of fuel cells [1,2].

Item	PEFC	AFC	PAFC	MCFC	SOFC
Ion conductor	H^+	OH^-	H^+	CO_3^{2-}	O^{2-}
Operating temperature	$40-80°C$	$65-220°C$	$205°C$	$650°C$	$600-1000°C$
Electrical efficiency	45%–60%	40%–60%	55%	60%–65%	55%–65%
Electrolyte	Hydrated polymeric ion exchange membranes	Mobilized or immobilized potassium in asbestos matrix	Immobilized liquid phosphoric acid in SiC	Immobilized liquid molten carbonate in $LiAlO_2$	Metal ceramic
Catalyst	Platinum	Platinum	Platinum	Made of electrode	Made of electrode
Interconnect	Carbon-metal	Metal	Graphite	Stainless steel	Ni, Ceramic

Besides, SOFCs possess other advantages [4]:

- Due to high-temperature operation, it has the highest efficiency compared with other fuel cells.
- The heat produced can be used in other parts of the fuel cell system.
- It is possible to use a wide range of fuel types (such as natural gas, hydrogen, and carbon monoxide) directly in the cell.
- Not required for fuel converters to use different fuels.
- Due to solid structure, the corrosion problem is low.

1.1 SOFC operation

In an SOFC, oxygen ions react with hydrogen or carbon monoxide (if present) and produce water and carbon dioxide. In Fig. 1, the principle of SOFC operation with hydrogen and carbon monoxide fuels is shown. At first, the fuel enters the anode channel and an electrochemical reaction is performed on the anode-electrolyte interface, Eventually, the products go to outside fuel channel with unreacted gases [5]. As shown in Fig. 1, if pure hydrogen fuel is used, the product includes only water, and if the fuel contains hydrogen and carbon monoxide, the product will contain water and carbon dioxide.

If H_2fed is used as fuel, reactions include:

$$Anode : H_2 \longrightarrow 2H^+ + 2e^-$$

$$Cathode : \frac{1}{2}O_2 + 2e^- \longrightarrow O^{2-}$$

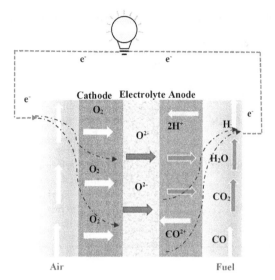

Fig. 1
Description of solid oxide fuel cell performance.

$$Anode : O^{2-} + 2H^+ \rightarrow H_2O$$

$$Overall : H_2 + \frac{1}{2}O_2 \longrightarrow H_2O \tag{1}$$

If CO is fed to the SOFC, reactions include:

$$Anode : CO \longrightarrow CO^{2+} + 2e^-$$

$$Cathode : \frac{1}{2}O_2 + 2e^- \longrightarrow O^{2-}$$

$$Anode : O^{2-} + 2CO^{2+} \rightarrow CO_2$$

$$Overall : CO + \frac{1}{2}O_2 \longrightarrow CO_2 \tag{2}$$

1.2 Fuel options for SOFC

The main advantage of SOFC is its flexibility for using a variety of fuels. If methane and hydrocarbons are used as fuel, the reforming and water gas shift reactions occur, which are shown by Eqs. (3), (4), respectively.

Steam reforming of hydrocarbons:

$$C_xH_y + xH_2O \leftrightarrow xCO + \left(x + \frac{y}{2}\right)H_2 \tag{3}$$

Table 2 SOFC categories based on various parameters.

Various parameters	Types
Electrolyte conduct	PC-SOFC (intermediate temperature)
	SOFC (high temperature)
SOFC design [6]	Planar
	Tubular
Type of scale	Cell level
	Stack level
	System level
Type of reformer [6]	External reforming
	Direct internal reforming
	Indirect internal reforming
Type of Modeling	White box
	Gray box
	Black box

Water-gas shift:

$$CO + H_2O \leftrightarrow CO_2 + H_2 \tag{4}$$

In Table 2, types of SOFC based on various parameters such as temperature, type of design, scale, electrolyte conduct, reformer, and modeling approach are shown.

The development of PSOFCs has progressed slower due to the sealing problem, which in recent years, the use of low temperature design has increased and the structure has improved. Recently, a lot of research has been done on the planar design of SOFC.

1.3 SOFC geometry

Fig. 2 shows planar and tubular geometries, which are examined and compared in the following section. PSOFC is composed of flat layers on top of each other. They are mainly divided into radial and planar. The reactant gases are directed to the center of the porous structure. The movement of gases within the channels is constrained by the electrodes and interconnections.

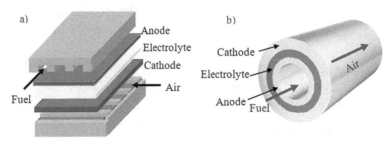

Fig. 2
Type of common SOFC designs (A) PSOFC [7] (B) TSOFC.

Table 3 Features of PSOFC and TSOFC [7].

Type of SOFC	Advantages	Disadvantages
Planar	• High power density • Simple structure that can be changed to other geometric shapes	• Unchangeable the scale of the layers • Low strength against mechanical and temperature stresses • Due to the sharp edges, there is a sealing problem
Tubular	• Suitable replacement for conventional heat engines • High efficiency • Designed to produce power up to 2 MW	• Have high ohmic loss • Have high fabrication costs

However, one of the challenges of a PSOFC is fabrication of high strength construction. The ceramic thin films have small thickness and break due to stress. The main disadvantages of flat design are as follows:

• The fragility of planar SOFCs under the influence of mechanical and thermal stresses
• Need proper sealing, especially around the edges
• Low scalability of thin films

Table 3 lists the advantages and disadvantages of both designs.

The tubular structure consists of concentric cylinders stacked on top of each other. Tubular SOFCs have more advantages than planar cells. One of the important advantages of this type of cell is no loss of fuel from the edges of the cell components and the use of heat generated in the cell, which reduces the construction and operating costs of the cell. The most important disadvantage of TSOFC is the relatively long path of electrons compared with the plate structure, which leads to a more significant ohmic potential drop. To solve this problem, additional equipment must be installed on bipolar cells to eliminate the problem of current density and heat transfer. In this case, the complexity and construction costs of system increase. Also, lateral heat exchangers must be used to preheat the air and prepare it to react in the cell. However, the most important advantage of the PSOFC is its lower potential drop compared with the tubular structure. In other words, bipolar plates lead to shorter path of electrons, which reduces the ohmic potential drop and increases cell efficiency [1].

1.4 Investigation based on the type of reformer

As mentioned, one of the advantages of SOFC is the variety in the fuels it can consume. If the fuel is hydrocarbon, a reformer is used to produce hydrogen. In the external or primary reforming type, the reforming process is performed in a side reactor next to the set of cells,

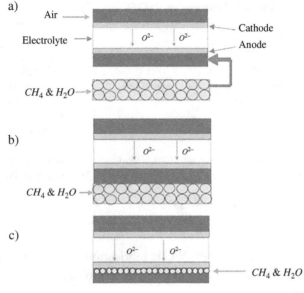

Fig. 3
(A) ER-SOFC, (B) IIR-SOFC, and (C) DIR-SOFC.

using the heat generated in the cell, or the reformer may be inside the stack, which then typically results in two direct or indirect states, which are shown in Fig. 3 (a) ER-SOFC, (b) IIR-SOFC, and (c) DIR-SOFC. In the IIR-SOFC, the fuel cell is in thermal contact with the reformer. The reformer uses the heat generated in the fuel cell to perform reforming reactions.

Hydrocarbon fuel enters the reactor with water vapor. After the reactions, the product of this reactor enters into the anode channel as fuel. The main problem of this type of SOFC is the preparation of water vapor. However, adding the peripheral equipment increases costs and complexity of the system.

In DIR-SOFC, hydrocarbon fuel enters the cell directly. The chemical and electrochemical reactions take place in electrode-electrolyte interface at the same time. Therefore, the catalyst must be made of a material that has the appropriate catalytic properties for both types of reactions [1].

2. SOFC modeling

Many studies have been done with the physical model approach. Actually, the researchers thoroughly analyzed the electrochemical reactions and ionic and electric properties by using of physical and analytical models. In fact, fuel cell modeling is primarily based on knowledge of the physical and chemical structures known as white-box, or first-principle models. This modeling view can be considered for the cell, stack, and system scale with different geometries, but can result in complex set of coupled equations, which need a lot of computational resources. There is another type of modeling based on empirical data. These models are called black-box models, which have applications such as control, monitoring, and optimization of systems. This approach is suitable for very complex systems. However, the strong reliance on experimental data makes it less generalizable, and another approach that called gray box is developed. The basis of gray-box modeling is somewhat experimental and slightly physical [4].

2.1 Physical models

Many studies have been reported on SOFC physical modeling, and the main purpose of those models is to examine the mass, energy, and momentum transport. However, in different types of models, the goal is to find the relationship between voltage and current density. Fuel cell models at the cell, stack, and system scale have been investigated.

2.1.1 SOFCs models at cell level

Serval studies have been presented with different level of accuracy and complexity to predict the performance of SOFCs. These SOFC models at the cell level have been divided into micro and macro scales. In macro models, transport phenomena, thermodynamics, and electrochemistry in three dimensions are considered to predict the cell performance. Also, it is assumed that electrochemical reactions take place at the electrode-electrolyte interface. In PSOFC macro modeling, a cell in solid structure and fluid flow in a channel are examined. The solid structure consists of porous electrodes containing the diffusion and reaction layers of the anode, cathode, and electrolyte. Fluid flow refers to the flow within the fuel and air channels. In solid structure, heat and charge transfer and chemistry are investigated. The heat, mass, and momentum transport phenomena are also used to calculate temperature, pressure, and distribution of the chemical species concentration. In general, mathematical models of SOFC that include heat transfer and electrochemical are considered in different layers including channels, porous electrodes, electrolyte, and interconnections. Their transport equations were coupled to a reliable electrochemical model. The chemical reactions take place at the interface of porous electrodes and electrolyte. During the process of electrochemical reactions, H_2O is produced at the anode, as well as O^{-2} at the cathode. In many studies, the models are examined at steady-state conditions. The time-dependent model is reported less than steady-state ones [8].

The SOFC system voltage is obtained via subtracting the irreversible losses from the Nernst voltage. As shown in Fig. 4, irreversible losses are mainly due to three sources, which include ohmic (η_{ohm}), concentration (η_{conc}), and activation (η_{act}) polarization. The losses are called polarization, overpotential, or overvoltage (η). The activation polarization plays more significant role in the total losses. It is more important in low- and medium-temperature fuel

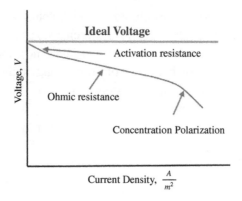

Fig. 4
Distribution of total voltage and standard voltage [9].

cells. This voltage is because of the resistance of the activation energy to perform the electrochemical reaction. However, the resistance decreases at higher temperatures and the reactions occur faster. In a fuel cell with electrolyte supported, activation polarization takes place at each positive and negative electrode. But, in the cathode, the current density is higher due to the smaller activation drop.

The activation polarization with lower current density gradually increases but rapidly at higher current density [10]. There are several relationships to calculate activation loss. In this study, the activation polarization is calculated by the Butler-Volmer relationship [1], which expresses relationship between current density and overpotential [11] as follows:

$$j = j_0 \left[exp \left(\frac{\alpha n F \eta}{RT} \right) - exp \left(-\frac{(1-\alpha)F\eta}{RT} \right) \right] \tag{5}$$

There is ohmic resistance due to oxygen ion transport in the electrolyte and electrical resistance of the electrode. Various ways have been suggested to reduce ohmic drop, including: reduce electrolyte thickness, using materials with high conductivity for electrodes, and implementation of an appropriate design based on the anode or interconnections.

The loss in concentration potential is due to the diffusion and mass transfer resistance of reactant gases. This resistance is very low while the fuel is introduced as pure gases. When the mixture of gases such as exhaust gases from the reforming system is used, the resistance is higher. It is better to use the high-temperature systems for controlling the resistances [1]. The overall voltage (V) is less than the reversible voltage E_{Th} calculated by

$$V = E_{Th} - \text{losses} = E_{Th} - \eta \tag{6}$$

$$V = E_{Th} - \eta_{act} - \eta_{ohm} - \eta_{conc} \tag{7}$$

The open-circuit voltage of an SOFC for a hydrogen electrochemical reaction is calculated as follows:

$$E = -\frac{\Delta \overline{g_f}}{nF} \tag{8}$$

$$\Delta \overline{g_f} = \Delta \overline{g_f}^0 - RTln\frac{\prod \alpha^{vi}_{product}}{\prod \alpha^{vi}_{reactants}} \tag{9}$$

So,

$$E_{Th} = V_{OC} + RTln\frac{\prod \alpha^{vi}_{product}}{\prod \alpha^{vi}_{reactants}} \tag{10}$$

$$\eta_{ohm} = jA\delta R_{ohm} \tag{11}$$

$$\eta_{conc} = c\ ln\frac{j_L}{j_l - j} \tag{12}$$

In Eq. (12), and c represents an approximate value of [12]:

$$c = \frac{RT}{nF}\left(1 + \frac{1}{\alpha_i}\right) \tag{13}$$

The mathematical models are presented below show mass, energy, and momentum transfer for air and fuel flow in SOFC channels. The chemical reaction takes place along the channels. The electrochemical reaction occurs at triple-phase boundary (TPB). The electrochemical reactions products are then returned to the channels and exited from the system with unconverted gases.

Continuity equation is:

$$\frac{\partial \rho}{\partial t} + \nabla \cdot \left(\rho \cdot \vec{v}\right) = S_y \tag{14}$$

Due to the very high temperature of SOFCs, all fluids are in the form of gases, so it can be assumed that the ideal gas law is valid:

$$\rho = \frac{PM}{RT} \tag{15}$$

The velocity field and pressure drop can be calculated from the following equation:

$$\frac{\partial\left(\rho.\vec{v}\right)}{\partial t} + \nabla \cdot \left(\rho\vec{v}\vec{v}\right) = \rho\vec{f} + \nabla \cdot \vec{\tau} \tag{16}$$

where f is sum of the body forces. However, none of these forces play a significant role in the fluid motion through the SOFC channels and can be neglected [5].

$$\frac{\partial\left(\rho\cdot\vec{v}\right)}{\partial t}+\nabla\cdot\left(\rho\,\vec{v}\,v\right)=\nabla\cdot\vec{\tau} \tag{17}$$

Assuming the fluid is a Newtonian fluid, the stress tensor is calculated as follows:

$$\vec{\tau}=\mu\cdot\left[\nabla\vec{v}+\left(\left(\nabla\vec{v}\right)^{T}\right)\right]-\left[p+\left(\mu v-\frac{2}{3}\mu\right)\nabla\cdot\vec{v}\right]\vec{I} \tag{18}$$

Mass transfer of chemical species in the channel is given as follows:

$$\frac{\partial(\rho y_\alpha)}{\partial t}+\nabla\cdot\left(\rho y_{\vec{v}\alpha}\right)+\nabla\cdot\vec{j_\alpha}=S_{y_\alpha} \tag{19}$$

S_y stands for the term of reaction for y component. To study the chemical species diffusion, there are several relationships that are examined. At first, the basic constitutive equation was proposed by Fick that is adapting the Fourier coefficient in heat transfer [13]. Fick described the typical binary mixture diffusion. For an isobaric and isothermal system, it is possible to write the diffusion of the ordinary type in a one-dimensional for species α and β:

$$\vec{j_\alpha}=-D_{\alpha\beta}\frac{\partial C_\alpha}{\partial x} \tag{20}$$

Groot suggested relation to the systems that are not isothermal and isobaric [14]:

$$\vec{j_\alpha}=-cD_{\alpha\beta}\nabla x_\alpha \tag{21}$$

In general, multicomponent diffusion is described by Stefan-Maxwell model. These equations are simplified as follows that assume a constant pressure system with negligible thermal diffusion [15]:

$$j=-\rho w_i\sum D_i^{eff}\left(\nabla x_i+(x_i-w_i)\frac{\nabla P}{P}\right) \tag{22}$$

The energy transfer is expressed as follows [16]:

$$\sum_{\forall\alpha}(h_\alpha y_\alpha)\frac{\partial\rho}{\partial t}+\rho C_p\frac{\partial T}{\partial t}+\nabla\cdot\left(\rho C_p\,\vec{v}\,T\right)-T\cdot\nabla\left(\rho C_p\,\vec{v}\right)-\nabla\cdot(\lambda\nabla T)+\nabla\cdot\sum_{\forall\alpha}\left(h_\alpha\vec{j}_\alpha\right)=0 \tag{23}$$

In micro-modeling, it is assumed that electrochemical reactions are occurred within the porous volumetric structure of the electrodes. The purpose of this type of modeling is to predict the chemical and electrochemical performance in the electrode microstructures. For micro-modeling, microstructures must be modeled to calculate the physical properties of porous electrodes. Because of simplicity and required lower computational resources, the random packing sphere model is recommended. In this model, the electrode is assumed as a random packing of electron conducting and ion conducting spheres with single size. To estimate the microstructure properties of electrode, the particle coordination number and the percolation theories are employed [17,18].

Overall, there has been a lot of progress in the micro-modeling of the SOFC. However, for a porous composite, the type of modeling must be generalized to be able to accurately predict the performance of the fuel cell. Conservation of mass, momentum, and energy as well as electrochemical equations must be considered to calculate the voltage and current density. The electrode model considers the motion, mass of species, and heat transfer in the SOFC electrodes. Also, the system is assumed to be homogeneous and isotropic. The microstructure parameters include ε porosity, τ tortuosity factor, d_p mean pore diameter, and κ permeability are used for describing electrode properties. Mass transfer of the components is in the electrode as follows:

$$\frac{\partial(\varepsilon \rho y_\alpha)}{\partial t} + \nabla \cdot (\rho y_\alpha \langle v \rangle) + \nabla \cdot \vec{j_\alpha} = S_{y_\alpha} \tag{24}$$

Therefore, v is the velocity of the fluid through the electrode, which can be obtained by Dupuit-Forchheimer equation $\langle v \rangle = \varepsilon \vec{v}_f$ being $\vec{v}_f$: the mean velocity of the fluid through the porous medium; and j_α, combined ordinary and Knudsen diffusion occur in porous composites such as anode and cathode layers and usually occur at the same time for binary and multicomponent mixtures. Therefore, we combine both influences of diffusion in Fick's law based on the definition of effective diffusion coefficient.

$$\frac{1}{D_\alpha^{eff}} = \frac{1}{D_{am}^{eff}} + \frac{1}{D_{K\alpha}^{eff}} \tag{25}$$

$$D_{am}^{eff} = \frac{\varepsilon}{\tau} D_{am} \tag{26}$$

$$D_{K\alpha}^{eff} = \frac{\varepsilon}{\tau} D_{K\alpha} \tag{27}$$

The fundamental Fickian equation for diffusion flux in porous media is [19]:

$$\vec{j_\alpha} = -\rho D_\alpha^{eff} \nabla y_\alpha \tag{28}$$

The dusty gas model (DGM) is first proposed by Krishna and Wesselingh [20]. This model was introduced as the most appropriate method for expressing gas diffusion modeling in a porous medium. This method was later used to calculate the gas diffusion in the electrodes by Suwanwarangkul et al. [21]. In general, thermo diffusion and effects of external forces are neglected.

$$-\frac{1}{RT}\nabla\cdot p_\alpha = \sum_{\beta\neq\alpha}\frac{x_\beta\vec{N}_\alpha - x_\alpha\vec{N}_\beta}{D_{\alpha\beta}^{eff}} + \frac{\vec{N}_\alpha}{D_{K\alpha}^{eff}} + \frac{1}{D_{K\alpha}^{eff}}\frac{p_\alpha}{RT}\frac{B_0}{\mu}\nabla p \tag{29}$$

The dusty gas model is less used than the above Fickian law because of its complexity. For the simplicity of the model [22], it is assumed that the viscous flow in the DGM should be negligible ($\nabla p = 0$) and use the following equation.

$$-\frac{p}{RT}\nabla\cdot x_\alpha = \sum_{\beta\neq\alpha}\frac{x_\beta\vec{j}_\alpha - x_\alpha\vec{j}_\beta}{D_{\alpha\beta}^{eff}} + \frac{\vec{j}_\alpha}{D_{K\alpha}^{eff}} \tag{30}$$

The enthalpy conservation equation in a porous medium, assuming a thermal equilibrium between the electrode interface and the fuel flow as [23]:

$$\frac{\partial[\epsilon\rho_f h_f + (1-\epsilon)\rho_s h_s]}{\partial t} + \nabla\cdot(\rho_f h_f\langle v\rangle) + \nabla\cdot\left[\epsilon\vec{q}_f + (1-\epsilon)\vec{q}_s\right] = Q + w \tag{31}$$

Assuming that the volumetric heat sources due to heat and Joule radiation inside the electrodes are ignored $Q\approx0$. Joule heat is considered only inside the electrolyte, it also avoids radiation, so the electrode surfaces are assumed to be opaque [24]:

$$\frac{\partial[\epsilon\rho_f h_f + (1-\epsilon)\rho_s h_s]}{\partial t} + \nabla\cdot(\rho_f h_f\langle v\rangle) + \nabla\cdot\left[\epsilon\vec{q}_f + (1-\epsilon)\vec{q}_s\right] = w \tag{32}$$

The diffusive heat fluxes within the SOFC electrodes are

$$\vec{q}_f = -\lambda_f\nabla T + \sum_{\forall\alpha}\left(h_\alpha\vec{j}_\alpha\right) \tag{33}$$

$$\vec{q}_s = -\lambda_s\nabla T \tag{34}$$

where the effective thermal conductivity of the porous media can be expressed based on the upper Wiener bound [25].

$$\lambda^{eff} = \epsilon\lambda_f + (1-\epsilon)\lambda_s \tag{35}$$

Considering the following relationship:

$$\rho_f h_f\langle\vec{v}\rangle + \sum_{\forall\alpha}\left(h_\alpha\vec{j}_\alpha\right) = \sum_{\forall\alpha}\left[h_\alpha\left(\vec{N}_\alpha W_\alpha\right)\right] \tag{36}$$

$$\frac{\partial[\epsilon\rho_f h_f + (1-\epsilon)\rho_s h_s]}{\partial t} + \nabla\cdot\sum_{\forall\alpha}\left[h_\alpha\left(\vec{N}_\alpha W_\alpha\right)\right] - \nabla\cdot\left[\lambda^{eff}\nabla T\right] = w \tag{37}$$

The term w is the heat of reaction in the cathode and anode electrodes, which is calculated as:

$$w = \begin{cases} \sigma_{el} \nabla \varphi_{el} \cdot \nabla \varphi_{el} & \text{in CDL, ADL} \\ \sigma_{el} \nabla \varphi_{el} \cdot \nabla \varphi_{el} + \sigma_{io} \nabla \varphi_{io} \cdot \nabla \varphi_{io} + J\eta + \sum_k (-\Delta H_{elec,k}) R_{elec,k} & \text{in ARL, CRL} \end{cases}$$

(38)

The performance of porous materials has been extensively studied in view of structure and hydrology. The Darcy equation has an empirical nature that expresses the forces due to the pressure gradient and the friction resistance of porous materials [26]. Although Darcy equation is used for a porous medium, but it cannot be used for nonslip walls condition. This equation is corrected by adding the Brinkman term.

$$\nabla P = -\frac{\mu_g \varepsilon}{\kappa} \vec{v}_f + \nabla \cdot \left(\rho_g \nabla \right) \vec{v}_f$$

(39)

The electrolyte is nonporous, dense solid. However, due to the flow of oxygen ions (O^{2-}), there is mass transfer and charge and heat dissipation. In this model, by considering the charge balance, the effect of ionic mass transfer as well as heat transfer can be calculated. The heat generated in the electrolyte is due to the transport of oxygen ions along the electrolyte. To analyze the movement of O^{2-} from the electrolyte-cathode to the interface of the electrolyte and the anode electrode, the transfer equation can be expressed as follows:

$$\frac{\delta \tilde{\rho}}{\delta t} + \nabla \cdot \vec{j} = S_{io}$$

(40)

According to Ohm's law [8]:

$$\vec{j}_{io} = -\tilde{\sigma}_{io} \nabla \phi_{io}$$

(41)

With placement of Eq. (41) into Eq. (40):

$$\frac{\delta \tilde{\rho}}{\delta t} - \nabla \cdot (\tilde{\sigma}_{io} \nabla \phi_{io}) = S_{io}$$

(42)

Temperature distribution is calculated in the electrolyte only by the conduction mechanism and the radiation (very thin electrolytes) and convection (it is a dense environment) are ignored [24]:

$$\frac{\partial (\rho h)}{\partial t} + \nabla \cdot \vec{q} = Q + w$$

(43)

where $w=0$, because the electrochemical reaction occurs only in the porous anode electrode. The conduction flux is calculated from the Fourier's law $\vec{q} = -\lambda \nabla T$, and the energy source is calculated by Joule heating, $Q = i^2 \tilde{\sigma}_{io}^{-1}$. By applying assumptions, the equation will eventually be as follows:

$$\frac{\partial (\rho h)}{\partial t} - \nabla \cdot (\lambda \nabla T) = i^2 \tilde{\sigma}_{io}^{-1}$$

(44)

The interconnects are assumed to be dense, so only Ohm's law is used to calculate the electron conservation of charge as [27].

$$\vec{j}_{el} = -\tilde{\sigma}_{el}\nabla\phi_{el} \tag{45}$$

$$\frac{\delta\tilde{\rho}}{\delta t} - \nabla \cdot (\tilde{\sigma}_{el}\nabla\phi_{iel}) = S_{el} \tag{46}$$

2.1.2 Comparison of SOFC at cell level

Some models shown in Table 4 are compared from the modeling perspective. This table examines fuel cell models at the cell scale. There is a focus on the DIR-SOFC. One major benefit of the SOFC is the flexibility in the fuel. Natural gas (methane) is used as the dominant fuel for SOFC. In many studies, the model is three-dimensional, but some parts can be described as one-dimensional model, such as PEN and fuel-air duct. Radiation is ignored in most cases. In most models, the effective diffusivity is estimated by Fick's law. Also, most of the research on SOFC is reported for flat structure, despite its disadvantages, it has more power density than tubular structure and is more interesting for researchers. Some key points in reported works are summarized as follows:

- There is a need to study more on electrochemical aspects and improve models in this area.
- In most studies, SOFC is considered to be self-supported. The high thickness of layers can result in high ohmic loss. This issue has not received enough attention and needs to be investigated in more detail.
- There are many reported works on modeling on SOFC with co-flow configuration. It is required to discuss other configurations.

2.1.3 SOFC models at the stack level

The SOFC stack is a set of single cells that are arranged sequentially or in parallel. By examining a single cell, the overall performance of the stack can also be predicted. The total computational cost of single cells is equal to the computational cost of a stack. This level model is extensively performed in SOFC systems to evaluate the performance of fuel cell. Chan et al. [39] presented a stack model based on the 0-D cell model. They calculated the amount of heat transfer in a fuel cell stack using external insulation. Braun et al. [40] presented a stack model based on 1-D cell model. They did not consider heat transfer. However, they added the expression of heat loss (equivalent to 3% of the additional heating value of the input fuel) to the energy level inside the storage medium. Colpan et al. [41] developed a two-dimensional SOFC at the stack level. However, heat transfer and its effect on SOFC performance were not investigated.

Table 4 Comparison of SOFC at cell level.

Ref.		[28]	[29]	[30]	[31]	[32]	[33]	[34]	[35]	[36]	[37]	[38]
Type of reformer	External	✓		✓				✓	✓	✓	✓	✓
	DIR		✓		✓	✓	✓					
	IIR											
Fuel	Hydrogen	✓	✓	✓	✓	✓	✓	✓	✓	✓	✓	✓
	Methane											
	Gas mixture											
Flow configuration	Co-flow	✓	✓	✓	✓	✓	✓	✓	✓	✓	✓	✓
	Counter-flow	✓			✓						✓	✓
	Cross-flow				✓							
Dimensional modeling	0-D											
	1-D											
	2-D								✓	✓	✓	✓
	3-D	✓	✓	✓	✓	✓	✓	✓				
Scale of cell modeling	Micro	✓	✓	✓		✓	✓	✓				✓
	Macro	✓	✓	✓	✓	✓	✓	✓	✓	✓	✓	✓
Heat transfer considerations	Convection	✓	✓	✓	✓	✓	✓	✓	✓	✓	✓	✓
	Radiation	✓	✓	✓				✓	✓		✓	✓
Type of support	Self-supporting	✓	✓	✓	✓	✓	✓		✓	✓	✓	✓
	External-supporting							✓				
Diffusion model	Fick	✓		✓		✓	✓			✓	✓	✓
	Maxwell		✓		✓				✓			
	Dusty-gas							✓				
Cell design	TSOFC	✓	✓	✓	✓			✓		✓		✓
	PSOFC					✓	✓	✓	✓		✓	
Dependence to time	Steady-state	✓	✓	✓	✓	✓	✓	✓	✓	✓	✓	✓
	Transient		✓					✓				
Thermomechanical modeling	Yes											
	No	✓			✓	✓	✓		✓	✓	✓	✓
Validation with experimental data	Yes	✓	✓		✓	✓	✓			✓		✓
	No		✓	✓				✓	✓		✓	

2.1.4 SOFC models at system level

In the recent years, attention has been paid to combine energy and power systems, which along with high electrical efficiency, hot water vapor can also be used as an output [42,43]. The system under consideration is a CHP system used for industrial and residential applications. The concept of SOFC-based poly generation systems is presented and shown in Fig. 5.

Natural gas, pure hydrogen, and biogas are the input fuels that enter at high temperatures. Integrating the fuel cell with equipment such as turbines helps increase efficiency. It can also be seen in the concept of the poly-generation, cooling and heating are the most common co products. ·However, other products such as carbon dioxide and hydrogen are also reported in numerous studies [44].

Fig. 6 shows the whole configuration of a combined heat and energy system whose fuel is biogas. The components of the system include a prereformer, a postburner, a boiler, and SOFC stack.

It is assumed that the input biogas, which includes sulfur and different elements, is refined (purification in this study is not considered). The hot outlet of the postburner used for the heating of the biogas and then the feed is mixed with the high-temperature steam. Then, the mixture is directed to a prereformer. Finally, the required fuel for SOFC is supplied during the chemical reaction. The air blower pressurizes the inlet air, pushes through heater, and then directs it into the fuel cell. Electricity and hot water vapor are generated in the stack during the electrochemical reaction [45]. The anode output of the SOFC stack includes carbon monoxide,

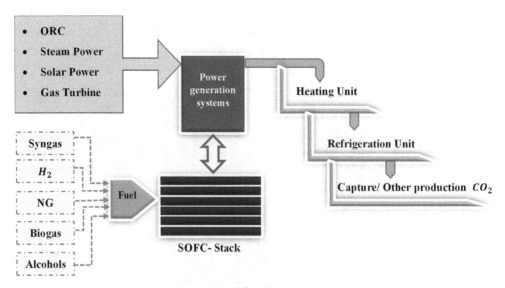

Fig. 5

General concept of high-temperature SOFC-based poly-generation systems [44].

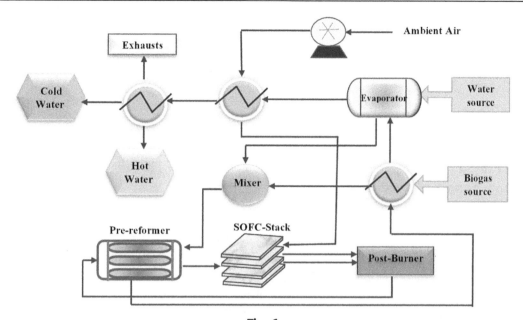

Fig. 6

The combined heat and power system with biogas feed [45].

hydrogen, and nonconverted gases. Within the postburner, it is completely oxidized. The exhaust gases are introduced into the reformer for the required heat of the reforming reaction, then the reformer output is added into the converter at a lower temperature. Heat exchangers and evaporators are used to preheat the fuel used. Finally, the burner exhaust enters to the boiler to produce hot water. The effect of water pump is ignored [46]. This model consists of a three-dimensional domain at stack level and a 0-D BOP component model. The BOP components include an air blower, preformer, postburner, several heat exchangers, and pipes. The electrochemical modeling is performed between the electrolyte and the electrode at the TPB. The cell-level model includes the mass transfer, charge transfer, and pressure drop.

For prereformer modeling, the steam reforming (SR) and water-gas shift reactions are considered [47].

$$CH_4 + H_2O \leftrightarrow CO + 3H_2 \quad \Delta H1 = 206.1 \text{ kJ/kmol} \tag{47}$$

$$CO + H_2O \leftrightarrow CO_2 + H_2 \quad \Delta H2 = -41.15 \text{ kJ/kmol} \tag{48}$$

In thermodynamic equilibrium, the species molar fractions are related to the equilibrium constant,

$$K_{eq,SR} x_{CH_4} x_{H_2O} = \left(\frac{P}{P_0}\right)^2 x_{CO} x_{H_2}^3 \tag{49}$$

$$K_{eq,WGS} x_{CO} x_{H_2O} = x_{CO_2} x_{H_2} \tag{50}$$

The heat required is provided by the postburner. The prereformer energy balance is calculated by [45]:

$$\dot{m}_{in} \sum_i y_{i,in} h_{i,in} + Q = \dot{m}_{out} \sum_i y_{i,out} h_{i,out} \tag{51}$$

$$\dot{m}_H (h_{H,in} - h_{H,out}) = Q \tag{52}$$

where Q is heat that is provided by the hot exhaust.

The exhaust gases from the SOFC anode are completely ignited in the burner during the following chemical reactions:

$$CH_4 + 2O_2 \leftrightarrow CO_2 + 2H_2O \tag{53}$$

$$2CO + O_2 \leftrightarrow 2CO_2 \tag{54}$$

$$2H_2 + O_2 \leftrightarrow 2H_2O \tag{55}$$

The required oxygen is provided during combustion from the cathode (N) and the rest of the required air is provided from the environment (O). Therefore, the output molar flow and temperature can be calculated [46].

$$\sum_{i,H} F_{i,H} \cdot h_{i,H} + \sum_{i,N} F_{i,N} \cdot h_{i,N} + \sum_{i,O} F_{i,O} \cdot h_{i,O} - (F_{CO,H} \cdot \Delta h_{r,CO}) - (F_{CH_4,H} \cdot \Delta h_{r,CH_4})$$
$$- (F_{H_2,H} \cdot \Delta h_{r,H_2}) = \sum_{i,l} F_{i,l} \cdot h_{i,l} \tag{56}$$

The following heat equation is used in the heat exchanger to calculate the amount of heat required for the preheater, air inlet gas, and biogas [46]:

$$\sum_{i,AC} F_{i,AC} \cdot \Delta h_{i,AC} + \sum_{i,LB} F_{i,LB} \cdot \Delta h_{i,LB} = \eta_{HC} \cdot \left(\sum_{i,GH} F_{i,GH} \cdot \Delta h_{i,GH} + \sum_{i,MN} F_{i,MN} \cdot \Delta h_{i,MN} \right) \tag{57}$$

$F_{i,AC}$ and $\Delta h_{i,AC}$ are the molar flow rate of the i species from path A to C and enthalpy changes from path A to C.

Table 5 compares products of systems and power scales.

To evaluate the SOFC system, it is necessary to calculate fuel utilization, air excess ratio, and CHP efficiency, defined as [50]:

$$U_f = \frac{jA}{F(8\dot{n}_{a,CH_4,in} + 2\dot{n}_{a,H_2,in} + 2\dot{n}_{a,CO,in})} \tag{58}$$

$n_{a,i,in}$ is the anode input flow rate of species i.

Table 5 Comparison products and power scales for system level of SOFC.

Source	Product			Power scale
	Electricity	Heating	Cooling	
Yan et al. [48]	Tubular SOFC + GT+ ORC Kalina power plant		Stored LNG cryogenic energy	SOFC ~3100kW, turbine generation power ~600 kW, ORC ~150 kW, Kalina power plant ~130 kW
Baghernejad et al. [49]	SOFC + steam turbine + gas turbine	Heat exchanger	Libr-water single effect absorption chiller	SOFC ~100 kW
Wang et al. [45] Kazempoor et al. [50]	Planar SOFC HT-DIR SOFC IT-DIR SOFC	Heat exchanger, boiler, postburner Water heat exchanger, RAP, gas-to-gas heat exchanger, burner, thermally driven chiller		SOFC AC Power~180 kW, HT-DIR SOFC ~4.03–6.4 kW IT-DIR SOFC ~1.4–2.4 kW
Bao et al. [51]	Tubular SOFC+ (HP) Turbine+ (LP) turbine	Air supply tube, burner, compact heat exchangers		Turbine generation power ~21 kW SOFC stack power ~171.67 kW
Corigliano et al. [46]	IID-SOFC	Heat exchangers		Not available (N/A)
Tippawan et al. [52]	SOFC	Libr-water double effect absorption chiller	Libr-water double effect absorption chiller	Not available (N/A)
Fong et al. [53]	SOFC + grid	Hot water generator + heater	Libr-water single effect absorption chiller + vapor compression chillers	SOFC ~150 Kw (Partial Load), ~600 kW (Full Load)
Al Moussawi et al. [54]	SOFC	Heat Exchanger, burner	Single-effect absorption chiller	SOFC electrical only ~211.844 (MW) Annul heating load coverage 41.56 MW Annual cooling load coverage ~13.02 (MW)
Elmer et al. [44]	Planar SOFC + grid	SOFC auxiliary boiler liquid desiccant system	Liquid desiccant system	SOFC ~1.5–2 kW; Heating ~500–1000 kW; Cooling ~330 – 650 kW

$$\eta_{elec} = \frac{P_{SOFC,AC} - P_{Blowers}}{\eta_{f,ent}\overline{LHV}_{biogas}} \tag{59}$$

$$\lambda_{cell} = \frac{n_{O_2,ca,in}}{(2\dot{n}_{a,CH_4,in} + 0.5\dot{n}_{a,H_2,in} + 0.5\dot{n}_{a,CO,in})} \tag{60}$$

$$\eta_{CHP,syn} = \frac{P_{SOFC,AC} - P_{Blowers} + Q_{heat}}{\eta_{f,ent}\overline{LHV}_{biogas}} \tag{61}$$

2.2 Gray-box model

The basis of the gray-box modeling is a combination of experimental data and additional knowledge about the system. In fact, this approach is used when experimental data are not sufficient to examine the system. So, the lack of information in the data can be replaced by knowledge of the system. In many studies, due to mode simplification and linearization with the application of hypotheses, the real system does not match the system modeled. In such cases, there is a significant error. Therefore, it is useful to use an optimal identification method that can describe the system using an experimental data-based model [55].

2.2.1 Basic modeling

The basis of gray-box modeling is mathematical relations. Before modeling, the limitations of the system must be examined and analyzed. Those limitations should be considered that they are correct. Basic modeling is describing the process using a set of mathematical equations. This description may result in different forms of equations: logical expressions, differential equations, and algebraic equations [56].

Moradi and et al. [57] reported a one-dimensional first principle model of the SOFC in MATLAB and a simulated gray-box model of SOFC system level in Aspen-Plus. The one-dimensional model described the temperature and concentration distributions of chemical components and the voltage distribution. By validating against experimental data, it was observed that there is a relative error of 1%–79%, which increased with current density. This error difference was because of the temperature distribution. The gray-box model was developed by considering a series of system-scale blocks. In this approach, the superposition method is used. The results showed that the max temperature in the one-dimensional first principle model is higher than that of the gray-box model. The gray-box and black-box models are useful for simulating an SOFC on a system level because they are simpler, although they should be used according to the internal temperature distribution.

2.2.2 Conducting experiments

A route for better identification of the system under study is the measurement of the input and output signals of the system. In the first step, experiments are designed depending on the conditions of the system. Input signals shall affect the output in a meaningful range. Pseudo random binary signal is the among the common inputs for response estimation [56].

Goodwin [58] presented a design procedure for a nonlinear system to estimate unknown parameters. This method is appropriate to minimize the variance. For this method, the limitations in the inputs and states are among the assumptions used. Appropriate criteria for designing experiments to be used in gray-box modeling have been proposed by Melgaard et al. [59]. One of the important factors is to stimulate the process sufficiently, meaning that this test affects all the different states of the system.

2.2.3 Model calibration

Calibration is done in order to find a suitable and simple model. The model should be able to describe the experimental data in an acceptable and enough accurate way [60]. The estimation of unknown parameters is based on likelihood function that needs to be optimized [60]. Throughout optimization, the standard deviation, SD, of prediction error is calculated.

$$W(N(\varphi)b) = \frac{1}{2} \sum_{t=1}^{N} \left[\log\left[det\, P_r(t|N(\varphi), b) \right] + \text{error}(t|N(\varphi), b)^T\, P_r\,(t|N(\varphi), b)^{-1}\, \text{error}(t|N(\varphi), b) \right]$$

$$(62)$$

The model average error is calculated as:

$$\text{Error}(t|N(\varphi), b) = z(t) - \widehat{z}\,(t|N(\varphi), b) \tag{63}$$

$$S.D = \sqrt{\frac{1}{n-1} \sum_{t=1}^{n} \left[P_r\,(t|N(\varphi), b) \right]^2} \tag{64}$$

The variable $P_r\,(t|N(\varphi), b)$ is the covariance matrix of the prediction errors, $N(\varphi)$ is the model, φ is the parameter vector, and b is data sequence.

2.2.4 Model validation

Model validation is defined as the procedure where a prepared model is assessed with a testing dataset. The gray-box modeling technique is used for different objectives such as system control, monitoring, as well as system optimization. If the model fails to predict the requested output, the model should be simplified, if complex models are used, or increase complexity if a rough simple model is used. Out of sampling testing or rotation estimation are among the techniques for model validation. There is a lot of work to be done on model validation [61–63].

2.3 Black-box model

This approach is based on an implicit relationship between input and output data. The studied models have been presented according to the laws of physics, which are suitable for examining model states. For optimization and control purposes, use of physical models is not appropriate due to their complexity. The empirical models are shown in Fig. 7, the nature of them is a biologically inspired approach (artificial intelligence). This approach is based solely on empirical data and ignores the phenomenological or constitutional approach. The black-box approach is derived from nature patterns such as genetic algorithms or biological processes [64].

This major flaw prompted some researchers [65–68] to try to develop a model based on experimental data. The black-box approach, which relies on artificial intelligence, is used to describe and predict the performance of nonlinear dynamic systems. [69]. However, a model requires a lot of data as well as a long time to find the accurate data needed to express and introduce the system.

2.3.1 Artificial neural netwroks

The artificial neural networks (ANN) are based on the idea of using data (measured in laboratory or collected from plant), when they may not be defined according to a specific algorithm. This approach has a network of node layers, an input layer, an output layer, and one

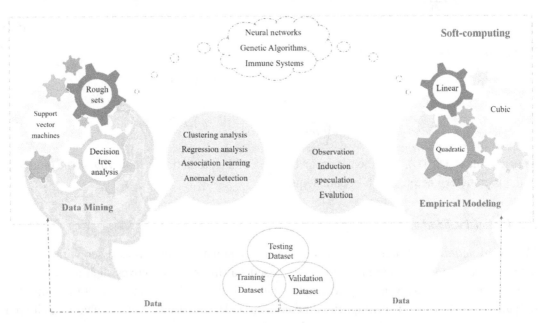

Fig. 7
Field of artificial intelligence.

or more hidden layers, each with own nodes. Validation is done using data not used in training or physical models.

Arriagada et al. [65] reported the application of a black-box model with ANN approach, in order to evaluate and investigate the SOFC system. The network in their study consists of two layers (see Fig. 8). Each output is determined by the implicit relationship between the input and the hidden layer.

The signals are in the input layer and the neurons are in the hidden layer. By comparing the results from the experimental study with the white-box model, the mean error value was less than 1%. In order to use in the system performance study, it seems that this model is very fast and easy in addition to high accuracy.

In another study by Song et al. [70], 30 pieces of SOFC stack were constructed and tested experimentally at different furnace temperatures and BP neural networks, support vector machine (SVM), and random forest (RF) are both used to predict the stack performance. Many assessments criteria used together with R^2, RMSE, MAE, coaching and take a look at time. The results show that the fitting errors of the three algorithms are all around 5%.

Milewski et al. [71] used an ANN approach to investigate and develop the system without providing an algorithm. They examined two sets of flow cell parameters include working conditions and cell architecture conditions. The backpropagation error algorithm was used for an ANN method. The results show that this model is well adapted to the system conditions. The error rate for anode porosity in this method is 0.09%–6.2%.

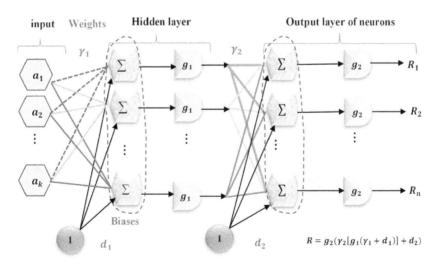

Fig. 8
Feed-forward two-layer neural network [65].

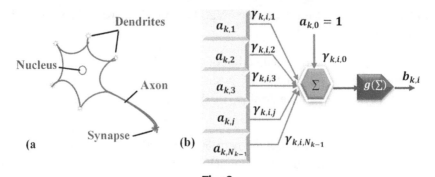

Fig. 9
(A) Neuron scheme and (B) its mathematical model [71].

An ANN is a black-box model that expresses the relationship between output data and input data. ANN uses no physical connection to learn the internal relations of the body, as well as to predict the behavior of the system. ANN consists of neurons that cluster in layers. Information is received by the dendrite and the active function (by the kernel). Modified information is then transmitted from axons and synapses (Fig. 9). In the first layer, the input values enter the nerve cells. Then they are upgraded by special weights ($\gamma_{k,i,1}$). Bias ($a_{k,1}$) is added to the sum ($s_{k,i}$). Special weights lead to simplicity of operation. It is then repeated for all inputs. In the end, the output $y_{k,\ i}$, which is the response of neurons, is obtained through the feature of activating neurons. The activation function is used only in the first layer. Hyperbolic tangent and sigmoid are used for subsequent layers. And for sharp outputs, the linear transmission function is used. Backpropagation was selectively changed due to the ANN learning method [72]. The equations describing the system are given below.

$$S_{k,i} = \sum_{j=0}^{N_{k-1}} \gamma_{k,i,j} \cdot a_{k,j} \tag{65}$$

$$b_{k,i} = g(S_{k,i}) \tag{66}$$

$$\varepsilon_{L,i} = d_{L,i} - b_{L,i} \tag{67}$$

$$\delta_{k,i} = \varepsilon_{k,i} \frac{\partial g(s_{k,i})}{\partial(s_{k,i})} \tag{68}$$

$$\varepsilon_{k,i} = \sum_{m=1}^{N_{k+1}} \delta_{k+1,m} \cdot \gamma_{k+1,m,i}, \quad \text{for } k = 1, 2, ..., L-1 \tag{69}$$

$$\gamma_{k,i,j}^{n+1} = \gamma_{k,i,j}^{n} + 2 \cdot \eta \cdot \delta_{k,i} \cdot a_{k,j} + \alpha \left(\gamma_{k,i,j}^{n} - \gamma_{k,i,j}^{n-1} \right) \tag{70}$$

where η is learning rate; α is momentum parameter. Parameters are used in the variables shown in Fig. 8.

2.3.2 Radial basis function neural networks

Another neural network approach is radial basis function neural networks (RBFNN). Its function is by using the initial radial functions. The output is linear. It is derived from input parameters and neurons. This method is used to approximate time series and optimal system control. As shown in Fig. 10, there are three layers in which neurons are activated in the middle layer using the radial basis function. A hidden layer consists of an array of computational units called hidden nodes (see Fig. 10) [73]. The middle layers are made up of hidden nodes for simplicity in calculations, which include an intermediate vector and an input length vector. The Euclidean distance is defined as follows: $\|x_i(t) - u_i(t)\|$. The Gauss activation function is mainly used as a nonlinear subset for the RBFNN method. It should be noted that the system is sensitive to initial values. If not selected correctly, the validity of the model will decrease [74].

Huo et al. [75] used a Hammerstein model to evaluate the performance a stack-level SOFC for loading voltage at DC current and two control loops for estimating output voltage and fuel utilization. The model consists of two parts, static and dynamic, which are developed with RBFNN and ARX, respectively. The fuel utilization was kept constant. It is also used to control the voltage of the MPC model. Finally, the simulation results suggest the model to control the voltage, while the nonlinear MPC controller is proven to control the potential. It is concluded that the MPC controller is very desirable for voltage control.where, $n = [n_1, n_2, ..., n_n]^T$ is the input, and $\gamma = [\gamma_1, \gamma_2, ..., \gamma_q]^T$ is the neural network weight. As mentioned k_i, the nonlinear function, obtained from the Gaussian activation method [74]:

$$g_i = exp\left[-\frac{(n - d_i)^T (n - d_i)}{2e_i^2}\right] \quad (i = 1, 2, ..., p) \tag{71}$$

where $d_i = (d_{i1}, d_{i2}, ..., d_{ij})^T, j = 1, 2, ..., n$, is the center of the ith RBFNN hidden unit, and e_i is width of the hidden unit. The output is shown as a weight function sum of p:

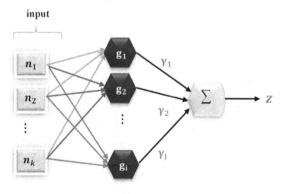

Fig. 10
RBF neural network [74].

$$z_p(t) = \sum_{i=1}^{p} \gamma_i g_i = \sum_{i=1}^{p} \gamma_i \, exp\left[-\frac{(n-d_i)^T(n-d_i)}{2e_i^2}\right] \tag{72}$$

Let $z(t)$ show the objective function in time. Network error in the form:

$$Error(t) = z(t) - z_p(t) \tag{73}$$

With the help of the following function, the cost can also be calculated:

$$C(t) = \frac{1}{2}[error(t)]^2 \tag{74}$$

By minimizing the cost function, the desired output can be obtained, the changes of the output parameters are as follows:

$$\gamma_i(t) = \gamma_i(t-1) + \eta\Delta\gamma_i + \alpha(\delta_i(t-1) - \gamma_i(t-2)) \tag{75}$$

$$d_{ij}(t) = d_{ij}(t-1) + \eta\Delta d_{ij} + \alpha(d_{ij}(t-1) - d_{ij}(t-2)) \tag{76}$$

$$e_i(t) = e_i(t-1) + \eta\Delta e_i + \alpha(e_i(t-1) - e_i(t-2)) \tag{77}$$

where α is the momentum term, and η is the learning rate, $\alpha \in [0,1], \eta \in [0, 1]$. The terms $\Delta\gamma_i$, Δc_{ij}, and Δb_i are defined as:

$$\Delta\gamma_i = \frac{\partial C}{\partial w_i} = \left(z(t) - z_p(t)\right)g_i = \left(z(t) - z_p(t)\right) exp\left[-\frac{(n-d_i)^T(n-d_i)}{2e_i^2}\right] \tag{78}$$

$$\Delta\gamma_i = \frac{\partial C}{\partial w_i} = \left(z(t) - z_p(t)\right)g_i = \left(z(t) - z_p(t)\right) exp\left[-\frac{(n-d_i)^T(n-d_i)}{2e_i^2}\right] \tag{79}$$

$$\Delta e_i = \frac{\partial C}{\partial e_i} = \left(z(t) - z_p(t)\right)\gamma_i g_i \frac{\|n_j - d\|^2}{e_i^3} \tag{80}$$

Choosing the appropriate initial value is very important in Eq. (78)–(80).

2.3.3 Least squares support vector machine

In a reported work by Suykens [76], a general SVM modification was proposed. Using multiobjective functions can help model credibility and reduce risk. It also has the ability to achieve the appropriate solution using the function matrix [77].

A nonlinear SOFC model was presented in Huo et al. [68], based totally on least squares support vector machine (LS-SVM). In this model, the fuel utilization and cell current are selected as two inputs and also the cell voltage as output. Cell-level data were analyzed in steady-state flow regime. According to the results, the LS-SVM model is suitable for stack description due to its simplicity and high speed in calculations. The system performance has also been compared with the RBFNN method, and LS-SVM has a high reliability.

Here, another LS-SVM algorithm that is based on the nonlinear system has been briefly described [76]:

First, datasets are assumed to be:

$$(n_1, z_1), ..., (n_N, z_N) \in K^m \times K \ (1) \tag{81}$$

The nonlinear function $\psi(\cdot)$ is used to write the original input space R^n to high-dimensional feature space $\psi(n) = (\varphi(n_1), \varphi(n_2), ..., \varphi(n_N))$. Then the linear decision function $z(n_i) = w^T \phi(n_i) + c$ is constructed in higher-dimensional feature space. Risk bound can be limited by the optimization formula. According to the structural risk minimization principle, it can be written as:

$$\min_{w,b,e} J(w, e) = \frac{1}{2} w^T w + \gamma \sum_{i=1}^{N} e_i^2, \quad \gamma > 0 \tag{82}$$

Therefore, the constraints of equality are as follows:

$$z_i = w^T \varphi(n_i) + c + e_i, \quad i = 1, ..., N \tag{83}$$

Lagrangian method is devised as:

$$\text{La}(w, c, e, \alpha) = J(w, e) - \sum_{i=1}^{N} \alpha_i \left(w^T \varphi(n_i) + c + e_i - z_i \right) \tag{84}$$

where $\alpha_i \ (i = 1, ..., N)$ are Lagrange multipliers. The optimal conditions are obtained:

$$\frac{\partial L}{\partial w} = 0 \rightarrow w = \sum_{i=1}^{N} \alpha_i \varphi(n_i), \quad \frac{\partial L}{\partial b} = 0 \rightarrow \sum_{i=1}^{N} \alpha_i = 0 \tag{85}$$

$$\frac{\partial L}{\partial e_i} = 0 \rightarrow \alpha_i = \gamma e_i, \quad i = 1, ..., N \tag{86}$$

$$\frac{\partial L}{\partial \alpha_i} = 0 \rightarrow y_i = w^T \varphi(n_i) + c + e_i, \quad i = 1, ..., N \tag{87}$$

With solution

$$\begin{bmatrix} 0 & 1 & \cdots & 1 \\ 1 & T(n_1, n_1) + 1/\gamma & \cdots & T(n_1, n_N) \\ \vdots & \vdots & \ddots & \vdots \\ 1 & T(n_N, n_1) & \cdots & T(n_N, n_N) + 1/\gamma \end{bmatrix} \begin{bmatrix} c \\ \alpha_1 \\ \vdots \\ \alpha_N \end{bmatrix} = \begin{bmatrix} 0 \\ z_1 \\ \vdots \\ z_N \end{bmatrix} \tag{88}$$

The resulting final LS-SVM model of the nonlinear system is as follows:

$$z(n) = \sum_{i=1}^{N} \alpha_i T(n, n_i) + c \tag{89}$$

Eq. (89) is linear. That parameter α_i can be calculated using methods of solving the normal linear system. By context of Karush-Kuhn-Tucker (KKT), the parameter b is calculated, so the LS-SVM model for nonlinear system can be obtained. The function of kernel $K(x,x_i)$ is a symmetric function that satisfies the condition of Mercer. The general of kernel function also can be referred to the linear and polynomial function, the radial basis function:

$$\text{Linear}: T(n_1, n_2) = n_1^T n_2 \tag{90}$$

$$\text{Ploynomial}: T(n_1, n_2) = \left(n_1^T n_2 + 1\right)^P \cdot P \in N \tag{91}$$

$$\text{RBF}: T(n_1, n_2) = exp\left(\frac{-\|n_1 - n_2\|^2}{2\sigma^2}\right) \tag{92}$$

3. SOFC optimization

The design of fuel cells is a major challenge because many physical phenomena need to be solved, optimized, and checked for proper performance of the system. Also, due to the specific design objectives and the lack of detailed models, the amount of computational resources to solve the equations needs to be optimized. Optimization may be done for some reasons such as minimizing the cost, effects of pollutants, and temperature or to maximize the efficiency and function of fuel cells [78]. The goal of optimization is actually finding the best solution of all possible solutions [79]. Systematic optimization strategies for defining optimal solutions to one or more parameters with limitations specified to maximize or minimize the purpose of these solutions [80]. In the fuel cell research area, some research has been done to optimize SOFC and its implementations over the years. To improve the performance of SOFC, optimization seems to be an essential task that should be done in several aspects. The research process is in developing optimization for optimal variables, objective functions, constraints, methods, and their implementation.

3.1 Decision variables

As previously mentioned, SOFCs are divided into two flat and tubular categories. Based on this division of fuel cells, including several structures such as micro and macro affects optimized parameters. In the design of microstructure, layer thickness, including electrode, anode and cathode, grain size, and porosity distribution are optimized [81]. GE analyzed the effects of micro structured design parameters such as weight fraction, empty fraction, particle size ratio, particle size and density for optimization of tubular SOFCs [82]. In another research [83], it has been proven that the distribution of porosity and particle size for electrodes can have effect on the performance of planar SOFCs.

3.2 Objective function

An important and significant part of the optimization strategy is the objective function that can be minimized or maximized and described below:

$$
\begin{aligned}
&\text{(a)} \quad \text{maximized or minimized}: O(t) \\
&\text{(b)} \quad \text{w.r.t } t_k \text{ for } k = 1,2,\ldots n \\
&\text{(c)} \quad \text{subject to}: R_j(t) = 0 \text{ for } j = 1,2,\ldots m \\
&\text{(d)} \quad W_p(t) \le 0 \text{ for } p = 1,2,\ldots z \\
&\text{(e)} \quad t_{\text{Lower}} \le t \le t_{\text{Upper}}
\end{aligned} \tag{93}
$$

In the above equations, $O(t)$ is object function to minimize or maximize. The $R_j(t) = 0$ are equality constraints, while $W_p(t) \le 0$ are inequality constraints compared with decision variable t. The t_{Lower} and t_{Upper} equations represent a lower and upper limit of decision variable [78]. The objective function for optimization strategy of SOFCs is based on thermodynamic, economic, and environmental aspects. The target function can be single or combination between several aspects that depend on the decision variable. Also, for simple systems, optimization problem is solved for one objective function but for complex systems, several objective functions are considered.

Table 6 summarizes the objective functions used to optimize the fuel cell system. These mathematical equations have been raised based on various thermodynamics, environmental, and economic issues. From thermodynamics point of view, the goal is maximizing the output voltage and efficiency. However, from economic and environmental point of view, the goals are to minimize the cost of system and the release of unwanted gases.

Table 6 A summary of the objective functions to optimize the SOFC system.

Ref.	Optimization area	Objective function
Thermodynamics		
[84] [85]	Maximizing OCV Maximizing power of SOFC	$V = V_n - V_{ohm} - V_{act} - V_{con}$ $P_{cell} = M \times \sum_j \frac{V_{cell}}{V_{cell}}$
Economics		
[86]	Minimize capital cost	$C.C = cos\, t_{Mbase} \left(\frac{M_{unit}}{M_{base}} \right)^{\beta}$
[87]	Minimize fuel cost	$C.F.E = \dfrac{1}{\left[\left(\frac{0.55}{city_{FE}} \right) + \left(\frac{0.45}{hwy_{FE}} \right) \right]}$
Environmental		
[88]	Maximizing C_2 production	$O.F = \dfrac{2\left[F_{C_2H_6} + F_{C_2H_4} \right]}{F_{CH_4} - F_{CH_4}}$

3.3 Constraints

Limitations are the cause of creating optimization constraints that can be related to process safety, construction materials, or manufacturing capabilities. As example, in a study [64], a hybrid system of SOFC has been investigated. It was concluded that to maximize the system's efficiency in proportion to turbine pressure changes, cell temperature should be fixed at $800°C$ and gas turbine at a maximum of $1100°C$. In another example, Wan et al. [89] optimized a PSOFC on the cell scale. In that study, the cell volume and thickness of interconnect, anode and cathode layers are considered as limits, which ultimately reach the maximum output power.

3.4 Methods for optimization

In a literature review by Ramadhani [81], fuel cell optimization was investigated in three methods that include determining, random, and metaexploration.

The determining method of repeated measurements used to optimize with mathematical calculations. Definitive optimization using simplex algorithm, numerical calculations, and linear programming have been done to obtain optimal variables for SOFC parameters. In another research by Oh et al. [90], a numerical simulation for SOFC is performed on a stack scale, which aims to optimize and control a class of 5kW fuel cells. This proposed method also used to optimize fuel flow and generator power to respond to load changes.

A random method is mainly used for complex cases that goals, decision-making variables, and restrictions are unclear. But due to solving complex problems for the accurate solution, it faces a computational cost issue [81]. Several randomized methods have been proposed in various studies that include Lagrange [91], Pareto optimal [92], Real-Time Optimization (RTO) [93], and Newton's Iteration [94].

The metaexploration approach is used to improve the random method with approximate solutions. This method improves search speed, and solutions are obtained in less time with high accuracy. This is called intelligent optimization method that can be controlled and evaluated with random solutions. Lopez et al. [95] evaluated a GT-SOFC. The purpose of this research is to optimize the location of the supply source and current value, which was performed using the Particle Swarm Optimization (PSO) method. They also used the genetic method. By comparing these two methods, they concluded that the PSO method is easier than genetic algorithm due to less relative parameters. The duration of review in PSO method is also less.

Table 7 is a summary of review papers listed on the optimization methods. In this table, the decision variables, the objective functions, optimization methods, and the SOFC scale are proposed.

Table 7 A review of studies in the field of optimization of solid oxide fuel cell.

Ref.	Decision variables	Objective function	Optimization method	SOFC scale
[82]	Geometry	Triple-phase boundary length	Micro-structure analysis using Home-built Mathematica	SOFC at cell level
[86]	Geometry, operating Condition	Minimize capital cost of system	Deterministic and Stochastic	SOFC-PEM hybrid system
[96]	Operating Condition	Minimize the cost of the system annually and maximize efficiency	Determining method using MATLAB	CHP-SOFC
[97]	Design, operating condition	Maximizing the exergy efficiency and minimizing the total cost rate (TCR)	Metaexploration approach (GA) in Aspen-Plus	SOFC-GT
[98]	Geometry	Minimize overpotential and degradation rate of the cathode	Metaexploration approach combination of ANN and GA	Electrode microstructures.
[99]	Geometry	Maximum cell current density	Thermodynamic examination in COMSOL	Stack
[100]	Operating Condition	Minimize electricity cost and annualized cost	Pareto optimal front using BELSIM-VAL	SOFC-GT
[101]	Design, operating condition	Maximize fuel utilization factor, nominal net power	Thermodynamic aspect:	Single SOFC

4. Conclusion

In this chapter, the modeling, simulation, and optimization of gas-powered SOFC systems are discussed. In general, for this purpose, the main focus is on the white-box, gray-box, and black-box approaches. At first, this type of fuel cell is compared with other different types with potential industrial applications. The most important advantage of SOFC is the variety in fuel type and high efficiency.

The white-box modeling approach is based on the physical relationships of transport phenomena. In the white-box modeling section, SOFCs at the cell, stack, and system level were examined. The SOFC is divided into external and internal categories based on reformer type. In the external or primary reforming type, the reforming process is performed in a side reactor next to the set of cells, using the heat generated in the cell. Also, the reformer may be inside the stack, which typically has two direct or indirect states. At cell level, SOFCs have been studied in micro and macro scales. In macro scale, transport phenomena, thermodynamics, and electrochemistry in three dimensions are considered to predict cell performance. In this scale, electrochemical reactions take place at the electrode-electrolyte interface, where in micro scale, they are considered within the porous volumetric structure of the electrodes.

The SOFC stack model is a set of single cells that are arranged sequentially or in parallel. By examining a single cell, the overall performance of the stack can also be predicted. Also, combined energy and power systems are studied at the system level. Fuel cell systems are modeled by gray and black-box approach. The gray-box model is based on a combination of experimental data and additional knowledge about the system. In many studies, due to modeling simplification and linearization with the application of hypotheses, the real system does not match the modeling system. In black-box approach, the basis of modeling is an implicit relationship between input and output data. The black-box approach, which relies on artificial intelligence, is used to describe and predict the performance of nonlinear dynamic systems. In this section, ANN, RBFNN, and LS-SVM models are studied. The design of fuel cells is a major challenge because many physical phenomena need to be solved and optimized and proper performance of the system should be checked.

To achieve the high performance of an SOFC, it is required to optimize the structure and the operating conditions. A short discussion has been given in this chapter about method and tools for optimization, decision variables, objective functions, and constraints. It was shown that, due the high complexity of SOFC and the lack of knowledge for accurate detailed modeling, intelligent optimization can help for better design and higher performance.

Abbreviations and symbols

0	pure material
0-D	zero-dimensional
1-D	one-dimensional
2-D	two-dimensional
3-D	three-dimensional
A	area (m^2)
A	as subscript: air, species
Aβ	species A in species
Act	activation
ADL	anode diffusion layer
AFC	alkaline fuel cell
Am	species a in mixture
ANN	artificial neural network
ARL	anode reaction layer
ASR	area-specific resistance
α	charge transfer coefficient; activity values
BOP	balance of plant
β	constant parameter
β́	defined parameter (J/mol)

C	concentration (mol/m^3)
CDL	cathode diffusion layer
CHP	combine heat and power
CRL	cathode reaction layer
Conc	concentration
C_p	heat capacity at constant pressure (J/mol K)
DC	direct current
D_{kl}	Knudsen diffusivity (m^2/s)
D_{l-m}	molecular diffusivity of l through m, (m^2/s)
D^{eff}	effective diffusivity (m^2/s)
DGM	dusty gas model
DIR	direct internal reforming
d	diameter (m)
Δ	average
δ	thickness (m)
$\acute{\delta}$	defined parameter (J/mol)
E	activation energy (kJ/mol)
ER	external reforming
E_{Th}	reversible voltage (V)
E_{rev}^0	standard voltage (V)
Eff	effective
El	electron
ε	porosity
η	polarization (V); efficiency (V)
F	Faraday's constant molar flow rate mol/s (C/mol)
F	fluid
FE	fuel economy
G	as subscript: gas
G	molar specific Gibbs free energy (J/mol)
$\acute{\gamma}$	defined parameter (J/mol)
Υ	velocity (m/s)
H	hot exhaust
H, h	molar specific enthalpy (J/mol)
Hwy	highway
I	species
IIR	indirect internal reforming
Io	ionic
j	current density (A/m^2)
J	ordinary diffusion flux (mol/s)
j_0	exchange current density (A/m^2)
K_{eq}	equilibrium constant

KKT	Karush – Kuhn – Tucker
κ	permeability
$\overline{LHV}$	lower heating value (J/mol)
LS-SVM	least squares support vector machine
λ	thermal conductivity coefficient (J/mol K s); air ratio (J/mol K s)
MCFC	molten carbonate fuel cell
MSE	mean squared error
MW	molecular weight (gr/mol)
μ	dynamic viscosity (kg/m s)
N	number fraction of electron- or ion-conducting particles
O	standard pressure
OCV	open circuit voltage
Ohm	Ohmic
PAFC	phosphoric acid fuel cell
PCFC	protonic ceramic fuel cell
PC-SOFC	proton-conducting solid oxide fuel cell
PEFC	polymer electrolyte fuel cell
PEN	positive electrode-electrolyte-negative electrode
PRBS	pseudo random binary signal
PSO	particle swarm optimization
PSOFC	planar solid oxide fuel cell
Φ	electric potential (V)
R	universal gas constant (J/mol K); polarization resistance ($/m^2$)
R^2	coefficient of determination
RBFNN	radial basis function neural network
RF	random forest
RMSE	root mean square error
P	pressure (bar)
ρ	density (kg/m^3)
$\tilde{\rho}$	volumetric charge density
S	solid structure
SD	standard deviation
SOFC	solid oxide fuel cell
SR	steam reforming reaction
SVM	support vector machine
$\tilde{\sigma}$	ionic conductivity (A/V m)
T	temperature (K)
TSOFC	tubular solid oxide fuel cell
tot	total
τ	tortuosity, stress tensor
U_{eff}	effective overall heat transfer coefficient (J/mol K)

U_f	fuel utilization ratio
V	Voltage (V)
Vi	stoichiometric coefficients
W	electric power (W)
W	mass fraction; weight
WGS	water-gas shift reaction
X	mole fraction in liquid
Y	mole fraction in gas

References

[1] J. Larminie, A. Dicks, M.S. McDonald, Fuel Cell Systems Explained, vol. 2, J. Wiley, Chichester, 2003.

[2] EG&G Technical Services, Inc., Handbook, F.C., Albuquerque, NM, DOE/NETL-2004/1206, 2004, pp. 1–10.

[3] C.O. Colpan, I. Dincer, F. Hamdullahpur, A review on macro-level modeling of planar solid oxide fuel cells, Int. J. Energy Res. 32 (4) (2008) 336–355.

[4] K. Wang, et al., A review on solid oxide fuel cell models, Int. J. Hydrogen Energy 36 (12) (2011) 7212–7228.

[5] M.G. Camprubı, Multiphysics Models for the Simulation of Solid Oxide Fuel Cells, Citeseer, 2011.

[6] M. Ghassemi, M. Kamvar, R. Steinberger-Wilckens, Fundamentals of Heat and Fluid Flow in High Temperature Fuel Cells, Academic Press, 2020.

[7] S.A. Hajimolana, et al., Mathematical modeling of solid oxide fuel cells: a review, Renew. Sustain. Energy Rev. 15 (4) (2011) 1893–1917.

[8] S. Kakac, A. Pramuanjaroenkij, X.Y. Zhou, A review of numerical modeling of solid oxide fuel cells, Int. J. Hydrogen Energy 32 (7) (2007) 761–786.

[9] C. Spiegel, PEM Fuel Cell Modeling and Simulation Using MATLAB, Elsevier, 2011.

[10] S. Chan, Z. Xia, Polarization effects in electrolyte/electrode-supported solid oxide fuel cells, J. Appl. Electrochem. 32 (3) (2002) 339–347.

[11] J. Newman, K.E. Thomas-Alyea, Electrochemical Systems, John Wiley & Sons, 2012.

[12] R. O'hayre, et al., Fuel Cell Fundamentals, John Wiley & Sons, 2016.

[13] A. Fick, Ueber diffusion, Ann. Phys. 170 (1) (1855) 59–86.

[14] S.R. De Groot, S.R. De Groot, Thermodynamics of Irreversible Processes, vol. 242, North-Holland, Amsterdam, 1951.

[15] J. Solsvik, H.A. Jakobsen, Modeling of multicomponent mass diffusion in porous spherical pellets: application to steam methane reforming and methanol synthesis, Chem. Eng. Sci. 66 (9) (2011) 1986–2000.

[16] Y. Funahashi, et al., Fabrication and characterization of components for cube shaped micro tubular SOFC bundle, J. Power Sources 163 (2) (2007) 731–736.

[17] M. Suzuki, T. Oshima, Co-ordination number of a multi-component randomly packed bed of spheres with size distribution, Powder Technol. 44 (3) (1985) 213–218.

[18] D. Bouvard, F. Lange, Relation between percolation and particle coordination in binary powder mixtures, Acta Metall. Mater. 39 (12) (1991) 3083–3090.

[19] T.X. Ho, et al., Modeling of transport, chemical and electrochemical phenomena in a cathode-supported SOFC, Chem. Eng. Sci. 64 (12) (2009) 3000–3009.

[20] R. Krishna, J. Wesselingh, The Maxwell-Stefan approach to mass transfer, Chem. Eng. Sci. 52 (6) (1997) 861–911.

[21] R. Suwanwarangkul, et al., Performance comparison of Fick's, dusty-gas and Stefan–Maxwell models to predict the concentration overpotential of a SOFC anode, J. Power Sources 122 (1) (2003) 9–18.

[22] C.K. Ho, S.W. Webb, Gas Transport in Porous Media, vol. 20, Springer, 2006.

[23] D.L. Damm, A.G. Fedorov, Local thermal non-equilibrium effects in porous electrodes of the hydrogen-fueled SOFC, J. Power Sources 159 (2) (2006) 1153–1157.

[24] K. Daun, et al., Radiation heat transfer in planar SOFC electrolytes, J. Power Sources 157 (1) (2006) 302–310.

[25] F. Tong, L. Jing, R.W. Zimmerman, An effective thermal conductivity model of geological porous media for coupled thermo-hydro-mechanical systems with multiphase flow, Int. J. Rock Mech. Min. Sci. 46 (8) (2009) 1358–1369.

[26] B. Haberman, J. Young, Three-dimensional simulation of chemically reacting gas flows in the porous support structure of an integrated-planar solid oxide fuel cell, Int. J. Heat Mass Transf. 47 (17–18) (2004) 3617–3629.

[27] J. VanderSteen, et al., Mathematical modelling of the transport phenomena and the chemical/electrochemical reactions in solid oxide fuel cells: a review, in: Canadian Hydrogen and Fuell Cells Conference, 2004.

[28] B. Huang, Y. Qi, M. Murshed, Solid oxide fuel cell: perspective of dynamic modeling and control, J. Process Control 21 (10) (2011) 1426–1437.

[29] M. Ilbas, B. Kumuk, Numerical modelling of a cathode-supported solid oxide fuel cell (SOFC) in comparison with an electrolyte-supported model, J. Energy Inst. 92 (3) (2019) 682–692.

[30] S. Hosseini, K. Ahmed, M.O. Tadé, CFD model of a methane fuelled single cell SOFC stack for analysing the combined effects of macro/micro structural parameters, J. Power Sources 234 (2013) 180–196.

[31] Z. Zhang, et al., Three-dimensional CFD modeling of transport phenomena in multi-channel anode-supported planar SOFCs, Int. J. Heat Mass Transf. 84 (2015) 942–954.

[32] A. Raj, A.P. Sasmito, T. Shamim, Numerical investigation of the effect of operating parameters on a planar solid oxide fuel cell, Energ. Conver. Manage. 90 (2015) 138–145.

[33] J. Shi, X. Xue, CFD analysis of a novel symmetrical planar SOFC design with micro-flow channels, Chem. Eng. J. 163 (1–2) (2010) 119–125.

[34] R. Bove, S. Ubertini, Modeling solid oxide fuel cell operation: approaches, techniques and results, J. Power Sources 159 (1) (2006) 543–559.

[35] K. Yingwei, et al., One-dimensional dynamic modeling and simulation of a planar direct internal reforming solid oxide fuel cell, Chin. J. Chem. Eng. 17 (2) (2009) 304–317.

[36] X. Luo, K. Fong, Development of 2D dynamic model for hydrogen-fed and methane-fed solid oxide fuel cells, J. Power Sources 328 (2016) 91–104.

[37] P. Aguiar, D. Chadwick, L. Kershenbaum, Modelling of an indirect internal reforming solid oxide fuel cell, Chem. Eng. Sci. 57 (10) (2002) 1665–1677.

[38] P. Dokmaingam, et al., Modeling of IT-SOFC with indirect internal reforming operation fueled by methane: effect of oxygen adding as autothermal reforming, Int. J. Hydrogen Energy 35 (24) (2010) 13271–13279.

[39] S.H. Chan, O.L. Ding, Simulation of a solid oxide fuel cell power system fed by methane, Int. J. Hydrogen Energy 30 (2005) 167–179.

[40] R. Braun, S. Klein, D. Reindl, Evaluation of system configurations for solid oxide fuel cell-based micro-combined heat and power generators in residential applications, J. Power Sources 158 (2) (2006) 1290–1305.

[41] C.O. Colpan, et al., Effect of gasification agent on the performance of solid oxide fuel cell and biomass gasification systems, Int. J. Hydrogen Energy 35 (10) (2010) 5001–5009.

[42] Y. Yi, et al., Fuel flexibility study of an integrated 25 kW SOFC reformer system, J. Power Sources 144 (1) (2005) 67–76.

[43] S. Farhad, F. Hamdullahpur, Conceptual design of a novel ammonia-fuelled portable solid oxide fuel cell system, J. Power Sources 195 (10) (2010) 3084–3090.

[44] A. Mehr, et al., Polygeneration systems based on high temperature fuel cell (MCFC and SOFC) technology: system design, fuel types, modeling and analysis approaches, Energy (2021) 120613.

[45] Y. Wang, et al., Analysis of a biogas-fed SOFC CHP system based on multi-scale hierarchical modeling, Renew. Energy 163 (2021) 78–87.

[46] O. Corigliano, P. Fragiacomo, Numerical modeling of an indirect internal CO_2 reforming solid oxide fuel cell energy system fed by biogas, Fuel 196 (2017) 352–361.

[47] C. Bao, et al., A multi-level simulation platform of natural gas internal reforming solid oxide fuel cell–gas turbine hybrid generation system: part I. Solid oxide fuel cell model library, J. Power Sources 195 (15) (2010) 4871–4892.

[48] X. Yan, et al., A theoretical strategy of pure carbon materials for lightweight and excellent absorption performance, Carbon 174 (2021) 662–672.

[49] A. Baghernejad, M. Yaghoubi, K. Jafarpur, Optimum power performance of a new integrated SOFC-trigeneration system by multi-objective exergoeconomic optimization, Int. J. Electr. Power Energy Syst. 73 (2015) 899–912.

[50] P. Kazempoor, V. Dorer, F. Ommi, Modelling and performance evaluation of solid oxide fuel cell for building integrated co-and polygeneration, Fuel Cells 10 (6) (2010) 1074–1094.

[51] C. Bao, N. Cai, E. Croiset, A multi-level simulation platform of natural gas internal reforming solid oxide fuel cell–gas turbine hybrid generation system–part II. Balancing units model library and system simulation, J. Power Sources 196 (20) (2011) 8424–8434.

[52] P. Tippawan, A. Arpornwichanop, I. Dincer, Energy and exergy analyses of an ethanol-fueled solid oxide fuel cell for a trigeneration system, Energy 87 (2015) 228–239.

[53] K. Fong, C. Lee, Investigation on zero grid-electricity design strategies of solid oxide fuel cell trigeneration system for high-rise building in hot and humid climate, Appl. Energy 114 (2014) 426–433.

[54] H. Al Moussawi, F. Fardoun, and H. Louahlia, 4-E based optimal management of a SOFC-CCHP system model for residential applications, Energ. Conver. Manage. 151 (2017) 607–629.

[55] J. Řehoř, V. Havlena, A practical approach to grey-box model identification, IFAC Proc. Vol. 44 (1) (2011) 10776–10781.

[56] B. Sohlberg, Grey box modelling for model predictive control of a heating process, J. Process Control 13 (3) (2003) 225–238.

[57] R. Moradi, et al., Comparison between 1-D and grey-box models of a SOFC, in: E3S Web of Conferences, EDP Sciences, 2019.

[58] G. Goodwin, Optimal input signals for nonlinear-system identification, in: Proceedings of the Institution of Electrical Engineers, IET, 1971.

[59] H. Melgaard, et al., Experiment design for grey-box models, IFAC Proc. Vol. 26 (2) (1993) 489–492.

[60] T. Bohlin, A Grey-Box Process Identification Tool: Theory and Practice, Technical Report IR-53-REG-0103, Department of Signals, Sensors and Systems, 2001.

[61] M. Gevers, et al., The role of experimental conditions in model validation for control, in: Robustness in Identification and Control, Springer, 1999, pp. 72–86.

[62] L. Lennart, System Identification: Theory for the User, vol. 28, PTR Prentice Hall, Upper Saddle River, NJ, 1999.

[63] J.-N. Juang, Applied System Identification, Prentice-Hall, Inc., 1994.

[64] J. Milewski, et al., Advanced Methods of Solid Oxide Fuel Cell Modeling, Springer Science & Business Media, 2011.

[65] J. Arriagada, P. Olausson, A. Selimovic, Artificial neural network simulator for SOFC performance prediction, J. Power Sources 112 (1) (2002) 54–60.

[66] U.K. Chakraborty, Static and dynamic modeling of solid oxide fuel cell using genetic programming, Energy 34 (6) (2009) 740–751.

[67] E. Entchev, L. Yang, Application of adaptive neuro-fuzzy inference system techniques and artificial neural networks to predict solid oxide fuel cell performance in residential microgeneration installation, J. Power Sources 170 (1) (2007) 122–129.

[68] H.-B. Huo, X.-J. Zhu, G.-Y. Cao, Nonlinear modeling of a SOFC stack based on a least squares support vector machine, J. Power Sources 162 (2) (2006) 1220–1225.

[69] K. Patan, Artificial Neural Networks for the Modelling and Fault Diagnosis of Technical Processes, Springer, 2008.

[70] S. Song, et al., Modeling the SOFC by BP neural network algorithm, Int. J. Hydrogen Energy 46 (38) (2021) 20065–20077.

[71] J. Milewski, K. Świrski, Modelling the SOFC behaviours by artificial neural network, Int. J. Hydrogen Energy 34 (13) (2009) 5546–5553.

[72] M.H. Beale, M.T. Hagan, H.B. Demuth, Neural Network Toolbox User's Guide, vol. 103, The MathWorks Inc, 1992.

[73] K. Warwick, R. Craddock, An introduction to radial basis functions for system identification. A comparison with other neural network methods, in: Proceedings of 35th IEEE Conference on Decision and Control, IEEE, 1996.

[74] X.-J. Wu, et al., Modeling a SOFC stack based on GA-RBF neural networks identification, J. Power Sources 167 (1) (2007) 145–150.

[75] H.-B. Huo, et al., Nonlinear model predictive control of SOFC based on a Hammerstein model, J. Power Sources 185 (1) (2008) 338–344.

[76] J.A. Suykens, J. Vandewalle, Least squares support vector machine classifiers, Neural. Process. Lett. 9 (3) (1999) 293–300.

[77] X.-J. Wu, et al., Nonlinear modeling of a SOFC stack based on ANFIS identification, Simul. Model. Pract. Theory 16 (4) (2008) 399–409.

[78] M. Secanell, J. Wishart, P. Dobson, Computational design and optimization of fuel cells and fuel cell systems: a review, J. Power Sources 196 (8) (2011) 3690–3704.

[79] J.J. Grefenstette, Optimization of control parameters for genetic algorithms, IEEE Trans. Syst. Man Cybern. 16 (1) (1986) 122–128.

[80] P.D. Zavattieri, E.A. Dari, G.C. Buscaglia, Optimization strategies in unstructured mesh generation, Int. J. Numer. Methods Eng. 39 (12) (1996) 2055–2071.

[81] F. Ramadhani, et al., Optimization strategies for solid oxide fuel cell (SOFC) application: a literature survey, Renew. Sustain. Energy Rev. 76 (2017) 460–484.

[82] X.M. Ge, Y.N. Fang, S.H. Chan, Design and optimization of composite electrodes in solid oxide cells, Fuel Cells 12 (1) (2012) 61–76.

[83] J. Shi, X. Xue, Microstructure optimization designs for anode-supported planar solid oxide fuel cells, J. Fuel Cell Sci. Technol. 8 (6) (2011).

[84] T.S. Lee, J. Chung, Y.-C. Chen, Design and optimization of a combined fuel reforming and solid oxide fuel cell system with anode off-gas recycling, Energ. Conver. Manage. 52 (10) (2011) 3214–3226.

[85] J. Shi, X. Xue, Optimization design of electrodes for anode-supported solid oxide fuel cells via genetic algorithm, J. Electrochem. Soc. 158 (2) (2010) B143.

[86] L. Tan, C. Yang, N. Zhou, Synthesis/design optimization of SOFC–PEM hybrid system under uncertainty, Chin. J. Chem. Eng. 23 (1) (2015) 128–137.

[87] K. Wipke, T. Markel, D. Nelson, Optimizing energy management strategy and degree of hybridization for a hydrogen fuel cell SUV, in: Proceedings of 18th Electric Vehicle Symposium, 2001.

[88] M.R. Quddus, Y. Zhang, A.K. Ray, Multi-objective optimization in solid oxide fuel cell for oxidative coupling of methane, Chem. Eng. J. 165 (2) (2010) 639–648.

[89] H. Wen, J. Ordonez, J. Vargas, Single solid oxide fuel cell modeling and optimization, J. Power Sources 196 (18) (2011) 7519–7532.

[90] S.-R. Oh, J. Sun, Optimization and load-following characteristics of 5kw-class tubular solid oxide fuel cell/ gas turbine hybrid systems, in: Proceedings of the 2010 American Control Conference, IEEE, 2010.

[91] D.F. Cheddie, Thermo-economic optimization of an indirectly coupled solid oxide fuel cell/gas turbine hybrid power plant, Int. J. Hydrogen Energy 36 (2) (2011) 1702–1709.

[92] M. Burer, et al., Multi-criteria optimization of a district cogeneration plant integrating a solid oxide fuel cell– gas turbine combined cycle, heat pumps and chillers, Energy 28 (6) (2003) 497–518.

[93] J. Ki, Integrated Modeling Approach for Solid Oxide Fuel Cell-Based Power Generating System, Mechanical Engineering, University of Texas at Arlington, 2013.

[94] L. Elliott, W. Anderson, S. Kapadia, Solid oxide fuel cell design optimization with numerical adjoint techniques, J. Fuel Cell Sci. Technol. 6 (4) (2009).

[95] P.R. Lopez, et al., Optimization of biomass fuelled systems for distributed power generation using particle swarm optimization, Electr. Pow. Syst. Res. 78 (8) (2008) 1448–1455.

[96] E. Riensche, U. Stimming, G. Unverzagt, Optimization of a 200 kW SOFC cogeneration power plant: part I: variation of process parameters, J. Power Sources 73 (2) (1998) 251–256.

[97] Z. Hajabdollahi, P.-F. Fu, Multi-objective based configuration optimization of SOFC-GT cogeneration plant, Appl. Therm. Eng. 112 (2017) 549–559.

[98] Z. Yan, et al., Modeling of solid oxide fuel cell (SOFC) electrodes from fabrication to operation: microstructure optimization via artificial neural networks and multi-objective genetic algorithms, Energ. Conver. Manage. 198 (2019) 111916.

[99] W. Kong, et al., Optimization of the interconnect ribs for a cathode-supported solid oxide fuel cell, Energies 7 (1) (2014) 295–313.

[100] S. Sharma, et al., Multi-objective Optimization of Solid Oxide Fuel Cell–Gas Turbine Hybrid Cycle and Uncertainty Analysis, University of Ljubljana, 2016.

[101] H. Wen, J. Ordonez, J. Vargas, Optimization of single SOFC structural design for maximum power, Appl. Therm. Eng. 50 (1) (2013) 12–25.

Index

Note: Page numbers followed by *f* indicate figures and *t* indicate tables.